SELECTION OF ENGINEERING MATERIALS

SELECTION OF ENGINEERING MATERIALS

GLADIUS LEWIS
Department of Mechanical Engineering
Memphis State University

Prentice Hall
Englewood Cliffs, New Jersey 07632

Library of Congress Cataloging-in-Publication Data

Lewis, Gladius.
Selection of engineering materials / Gladius Lewis. p. cm.
Bibliography: p.
Includes index.
ISBN 0-13-802190-2
1. Materials. I. Title.
TA403.L45 1990
620.1'1--dc19 88-31558
CIP

Editorial/production supervision and
interior design: Jennifer Wenzel
Cover design: Photo Plus Art
Manufacturing buyer: Mary Noonan
Page layout: Richard Dombrowski
See page 565 for list of trademarks

Printed in the United States of America
10 9 8 7 6 5 4 3 2 1

ISBN 0-13-802190-2

PRENTICE-HALL INTERNATIONAL (UK) LIMITED, *London*
PRENTICE-HALL OF AUSTRALIA PTY, LIMITED, *Sydney*
PRENTICE-HALL CANADA INC., *Toronto*
PRENTICE-HALL HISPANOAMERICANA, S.A., *Mexico*
PRENTICE-HALL OF INDIA PRIVATE LIMITED, *New Delhi*
PRENTICE-HALL OF JAPAN, INC., *Tokyo*
SIMON & SCHUSTER ASIA PTE. LTD., *Singapore*
EDITORA PRENTICE-HALL DO BRASIL, LTDA., *Rio de Janeiro*

In this work, when it shall be found that much is omitted, let it not be forgetten that much likewise is performed.

Samuel Johnson,
Preface to the two-volume, Dictionary of the English Language

CONTENTS

PREFACE

My main acknowledgments go to the many students with whom I have interacted during the past decade on three continents while teaching and researching in the field of engineering materials and their applications. I appreciate, among many other things, their insight, probing questions, insatiable thirst for knowledge, and remarkable initiative. Many thanks also go to the following persons for their help in the "nut and bolts" aspects of converting a handwritten manuscript to a typewritten, illustrated one: Barbara Moore and Ann Hester of the Department of Mechanical Engineering of the University of Alabama in Huntsville, Alabama; and Billie Bulmanski, Beth Caldwell, and Ann Brock of Memphis State University in Memphis, Tennessee. In addition, I am grateful for the permission given by various individuals and organizations to use a number of diagrams, photographs, and data sets. All such materials are clearly acknowledged in the text. An acknowledgments list of all proprietary products and the patent-holding companies is provided on page 565. Finally, I acknowledge the use of an enormous body of papers and documents published elsewhere. All information taken from these sources is clearly identified in the text, by means of numbers in brackets, corresponding to a reference list at the end of each chapter. My thanks go to all the authors of these referenced works.

INTRODUCTION

The most important decision in engineering is the proper selection of materials. That this is so is highlighted by the fact that all decisions regarding the manufacturing of engineering structures and components are predicated on proper selection of materials. Indeed, the combination of a proper material selection methodology and sound economic judgment is the cornerstone of the ability to transform an engineering idea from concept to implementation. The boldness or, as some may argue, the dogmatic nature of these remarks is meant to underscore the pivotal position that a course in the selection of engineering materials should occupy in the curricula of engineering degree programs. This is not the case at the present time.

Too many of today's engineering students and tomorrow's engineers are being wrongly or badly educated in this subject as a result of errors of omission and commission in the curriculum. Insufficient efforts are being made either to disabuse the minds of engineering students completely or to point out to them the limitations of the "old" approach to the selection of materials—that the process is largely empirical. In essence, that approach stressed the ability of engineers to draw on their fund of experience; but invariably, they tend to play it safe and prescribe the few materials that have been tested and proved or with which the engineers are most familiar. In addition, at the college level, courses in such pertinent fields as materials science, mechanics of materials, and engineering design are not coordinated well enough for their symbiotic nature to be appreciated, thereby thwarting a cross-fertilized, coherent, holistic treatment.

In the past two decades or so, there has been a steadily growing realization that for optimum selection of materials, the engineer must be knowledgeable about the nature, properties, merits, and demerits of the widest spectrum of currently available materials as well as materials with near-term availability potential, in addition to being well grounded in the fundamentals of the mechanical aspects of design, environmental degradation of materials, manufacturing processes, quality control, salvage and recycling potential of materials, and so on. It is therefore necessary to present the material in this book as a synthesis of ideas from these and other subject areas. It is gratifying to note that in a small number of engineering colleges, such approaches are being taken in the teaching or development of the curriculum in this field.

This book is a contribution to the emerging "new approach" to the teaching and practice of material selection. The approach is, for a given application, to establish the link between the functional and material requirements and hence narrow the choice to a few candidate materials. On the basis of established criteria, either an unambiguous choice is made, or the compromises and/or dilemmas associated with the selection process are highlighted, leaving the final choice to the reader. It is useful to note here that when candidate materials are selected on the basis of so-called sizing material properties (properties that affect the final dimensions of the component, such as modulus of elasticity, fracture toughness, and density), the final choice is comparatively easier than when such selection is to be made on the basis of so-called discriminating material properties (properties that have little or no effect on the size of the component but are nonetheless important, such as ductility, oxidation resistance, and maximum reuse temperature).

The material in this book is divided into four parts. In Part 1, some 30 case studies are examined, running the gamut of engineering applications from a vaulting pole to a flywheel as an energy storage unit in a ground vehicle and from the thermal protection system of the U.S. space shuttle to the wheels of a skateboard. The objective in this part is to present as well rounded a study as possible: definition or description of the case, establishment of service conditions, presentation of equations governing the mechanics of the system (where applicable), considerations relevant to the selection of suitable materials, candidate material choices, and a final evaluation of these materials, culminating in a recommendation.

Part II consists of a series of 58 questions and suggested answers on a wide range of materials applications, from the driveshaft of an automobile to a gas-cooled nuclear reactor and from a pyroelectric transducer to an ovenproof casserole. The objective in this part is to illustrate, by way of these suggested answers, both style and format for the presentation of information and ideas. It is not intended that these answers be taken as definitive; rather, it is hoped that they will be used as "springboards" for developing answers independently.

In Part III, the latest developments in the field of engineering materials and their applications are presented in 16 chapters. The objective in this part is to give as comprehensive and panoramic a survey as possible. The chapters therefore cover trends in metals, ceramics, polymers, and composite materials. Of course, in a field as volatile as this one (as is evidenced in the explosion of developments in superconducting materials since February 1987), it is neither possible nor necessary to make the material in Part III encyclopedic. The best that can be hoped for is to capture the essence of research and developments results at the time of writing. (In the 24 months of preparation of the manuscript, some portions of Part III had to be updated no less than 10 times!)

The guiding principle in the presentation in Part III is a recognition of the key links between properties and uses of materials. Thus much space is devoted to details about the latest developments in materials from the viewpoint of their properties, advantages, and disadvantages. The intention is to present the information about these aspects, taken from as many sources as possible, in summarized form, thereby sparing the reader an agonizing journey through hundreds of journals and other publications.

Part IV presents 45 exercises in material selection for the student to tackle (ideally, under the guidance of a discriminating instructor).

It is useful to insert a number of caveats here, as far as an assessment of this book is concerned. First, it should be appreciated that access to detailed properties of many nonmetallic materials is not easy. As for raw material and fabrication cost data, these are mostly nonexistent in the open literature. To obtain them from vendors and/or user companies requires both Jobian patience and Herculean tenacity! Although various organizations are now earnestly trying to correct this deficiency in information sources (see Appendix A), much more needs to be known.

It is against this background of difficulty of obtaining cost data that a decision was made to forgo any exposition on the economics of materials selection. It is clearly an important element in any methodology (very rarely do the most economic solution and the best engineering solution to a given material selection problem coincide!). Also, it is important to recognize the fact that material properties differ in importance, depending on the application. Thus material property weighting factors must be assigned. This is by no means a trivial task and is clearly outside the ambit of this book. To underscore the fact that both cost and weighting factors are important in final material selection, a highly abbreviated and rudimentary treatment of them, for two named applications, is given in Appendix B. It is hoped that the reader with access to detailed cost and weighting factor information will apply these to any case of interest. (College professors using this book in a course might regard this exercise as *de rigueur* for their students).

The second caveat is to do with the intended readership. It is anticipated that at the college level, this book will be used in a capstone course at the senior undergraduate or first-year graduate level. These students would have the necessary background and preparation in subjects such as physics, chemistry, mathematics, mechanics of materials, materials science, thermofluids science, and electromagnetic theory. In other words, a certain minimum knowledge of relevant concepts in these courses is assumed.

The third caveat is, in fact, an extension of the second. This book is not intended to supplant existing texts on materials. Its main thrust is not the science of materials (which would, presumably, have been thoroughly studied before embarking on a capstone course) but the applications of materials that are predicated on the science. In one sense, this book is intended to complement or supplement some existing texts; in another sense, it is a "first-generation" resource book on materials and their applications. That there are so few texts that this book can complement or supplement (even though "Choice of Materials" or "Selection of Materials" is a featured course in the curricula of many engineering colleges—see Appendix C) provides its *raison d'être*.

The fourth caveat is to do with the amount of detail given in the presentation of the engineering aspects of some of the case studies. This is deemed necessary in order to reinforce the notion that these are real engineering projects, not merely academic exercises. In addition, the inclusion of these details helps to "educate and inform" the reader who has no knowledge about these case studies or whose familiarity with them is, at best, perfunctory.

Finally, it should be appreciated that within the context and constraints of a book of this type, some aspects of material behavior in service, important as they are, will be treated only briefly. Examples of such aspects are fatigue, fracture, creep, corrosion, oxidation, wear, and erosion. Since the book is not intended to be exhaustive or even comprehensive but rather representative, and being cognizant of the intended readership, it is suggested that details of these aspects be obtained from other texts (see Appendix D).

It is my hope that this book will provide a real service for college students and will find a wider audience in the engineering community of practicing material engineers, machine element designers, and all people involved in the decision-making process regarding use of materials in the manufacturing and production engineering industries.

Gladius Lewis
Memphis, Tennessee

Part I

Case Studies

FLYWHEEL ENERGY STORAGE SYSTEM IN AN ELECTRIC VEHICLE FOR THE POSTAL SERVICE

CHAPTER 1

1.1 INTRODUCTION

A flywheel is, quite simply, a device for storing kinetic energy. It has a number of attractive features, including (a) excellent flexibility (it can be made of any size and can be put anywhere in an electrical or mechanical system), (b) a variety of energy interfaces (the input or output can be electrical, hydraulic, mechanical, or any combination thereof), (c) relative insensitivity to the number and frequency of charge and discharge cycles, (d) unlimited depth of discharge, (e) simplicity, (f) little maintenance, (g) infinite shelf life, and (h) benign effect on the environment. The disadvantages of a flywheel include (a) relatively poor energy storage per mass, (b) poor efficiency (short rundown time), and (c) an ever-present hazard of catastrophic failure (leading to a substantial derating of steel flywheels to a small fraction of their theoretical maximum performance).

Flywheels have found a wide range of applications, including use in automobiles, urban subway trains, mass transit buses, earth-moving vehicles, wind power–generating machines, switching locomotives, large road scrapers, electric welding units, and helicopter high-speed hoists, as well as attitude control systems in spacecraft and satellites, power sources for communications satellites, plasma physics experiments, and telephone exchanges and backup power sources for computers and other equipment that require uninterrupted power in order to avoid very expensive consequences.

A priori, it could be anticipated that since the objective is to store the maximum kinetic energy per unit mass and since kinetic energy is directly proportional to mass density, storage ability will be dependent only on mass density. This, however, would be an incomplete consideration. It is important to note that the maximum speed of the flywheel is determined by some failure criterion, typically expressed in terms of stress due to centrifugal body forces, which is in turn dependent on mass density. For a flywheel with given dimensions, the kinetic energy per unit mass is dependent on the density of the material. Thus the design criterion of maximizing the kinetic energy per unit mass favors the use of the lightest possible material, provided that this material can withstand reasonably high stresses up to failure.

1.2 OPTIMIZATION ANALYSIS

The objective in flywheel design being to store the maximum amount of kinetic energy per unit mass of the flywheel, the two criteria of performance are thus stored kinetic energy and flywheel mass. In analyzing for the optimum material, the discussion is limited to thin flywheels so that plane stress conditions may be assumed.

For such a case, Christensen and Wu [1] have shown that the stored kinetic energy per unit mass (U/M) for a flywheel made of a composite material is given by

$$\frac{U}{M} = \left[\frac{1 + (\nu_{\theta r}/\nu_{r\theta}) + 2\nu_{\theta r}}{1 - \nu_{r\theta}\nu_{\theta r}}\right]\frac{E_r e}{2\rho} \tag{1.1}$$

where $\nu_{\theta r}$ is the Poisson's ratio of the material that characterizes contraction in the r direction due to tension in the θ direction, $\nu_{r\theta}$ is the Poisson's ratio of the material that characterizes contraction in the θ direction, E_r is the radial modulus of elasticity of the material, ρ is the density of the material, and e is the strain in the material in either the radial or hoop direction. Defining S as an appropriate elastic failure strength of the material, thus making $S = E_r e$, makes equation (1.1) become

$$\frac{U}{M} = \left[\frac{1 + (\nu_{\theta r}/\nu_{r\theta}) + 2\nu_{\theta r}}{1 - \nu_{r\theta}\nu_{\theta r}}\right]\frac{S}{2\rho} \tag{1.2}$$

Equation (1.2) needs to be modified by the shape factor k_s. This factor reflects the effects of the specific configuration of the flywheel (Table 1.1). Thus equation (1.2) becomes

$$\frac{U}{M} = \left[\frac{1 + (\nu_{\theta r}/\nu_{r\theta}) + 2\nu_{\theta r}}{1 - \nu_{r\theta}\nu_{\theta r}}\right]\frac{Sk_s}{2\rho} \tag{1.3}$$

TABLE 1.1 FLYWHEEL CONFIGURATIONS AND SHAPE FACTORS

Flywheel Configuration	Mass Distribution	Shape Factor, k_s
Isotropic Materials		
Constant stress shaped disc (no hole)		1.0
Flat disc		0.6
Flat disc (with hole)	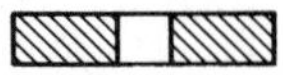	0.3
Modified constant stress disc (no hole)		0.9
Anisotropic Materials		
Thin rim		0.5
Thin rod	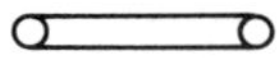	0.3

For an isotropic material, $\nu_{r\theta} = \nu_{\theta r} = \nu$, the Poisson's ratio of the material. Thus

$$\frac{U}{M} = \frac{Sk_s}{(1 - \nu)\rho} \tag{1.4}$$

The choice of materials for the flywheel may be made on the basis of equation (1.3) or (1.4). First, though, one important clarifying comment is necessary. Since the normal mode of operation of a flywheel is rotational, leading to fluctuating stresses (due to the fluctuating speed), the relevant failure model is fatigue. Thus in equations (1.3) and (1.4), put $S = FL$, the fatigue limit or endurance strength* of the material in tension-tension loading. Fortunately, the best fatigue properties in filamentary composite materials are usually obtained in this loading mode, with the load being along the fiber direction, which is the situation in most flywheel designs. Some fatigue loading data are presented in Figs. 1.1 and 1.2.

A second important requirement in a material to be used to manufacture a flywheel is thus a high value of FL or, more realistically, a high value of the specific fatigue or endurance strength, FL/ρ.

A third important requirement is a high value of the stored kinetic energy per unit mass per unit total cost. The total cost is made up of the raw material and fabrication costs. The latter costs, however, are related to the selected con-

*For ferrous materials, use fatigue limit; for all other materials, use endurance strength at 10^6 cycles.

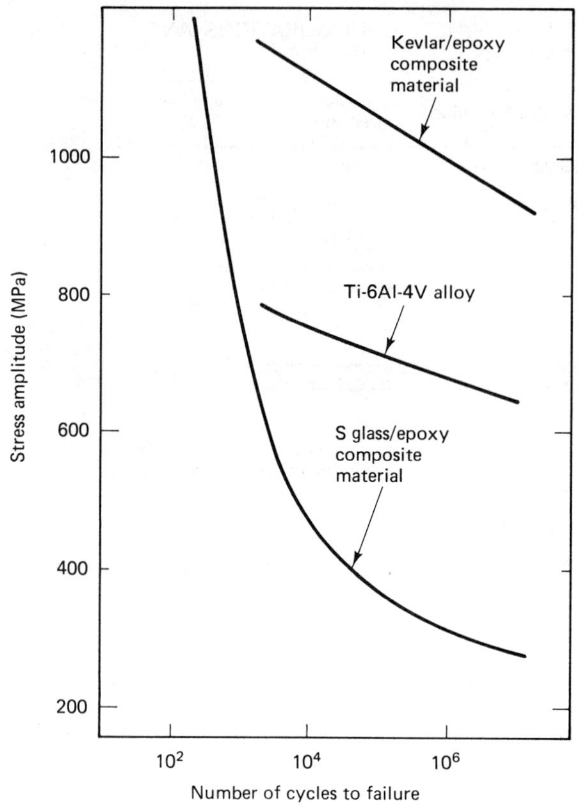

Figure 1.1 Tension-tension fatigue failure curves for Kevlar/epoxy, a titanium alloy and glass/epoxy materials.

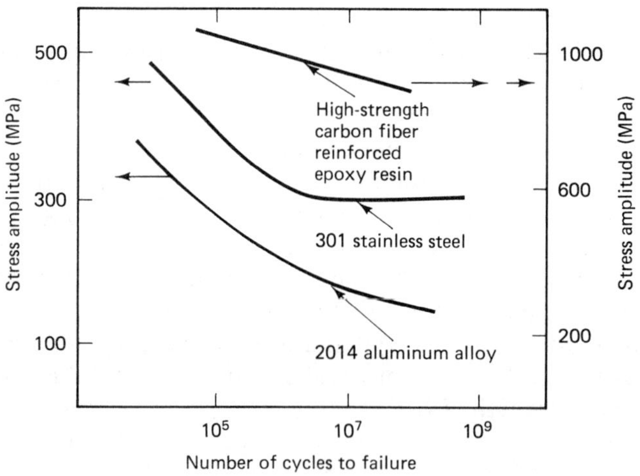

Figure 1.2 Tension-tension fatigue failure curves for carbon/epoxy, a stainless steel and an aluminum alloy.

figuration of the flywheel and its shape (radial reinforcements, circumferential reinforcements, circumferentially wound ring, etc.). However, fabrication methods and selection of "best" shapes lie in the realm of practical design and hence outside the scope of this book. Thus only raw material cost per unit stored kinetic energy per unit mass will be considered.

1.3 CANDIDATE MATERIALS

Until recently, the candidate materials for flywheels were almost invariably metallic (alloys of iron, aluminum, or titanium). Since about 1975, however, composite materials (epoxy resin reinforced with fibers of carbon, graphite, Kevlar 49, or glass) have been considered suitable for this application. Thus all of the following are candidate materials:

- 301 stainless steel
- 2014 aluminum alloy
- Ti-6Al-4V alloy
- Carbon fiber reinforced epoxy resin
- S-glass fiber reinforced epoxy resin
- Kevlar 49 fiber reinforced epoxy resin

Table 1.2 shows that the most promising material is the Kevlar 49 fiber reinforced epoxy resin. It has very high values of both energy density (U/M) and specific endurance strength (FL/ρ), although its energy cost, at the present time, is rather high.

In general, composite materials possess the following advantages for use in flywheel construction:

- High specific stiffness
- High specific strength—they can thus store high quantities of energy per unit mass
- Good potential for satisfactory long-term performance
- Fragmentability—when they fail, it is easier to contain the various fragments because they tend to break into small pieces
- Excellent potential for low manufacturing cost
- Excellent potential for tailorability—they can be made into filament-wound rings, laminated discs, etc.

Notwithstanding this catalog of attractive features, a number of considerations are specific to the use of filamentary composite materials for flywheels (and may, in fact, compromise their selection).

1. A generalized performance index, expressed as

$$\left(\frac{U}{M}\right)_{\text{max}} = \psi \cdot \min \frac{k_{si} S_i}{\rho} \tag{1.5}$$

is necessary.

In equation (1.5), i equals the ith mode of failure of the total number possible for a given design concept; k_{si} is the shape factor incorporating geometry, material orthotropy, type of fabrication, and failure mode; S_i is the base strength of the material for the ith failure mode; and ψ incorporates the Poisson's ratio and density of the material.

TABLE 1.2 MECHANICAL PROPERTIES, ENERGY DENSITY, AND ENERGY COST FOR CANDIDATE MATERIALS

Material	E_{rr} (GPa)	$E_{\theta\theta}$ (GPa)	FL after 10^6 Cycles (MPa)	ρ (kg m^{-3})	$\nu_{r\theta}$	$\nu_{\theta r}$[a]	k_s	U/M (Wh kg^{-1})	Material Cost (\$ kg^{-1})	Energy Cost[b] (\$ Wh^{-1})	FL/ρ (10^3 m)
301 stainless steel	193	193	300[c]	8027	0.28	0.280	0.9[d]	12.98	3.5	0.27	3.81
2014 Al alloy	73	73	200[c]	2796	0.35	0.350	0.9[d]	27.51	2.5	0.09	7.29
Ti-6Al-4V alloy	116	116	700[e]	4540	0.31	0.310	0.9[d]	55.86	15.0	0.27	15.72
High-strength graphite fiber reinforced epoxy resin[g]	8.8	124	996[c]	1570	0.27	0.019	0.5[f]	49.08	130.0	2.65	64.67
S-glass fiber reinforced epoxy resin[g]	21	54	300[e]	1997	0.25	0.097	0.5[f]	16.91	5.0	0.29	15.31
Kevlar 49 fiber reinforced epoxy	4.5	76	1000[e]	1366	0.33	0.020	0.5[f]	56.32	50.0	0.89	74.62

[a] $\nu_{\theta r} = \nu_{r\theta} E_{rr}/E_{\theta\theta}$.

[b] Energy cost = material cost/(U/M).

[c] From Fig. 1.2.

[d] Modified constant stress disc; no hole.

[e] From Fig. 1.1.

[f] Multidrum thin-rim design.

[g] Volume fracton of fiber = 0.60.

This index is necessary to account for the greater complexity of composite materials arising from such factors as the anisotropy of the elastic and strength properties and their heterogeneity. Also, fiber composite materials have low strengths in failure modes that are not controlled by the fiber properties, and the time-dependent properties of these materials can differ significantly from the static values.

2. Material anisotropy leads to material-geometry coupling, which must be taken into account in the design optimization. This could mean, for example, tailoring the material properties (for a given design) or tapering the flywheel profile (for a given material system) in order to achieve the maximum kinetic energy density possible for that flywheel.

3. Composite material strength properties (transverse tensile) controlled by the matrix can limit the flywheel performance to a level below the theoretical maximum kinetic energy density.

4. Definition of the time-dependent properties of fiber composite materials is essential. A body of reliable data on this aspect is not currently available since much of it has only recently been generated. A good estimate can be obtained by adjusting the material strength values to reflect the effects of cyclic loading and stress rupture. In effect, the flywheel can be designed to a specified minimum lifetime, say, 10 years or 500,000 cycles.

1.4 FLYWHEEL CONFIGURATIONS

A number of configurations for the flywheel using composite materials have emerged. The first example is the laminated circular brush rotor, which employs a large number of thin rods of unidirectional composite material (Fig. 1.3). This configuration has the advantage of intrinsic simplicity and safety, but its principal draw-

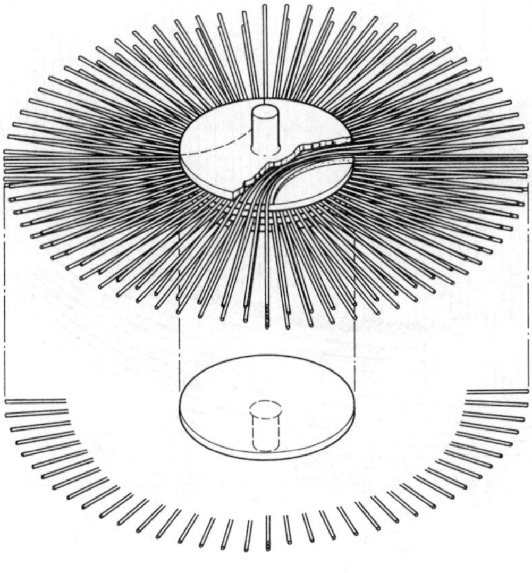

Figure 1.3 Laminated circular brush rotor.

backs are its relatively low volume efficiency and a somewhat higher rotational speed requirement. A second configuration (Fig. 1.4) is the multiring type, which does not have the disadvantages of the laminated circular brush rotor and provides some additional improvement in performance in terms of stored kinetic energy per unit volume of material. A third configuration (Fig. 1.5) has as its main at-

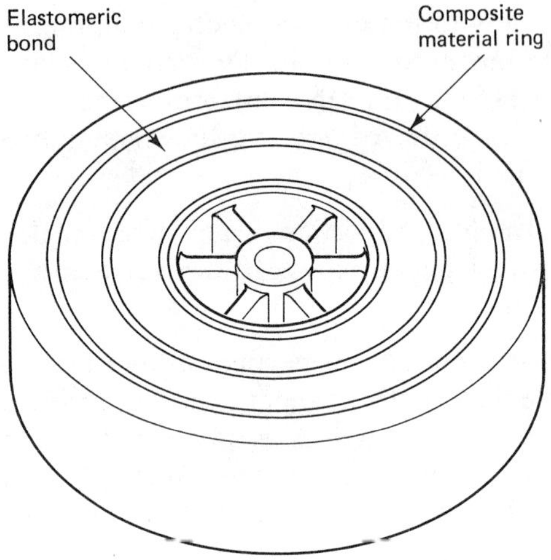

Figure 1.4 Multiring flywheel design.

Figure 1.5 The pseudoisotropic flywheel configuration.

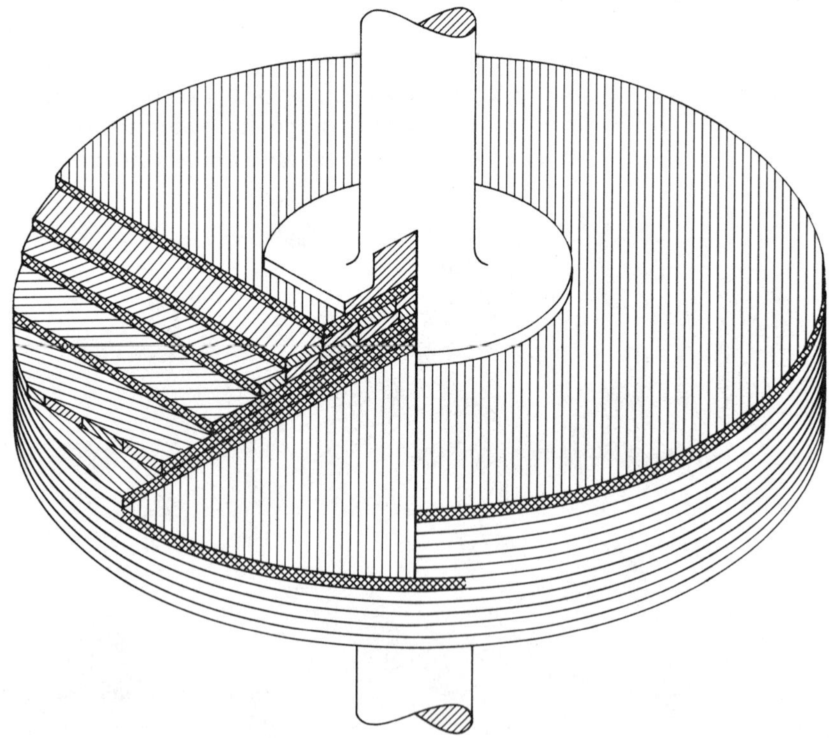

TABLE 1.3 SOME ADVANCED FLYWHEEL DESIGNS

Concept (Manufacturer)	Materials	Demonstrated Energy Density (Wh kg^{-1})	Comments
Multiple thin ring (Garrett Air Research)	S2 glass-epoxy inside rings; Kevlar 29-epoxy interior rings; Kevlar 49-epoxy outside rings	60	This concept is probably the best for Kevlar systems. Arguably, the most successful concept to date.
Solid disc (General Electric/CR&D)	E-glass-epoxy quasi-isotropic laminated flat disc with attachment holes	20	Early rotors were not made from high-quality laminates.
Bare filament (Rabenhorst, Johns Hopkins University/Applied Physics Lab.)	Kevlar fiber (hoop-wound bare filament)	100	Interesting low-cost system with excellent performance.
Thick ring (Oak Ridge/Y-12)	Kevlar-epoxy (hoop-wound)	20	Kevlar is probably not a good choice because transverse property requirements are important.

tributes high volume effectiveness and low manufacturing cost but is limited by a rather moderate value of maximum theoretical stored energy density of 66 Wh kg^{-1}. Some details regarding a few examples of advanced flywheel designs are presented in Table 1.3.

REFERENCE

1. R. M. Christensen and E. M. Wu, "Optimal design of anisotropic (fiber-reinforced) flywheels," *J. Composite Materials 11*, pp. 395–404, 1977.

WINGS OF A HIGH-PERFORMANCE FIGHTER JET

CHAPTER 2

2.1 INTRODUCTION

It is generally acknowledged among students of strategic defense that air power remains one of the most effective deterrents to war at every level. Thus there is constant competition between the superpowers of NATO and the Warsaw Pact to gain and maintain a technological edge in the field of fighter aircraft. The Soviets have three advanced all-weather fighters—the MiG-29, the Sukhoi Su-27 Flanker, and the MiG-31 Foxhound—in service now or soon to become operational. On its side, NATO has the U.S. F-16 Falcon (Fig. 2.1), the F-15 Eagle, the F-18 Hornet, and Western Europe's Tornado.

2.2 BUCKLING OF A THIN RECTANGULAR PLATE

The stress analysis of the upper wing surface of an aircraft may be simplified by approximating it as a long, thin rectangular plate of constant thickness uniformly compressed, at its midplane in one direction, by a uniformly distributed force per unit length, N_x, arising from the wind loading applied to simply supported opposite edges, with the unloaded edges being free (Fig. 2.2). The governing condition for the operation of such a structure is its buckling at some critical value of N_x.

Figure 2.1 The U.S. Air Force's F-16 in a combat exercise. (Courtesy of *Aerospace America*)

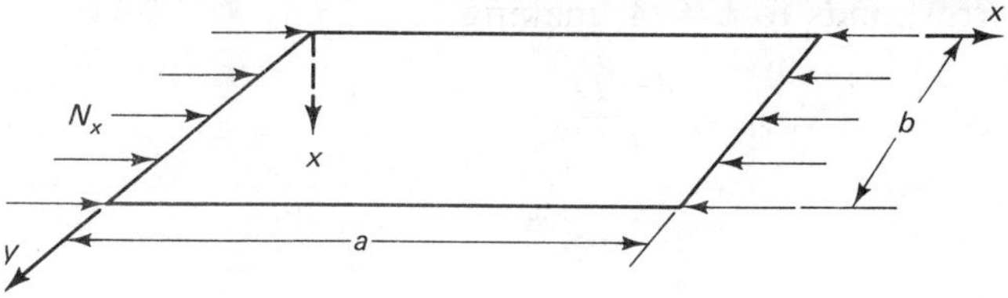

Figure 2.2 Simply supported laminated rectangular plate under uniform uniaxial in-plane compression.

N_x is given as [1]

$$N_x = \frac{\pi^2 D}{b^2}\left(\frac{mb}{a} + \frac{n^2 a}{mb}\right)^2 \tag{2.1}$$

where D is the flexural rigidity of the plate $= Et^3/12(1 - \nu^2)$ (E and ν are the modulus of elasticity and Poisson's ratio, respectively, of the plate material, and t is the plate thickness), m and n are coefficients in the Fourier series used to describe the deflection of any point on the plate,* and the plate is simply supported along edges $x = 0$, $x = a$, $y = 0$, and $y = b$ (Fig. 2.2). N_x has its lowest value when $n = 1$, indicating that, on buckling, there can be a number of halfwaves in the direction of compression but only one halfwave in the orthogonal direction. The critical value of N_x is thus

$$(N_x)_{cr} = \frac{\pi^2 D}{b^2}\left(\frac{mb}{a} + \frac{a}{mb}\right)^2 = k\,\frac{\pi^2 D}{b^2} \tag{2.2}$$

The factor k thus has a magnitude that depends on the number of halfwaves m and the aspect ratio of the plate r ($= a/b$). The values of $(N_x)_{cr}$ and m can be

*m and n are, in fact, the number of halfwaves in the x and y directions, respectively.

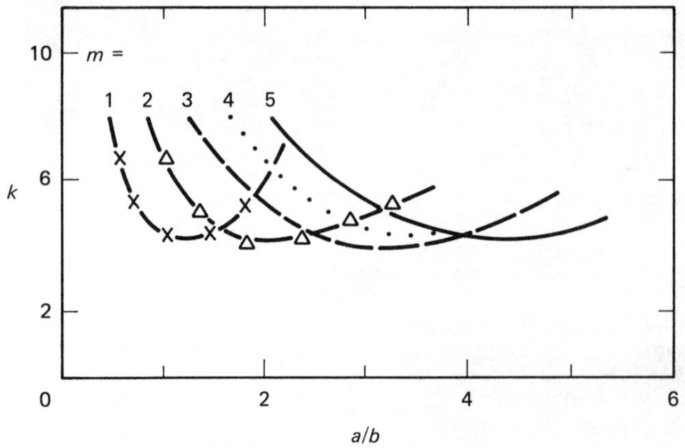

Figure 2.3 The buckling modes of a thin rectangular plate under compressive loading.

obtained, for any value of r, by taking the value of the ordinate of Fig. 2.3, which gives the smallest value of k for that value of r.

The minimum value of $(N_x)_{cr}$ occurs when

$$\frac{d(N_x)_{cr}}{d(mb/a)} = \frac{2\pi^2 D}{b^2}\left(\frac{mb}{a} + \frac{a}{mb}\right)\left[1 - \frac{1}{(mb/a)^2}\right] = 0 \tag{2.3}$$

that is, $m = a/b$. This corresponds to $k = 4$, making

$$(N_x)_{cr} = \frac{4\pi^2 D}{b^2}$$

Recalling that

$$D = \frac{Et^3}{12(1 - \nu^2)}$$

yields

$$(N_x)_{cr} = \frac{4\pi^2 Et^3}{12b^2(1 - \nu^2)}$$

The critical or buckling stress $\sigma_{cr} = (N_x)_{cr}/t$ becomes, in this case,

$$\begin{aligned}\sigma_{cr} &= \frac{4\pi^2 E}{12\,(1 - \nu^2)}\left(\frac{t}{b}\right)^2 \\ &= \frac{\pi^2 E}{3\,(1 - \nu^2)}\left(\frac{t}{b}\right)^2\end{aligned} \tag{2.4}$$

The buckling load $P_{cr} = \sigma_{cr} tb$ or

$$P_{cr} = \frac{\pi^2 E}{3\,(1 - \nu^2)}\left(\frac{t}{b}\right)^2 tb \tag{2.5}$$

or

$$t^3 = \frac{P_{cr} 3b(1 - \nu^2)}{\pi^2 E} \tag{2.6}$$

Let us define the structural efficiency of the wing η as

$$\eta = \frac{\text{critical load on wing}}{\text{weight of wing}} = \frac{P_{cr}}{tb\rho} \tag{2.7}$$

Thus

$$\begin{aligned} \eta &= \frac{\pi^2 E t^3}{3(1 - \nu^2)b \cdot tb\rho} \\ &= \frac{\pi^2}{3(1 - \nu^2)b^2} \left(\frac{E}{\rho}\right) t^2 \end{aligned} \tag{2.8}$$

Solving for t from equation (2.6) and inserting the result in eq. (2.8) gives

$$\begin{aligned} \eta &= \frac{\pi^2 E}{3(1 - \nu^2)b^2\rho} \left[\frac{3P_{cr}(1 - \nu^2)b}{\pi^2 E}\right]^{2/3} \\ &= \text{constant} \left(\frac{E}{1 - \nu^2}\right)^{1/3} \left(\frac{1}{\rho}\right) \left(\frac{P_{cr}}{b^2}\right)^{2/3} \end{aligned}$$

(P_{cr}/b^2) is the maximum compressive wing loading; it has a value of 12 MPa for low–aspect ratio wings, such as those on double Delta jets, and 30 MPa for high–aspect ratio wings, such as those on the Tornado [2].

Thus, for a given wing loading, the structural efficiency of the upper wing surfaces is dependent on the material parameter given by

$$\frac{1}{\rho}\left(\frac{E}{1 - \nu^2}\right)^{1/3}$$

2.3 FATIGUE RESISTANCE AND FRACTURE TOUGHNESS

As far as the lower wing surfaces are concerned, there are two important requirements for suitable materials. One is high fatigue or endurance strength; this is very important because in a fighter jet aircraft, in addition to fluctuating stresses encountered during takeoff, climb, approach, and landing, very high cyclical stress is generated during the numerous and rapid maneuvers that the craft is put through. The second requirement is high fracture toughness: The combination of tensile stress and flaws in these surfaces can limit the life of the wing.

2.4 CANDIDATE MATERIALS

The demand on the performance of fighter aircraft has been increasing. In the future, these aircraft will be expected to operate over broader combat envelopes, that is, a combination of higher altitudes and faster speeds (Fig. 2.4). (Fortunately, reliability and maintainability, as measured by the ratio of design service life to flight years, continue to improve; see Fig. 2.5.) These demands and expectations may be met by a reduction in structural weight, for a given set of design constraints, achieved through a series of advanced concepts (Table 2.1). The most dramatic

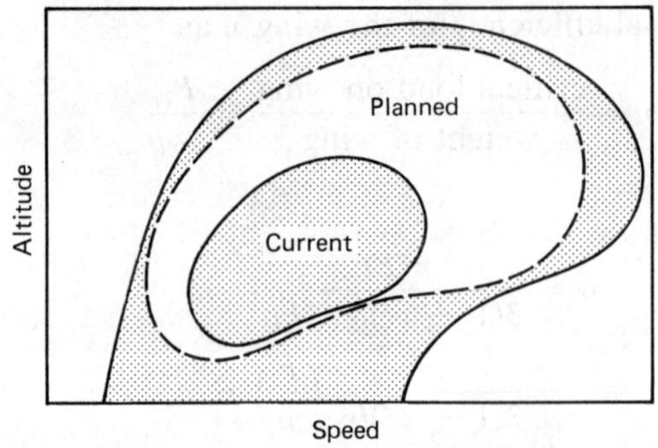

Figure 2.4 Flight envelopes of fighter aircraft.

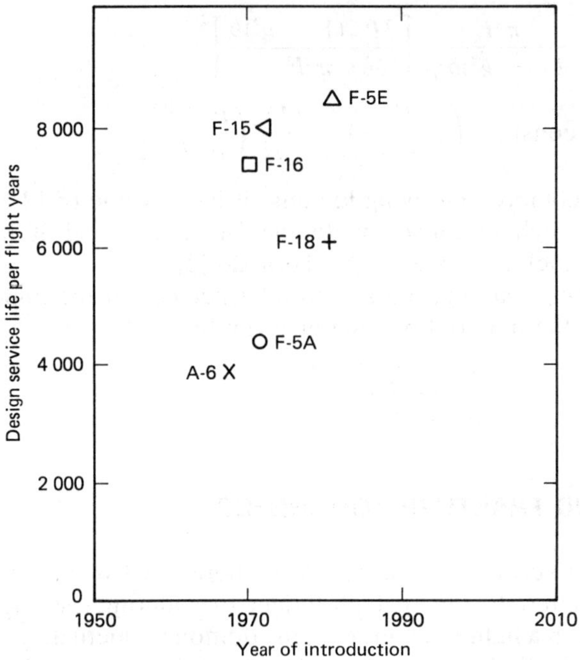

Figure 2.5 Design service life of U.S. fighter aircraft.

TABLE 2.1 TRENDS IN AIRCRAFT STRUCTURAL WEIGHT REDUCTION

	Average weight reduction (%) due to			
Year	Advanced structural concepts	Automated design	Active load control and flutter suppression	New and improved materials
1970		—	—	5.0
1980	2.5	1.0	2.5	10.0
1990	3.0	2.5	3.0	15.0
2000	3.0	2.5	3.0	20.0
2010	3.0	3.0	3.0	20.0

Source: M. M. Ratwani, "Impact of composite materials on advanced fighters," *SAMPE Quart.* *17* (2), p 21, 1986.

weight reduction is obtained through the use of new and improved materials. In this category, filamentary composite materials are currently the most promising candidates (it is anticipated that future work will focus on improved versions of these materials as well as metal-matrix composite materials). This is so because of their very high specific stiffness and specific strength. In fact, there has been a substantial growth in the use of these materials in a multitude of applications in fighter jet aircraft (see Tables 2.2 and 2.3 and Figs. 2.6, 2.7, and 2.8).

In a study of future tactical aircraft in the United States, it was found that significant size and cost reductions are possible by using composite materials to account for up to 75% of the structural weight (Table 2.4).

The wings of the prototype of the X-29 aircraft are made from laminated assemblies of a graphite-epoxy unidirectional tape. The X-29, with its forward-

TABLE 2.2 APPLICATION OF FILAMENTARY COMPOSITE MATERIALS IN FIGHTER JET AIRCRAFT

Aircraft (popular name)	Weight of Composite as a Proportion of Total Structural Weight (%)	Estimated Weight Saving Accruing from Use of Composites (%)
F14 (Tomcat)	1	20
F15 (Strike Eagle)	1	25
F16 (Falcon)	2	25
F18 (Hornet)	10	35
Advanced Harrier (AV-8B)	30	23
French experimental combat aircraft (Rafale)	25	N.A.

N.A., data not available as of June 1988.

TABLE 2.3 WEIGHT AND COST SAVINGS FROM USE OF ADVANCED COMPOSITE MATERIALS IN SOME COMPONENTS OF THE U.S. B1 BOMBER PROTOTYPE

Application	Weight savings (%)	Cost savings (%)
Avionic access door[a]	12	2
Dorsal longeron[a]	44	Data not available
Flap[b]	15	Data not available
Horizontal box[c]	14	20
Slat[d]	15	40
Vertical box[a]	15	15
Weapons bay door[a]	12	10

Sources: (a) Rockwell International, El Segundo, California.
(b) Northrop Corp., Hawthorne, California.
(c) Grumman Corp., Bethpage, New York.
(d) Lockheed Missiles and Space Company, Inc., Sunnyvale, California.

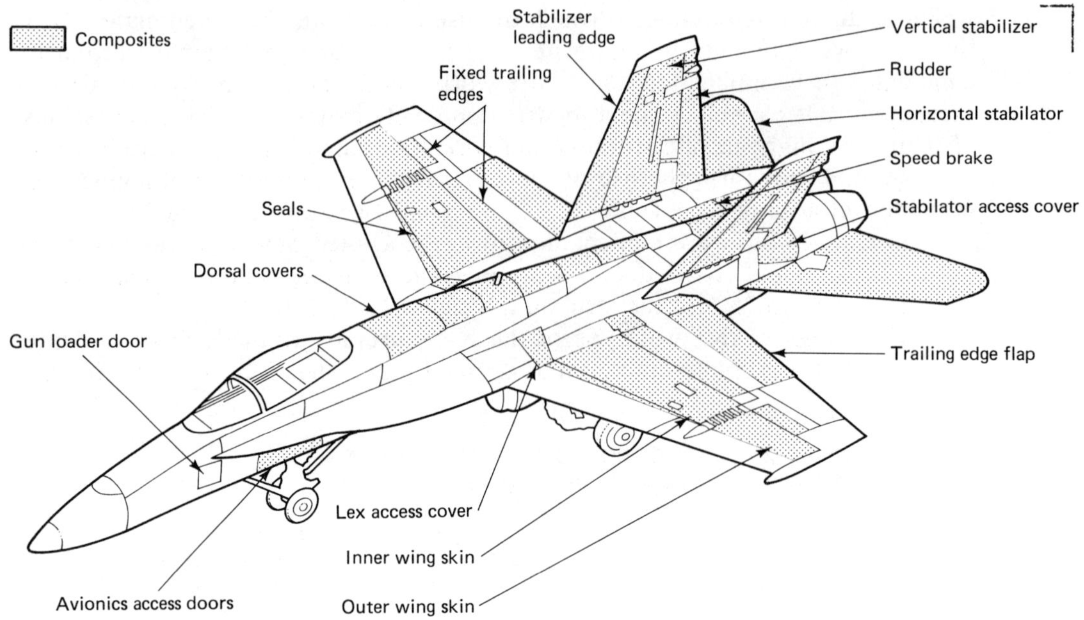

Figure 2.6 Distribution of filamentary composite materials on a U.S. Air Force F/A-18 fighter jet. (Courtesy of McDonnell Douglas)

Figure 2.7 Use of boron fiber reinforced epoxy composite material in the stabilizers and rudder on a U.S. Air Force F-15 fighter jet. (Courtesy of McDonnell Douglas)

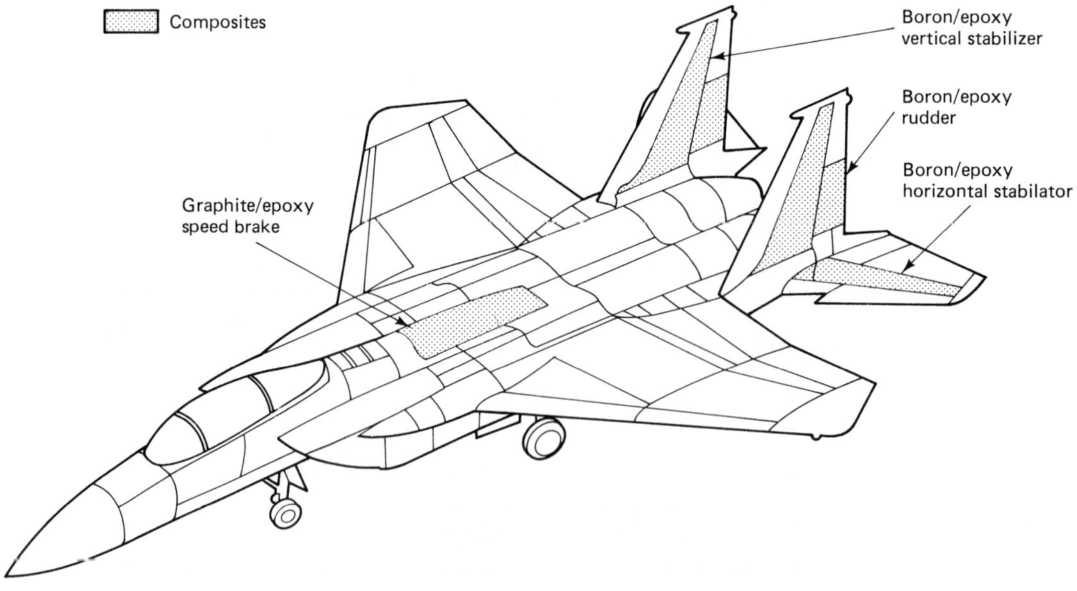

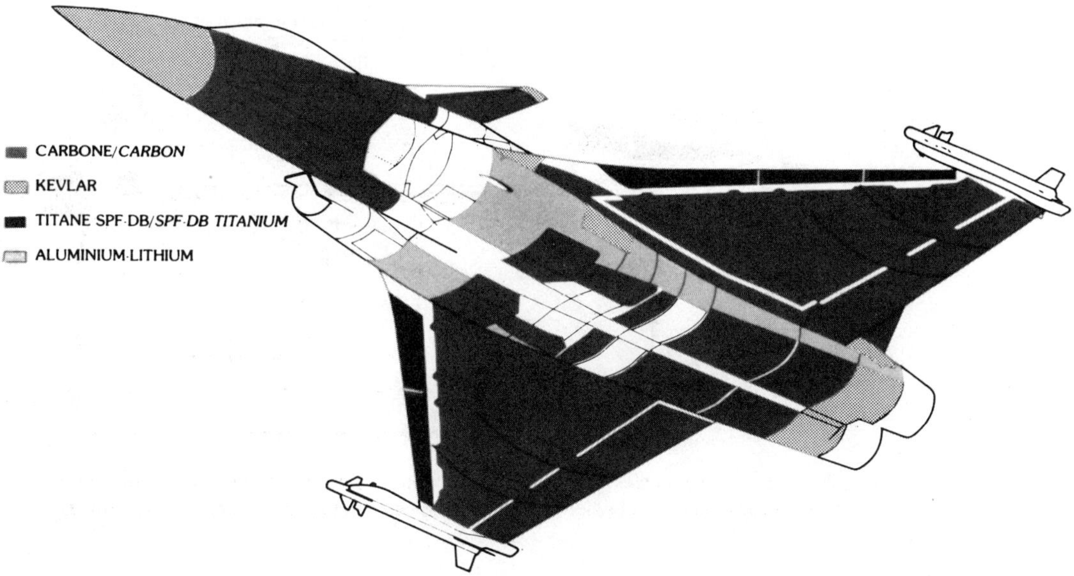

Figure 2.8 Use of filamentary composite and other advanced materials in the French Rafale fighter jet aircraft. (Courtesy of *Aerospace America*)

TABLE 2.4 WEIGHT COMPARISON OF SOME ADVANCED AIRCRAFT DESIGNS

	Take off gross weight (t) due to			
Type of aircraft	Useful load	Fuel	Systems	Structures
Advanced design composite	2.2	5.5	4.5	4.5
Advanced metal	2.2	7.0	5.0	8.0
Composite substitution	2.0	7.0	5.5	7.0

Source: J. F. Young, and K. S. Shane (ed.), *Materials and Processes Part A: Materials,* 3rd ed., New York: Marcel Dekker, 1985, p. 789.

swept wings and canards (Fig. 2.9), is the latest of the U.S. military aircraft (during 1988 it was in the advanced stages of research and development).

The candidate materials for the wings are thus the traditional ones (alloys of aluminum, iron, or titanium) as well as filamentary composites and the new aluminum-lithium alloys.

The maximum value of the structural efficiency of the upper wing surfaces, for a given wing loading, has been evaluated for the various candidate materials, and the results are presented in Table 2.5. The best results are obtained with the filamentary composite materials; high-modulus graphite and Kevlar 49 fiber reinforced epoxy resins are the strongest contenders.

A comparison of the candidate materials for the lower wing surfaces is given in Table 2.6. Again, filamentary composite materials are superior to metallic materials.

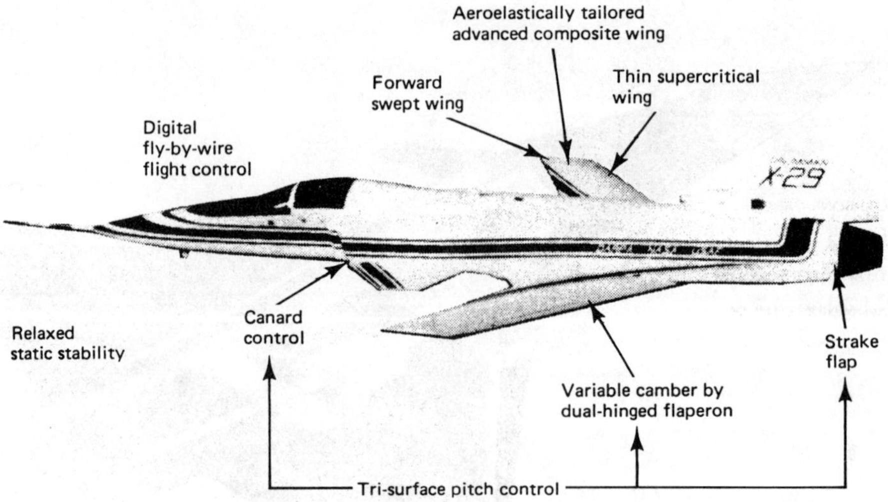

Figure 2.9 Some technologies incorporated into the X-29 aircraft for structural weight reduction and improved control when flying at low speed at a high angle of attack. (Courtesy of *Aerospace America*)

TABLE 2.5 STRUCTURAL EFFICIENCIES OF CANDIDATE MATERIALS FOR THE UPPER SURFACES OF THE WINGS OF A FIGHTER JET AIRCRAFT

						Structural efficiency ($N^{-2/3}$ $m^{10/3}$ s^{-2})	
Material	E_{xx} (GPa)	E_{yy} (GPa)	ν_{xy}	ν_{yx}[a]	ρ ($kg\ m^{-3}$)	η_{xx}[b]	η_{yy}[c]
301 stainless steel	193	193	0.28	0.280	8027	0.73	0.73
2014 Al alloy	73	73	0.35	0.350	2796	1.55	1.55
Ti-6Al-4V alloy	116	116	0.31	0.310	4540	1.10	1.10
Al-Li alloy	79	79	0.30	0.300	2685	1.64	1.64
High-modulus graphite fiber reinforced epoxy resin[d]	207	6	0.25	0.007	1412	4.24	1.28
High-strength graphite fiber reinforced epoxy resin[d]	141	8	0.27	0.015	1570	3.37	1.26
Kevlar 49 fiber reinforced epoxy resin[d]	86	7	0.33	0.027	1366	3.33	1.39
Carbon fiber reinforced epoxy resin[d]	138	7	0.30	0.015	1661	3.19	1.14
Boron fiber reinforced epoxy resin[d]	207	21	0.30	0.030	1990	3.05	1.37
Boron fiber reinforced 6061 Al alloy[d]	262	159	0.30	0.182	2600	2.52	2.09

[a] $\nu_{yx} = \nu_{xy}E_{yy}/E_{xx}$

[b] $\eta_{xx} = [E_{xx}/(1 - \nu_{xy}^2)]^{1/3}/\rho$

[c] $\eta_{yy} = [E_{yy}/(1 - \nu_{yx}^2)]^{1/3}/\rho$

[d] Volume fraction of fiber = 0.60.

TABLE 2.6 PROPERTIES OF CANDIDATE MATERIALS FOR THE LOWER SURFACES OF THE WINGS OF A FIGHTER JET AIRCRAFT

Material	ρ (kg m^{-3})	Fatigue or Endurance Strength, FL[a] (MPa)	Fracture Toughness, K_{IC} (MPa m$^{0.5}$)	FL/ρ (10^3 m)	K_{IC}/ρ (10^3 m$^{1.5}$)
301 stainless steel	8027	300	100	3.81	1.27
2014 Al alloy	2796	200	50	7.29	1.82
Ti-6 Al-4V alloy	4540	700	65	15.72	1.46
Al-Li alloy	2685	278[b]	33	10.55	1.25
High-modulus graphite fiber reinforced epoxy resin	1412	800	12[c]	57.75	0.87
Kevlar 49 fiber reinforced epoxy resin	1366	1000	10[c]	74.62	0.75
Carbon fiber reinforced epoxy resin	1661	530	40	32.53	2.45
Boron fiber reinforced epoxy resin	1990	1070	46	54.81	1.80

[a] At 10^6 cycles (for nonferrous materials).

[b] Assumed to be 50% of ultimate tensile strength.

[c] Normal to crack direction.

REFERENCES

1. C. T. Wang, *Applied Elasticity*. New York: McGraw-Hill, 1953.
2. R. J. Sellars and G. Terry, "Sophisticated aircraft structure developments: Combat aeroplanes," *Aeronautical J. 85* (847), pp. 334–342, 1981.

RADAR CROSS-SECTION REDUCTION

CHAPTER 3

3.1 INTRODUCTION

The emergence of increasingly accurate and sophisticated detection systems can compromise the effectiveness with which many weapons systems carry out their mission. Many scientific and engineering studies are thus being carried out aimed at ensuring the success of these missions by reducing the detectability of the weapons systems through a decrease in their radar cross section.

This is the rationale behind the exotic fighter aircraft expected to become part of the defense aircraft fleet of the United States by 1990, the bomber called Stealth (Fig. 3.1). Reduction of radar signature in this aircraft is effected through three technically advanced methods. One is the reduction of radar cross section through (a) shaping: the use of the flying wing, which, in its purest form, has neither fuselage nor tail and does away with all reflective edges, corners, and boxy surfaces, and (b) the use of radar-absorbing filamentary composite materials. The second method is the introduction of innovative aerodynamic concepts; for example, the aircraft has few control surfaces—the roles of ailerons, elevators, rudders, and perhaps flaps will be filled by vector nozzles that will guide the craft by directing the flow of the jet engine's exhaust. The third method is the use of highly sensitive and sophisticated electronic countermeasures.

Figure 3.1 The B-2 "STEALTH" bomber of the United States Air Force (Courtesy of the United States Air Force).

3.2 RADAR CROSS SECTION (RCS)

When an electromagnetic wave impinges on an object, energy is dispersed in all directions. The spatial distribution of this energy depends on the size, shape, and composition of the object and on the frequency and nature of the incident wave. This distribution, called scattering, is characterized by a cross section σ. Thus in the case where the object is impinged on by radar, we have monostatic scattering (i.e., the source and the receiver of the electromagnetic wave are located at the same point). It is usual to display the data in logarithmic form and specify RCS in terms of dB_{sm}, defined as "decibels above a square meter" such that $dB_{sm} = 10 \log \sigma$, where σ is in m^2.

The RCS of the simplest of three-dimensional shapes, the sphere, is presented in Fig. 3.2 as a function of the electrical circumference, $ka = 2\pi a/\lambda$, where λ is the frequency of the wave and a is the radius of the sphere. For other simple shapes, the maximum values of σ are given in Table 3.1.

The RCS of a complex body, such as an aircraft or a ship, is obviously more complicated because it contains several significant "scattering centers" and many

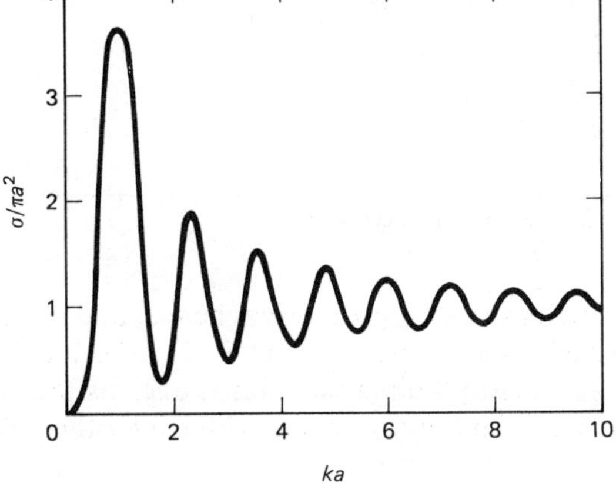

Figure 3.2 The radar cross-section of a metallic sphere, as a function of its electrical-circumference.

TABLE 3.1 RADAR CROSS SECTION (σ) OF SOME SIMPLE GEOMETRIC SHAPES

Geometric Shape	Formula for σ[a]
Square trihedral corner retro-reflector	$12\pi a^4/\lambda^2$
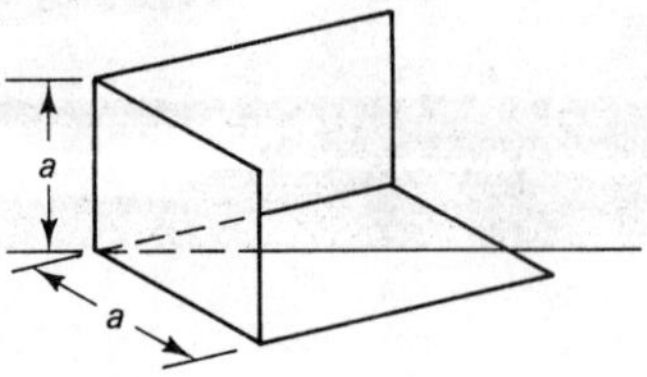	
Right dihedral corner reflector	$8\pi a^2 b^2/\lambda^2$
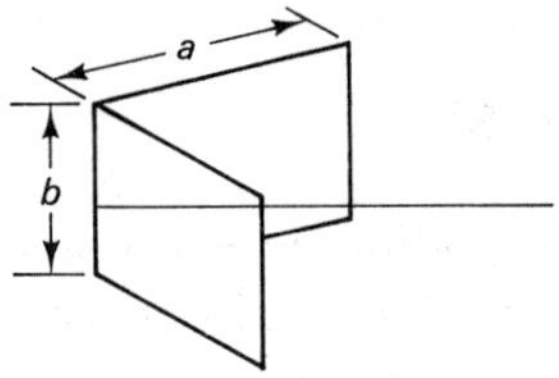	
Flat plate	$4\pi a^2 b^2/\lambda^2$
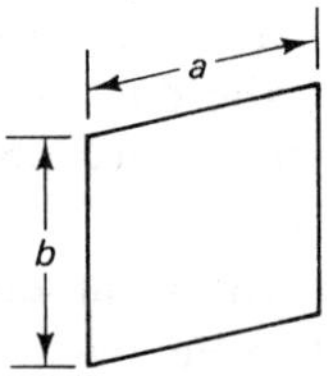	
Cylinder	$2\pi a b^2/\lambda$
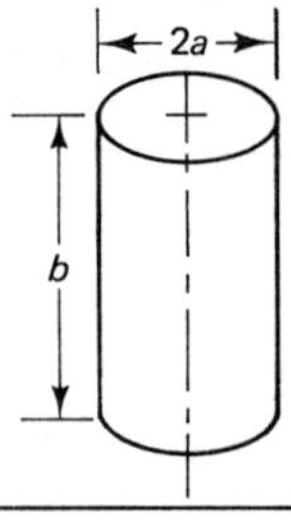	

[a] λ is the wavelength of the incident wave.

more less significant ones, such as seams, ports, rivet heads, and stiffening plates. Because of this multiplicity of "scatterers," the net RCS pattern exhibits a rapid scintillation with aspect angle due to the mutual interference as the contributions from the various scatterers go in and out of phase with each other. See Figs. 3.3 and 3.4 for the RCS of a commercial airliner.

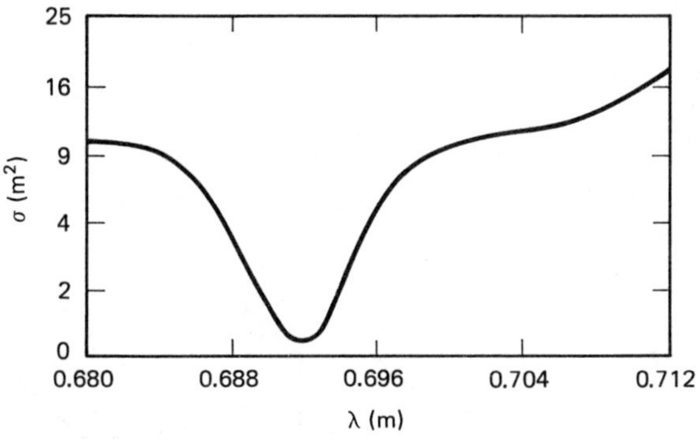

Figure 3.3 The radar cross section, as a function of wavelength, λ, for the nose-on aspect of a commercial aircraft.

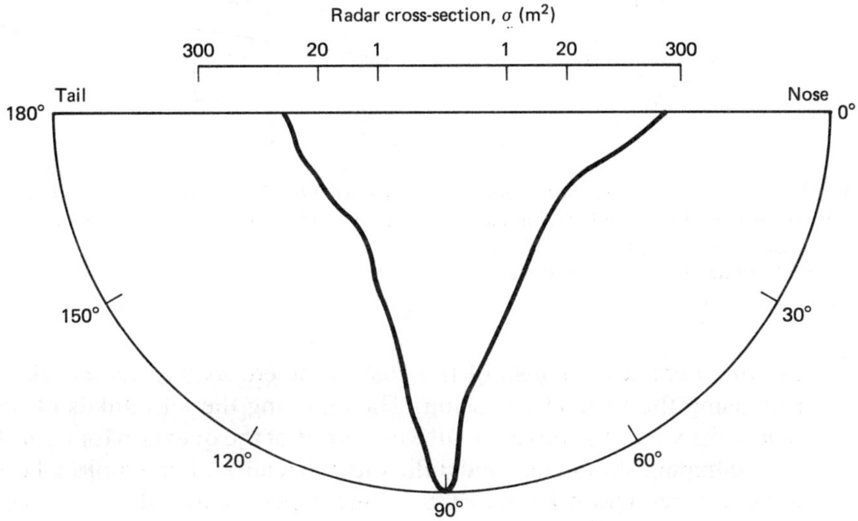

Figure 3.4 The radar cross-section of a commercial aircraft, as a function of aspect angle.

3.3 RADAR CROSS-SECTION REDUCTION (RCSR)

There are four basic techniques for reducing radar cross section.

Shaping

The objective here is to orient the target surfaces and edges so as to deflect the reflected energy away from the radar source. In the case of an aircraft, large cross sections may be "shifted" out of the forward sector toward the broadside sectors by, for example, sweeping airfoils back at sharper angles or using highly swept wings (Fig. 3.5), or else fin shapes may be changed (Fig. 3.6). For ships and ground transportation vehicles, internal dihedral and trihedral corners may be avoided by bringing intersecting surfaces together at acute or obtuse angles.

Shaping is a more useful tool in the design stage but only if RCS threat sectors can be specifically identified. This method is less useful for retrofitting

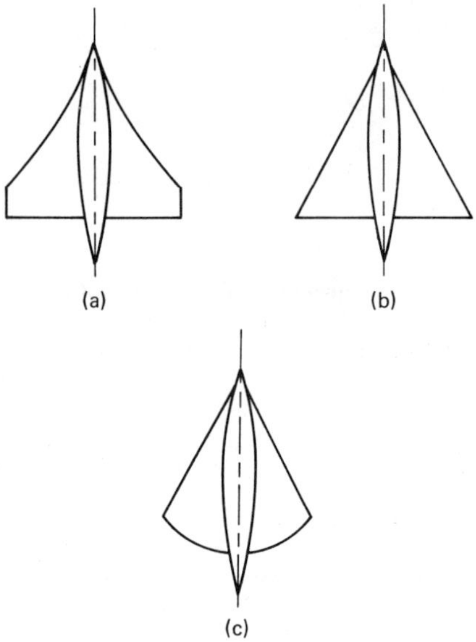

Figure 3.5 Radar cross-section reduction by shaping: (a) curved leading edge at a greater sweep angle; (b) delta wing; (c) modified delta wing with rounded corners.

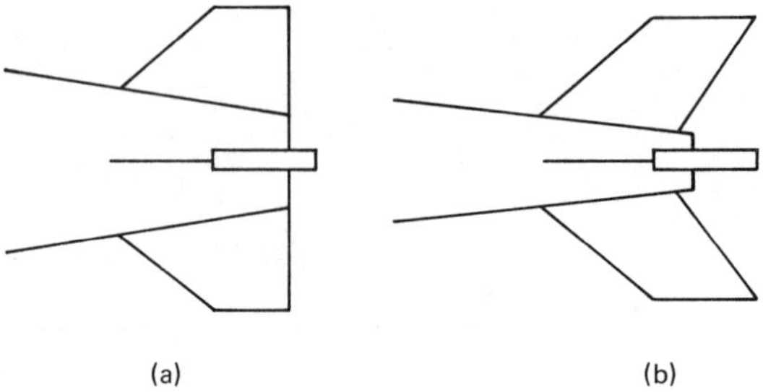

Figure 3.6 Radar cross-section reduction by improved fin design: (a) bad; (b) good.

existing vehicles because of the cost involved, as well as the likelihood of compromising the vehicle's mission. Thus moving the jet intakes of an aircraft from below the wings to above the fuselage is out of the question for operational aircraft.

Shaping should be used judiciously because of an empirical observation that for every reduction in the RCS at one aspect angle, there is a concomitant enhancement of RCS at another aspect angle.

The following are some of the shaping innovations reportedly* incorporated in the STEALTH bomber which lead to an estimated RCS of 0.5 m^2†:

- sharply swept back leading edges
- wingspan of about 55m
- vertical surfaces tilted inward or completely eliminated (for example, no vertical tail surface)
- rivets on the wing skin eliminated (the trailing edge is flexible enough to move vertically).
- the jet engines have S-shaped intakes which are located above the wings (to reduce possible stalling and cutoff of the engines, due to a restricted amount of inlet air at high attack angles such as at take off and landing, boundary air control or auxiliary air inlets under the wings may be used).

*Based on the artist's impression drawing of the bomber provided by the U.S. Air Force.

†Author's estimate based on information gleaned from the limited information in the open literature.

Passive Cancellation (Impedance Loading)

This involves adjusting the amplitude and phase of an echo source so as to cancel out another echo source. This method is impractical for all but the simplest of objects.

Active Cancellation (Active Loading)

In this method, the target emits radiation whose amplitude and phase cancel the reflected energy. This is obviously an ambitious technique since it implies that the target must be "clever" enough to sense the angle of arrival, intensity, frequency, and waveform of the incident wave; "know" its own echo characteristics for that particular wave and angle of arrival; fast enough to generate the required waveform and frequency; and be versatile enough to adjust and radiate a pulse of the proper amplitude and phase.

Radar-absorbing Materials (RAMs)

This method relies on the ability of the materials to absorb as much as possible of the incident electromagnetic energy, thereby reducing the net energy available for reflection back to the target. In order to absorb energy, the materials must behave in such a way that the induced currents are in phase with the incident fields.

Many RAMs are manufactured with carbon providing the loss mechanism and the dissipation of energy taking place by conversion of the electromagnetic energy into heat. Another class of materials is dielectrics, whose refractive indices are complex numbers, with the imaginary part giving rise to the loss. The molecules in such materials are essentially small dipoles with tendencies to orient themselves along the incident field. Conduction losses are lumped together with the imaginary part of the refractive index for engineering convenience. Also, magnetic and electric losses can occur because the refractive index is a function of both the magnetic permeability and the electric permittivity. These lossy materials may include ferrites or carbonyl iron in addition to or in place of carbon.

1. One type of radar absorber, the Salisbury screen, is a sheet of porous material impregnated with graphite (the load) placed a quarter of a wavelength from a metallic backing plate. From transmission theory, this setup creates an open circuit at the load. Thus the incident wave "sees" free space and hence there is no reflection. If the backing plate is at a distance other than a quarter of a wavelength from the load, the wave "sees" a finite impedance in parallel with the impedance of the resistive sheet, resulting in a net complex impedance that is different from that of free space. The result is that part of the incident wave is reflected, with a magnitude that increases with the frequency with which their incident moves away from the quarter wavelength spacing. Thus a Salisbury screen is a narrowband device whose bandwidth may be increased by placing several sheets, one behind the other, separated by spacers. The screen has been used in various commercial RAMs. However, the rapid oscillations for large spacings makes it rather ineffective over a wide frequency range. For increased mechanical rigidity, polymers or higher-density foam may be used as the spacers.

2. Another type of radar absorber, the Dallenbach layer, consists of a layer of a homogeneous lossy material backed by a metal plate. The reflection at the surface of the material is due to the impedance change "seen" by the wave at the junction between the metal and the lossy material. Thus if a material can be found whose impedance relative to free space equals unity (i.e., $\epsilon_r = \mu_r$),* there will be no reflection at the surface. Attenuation will then depend on the loss properties of the material (i.e., ϵ'' and μ_r'')* and the electrical thickness. However, materials with appropriate dielectric and magnetic properties to act as a matched RAM over any appreciable frequency range are uncommon, and hence the best approach is to optimize the loss at a given frequency by using available materials.

For a single material layer backed by a conducting plate, the reflection coefficient, R, is given by

$$R = \frac{\sqrt{(\mu_r/\epsilon_r)}\tanh(-ikt) - 1}{\sqrt{(\mu_r/\epsilon_r)}\tanh(-ikt) + 1}$$

where t is the thickness of the lossy material. The variation of R with t/λ (where λ is the wavelength) is presented in Figs. 3.7 and 3.8 for predominantly electric and predominantly magnetic loss materials, respectively. For the former type of

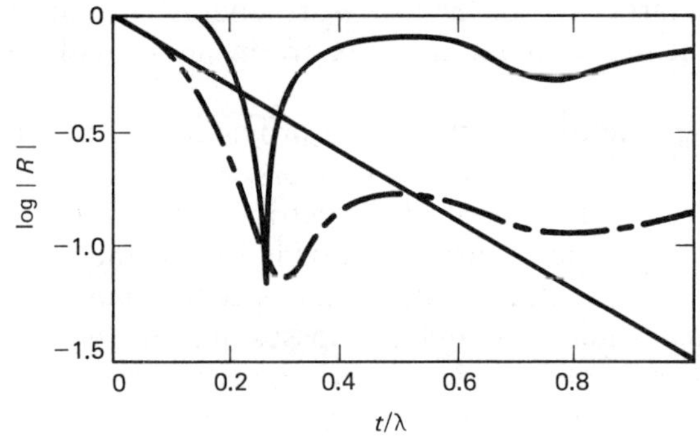

Figure 3.7 Reflectivity curves for predominantly electric materials. Solid line is for $|\varepsilon_r| = 16$, $|\mu_r| = 1$, $\delta_\varepsilon = 20°$, and $\delta_\mu = 0°$. Dashed curve is for $|\varepsilon_r| = 25$, $|\mu_r| = 16$, $\delta_\varepsilon = 30°$, and $\delta_\mu = 20°$. Diagonal line is for $|\varepsilon_r| = |\mu_r| = 4$ and $\delta_\varepsilon = \delta_\mu = 15°$.

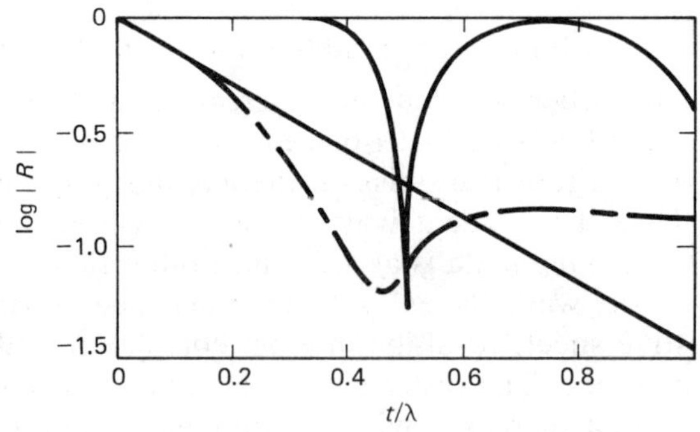

Figure 3.8 Reflectivity curves for predominantly magnetic materials. Solid line is for $|\mu_r| = 16$, $|\varepsilon_r| = 1$, $\delta_\mu = 10°$, and $\delta_\varepsilon = 0°$. Dashed curve is for $|\mu_r| = 25$, $|\varepsilon_r| = 16$, $\delta_\mu = 20°$, and $\delta_\varepsilon = 30°$. Diagonal line is for $|\varepsilon_r| = |\mu_r| = 4$ and $\delta_\varepsilon = \delta_\mu = 15°$.

*Relative permittivity: $\epsilon_r = \epsilon_r' + i\epsilon_r''$
Relative permeability: $\mu_r = \mu_r' + i\mu_r''$
Electric loss tangent: $\delta = \tan^{-1}(\epsilon_r''/\epsilon_r')$
Magnetic loss tangent: $\delta_m = \tan^{-1}(\mu_r''/\mu_r')$
Wave number in the material: $k = k_0\sqrt{\mu_r\epsilon_r}$

where $k_0 = \omega\sqrt{\mu_0\epsilon_0}$ is the free space number and ω is the radian frequency.

material ($\mu_r \simeq 1$), the best RCSR performance is obtained when the material has an electrical thickness of about $\lambda/4$, while for a pure magnetic absorber ($\epsilon_r \simeq 1$), the optimum thickness is about $\lambda/2$.

3. The Gauss absorber [1], which is another single-layer type of absorber, consists of filaments of radar absorbing chaff (metallic strands with diameters and lengths typically 50 nm and 50 μm, respectively) in a solid binder of near unity dielectric constant. Resistive dissipation in the filaments leads to attenuation of the incident wave. It has been calculated [1] that for a 20-mm-thick application, RCSR exceeds 30 dB from 10 GHz to 100 GHz (when the filaments are in a regular array in the matrix) and ranges from 13 dB at 10 GHz to more than 30 dB at 20 GHz to 100 GHz (when the filaments are randomly dispersed in the matrix).

4. The Jaumann and graded dielectric absorbers represent two of the improvements of the single-layer absorbers through the use of multiple layers. A Jaumann absorber is obtained when additional resistive sheets and spacers are added to a Salisbury screen. For maximum performance, the front sheets should have a high resistivity and the back sheets should have a low resistivity. The bandwidth is a function of the number of sheets used, i.e. thickness (Table 3.2).

TABLE 3.2 BANDWIDTH OF JAUMANN ABSORBERS

Number of Sheets[a]	Fractional Bandwidth
1	0.27
2	0.55
3	0.95
4	1.16

[a] Each sheet is 7.6 mm thick.

A graded dielectric can be used to match the impedance between free space and a perfect conductor. The optimum method for designing such an absorber is to determine analytically the required permeability μ and permittivity ϵ as a function of distance into the material and to limit the modulus of the reflection coefficient frequency range, subject to considerations of incidence angle and thickness. Unfortunately, there is not yet a fully theoretical solution to this problem [2]. An engineering approach has been to assume a model for μ and ϵ as a function of distance into the absorber z and the layer thickness l and solve for the resulting reflection coefficient, R. A number of such models have yielded the results given in Table 3.3. The minimum thickness is of the order of $\lambda/3$ suggesting that, even in the ideal case, an absorber nearly 50 mm thick would be required for a RCS of 20 dB down to 2 GHz.

Graded dielectric RAMs are made from discrete layers, with properties varying from layer to layer. One commercially available type is the AN series, for example AN-74, marketed by Emerson and Cuming [3]. This is a three-layer foam absorber, about 20 mm thick, and is advertised to provide an RCSR of 20 dB down to a frequency of 3.5 GHz.

TABLE 3.3 SEVERAL GRADED DIELECTRIC RAM DESIGNS

Type of Variation	$\mu_r'(z)$	$\mu_r''(z)$	$\epsilon_r'(z)$	$\epsilon_r''(z)$	Minimum l/λ for $R \leq 0.1$
Ideal Jacobs	1	0	$(1 - z/l)^{-2}$	Small constant $\ll 1$	0.30
Finite Jacobs	1	0	$(1 - 0.95\, z/l)^{-2}$	0.50	0.42
Linear	1	0	1	$3\, z/l$	0.55
Exponential	1	0	1	$0.285 \exp(2.73\, z/l)$	0.35
Three-layer discrete approximation to exponential	1	0	1	0.58 for 0.344 l, 1.16 for 0.359 l, 3.48 for 0.297 l	0.33

Source: G. T. Ruck (ed.), *Radar Cross Section Handbook*. Vol 2. New York: Plenum, 1970, p. 628.

5. Another type of RAM employs the principle of reducing the reflection from the front face of a flat absorber by using a material whose intrinsic impedance is approximately equal to one. There are two common variants of this type of material. One, called the hair type, is formed by mixing cured animal hair with conducting carbon black in neoprene. Plessey [4] introduced one such material that gives advertised RCSRs of better than 15 dB in the 8–14 GHz range and better than 10 dB in the 6–100 GHz range. The other type uses light carbon-loaded foams. They come in a variety of shapes, such as pyramidal, aggregate sine wave front, and off-normal angle wedges, and give an RCSR of better than 50 dB, but thicknesses in excess of 10λ are needed in order to achieve this level of performance [5].

6. Circuit analog radar absorbing materials (CA RAMs) use lossy materials deposited in various geometric patterns (such as dipoles, triangles, and crosses) on a thin lossless film. The thickness of the deposit affects the effective resistance of the layer, while effective inductance and capacitance are functions of the geometry of the pattern. The layer can therefore be carefully selected to give required values of inductance, capacitance, and resistance, and its performance can be analyzed in terms of lumped parameters. Typically, a CA absorber panel is comprised of a number of layers separated by light spacers.

7. Magnetic materials depend primarily on magnetic rather than dielectric losses for their radar absorbing properties. Because of this, compounds of iron are most often used. Ferrites and carbonyl iron (a form of pure iron powder), are common examples. Ceramic materials containing ferrites are particularly useful for high-temperature applications provided the operating temperatures do not exceed the Curie point of the material. (For example, the Curie points of cubic spinel ferrites and hexagonal ferrites are about 600°C and 500°C, respectively.) The ferrites used are typically sintered in the form of small, rigid tiles, thus making application to a surface critically dependent on the bonding method used.

Alternatively, the magnetic material may be embedded in a flexible matrix of natural or synthetic elastomer and the whole piece bonded onto the surface to be shielded using an appropriate adhesive. Several companies have developed spray-on materials (so-called iron paints) in which the magnetic "dust" is suspended in an epoxy medium. Constant stirring is necessary in order to prevent the iron particles from settling at the bottom. This method has the advantage of being able to be applied to irregular surfaces with ease.

One major disadvantage associated with the use of magnetic materials is the problem of "streaking," the rusting of the material in an oxidizing environment due to the presence of some form of iron. At high temperatures, oxidation of carbonyl iron to nonmagnetic αFe_2O_3 is a serious problem since it leads to a loss of absorbing power. Also, these materials tend to be heavy (about 34 kg m^{-2}), although they provide significant low-frequency RCSR with a treatment only about 7 mm thick [6] (see Fig. 3.9).

8. Hybrid absorbers can be designed and fabricated using combinations of the materials discussed in categories 1–7 provided the electromagnetic properties of the various materials, over a wide frequency range, are known.

Although hybrid absorbers can provide significant performance in a reasonable thickness, problems in bonding, complexity in production and maintenance, and high cost may limit their widespread adoption.

9. A new class of materials, called retinyl Schiff base salts, was recently discovered [7]. Black in color (resembling graphite), they reportedly could absorb 80% of incident radar energy, and they are about 10% the weight of ferrites (for

Figure 3.9 Typical sintered ferrite absorber properties [8]. (reprinted with permission from Emerson and Cuming).

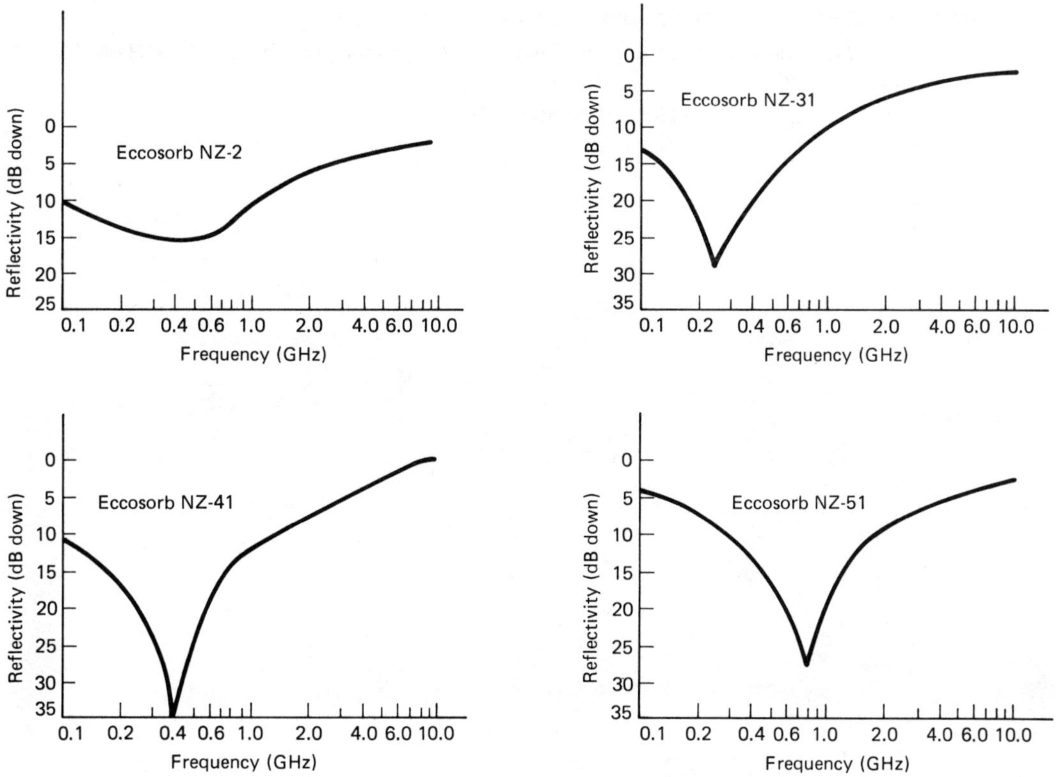

the same volume). Not much is known about the radar-absorbing potential of these materials as work on them is classified under United States Department of Defense regulations.

3.4 CANDIDATE MATERIALS

For reduction of radar cross section using the materials approach, some excellent candidate materials are sintered ferrites (e.g., Eccosorb NZ-2, Eccosorb NZ-31, Eccosorb NZ-41, and Eccosorb NZ-51) and graded dielectrics (e.g., AN-74).

REFERENCES

1. A. Gauss, "A new type of electromagnetic absorbing coating," Ballistic Research Laboratory Report AD A117472, June 1982.
2. G. T. Ruck (ed.), *Radar Cross Section Handbook,* vol. 2, chap. 8. New York: Plenum, 1970.
3. Emerson and Cuming, Inc., "Eccosorb AN-W high-loss dielectric microwave absorbers," Technical Bulletin 8-2-4A, January 1980.
4. Plessey (U.K.) Ltd., "Advance Information, External Netting Absorber (ENA)" April 1983.
5. W. H. Emerson, "Electromagnetic wave absorber: useful tools for engineers," Emerson and Cuming, Inc., Canton, MA, no date.
6. Emerson and Cuming, Inc., "Eccosorb NZ thin ferrite absorber for 50 MHz to 15 GHz," Technical Bulletin 8-2-17, May 1979.
7. R. R. Birge, Private Communication, 1988.

HEAT PIPE IN A COAL GASIFICATION PLANT

CHAPTER 4

4.1 INTRODUCTION

Coal gasification may be described as the chemical processes by which coal is reacted with air or a mixture of oxygen and steam to produce gaseous products. This system for producing synthetic fuels is very promising. Used on a small scale since the mid-18th century, it is now being extensively studied and evaluated by a number of countries (in the United States through the U.S. Energy Research and Development Agency, ERDA).

Coal gasification processes are currently commercialized in one of the following categories:

- Lurgi process (fixed-bed gasification; see Fig. 4.1)
- Koppers-Totzek (entrained flow gasification; see Fig. 4.2)
- Winkler (fluidized-bed gasification; see Fig. 4.3)
- Wellman-Galusha (revolving-grate gasification); see Fig. 4.4)

In a Lurgi fixed-bed gasifier, the coal bed rests on a rotary grate. The gasification agent, a blend of oxygen and steam, is led into the bed from the bottom upward. Due to the countercurrent configuration, a temperature profile, the peak of which comes close to the melting point of the ash, is obtained in the longitudinal

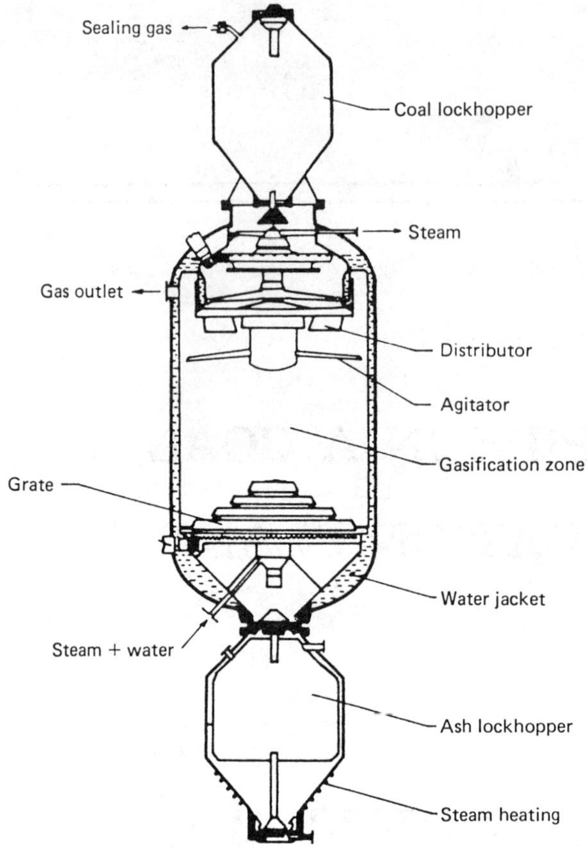

Figure 4.1 The fixed-bed Lurgi generator.

Figure 4.2 The Koppers-Totzek gasifier (model with two burner nozzles).

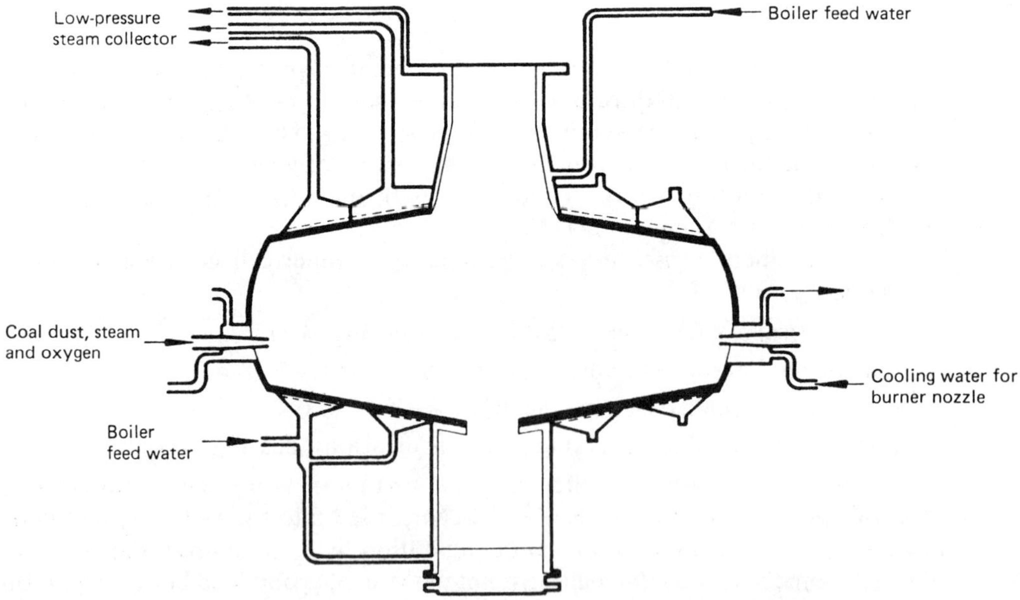

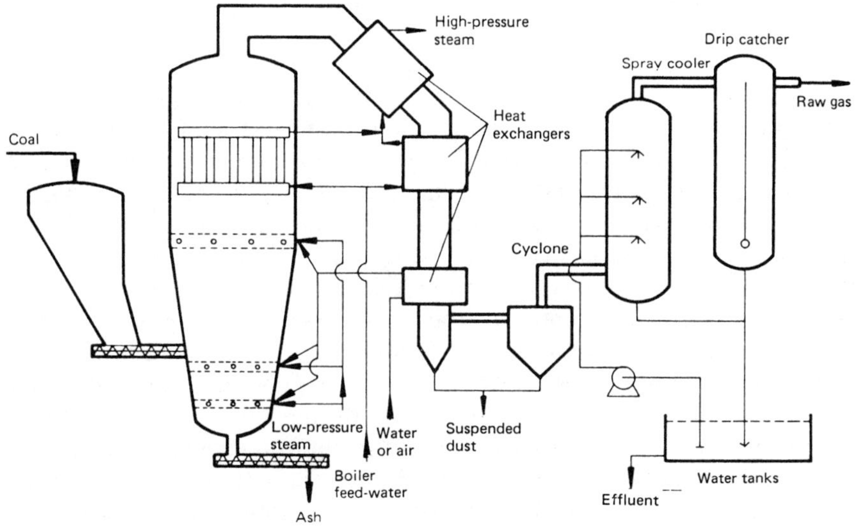

Figure 4.3 Schematic representation of the Winkler process.

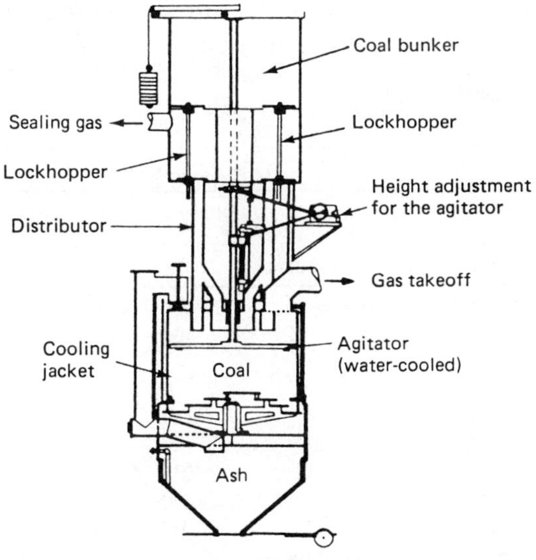

Figure 4.4 The Wellman-Galusha gas generator (with agitator).

axis of the reactor. The gas leaves the reactor at a temperature of about 400°C and entrains, besides the gaseous components, condensing components such as higher hydrocarbons and steam. In its basic form, this type of reactor is lined with refractory material. Two examples of plants using this process are in Sasolburg, South Africa, and Westfield, Scotland.

In the Koppers-Totzek gasifier, two burners are arranged opposite to each other in the lateral walls. The flames are deflected upward without the reactor walls being touched. While the slags are withdrawn by gravity from the reactor bottom, the gas leaves the reactor via the top. The reactor is fitted with a cooled refractory lining. The Shell-Koppers plant in Hamburg, West Germany, is a typical operational unit.

The fluidized bed is characterized by the movement of individual particles. These particles move in suspension within the reactor since they are borne by the gasification agent streaming through the fluidized bed from the bottom upward. A fluidized-bed temperature that is homogeneous over the whole reactor is lower than the melting point of the ash in order to avoid agglomeration. The impact effect of the particles in the suspension is negligible, so erosion in the reactor is not a problem. Examples of plants using this process are in Frechen, West Germany; Gorazde, Yugoslavia; and Madras, India.

Grate generators represent the earliest type of equipment used industrially for the gasification of solid fuels. Difficulties in ash removal, however, frequently disrupted or halted operation with this type of generator. The introduction of the revolving-grate generator around the turn of the 20th century constituted a major design improvement. A revolving-grate generator generally consists of a cylindrical shaft, usually between 1 m and 3 m in diameter. The walls are constructed as a water jacket or may be brick-lined. The shaft has a sturdy cover, which carries the charging equipment. The gasifying medium, air saturated with steam at a high temperature or an oxygen-steam mixture, is injected from below through the revolving grate.

Some operational data and process characteristics of all four types of gasification processes are given in Table 4.1. No matter which process is used, the

TABLE 4.1 OPERATIONAL DATA FOR COMMERCIAL COAL GASIFICATION PROCESSES

	Type of Process			
Operational Parameter	Lurgi	Koppers-Totzek	Winkler	Wellman-Galusha (with O_2 and steam)
Temperature (°C)	760–1400	1300	800–1000	750
Pressure (MPa)	3–10	0.1–4	0.1–1	0.1
Raw gas analysis (%)				
CO_2	27–32	7–12	13–25	11–16
CO	18–25	55–66	30–50	40–54
H_2	36–40	22–32	35–45	30–40
CH_4	9–10	0.1	0.5–2.0	0.4–0.9
C_nH_m	0.6	—	—	—
N_2	4.0	1.0	0.5–1.5	1.5–2.6
Calorific value (MJ m^{-3})	11.5	10.6–11.8	9–13	10.4
Specific yield of gas (m^3 t^{-1} of coal)	1400–1700	1650	1600	2434–2560
By-products	Tar, oil, naphtha, ammonia	None	Much dust, hardly any higher hydrocarbons	None
Coal requirements	Lignites, weakly coking coal	Any	Many lignites	Anthracite, coke
Coal size (mm)	6–40	<0.1	<8	2–70
Specific steam consumption (t)	1.0–1.4	0.24	0.88	1.0–1.4
Specific oxygen consumption (m^3)	220–300	540	350	540–560
Carbon conversion efficiency (%)	99	90–96	90	98
Gasification efficiency (%)	75–85	70–77	≤75	81–86
Thermal efficiency (%)	75	75–85	80	88–96

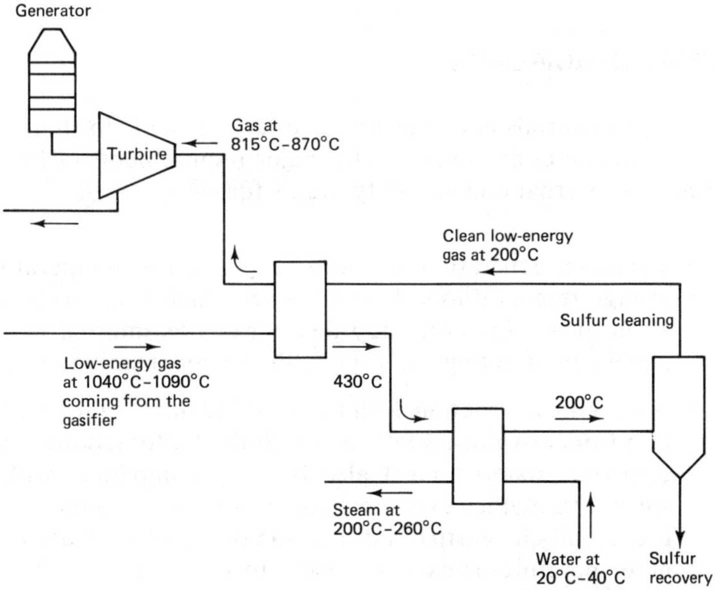

Figure 4.5 Sulfur cleaning system of low-energy gas with heat recovery.

requirements are quite severe (temperatures in excess of 750°C, pressures in excess of atmospheric, and corrosion and erosion by molten coal, slag, iron, etc.).

The chemical processes involved thus demand (a) extensive use of heat exchangers for energy conservation, (b) control of reaction temperatures for process efficiency, and (c) production of synthetics largely free from sulfur. Heat pipes offer unique advantages in each of these needs.

In the present example, we consider the selection of material for the heat pipe for the section of the coal gasification plant involved with sulfur cleaning of low-energy gas (Fig. 4.5). Heat pipe recovery units are used in two stages to cool hot gases to about 210°C and, following the sulfur scrubbing process (which removes hydrogen sulfide and other sulfer-containing compounds), heat the clean gases prior to use in the gas turbine.

The basic heat pipe, shown in Fig. 4.6, consists of a permanently closed container that encloses a small amount of vaporizable fluid.

Figure 4.6 Basic heat pipe.

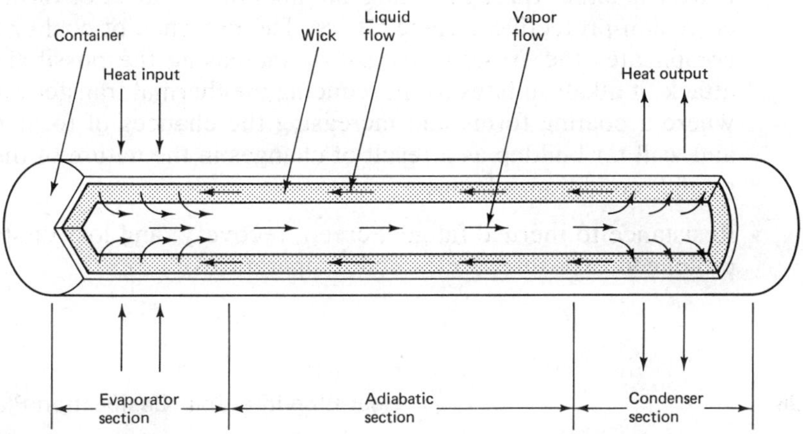

4.2 MATERIAL REQUIREMENTS

Envelope materials must meet certain requirements for both internal and external environmental compatibility. The major requirements of heat pipe envelope materials for internal compatibility are as follows:

- Minimum generation of hydrogen gas at low temperatures and oxygen gas at high temperatures. Requiring very high level vacuums for reliable operation, heat pipe efficiency deteriorates as internal gas pressure increases, adversely affecting the vaporization-condensation cycle.
- No chemical reaction with the working fluid. Fluid breakdown, gas desorption from envelope walls, and wall-fluid interactions can all create gas. The envelope material must also be fully compatible with the working fluid, subject to neither excessive corrosion nor contamination of the fluid. There is currently a dearth of information on envelope material–fluid compatibility at temperatures in excess of 500°C to allow selection of reliable combinations.
- Minimal attack of material. Degradation of envelope material can occur at elevated temperatures where liquid sodium, potassium, and cesium are often used as working fluids. For example, sustained exposure of pipes made of 304 stainless steel to liquid sodium, at 700°C, leads to segregation of nickel and chromium in the grain boundaries at the cold and hot ends of the pipe. Such a process reduces the cross sections of certain parts of the envelopes and thereby increases the corrosion susceptibility of the material.

The major requirements of heat pipe envelope materials for external compatibility are as follows:

- Good resistance to corrosion by hot gases and dilute acids. Corrosion may be general or specific (such as chloride stress corrosion cracking, hydrogen embrittlement, or corrosion fatigue). Thermal fatigue and corrosion fatigue may also occur.
- Resistance to erosion, abrasion, and scaling by coal particles in order to maintain the heat transfer efficiency. Erosion may damage heat pipe envelope materials either through pitting and material removal or through removal of corrosion-protective surface layers. The presence of coal or fly ash further complicates the erosion process by increasing the possibility of chemical attack, if alkali sulfates form, reducing the thermal transfer rates in all cases where a coating forms and increasing the chances of formation of scaling and wall tar buildup as a result of changes in the nature of the fly ash when gasifier conditions change.
- Resistance to thermal fatigue, creep, recovery, and loss of strength (due to prolonged exposure at elevated temperatures).

4.3 CANDIDATE MATERIALS

Of the operational conditions in coal gasification plants, corrosion, erosion, and scaling resistance to various products (Table 4.2) are probably the most important considerations in the material selection process. In this regard, a number of research groups evaluated the corrosion-erosion behavior of a host of materials in these plants.* The results (such as given in Tables 4.3 and 4.5) helped narrow

TABLE 4.2 TYPICAL COAL COMBUSTION PRODUCTS IN A COAL GASIFICATION PLANT

Composition	Mole Percent
Flue Gas	
H_2O	21
H_2	17
CO_2	16
CO	14
CH_4	12
NH_3	0.4
SO_2	0.06
H_2S	0.01
HCl	0.2
SO_3	0.002
N_2	Remainder
Fly Ash	
Fe_2O_3,	
SiO_2	

TABLE 4.3 CORROSION AND EROSION BEHAVIOR OF VARIOUS MATERIALS IN COAL GASIFICATION ENVIRONMENTS

Material[a]	Weight Loss Due to Corrosion ($mg\ cm^{-2}$)	Weight Loss Due to Erosion-corrosion ($mg\ cm^{-2}$)
Incoloy alloy 800	7.4	142.0
Inconel alloy 601	25.8	58.9
310 stainless steel	7.8	24.2
Crutemp alloy 25	4.3	23.3

[a]For composition, see Table 4.4.

Source: W. L. Silence, V. L. Hill, and E. J. Vesely, "The effects of environmental variables on the erosion-corrosion behavior of candidate materials for coal conversion plants," in V. L. Hill and H. L. Black (eds.), *The Properties and Performance of Materials in the Coal Gasification Environment*. Metals Park, Ohio, ASM, 1981, p. 638.

*Groups such as the Metals Properties Council, in a program sponsored by the U.S. Department of Energy and the Gas Research Institute [1] and Battelle Columbus Laboratories [2]; mostly pilot plants and laboratory test equipment were used.

TABLE 4.4 COMPOSITION OF ALLOYS (wt %)

Material	C	Mn	Si	Fe	Ni	Cr	Co	Al	Mo	Ti
Incoloy alloy 800	0.03	0.80	0.33	47.10	30.80	20.60	0.10	0.32	—	0.42
Inconel alloy 601	0.04	0.24	0.19	15.90	59.50	22.50	0.03	1.19	—	—
310 stainless steel	0.06	1.71	0.68	52.20	20.20	25.00	—	—	—	—
Crutemp alloy 25	0.07	1.50	0.60	47.20	24.80	25.00	—	—	0.40	—

TABLE 4.5 RESISTANCE OF CANDIDATE MATERIALS TO ENVIRONMENTAL ATTACK

Material	General Corrosion Resistance to Dilute Acids (HCl, H_2SO_4, and HNO_3)	Resistance to High-Temperature Sulfide Stress Corrosion Cracking	Resistance to Corrosion by Coal Ash
Inconel 600 (annealed/cold-worked)	F	G	G
Fe-25Cr-6.5Ni	F	G	F
Fe-20Cr-33Ni-4A1 (Incoloy 800)	G	G	G
Co-25Cr-10Ni	G	G	F
50Ni-50Cr alloys	G	G	G
304 stainless steel	F	P	F
302 stainless steel	F	P	P
Ferritic stainless steel	F	P	P
K Monel	P	P	P

F, fair; G, good; P, poor.

Source: Adapted from G. J. Ewell and A. Basiulis, "Heat pipe materials: Unique requirements for coal gasification processes," paper presented at the 23d national SAMPE symposium, Anaheim, Calif., 1978.

down the field of candidate materials to the highly alloyed nickel- and iron-base alloys. Note, however, that other materials, such as low-carbon steel coated with PTFE, heresite, or ceramics (such as BaO, Cr_2O_3, SiO_2, or Al_2O_3) can also be used in these plants.

REFERENCES

1. A. O. Schaefer (ed.), "A program to discover materials suitable for service under hostile conditions obtaining in equipment for the gasification of coal and other solid fuels," quarterly reports to the Energy Research and Development Agency, U.S. Department of Energy, by the Metals Properties Council, 1977–81; J. L. Blough, V. L. Hill, and B. A. Humphreys, "The performance of alloys in the Metal Properties Council (MPC) Gaseous Corrosion Study," in V. L. Hill and H. L. Black (eds.), *The Properties and Performance of Materials in the Coal Gasification Environment*. Metals Park, Ohio: American Society for Metals, 1981, pp. 225–255.

2. I. G. Wright, ''An overview of the erosion/corrosion problems and solutions in the coal gasification environment,'' in V. L. Hill and H. L. Black (eds.), *The Properties and Performance of Materials in the Coal Gasification Environment*. Metals Park, Ohio: American Society for Metals, 1981, pp. 795–817.

ELECTRONIC DEVICE APPLICATIONS

CHAPTER 5

5.1 INTRODUCTION

The unique electrical properties of semiconductors permit their use in devices to perform specific electronic functions. Advantages of semiconductor devices (also known as solid-state devices) include small size, low power consumption, and no warm-up time. Vast numbers of extremely small circuits, each consisting of numerous electronic devices, may be incorporated in a *chip*. It is generally agreed that the invention of semiconductor devices, which has given rise to miniaturized circuitry, is the greatest of all 20th-century inventions, being responsible for the advent and extremely rapid growth of a host of new industries, as well as the revolutionizing of other industries since about 1970. The range of applications of microelectronic circuits using chips is indicated in section 5.2, with the emphasis on some recent developments, as well as near-future ones, in the United States.

5.2 RECENT AND ANTICIPATED DEVELOPMENTS IN THE U.S.

Military Computers

The inexorable trend toward miniaturization of advanced weaponry parts is best illustrated by a rugged, compact computer developed by the Sandia National

Laboratory in New Mexico. About the size of a shoebox but as powerful as a mainframe computer for some applications, this new computer, weighing only 9 kg and called the Sandia airborne computer (Sandac/V), may be used in nuclear reentry vehicles, helicopters, and land vehicles.

Strategic Defense Initiative (SDI)

In essence, the SDI program (popularly known as "Stars Wars"), developed by the United States in response to a speech by President Reagan on March 23, 1983, seeks to investigate whether new technologies could provide the means for countering nuclear ballistic missiles during their trajectories (Fig. 5.1).

SDI has six basic components:

- Surveillance, acquisition and tracking
- Directed energy systems (such as neutral-particle beams, chemical and x-ray lasers) and kinetic energy systems ("smart rocks"), which are to attack missiles in the boost, midcourse, and/or terminal phases of their trajectories
- Conventional weapons
- Battle management, communications, and data processing
- Systems concepts
- Countermeasures and tactics

SDI has attracted enormous amounts of funding from the U.S. Department of Defense (Fig. 5.2), but the program has been criticized by both the scientific and engineering community and the political and defense community as being technically infeasible, politically dangerous, and economically wasteful [1]. However, because of budgetary commitments, SDI is likely to be a main engine of

Figure 5.1 Typical ballistic-missile trajectory.

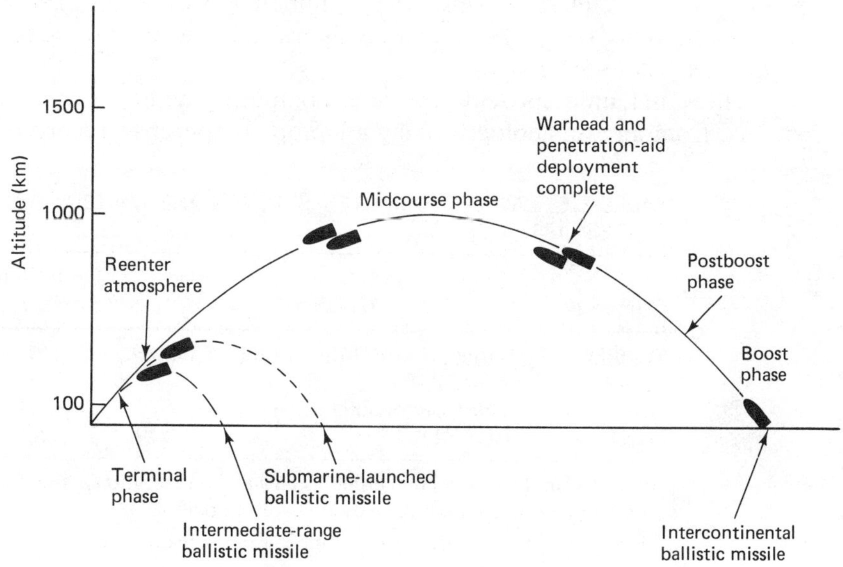

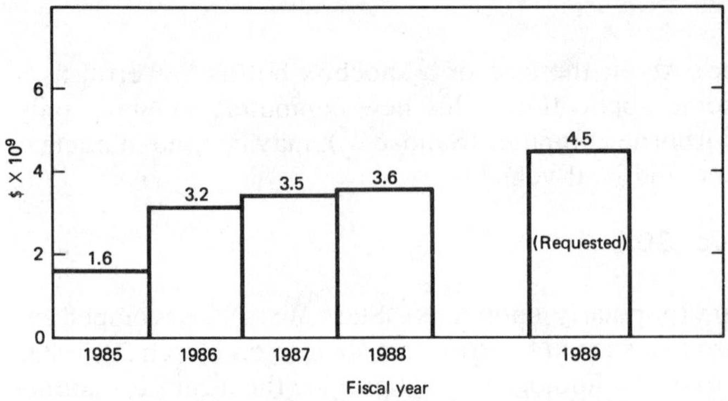

Figure 5.2 Budget authority for the United States' SDI ("Star Wars") program.

growth of the defense industries and their peripherals in the short run. SDI's impact on computer developments and innovations would be far-reaching and even revolutionary because all six of the basic components of the system are inextricably woven with advanced computing systems and techniques.

Computer Technology

The current thrust of research and development efforts in this area is in three directions: supercomputers, smaller but powerful personal computers, and workstations.

Supercomputers, such as the Cray X-MP/2, are being used in a wide variety of applications, such as the search for and exploitation of energy resources, design of airplanes, computational fluid dynamics, and tests of the crashworthiness of automobiles.

As far as advanced personal computers are concerned, major strides have been made in developments in both memory size and configuration. Personal computer models are now available with 256 kilobytes of random-access memory (RAM) (such as the IBM System/36), 512 Kb of RAM (IBM PCAT), 1 million Kb of RAM (Unisys PC/IT), and even 2 million Kb of RAM (Digital Equipment Corp.'s Microvax II). In the field of configuration, the latest trend is toward laptop computers (Table 5.1.).

These machines provide portable computing ability to people in all occupations, from an archaeologist using a laptop computer to record findings from a

TABLE 5.1 LAPTOP COMPUTER SALES IN THE UNITED STATES, 1987[a]

Company	Model(s)	Units Sold (10^3)
Toshiba	T1000, T1100 Plus, T1200, T3100	113
Zenith	Z181, Z183	63
IBM	The Convertible	51
NEC	EL, HD	25

[a]Mainstream MS-DOS-compatible clamshut configurations; top four producers, with a combined market share of about 85%.

Source: International Data Corporation, Framingham, Mass.

remote excavation site to a sales manager using one to perform sales analyses while away from his or her base office. The growth of the sales of these machines proceeds feverishly. Sales in 1986 were 307,000 machines at $750 million, up by 65% and 49%, respectively, from 1985; 1990 sales are projected to be 1,444,000 units at a value of $2.7 billion.

Workstations are full 32-bit computers, usually with sophisticated graphic displays, a keyboard, and a mouse and usually attached to a high-speed network. Consequently, workstation users have at their fingertips computing power exceeding that of personal computers by several orders of magnitude.

Medical Equipment

- Computerized axial tomography
- Image transmission systems for sending, for example, x-ray images over telephone lines to medical offices from a radiologist's office

Automobile Electronics

- Navigation systems, such as Chrysler's experimental system termed CLASS, which uses signals from NAVSTAR communications satellites and a digital laser disc to provide maps and other pieces of information on a cathode ray tube display
- Microprocessor-based spark ignition advance-angle meters

Communications

- Real-time conferencing systems that allow participants to view, edit, and process information jointly
- A system of telephones installed in a limited number of aircraft that use computers to direct radio beams between airplanes and a number of ground stations

General

- Television sets, video games, video cassette recorders, digital synthesizers, microprocessors, computer-aided design and manufacture, computer-enhanced photography (as was used to display the photograph of the tail of Halley's comet taken in March 1986 by cameras installed in the Vega 1 and 2 and Giotto spacecraft)
- Doppler radar, used to detect microburst wind shear

5.3 COMPETITION BETWEEN SILICON AND GALLIUM ARSENIDE

A vast majority of the developments in section 5.2 and others are based on silicon as the semiconducting material. However, solid-state physicists and electronic engineers now generally agree that in selected applications (particularly in the

TABLE 5.2 SOME ELECTRONIC PROPERTIES OF III–IV COMPOUNDS OF TECHNOLOGICAL INTEREST (AT 298 K)

Material	Band Gap (eV)	Mobility of Electrons ($cm^2\ V^{-1}\ s^{-1}$)	Gap Transition
GaAs	1.43	5,000–8,000	Direct
GaSb	0.68	5,000	Direct
GaP	2.25	300	Indirect
InAs	0.36	30,000	Direct
InSb	0.17	80,000	Direct
InP	1.27	4,500	Direct

fields of aerospace, defense, and supercomputing), future developments are predicated on chips that employ a better semiconducting material. Although a number of such alternatives are possible (see Table 5.2), the consensus is that gallium arsenide, GaAs, is the semiconducting material of the future. But notwithstanding its attractive physical electronic properties, there are some legitimate questions about its suitability for the mass manufacture of very large scale integrated (VLSI) circuits.

The perceived preference of GaAs over silicon is open to question and is being hotly debated among the experts. It is therefore useful to compare these materials on the basis of knowledge and technological practice currently available. The final material choice would inevitably be tied to the application and the levels of reliability and precision required.

The advantages of GaAs over silicon stem from its properties and may be summarized thus:

1. Currently, for the same power consumption, a typical GaAs device operates at a speed five times faster than emitter-coupled logic, the fastest silicon family. This statement applies to integration levels and fabrication processes that can be depended on for production purposes.

2. GaAs is radiation-hard [2]. This arises from the fact that the GaAs surface is stabilized by surface defect states (rather than passivated by a native oxide). On this question, though, there is some disagreement. Huge differences amounting to several orders of magnitude have been reported, but other measurements from an experimental microprocessor design suggest that the difference is barely noticeable [3].

3. GaAs is tolerant of temperature variations. Its operating range is from about −200°C to about 200°C. Its advantage over Si on this score is indisputable.

4. GaAs is well suited to the efficient integration of electronic and fiber-optic components. Should this potential of GaAs be fully exploited, its impact on the area of system design will be enormous.

A summary of the physical electronic properties of GaAs and silicon is given in Table 5.3.

TABLE 5.3 COMPARISON OF ELECTRONIC PROPERTIES OF GALLIUM ARSENIDE (GaAs) AND SILICON (Si) AT 300 K

Property	GaAs	Si
Band gap energy (eV)	1.43	1.12
Electron mobility ($cm^2\ V^{-1}\ s^{-1}$)	8600	1350
Hole mobility ($cm^2\ V^{-1}\ s^{-1}$)	250	480
Intrinsic carrier concentration (cm^{-3})	10^7	1.50×10^{10}
Electronic affinity (V)	4.07	4.01
Effective density of states in conduction band (cm^{-3})	4.70×10^{17}	2.80×10^{19}
Effective density of states in valence band (cm^{-3})	7.0×10^{18}	1.02×10^{19}
Electrical conductivity at 298 K ($\Omega^{-1}\ cm^{-1}$)	10^{-6}	4×10^{-6}
Effective mass of electrons (kg)	0.62×10^{-31}	3.00×10^{-31}
Effective mass of holes (kg)	5.10×10^{-31}	5.10×10^{-31}

The disadvantages of GaAs relative to silicon involve applicable technology and may be summarized thus:

1. GaAs wafers presently exhibit a greater "density of dislocations" compared to Si chips. Density of dislocations is the number of irregularities per unit area. This means that GaAs chips have (a) a smaller transistor count, (b) a smaller VLSI area, and (c) a lower yield. Current estimates for the transistor count for GaAs ICs are 20K–30K. (A GaAs memory chip with 102.3K transistors, reported in 1986, is believed to have been achieved at the expense of yield.) Silicon is expected to maintain well into the 1990s its advantage over GaAs as far as integration levels are concerned.

2. Substrates of GaAs are considerably more expensive than those of silicon (Table 5.4). In addition, since GaAs is intolerant of aluminum, gold interconnects have to be used, and the wafers can be damaged easily during fabrication of integrated circuits because GaAs is very brittle. These limitations are thought to be temporary, with some workers anticipating a drop in the comparative steady-state price of GaAs to only about 10 times that of silicon by 1996 [4].

3. The noise margin in GaAs is worse than in silicon, making it necessary, in many applications, to trade off chip area for higher reliability. There are also problems (generally agreed to be temporary) of sensitivity of GaAs to increased fan-in and fan-out.

4. The lack of passivation in GaAs eliminates the possibility of metal oxide semiconductor (MOS)–type circuits that form the basis for the widest segment

TABLE 5.4 TRENDS IN THE PRICES OF RAW WAFERS ($ m^{-2})

Year	GaAs	Si
1984	376.25	8.40
1987	268.75	7.42
1990	161.25	6.56

of present-day silicon VLSIs. This has led to a number of serious problems in the design of a GaAs microprocessor. Whereas the gate speed ratio of GaAs and silicon devices is about 6:1, the ratio of processor speeds (for compiled HLL code) is only 3:1 in favor of GaAs, and for the arithmetic coprocessor, the speed ratio is only 2:1 in favor of GaAs [3]. The conclusion is clear: Advances are needed in the area of GaAs microprocessor architecture and design. Some have occurred since 1985, including development of a 32-bit microprocessor using a direct-coupled FET logic enhancement/depletion-mode metal semiconductor field-effect transistor (DCFL E/D-MESFET) [5]; one-transistor dynamic random-access memory (DRAM) cells designed with AlGaAs/GaAs heterojunction field-effect transistor technology, known variously as MODFET, HEMT, or HIGFET technology [6]; and cache-based, multidistance, and pipelined memory design [7].

The electronic applications performance indices of GaAs and silicon are summarized in Table 5.5.

It is the opinion of many industry watchers and researchers in the field of semiconducting materials and devices that gallium arsenide is the semiconducting material of the future for a wide range of specialist and sophisticated devices, especially in SDI work, supercomputers, and communications. For example, in June 1986, Cray Research announced that GaAs circuits with 300 to 400 gates having switching times of 120 trillionths of a second were being fabricated for the next generation of its supercomputers. Also, in November 1987, Prisma, Inc. (Colorado Springs, Colo.), announced design plans for a GaAs-based 32-bit real-time supercomputer.

GaAs should begin appearing in affordable monolithic microwave integrated circuits (MMICs) for radar power amplifiers. One such MMIC chip is reported to have produced the exceptionally high output power of 2.2 W over a frequency band from 8 to 12 GHz with a 25 dB gain and is only 1/100 the size of a conventional hybrid radio frequency amplifier [8]. It is expected that the use of GaAs for these

TABLE 5.5 COMPARISON OF SOME ELECTRONIC APPLICATIONS PERFORMANCE INDICES OF GALLIUM ARSENIDE (GaAs) AND THE FASTEST SILICON (Si) INTEGRATED CIRCUITS

Index	GaAs (1 μm DCFL E/D-MESFET)[a]	Si (2 μm ECL)[b]
Transfer count per chip	20k–30k	30k–45k
Speed		
Gate delay	50–150 ps	150–250 ps
On-chip memory access (32 × 32)	0.5–2.0 ns	2–3 ns
Off-chip/on-package memory access (256 × 32)	4–6 ns	6–10 ns
Off-chip/off-package memory access	10–40 ns	20–60 ns
IC design		
Transistors per gate	1+ (fan-in)	3+ (fan-in)
Transistors per memory cell	6 static, 1 dynamic	6 static

[a]Direct-coupled FET logic enhancement/depletion-mode metal semiconductor field-effect transistor.

[b]Emitter-coupled logic.

TABLE 5.6 WORLDWIDE GaAs INTEGRATED CIRCUITS MARKET, BY MAJOR SUPPLIER ($ m)

Supplying region	1986	1987	1988	1989	1990	1992[a]
USA	103	156	226	338	506	700
Europe	6	28	50	90	163	300
Japan	15	30	60	121	243	500
Total	124	214	336	549	912	1500

Source: HTE Management, Inc., Scotts Valley, CA.
[a]Author's projections based on data from the source.

TABLE 5.7 WORLDWIDE GaAs INTEGRATED CIRCUITS MARKET, BY MAJOR MARKET SEGMENT ($ m)

Application	1984	1986	1988	1990	1992	1995[a]
Business	5	23	90	299	702	1200
Communications	10	43	149	460	1021	1600
Consumer	4	30	123	359	756	1250
Data processing	6	26	116	344	865	1400
Governments	61	179	395	681	1157	1800
Industrial and Instrumentation	8	41	170	494	1113	1750
Total	94	342	1043	2637	5614	9000

Source: Strategic Systems, Inc., Reading, PA, USA (1983)
[a]Author's projections based on data from the source.

and a host of other applications will increase dramatically through the mid 1990s, although estimates of this growth vary from the conservative (Table 5.6) to the overly optimistic (Table 5.7).

5.4 ALTERNATIVES

Recently, both computer scientists and electronic engineers have come to accept that chip technology is fast approaching the lowermost limits for the size of components produced from silicon, gallium arsenide, or any other metallic semiconductor. To bypass this size limitation, research attention is now turning to organic compounds similar to those found in living organisms.

The principal characteristic that an organic molecule must display for use in a molecular electronic device (MED) is the ability to exist in two characteristic states, such as different colors, that might represent the 0 and 1 of the binary code. Researchers have identified a number of such molecules, which can change color when exposed to light. Among them is a pigment called bacteriorhodopsin, which is part of the photosynthetic system of the purplish bacterium *Halobacterium halobium*. Isolated from the bacterium, the pigment can exist in two states:

one that absorbs green light (which can be used to store information, for example) and one that absorbs red light (which can be used to retrieve the information, for example).

Use of organic molecules in electronic devices has a number of advantages over metallic semiconductors. First, these molecules are less susceptible to electromagnetic interference and can thus be stacked one on top of the other, allowing many electronic components to be packed into a much smaller area. Second, the use of these molecules would considerably increase the speed of operation of electronic devices. Early indications are that only 5 ps are required for bacteriorhodopsin molecules to change colors. This is about 1000 times faster than the time in which information can be stored or retrieved in a magnetic device.

Work in the field of MEDs is being actively pursued in a small number of research centers in the United States, such as Carnegie-Mellon University, and Hughes Aircraft Corporation.

A second important recent development was the announcement in February 1987, by Hypres, Inc. (Elmsford, N.Y.) of the world's first commercial computer system, based on Jophenson junctions, which switch on and off 1000 times faster than semiconductor transistors. Called the Hypres PSP-1000, it is a modest machine currently used to analyze electronic signals, but it contains the key to a major breakthrough.

REFERENCES

1. B. Rensberger, "Star Wars splits experts into two camps," *Washington Post*, March 3, 1985, pp. A-1, A-18, A-19; R. Nimroody, "The trillion dollar brain drain," *Science 94* (8), 50–57, 1986.
2. "Aerospace spotlight," *Aerospace America 23* (5), p. 1, 1985.
3. W. Helbit, "RISC vs. CISC, GaAs vs. Silicon, Hardware vs. Software," lecture presented at the joint chapter meeting of the IEEE Microwave Theory and Techniques and IEEE Electron Devices societies, Princeton, N.J., January 1986.
4. M. R. Namordi, "Advances in GaAs technology," lecture presented to the Department of Electrical Engineering, Purdue University, West Lafayette, Ind., October 1984.
5. S. Karp and S. Roosild, "DARPA, SDI and GaAs" *Computer 19,* pp. 17–19, October 1986.
6. J. A. Cooper, "GaAs DRAM cell research is underway at Purdue," *Computer 19,* p. 11, October 1986.
7. V. Milutinovic, et al., "Issues of importance in designing GaAs microcomputer systems," *Computer 19,* pp. 45–56, October 1986.
8. H. C. Nathanson, M. C. Driver, and R. N. Thomas, "Gallium arsenide changes the shape of airborne radar," *Aerospace America 25,* p. 26, April 1987.

SHIELD OF A NUCLEAR REACTOR

CHAPTER 6

6.1 INTRODUCTION

The primary components of a typical nuclear fission reactor (Fig. 6.1) are (a) the nuclear fuel (usually uranium or plutonium); (b) the main structures (usually made from zirconium alloys, mild steel, stainless steels, or nickel alloys); (c) the moderator and reflector, in thermal reactors, or blanket and reflector, in fast reactors (made, in either case, from depleted uranium, thorium, beryllium, or graphite); (d) control elements (made from boron carbide, cadmium, hafnium, boric acid, or burnable absorbers); (e) coolant (helium, carbon dioxide, light water, heavy water, liquid sodium); and (f) safety systems (pressure suppression, emergency core cooling, and instrument monitoring systems) and shields.

6.2 RADIATION SHIELDING

We shall concentrate on the choice of materials for the shields. It is important to recognize that the character of adequate shielding depends on the purpose of the reactor (thermal, fast, research, power, etc.). Shield design is an integral part of reactor design. Notwithstanding this observation, the main function of the shield, irrespective of the type of reactor, is to prevent or reduce the passage of any type

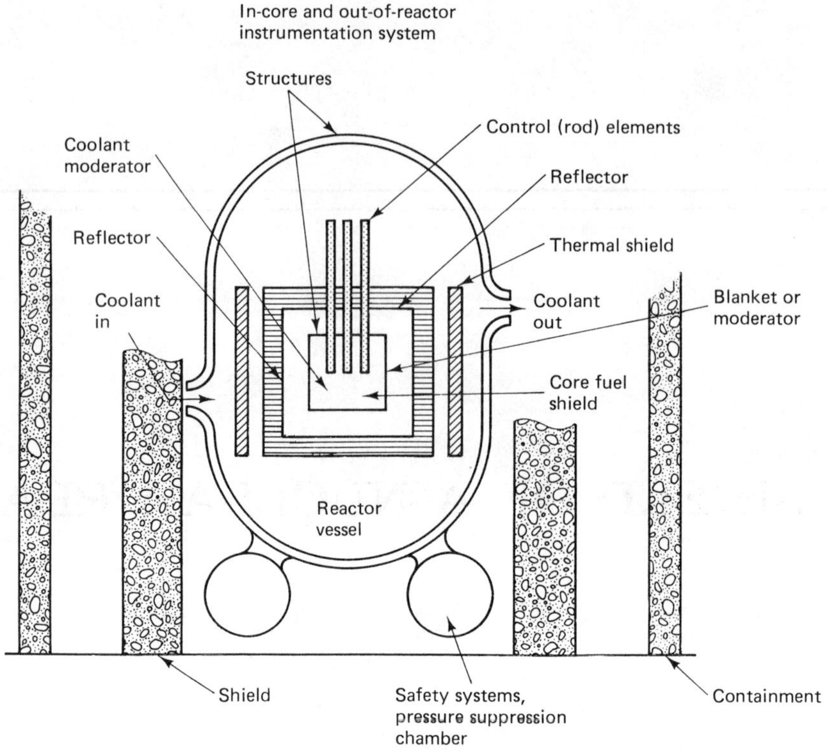

Figure 6.1 Schematic diagram of the components of a nuclear fission reactor.

of radiation—x rays, beta rays, protons resulting from the fission reactions, neutrons, and gamma rays—to the outside of the reactor, leading to radioactive contamination of the atmosphere as in the reactor accident at Chernobyl, USSR, in April 1986. Basically, there are three subsets of radiation shield functions:

- Thermal shielding, to protect the pressure vessel, coolant loop, and inner shield from the intense heat liberated upon absorption of the nuclear radiation
- Biological shielding, to protect personnel and safeguard public health
- Apparatus-instrument shielding, to protect electromagnetic apparatus and electronic instruments whose operation may be severely impaired when exposed to a radiation field

6.3 MATERIAL REQUIREMENTS

The candidate materials for the shields should have the following characteristics:

- High density, which allows for the use of a minimum thickness for maximum shielding. Both the relaxation length (the thickness needed to reduce outgoing radiation to 63.2% of the incoming radiation) and the halfwave thickness (the thickness needed to reduce outgoing radiation to 50% of the incoming radiation) are inversely dependent on shielding material density (see Table 6.1).

TABLE 6.1 RELAXATION (T_r) AND HALFWAVE ($T_{0.5}$) THICKNESSES FOR TWO TYPES OF CONCRETE HAVING A THICKNESS OF 200–800 mm.

Concrete	Density (kg m^{-3})	For 5 MeV Gamma Radiation T_r (mm)	For 5 MeV Gamma Radiation $T_{0.5}$ (mm)	For Neutron Radiation T_r (mm)	For Neutron Radiation $T_{0.5}$ (mm)
Ordinary	2400	130	90	110	76
Limonite	3300	110	76	90	63

Source: A. N. Komarovskii, *Shielding Materials for Nuclear Reactors*. Elmsford, N.Y.: Pergamon Press, 1961.

- Constancy of density
- Homogeneity
- High content of hydrogen (generally, chemically combined in the material) for the moderation of neutrons into thermal neutrons
- High content of heavy elements, for moderating fast neutrons and absorbing secondary radiation resulting from such moderation
- Negligible or no susceptibility to decomposition or weakening under the influence of radiation
- Good constructional strength and stability, because in most cases the radiation shield is a load-bearing member not only for the reactor but for the building itself as well (reactor roof, etc.)
- Minimum settling during the erection of the shield and subsequently
- High fire resistance
- Imperviousness to water, gases liberated during heating, and corrosive action of metals

6.4 CANDIDATE MATERIALS

These materials may be divided into three groups:

- Heavy or moderately heavy elements, which have the function of attenuating the gamma radiation and of slowing down fast neutrons to about 1 MeV by inelastic scattering in the high-energy region. Examples are lead, bismuth, tungsten, tantalum, iron, and ceramics.
- Medium-weight elements and compounds, which have the function of moderating and thermalizing neutrons by elastic scattering in the intermediate and low-energy regions. Examples include ferrophosphorus concretes and ferroboron concretes (in the latter material, minerals of limonite, $2Fe_2O_3{\cdot}3H_2O$, and boron compounds such as boric acid, borax, and colemanite, $C_{96}O_{11}{\cdot}5H_2O$, replace most of the sand and gravel in ordinary concretes).

TABLE 6.2 SUITABLE MATERIALS FOR RADIATION SHIELDING

Material	Comments
Heavy and Moderately Heavy Elements and Materials	
Lead	Most commonly used shield material for x rays because it is abundant and economical.
Bismuth	Associated with Pb in some alloys; suitable for low-temperature operation.
Tantalum, tungsten	High density; high melting point and thus suitable for high-temperature service.
Carbon and stainless steels	Inpurity contents such as cobalt and manganese are a nuisance, since these elements emit induced gamma radiation upon reactor shutdown.
Medium-Weight Materials	
Light water	Effective neutron shield material; efficient in removing the heat generated by nuclear radiation absorption.
Metal hydrides (e.g., ZrH_2, TiH_2, ThH_2, UH_2)	High melting points.
Special concretes (e.g., barite concrete, iron concrete, ferrophosphorus concrete, ferroboron concrete)	Good for neutron and gamma-ray shielding; high mechanical strength; various configurations of construction possible. (See Table 6.3 for further details.)

- Light elements, which have the function of slowing down and capturing fast neutrons without producing secondary gamma rays. Examples include boron, boron oxides (such as B_2O_3), boric acid, borax, and boral (a variable mixture of boron carbide and aluminum).

See Tables 6.2 and 6.3 for comments on some of these shielding materials.

Figures 6.2 and 6.3 show some typical results for the distribution of neutrons and gamma rays in various types of radiation shielding materials in a pressurized-water nuclear reactor operating at a power level of 100 MW (electrical).

Note that the fast neutron flux decreases regularly, whereas the thermal neutron flux increases in water due to the moderation effect, decreases in the steel layers due to neutron capture, and then increases as the fast neutrons are slowed down, moderated, and thermalized in the concrete. The primary gamma dose decreases in the steels, lead, and concrete, while the secondary gamma dose decreases slowly and then becomes appreciably higher than the primary gamma rays as a result of the absorption of the thermal neutrons and the emission of the secondary gamma rays in the shields.

TABLE 6.3 SOME CONCRETES FOR RADIATION SHIELDING

Type of Concrete	Composition (wt %)	Density (kg m^{-3})	Macroscopic Cross Section: Fast Neutron Removal (cm^{-1})	Macroscopic Cross Section: Gamma-Ray Absorption (cm^{-1})
Iron	Iron punchings, 57; limonite, 26; Portland cement, 13; water, 4	4500	0.160	0.130
Barite	Barite[a], 60; limonite, 22; Portland cement, 11; water, 7	3500	0.105	0.101
Ordinary	Cement, 10; coarse aggregate, 50; fine aggregate, 32; water, 8	2400	0.085	0.066

[a]Includes both barite ($BaSO_4$) and witherite ($BaCO_3$).

Source: B. A. Ma, *Nuclear Materials and Applications,* New York: Van Nostrand Reinhold, 1983, p. 385.

Figure 6.2 Fast and thermal neutron flux distributions in the radial direction of the shields.

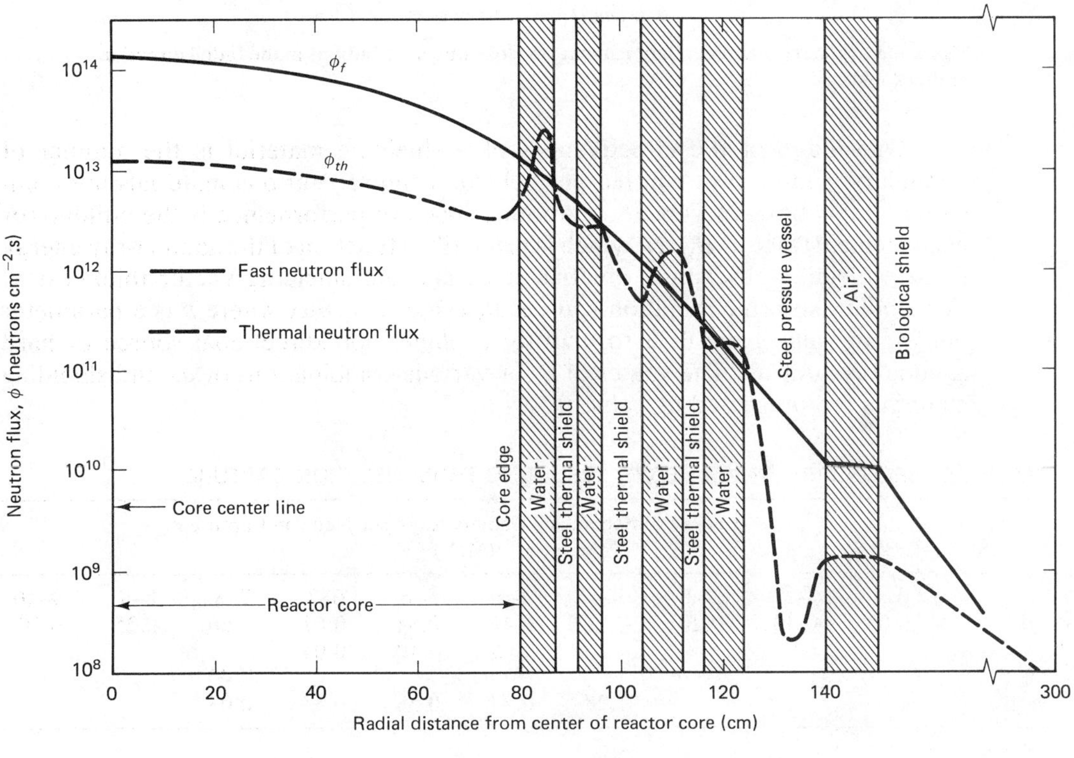

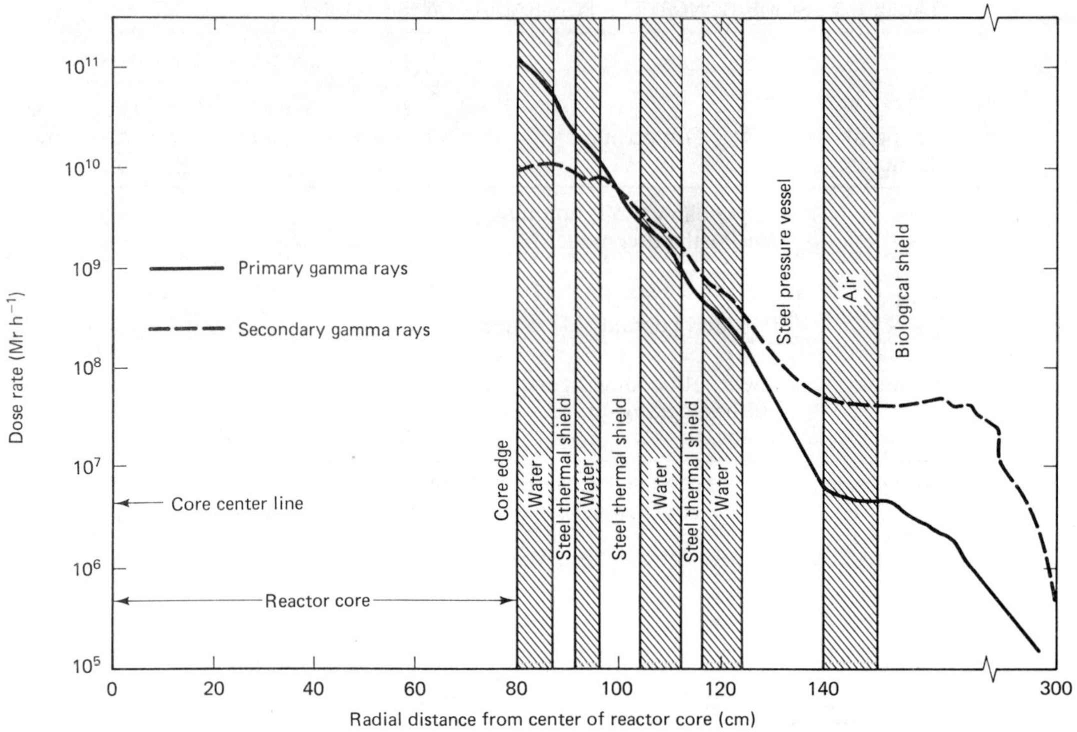

Figure 6.3 Primary and secondary gamma-ray dose rate distributions in the radial direction of the shields.

One index of the effectiveness of a shielding material is the number of secondary gamma rays emitted per neutron capture. Various materials are compared on this basis in Table 6.4. Another index of performance is the buildup (or enhancement) factor, $B(\mu x)$, which is primarily a function of the gamma-ray energy E, the shielding absorption coefficient μ, and the shielding vector thickness x. For a point-isotropic radiation source, $B(\mu x) = 1 + b\mu x$ where b is a parameter ranging in value from 0.33 to 0.50 for a plane monodirectional source of hard gamma rays in lead. The values of B for various combinations of μx and shielding materials are presented in Table 6.5.

TABLE 6.4 SECONDARY GAMMA RAYS RESULTING FROM NEUTRON CAPTURE

Element	Energy of Secondary Gamma Rays per Neutron Capture (MeV)									
	0–1	1–2	2–3	3–4	4–5	5–6	6–7	7–8	8–9	9–10
Fe	0.05	0.30	0.20	0.37	0.45	0.84	0.67	2.66	0.25	0.20
Ta					0.05	0.10	0.04			
Pb								7.40		
W				1.28	0.77	0.58	0.33	0.03		

TABLE 6.5 DOSE BUILDUP FACTOR FOR AN ISOTROPIC POINT SURFACE

Shielding Material	μx	Secondary (Capture) Gamma-Ray Energy (MeV) 0.5	4.0	10.0
Iron	1	1.98	1.45	1.20
	7	11.70	4.91	2.99
	20	55.60	16.00	12.40
Lead	1	1.24	1.27	1.09
	7	2.00	3.61	2.52
	20	2.73	16.30	39.20
Concrete (density	1	2.80	1.61	1.35
= 2350 kg	7	21.40	5.00	3.02
m^{-3})	20	125.00	13.20	6.32

The gamma (or neutron) flux after passage through a shield of thickness x is given by ϕ', with $\phi' = B(\mu x)\phi(0)\exp(-\mu x)$, where $\phi(0)$ is the flux before passage through the shield.

In addition to these indices, a large number of shield design charts, standards, and codes for commonly used shielding materials are available [1].

REFERENCE

1. N. M. Schaeffer (ed.), "Reactor shielding for nuclear engineers," U.S. Atomic Energy Commission Report TID-25291, National Technical Information Service, 1973; J. T. West, et al., "Developing an American Nuclear Society coupled cross section standard for concrete shielding," *Trans. Amer. Nucl. Soc. 28*, p. 642, 1978.

THERMAL SHOCK RESISTANCE

CHAPTER 7

7.1 INTRODUCTION

Thermal shock is the phenomenon whereby a body experiences instantaneous or stepwise changes in temperature due to rapid change in either external temperature or the rate of internal heat generation. As a result of these changes, internal thermal stresses are generated that may weaken the material or, in extreme cases, lead to fracture. When fracture occurs, it takes place after one cycle of change.

In the case of ductile metals and polymers, alleviation of these stresses may be achieved by plastic deformation. But the nonductile nature of most ceramics and glasses, under normal conditions, enhances the possibility of thermal shock. Rapid cooling of a brittle body is most likely to inflict thermal shock than rapid heating, since the induced surface stresses are tensile in the former case and crack formation and propagation from surface flaws are more probable when the imposed stress is tensile.

The applications in which thermal shock is an important design consideration include ovenware, laboratory ware, reentry space vehicles, components of internal combustion engines, heat exchangers, and radomes of aircraft and missiles.

7.2 THERMAL SHOCK RESISTANCE PARAMETERS

The capacity of a material to withstand thermal shock is termed its *thermal shock resistance* and is characterized by a number of thermal shock or damage resistance parameters, each of which is applicable to specific sets of heat transfer conditions.

Once the temperature distribution is known for a particular body shape, we can easily relate the thermal strain to the thermal stresses by one of Hooke's generalized laws for elastic stress and strain (in the absence of body forces). Thus we can write

$$\alpha(T_{av} - T) = \frac{\sigma_1 - \nu\sigma_2}{E}$$

where α is the coefficient of thermal expansion of the material, E is its modulus of elasticity, σ_1 and σ_2 are the principal thermal stresses, ν is the Poisson's ratio of the material, T is the temperature distribution in the body as a function of distance, x, from the surface and T_{av} is the average temperature in body thickness given by

$$T_{av} = \frac{1}{a}\int_{o}^{a} T(x)\,dx$$

where a is the thickness of the body. Assuming that isotropic conditions prevail and that the elastic and thermal constants are not seriously affected by temperature, we can write

$$\alpha(T_{av} - T) = \frac{\sigma(1 - \nu)}{E\alpha}$$

Dividing by T_o and considering the surface, we get an expression for a dimensionless stress factor, σ_s^*, given by

$$\sigma_s^* = \frac{T_{av} - T_s}{T_o} = \frac{\sigma_s(1 - \nu)}{T_o E\alpha}$$

where T_o is the initial temperature of the body above the surroundings, T_s is the surface temperature, and σ_s is the surface thermal stress.

For an infinite plate of finite thickness, σ_s^* versus a dimensionless time factor as a function of the Biot modulus β has been calculated (Fig. 7.1). Note that the Biot modulus is given by $\beta = ah/k$, where a is a typical dimension of the body in the direction of the heat flow or the characteristic heat transfer length of the body (it is usually taken to be the ratio of the volume to the surface area of the body or a multiple thereof), h is the surface heat transfer coefficient or the surface conductance, and k is the thermal conductivity of the material of the body. The maximum values of σ_s^* at each value of β are plotted in Fig. 7.2., for which a suitable fit is

$$\frac{1}{\sigma_{s\,max}^*} = A + \frac{B}{\beta} - C\exp\left(\frac{-D}{\beta}\right) \tag{7.1}$$

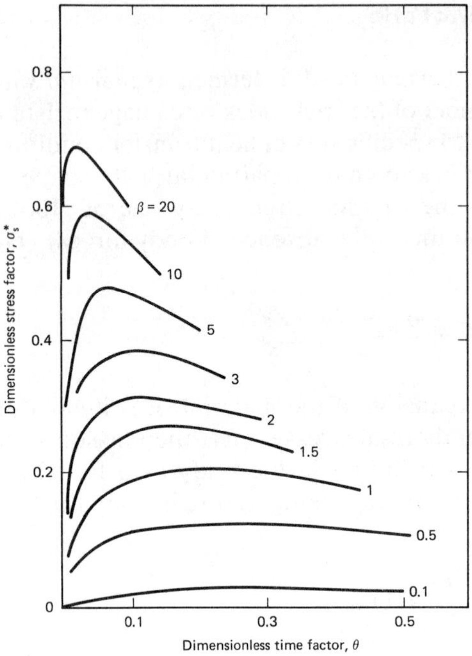

Figure 7.1 Variation of the dimensionless stress factor with the dimensionless time factor.

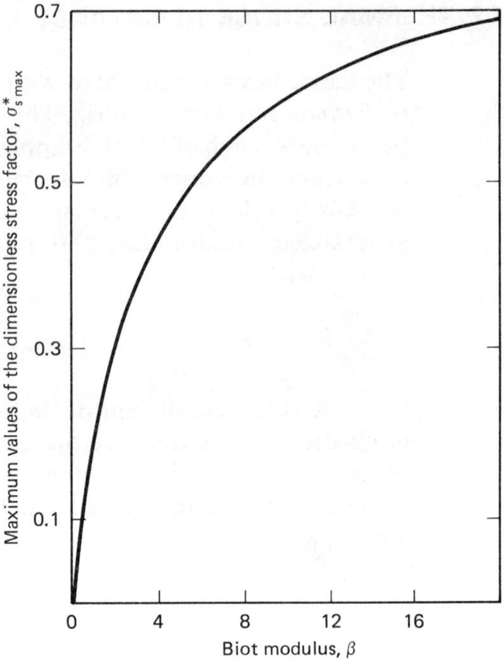

Figure 7.2 Variation of the maximum value of $\sigma^*_{s\,max}$ with Biot modulus.

For low values of β ($\beta < 10$),

$$\frac{1}{\sigma^*_{s\,max}} = A + \frac{B}{\beta}$$

When failure occurs, $\sigma^*_{s\,max} = \sigma_F$, the fracture strength of the material. In the case of brittle materials, which normally have large dispersion in strength values, it is arguable what an appropriate value for σ_F should be. It is suggested that an acceptable value might be related to the lower extreme of the strength distribution, with the problem being treated from a statistical viewpoint.

Equation (7.1) then becomes

$$T_{o\,max} = \left(\frac{A}{k} + \frac{B}{ah}\right)\left[\frac{(1 - \nu)\sigma_F k}{E\alpha}\right] \tag{7.2}$$

where $T_{o\,max}$ is the maximum temperature above the surroundings that the body can withstand without fracturing. In general, we can write

$$\Delta T_{max} = \left[\frac{(1 - \nu)\sigma_F k}{E\alpha}\right]\frac{A' + B'}{ah}$$

where A' and B' are numerical constants, which take different values for different shapes and sizes of various bodies, and ΔT_{max} is the maximum temperature change the body can withstand before fracture. Thus the first thermal shock resistance parameter R_1 becomes equal to ΔT_{max} and, for a given body under specified heat transfer conditions, becomes

$$R_1 = \frac{(1 - \nu)\sigma_F k}{E\alpha} \tag{7.3}$$

The physical interpretation and heat transfer conditions described by R_1 are these:

- The maximum allowable heat flux through a body under conditions of steady heat flow
- The maximum allowable temperature difference to which the body can be subjected under convective heat transfer, for small values of β
- The maximum allowable heat flux in a steady heat condition in a solid containing nonconducting holes or cavities

For values of β greater than 10, equation (7.1) reduces to

$$\frac{1}{\sigma^*_{s\,max}} = \mathrm{A}$$

and, at failure,

$$T_{o\,max} = \frac{\mathrm{A}\sigma_F(1 - \nu)}{E\alpha}$$

or, in general,

$$\Delta T_{max} = \frac{\mathrm{A}\sigma_F(1 - \nu)}{E\alpha}$$

Thus, for a given body under specified temperature change conditions, another relevant thermal shock resistance parameter R_2 becomes

$$R_2 = \frac{\sigma_F(1 - \nu)}{E\alpha} \tag{7.4}$$

This parameter has the following combination of physical interpretation and heat transfer conditions:

- The maximum allowable temperature difference in the body under conditions of steady heat flow
- The maximum allowable temperature difference to which the body can be subjected in a convective environment for large values of β
- The maximum allowable temperature gradient in steady-state heat conduction in a solid containing nonconducting holes or cavities

For the case where an opaque body is subjected to changes in the level of blackbody radiation, the relevant thermal shock resistance parameter may be derived on the basis of the recognized significance of the emissivity of the material of which the body is made [1].

The basic law for heat transfer by radiation [2] is

$$Q = \varepsilon S(T_2^4 - T_1^4) \tag{7.5}$$

where ε is the emissivity of the surface of the body, S is the Stefan-Boltzmann constant, and T_1 and T_2 are the temperature of the body and the surroundings, respectively.

The temperature rise at any radius r after any time t in a sphere of radius c with an arbitrary initial temperature subjected to a constant heat flux [3] is

$$T(r, t) = \frac{3Q_o t(TD)}{ck} + \frac{Q_o(5r^2 - 3c^2)}{10ck} - \frac{2Q_o c^2}{kr} \sum_{n=1}^{\infty} \frac{\sin (r\theta_n/c)}{\theta_n^2 \sin \theta_n} \exp\left[-\frac{(TD)\theta_n^2 t}{c^2}\right] \tag{7.6}$$

where θ_n, with $n = 1, 2, 3, \ldots$, represents the positive roots of the equation $\tan \theta = \theta$.

In equation (7.6), Q_o is the constant surface heat flux. The expressions for the radial and hoop stresses of a sphere subjected to a temperature gradient [4] are

$$\sigma_r = \frac{2\alpha E}{1 - \nu)}\left(\frac{1}{c^3}\int_o^r Tr^2\,dr - \frac{1}{r^3}\int_o^c Tr^2\,dr\right) \tag{7.7}$$

and

$$\sigma_h = \frac{\alpha E}{1 - \nu)}\left(\frac{1}{r^3}\int_o^r Tr^2\,dr + \frac{2}{c^3}\int_o^c Tr^2\,dr - T\right) \tag{7.8}$$

The maximum stress occurs at the center of the sphere, at $r = 0$. We can use the L'Hospital rule to evaluate the indeterminate integrals thus:

$$\lim_{r\to 0} \frac{1}{r^2}\int_o^r Tr^2\,dr = 0$$
$$\lim_{r\to 0} \frac{1}{r^3}\int_o^r Tr^2\,dr = \frac{T_o}{3} \tag{7.9}$$

where T_o is the temperature at the center of the sphere. The substitution of equations (7.6) and (7.9) in equations (7.4) and (7.8) gives the stresses

$$\sigma_r(o, t) = \sigma_h(o, t) = \frac{2\alpha E Q_o c}{k(1 - \nu)}\left\{\frac{1}{10} - \frac{2}{3}\sum_{n=1}^{\infty}\frac{1}{\theta_n^2 \sin \theta_n}\exp\left[-\frac{(TD)\theta_n^2 t}{c^2}\right]\right\} \tag{7.10}$$

The maximum stress is, from equation (7.10), given by

$$\sigma_{max} = \frac{Q_o \alpha E c}{5(1 - \nu)k} = \sigma_F \tag{7.11}$$

Assuming that $T_2^4 > T_1^4$, so that $(T_2^4 - T_1^4) \approx T_2^4$, equation (7.5) becomes

$$Q = T_2^4 \varepsilon S \tag{7.12}$$

We now combine equations (7.11) and (7.12) to give

$$\sigma_F = \left(\frac{\varepsilon S c \alpha E}{5(1 - \nu)k}\right) T_2^4$$

The influence of c, S, and ν is small compared to that of the other terms; thus

$$T_2 = \left(\frac{\sigma_F k}{\alpha E \varepsilon}\right)^{1/4}$$

This causes the thermal shock parameter R_3 to be given by

$$R_3 = \frac{\sigma_F k}{\alpha E \varepsilon}$$

This relation holds true for other geometric shapes such as cylinders and plates. R_3 is thus an index of the maximum allowable blackbody radiation to which an opaque material can be subjected without risking fracture due to thermal shock.

There are other thermal shock resistance parameters, as suggested by Hasselman [5]:

- R_4, an index of the resistance of the material to thermal shock by kinetic crack growth, given by

$$R_4 = \frac{E\gamma}{\sigma_F^2(1 - \nu)}$$

  where γ is the work of fracture
- R_5, which is indicative of the resistance of the material to thermal shock by quasi-static crack growth, given by

$$R_5 = \left(\frac{\gamma}{E\alpha^2}\right)^{0.5}$$

- R_6, which gives an indication of the maximum allowable rate of surface heating, given by

$$R_6 = \sigma_F(1 - \nu)\frac{TD}{E\alpha}$$

  where TD is the thermal diffusivity of the material
- R_7, which gives an indication of the minimum in the elastic energy at fracture available for crack propagation, given by

$$R_7 = \frac{E}{\sigma_F^2(1 - \nu)}$$

A summary of less used thermal shock resistance parameters, as suggested by Hasselman [5], is presented in Table 7.1.

TABLE 7.1 SOME CHARACTERISTICS OF INFREQUENTLY USED THERMAL SHOCK RESISTANCE PARAMETERS

Thermal Shock Resistance Parameter[a]	Designation[b]	Physical Interpretation/ Heat Transfer Conditions
$\dfrac{\sigma_F(1-\nu)}{\alpha\eta}$	R_{cr}	The maximum allowable rate of increase of temperature difference across a body undergoing thermal stress relaxation by creep.
$\dfrac{\sigma_F(1-\nu)k}{\alpha\eta}$	R'_{cr}	The maximum allowable rate of change of heat flux through a body undergoing thermal stress relaxation by creep.
$\left[\dfrac{\sigma_F(1-\nu)k}{\alpha E\varepsilon(1-F\lambda_o)}\right]^{0.25}$	R_{trans}	The maximum allowable blackbody temperature to which semitransparent material can be subjected, the absorption properties of which can be described as transparent for $\lambda < \lambda_o$ and opaque for $\lambda > \lambda_o$.
$-\log\left[\dfrac{\sigma_F^2}{E\gamma}\right]$	T	The minimum in the total numbers of fragments resulting from thermal stress fracture initiation.
$\left(\dfrac{\gamma k^2}{\alpha^2 E}\right)^{0.5}$	R'_{st}	1. The minimum allowable temperature difference required to propagate long cracks for mild thermal stress conditions. 2. The maximum allowable heat flux in a body containing cracks perpendicular to heat flow direction.

[a] η is the viscosity of the material, ε is its emissivity, $F\lambda_o$ is the fraction of the total energy from blackbody radiation below frequency λ_o, and λ is wavelength.

[b] As given by Hasselmann [5].

In the next two sections, we consider, in some detail, the selection of materials for thermal shock resistance in three applications: the orbiter of the United States' space shuttle, the support of a catalytic converter in an automobile, and an insulated combustion chamber in a turbocharged diesel engine.

7.3 THE SPACE SHUTTLE ORBITER THERMAL PROTECTION SYSTEM

With the inaugural flight of the U.S. Space Transportation System, popularly known as the space shuttle, in April 1981, interest in the design of aerodynamically heated manned reusable space vehicles surged. The space shuttle orbiter, the cargo-carrying element of the system, is launched, orbits, and reenters as a spacecraft. Once inside the lower atmosphere, it is maneuvered to land as an aircraft. The temperatures to which various parts of the orbiter are subjected during the critical stages of ascent and reentry (Figs. 7.3 and 7.4) limit the number of materials

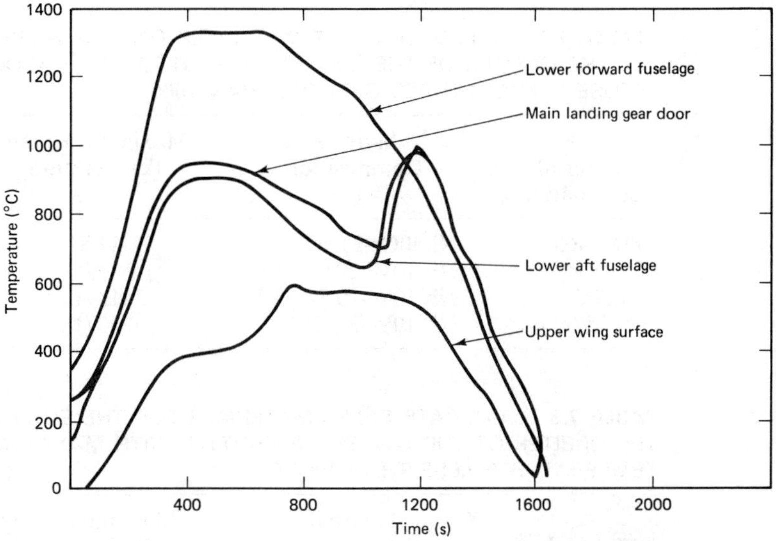

Figure 7.3 Entry surface temperatures at various sections of the U.S. space shuttle orbiter.

Figure 7.4 Approximate temperature distribution on the fuselage of the U.S. space shuttle orbiter during ascent and reentry.

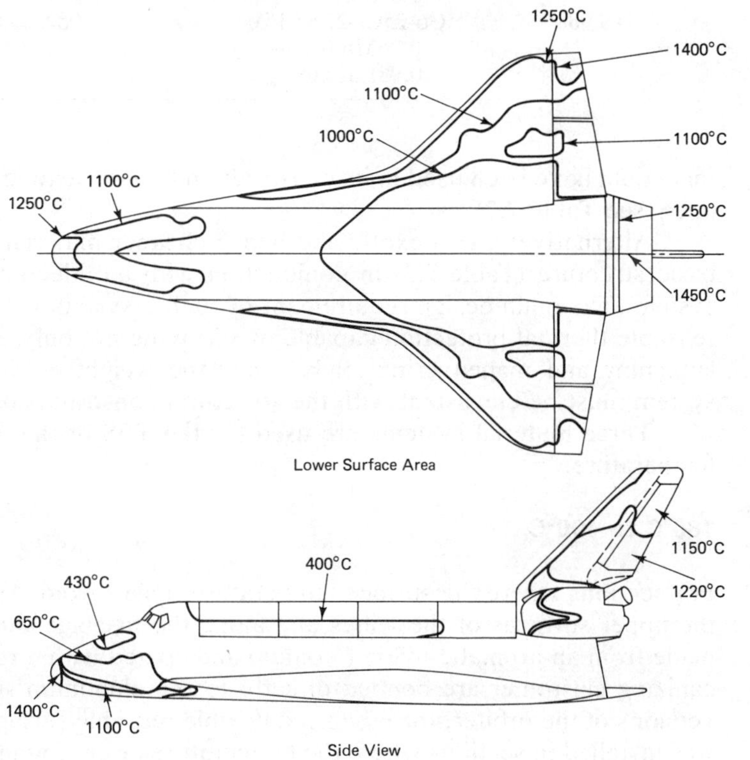

TABLE 7.2 CANDIDATE SHEET MATERIALS FOR THE SUBSTRUCTURE OF THE ORBITER OF THE U.S. SPACE SHUTTLE, WITH MAXIMUM REUSE TEMPERATURES GREATER THAN 1000°C

Material Designation	Material Composition (wt %)	Maximum Reuse Temperature (°C)	Cost ($ kg^{-1})
TD NiCr	Ni-20Cr-2ThO$_2$	1100	250
FS85	Nb-10W-28Ta-1Zr	1371	220
C129Y	Nb-10W-10Hf-0.2Y	1371	260
Cb-752	Nb-10W-2.5Zr	1371	260

TABLE 7.3 CANDIDATE SHEET MATERIALS FOR THE SUBSTRUCTURE OF THE ORBITER OF THE U.S. SPACE SHUTTLE WITH MAXIMUM REUSE TEMPERATURES LESS THAN 1000°C

Material Designation	Material Composition (wt %)	Maximum Reuse Temperature (°C)	Cost ($ kg^{-1})
7075-T73	Al-1.6Cu-5.6Zn - 2.5Mg-0.3Cr	150	2
Ti-6Al-4V	Ti-6Al-4V	482	33
René 41	Ni-19Cr-11Co- 9.75Mo-3.15Ti- 1.6Al-0.09C	871	40
HS 988	Co-25Cr-22Ni-15W- 0.75Mn-0.2Si- 0.08La-0.08	982	40

that could have been used in its construction to a few exotic and hence expensive ones (see Table 7.2).

Alternatively, less exotic and hence cheaper materials can be used for the basic structure (Table 7.3) in conjunction with an adequate thermal protection system. The main design requirement of such a system is to provide lightweight reusable thermal protection capable of surviving not only entry heating but also launching and maneuvering loads. Thus the weight of the thermal protection system must be consistent with the structural constraints of the entire orbiter.

Three material systems are used for the TPS of the orbiter, depending on temperature.

180°C to 400°C

For sections where the surface temperature ranges from 180°C to 400°C (such as the upper surfaces of the wings and along the fuselage sidewalls), felt blankets made from an aromatic nylon (Nomex) and coated with a room-temperature vulcanizing elastomer are bonded directly to the aluminum substrate. In the later versions of the orbiter, the advanced flexible reusable surface insulation (AFRSI) was installed in sections where the temperatures were found to exceed the design values, such as over 1.5 m^2 of the elevon cove of the space shuttle *Columbia*. AFRSI consists of silicaglass on the outside, microquartz felt in the middle, and a glass cloth on the inside.

400°C to 1260°C

For sections where the surface temperature is in the 400°C to 1260°C range (which represents some 75% of the surface area of the orbiter), the acceptable thermal protection system (TPS), in addition to being lightweight, must be made of a material that has excellent thermal shock resistance since these sections are subjected to very large temperature gradients during ascent and reentry of the orbiter.

Various thermal protection system concepts have been studied and used in the aerospace industry since about 1960, including ceramic insulations, detachable ablative heat shields (which had been used on earlier manned spacecraft such as the Apollo mission crafts), heat sinks, transpiration cooling systems, and radiative metallic structures. Only ceramic insulations, ablators, and radiative metallic structures could be regarded as serious options for a TPS on a structure as large as the orbiter (a heat sink would be heavy, and transpiration cooling would be too complex).

Use of ablators for the TPS presented two problems. First, ablators do not retain their aerodynamic surfaces through the reentry trajectory, necessitating costly refurbishment after each flight. Second, ablators such as those used on Apollo have nearly four times the density of ceramic insulators.

Hot radiative structures have been used extensively; for example, Inconel alloy X-750 was used on the X15 vehicle, and the Mercury and Gemini spacecraft included shingles made of René 41 on sidewalls to endure entry temperatures up to about 930°C. But this concept has several drawbacks. (a) It uses materials with high densities. (b) It has limited over-temperature capability. (c) Its fragile silicide coating is easily damaged, resulting in rapid oxygen embrittlement at high temperatures and, probably, catastrophic ignition above the melting point of its principal oxide (in the case of a columbiun metallic heat shield, for example, the principal oxide Cb_2O_5, and its melting point is 1500°C. (d) Design and manufacturing are complex. The design and thermal stress analysis of even the simplest of these structural panels is very labor-intensive. The problems of attaching panels to a typical substructure while maintaining the close tolerances necessary to avoid excessive heating from early boundary layer transition or plasma ingestion are formidable.

The final choice made in the space shuttle orbiter was to use aluminum for the basic structure together with a TPS consisting of (a) some 30,000 high-purity silica tiles* (LI 2000), each measuring 0.15 m by 0.15 m; (b) a room-temperature vulcanizing (RTV) elastomeric adhesive-sealant; and (c) felt strain-isolator pads (SIP).

The tiles are coated with a high-emittance layer of black borosilicate glass and functions as both an insulator and a radiator to limit heating of the structure.

*In later versions of the shuttle orbiter, high-purity silica was replaced by a material consisting of 80 wt % pure silica and 20 wt % proprietary alumino boria silicate fiber, called Nextel.

TABLE 7.4 COMPARISON OF THERMAL SHOCK RESISTANCE OF CANDIDATE MATERIALS FOR THE THERMAL PROTECTION SYSTEM OF A REUSABLE SPACE VEHICLE SUCH AS THE ORBITER OF THE U.S. SPACE SHUTTLE

Material[a]	Flexural Strength, σ_F (MPa)	ν	E (GPa)	α (10^{-6} K^{-1})	R_2 (K)
LI 900 silica (in-plane direction)	0.4	0.18	0.172	0.83	2298
Magnesium oxide, MgO	105	0.36	280	13.50	18
Alumina, Al_2O_3	280	0.26	370	8.80	64
Beryllium oxide, BeO	100	0.24	393	7.38	26

[a]Means in the 300 K–1260 K temperature range.

The strain-isolator pads provide for mechanical isolation of the tiles from the vehicle deflections and strains. The RTV bonds the tiles to the SIPs and the SIPs to the orbiter.

Table 7.4 presents a comparison of the thermal shock resistance of high-purity silica with other candidate materials for this application.

Higher Temperatures

For sections such as the wing leading edge and nose cone areas, where the temperatures climb, in some locations, above 1600°C, a coated reinforced carbon-carbon (RCC) composite material provides the needed thermal protection. Basically, RCC is an all-carbon material reinforced for strength and treated to resist oxidation at extreme temperatures. It is lightweight and rugged and can be fabricated readily into virtually any desired shape. At temperatures devastating to metals RCC actually increases in strength*, providing excellent thermal shock resistance and outstanding fatigue life.

The locations of various TPS materials in the space shuttle *Columbia* are shown in Fig. 7.5.

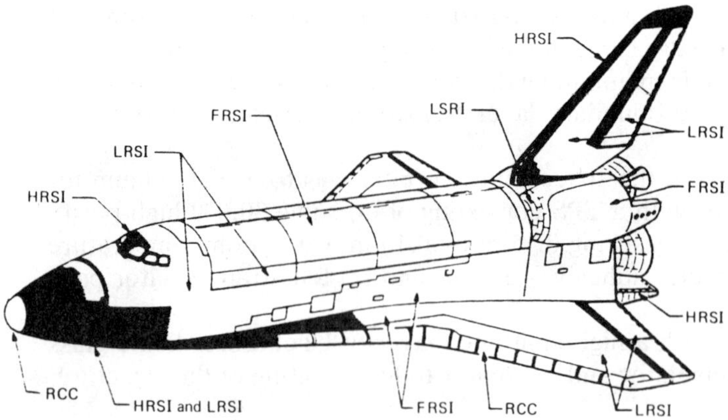

Figure 7.5 Locations of the thermal protection system in the U.S. space shuttle *Columbia*. RCC, reinforced carbon/carbon; FRSI, felt reusable surface insulation; LRSI, low-temperature reusable surface insulation; HRSI, high-temperature reusable surface insulation.

*See Chapter C15 in Part 3.

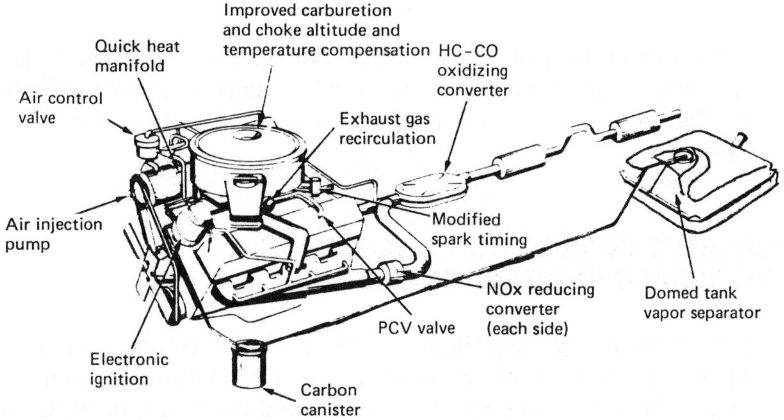

Figure 7.6 A dual catalytic converter system for an automobile.

7.4 CERAMIC SUPPORT FOR A CATALYTIC CONVERTER IN AN AUTOMOBILE

Automobile exhaust emission reduction levels set by the Clean Air Act of 1970 rely on the chemical conversion of nitrous oxides and unoxidized hydrocarbons to nitrogen, oxygen, water, and carbon dioxide. Such conversion depends on reactions catalyzed by platinum salts deposited on high-surface-area ceramic supports. A typical automobile dual catalytic converter system is shown schematically in Fig. 7.6. After engine ignition, the catalytic converters reach an operation temperature of 800°C to 900°C in 2 min, which places the ceramic support in a severe thermal shock mode.

Two approaches have been taken to combat possible critical stress problems induced by these conditions. First, the support could be made of a large number of small pellets so that each part is subject to only a fraction of the total temperature gradient and the thermal expansion of the support material is not critical. Second, monolithic honeycomb structures can be used, high thermal stresses being reduced by controlling (a) phase composition, to produce low thermal expansion; (b) porosity, to reduce modulus and crack propogation, etc.; and (c) web thickness, to increase heat transfer and thereby reduce thermal gradients.

Monolithic support materials are selected on the basis of their thermal shock resistance. Table 7.5 compares candidate materials for this application. In most

TABLE 7.5 COMPARISON OF THERMAL SHOCK RESISTANCE OF CANDIDATE MATERIALS FOR THE CERAMIC SUPPORT OF A CATALYTIC CONVERTER IN AN AUTOMOBILE

Material[a]	Flexural Strength, σ_F (MPa)	ν	E (GPa)	α (10^{-6} K^{-1})	R_2 (K)
Aluminum titanate, $Al_2O_3 \cdot TiO_2$	38	0.30	18	1.3	1137
Alumina, Al_2O_3	280	0.26	370	8.8	64
Mullite, $3Al_2O_3 \cdot 2SiO_2$	95	0.30	145	4.6	100
Spodumene, $LiO_2 \cdot Al_2O_3 \cdot 4SiO_2$	136	0.27	70	1.0	1418

[a]Means in the 300 K–1260 K temperature range.

of the currently available catalytic converters, the monolithic honeycomb structures are made of a laminate material, usually one layer of high cordierite content bonded onto a layer of low mullite content bonded onto a layer of low alumina content.

7.5 INSULATED COMBUSTION CHAMBER IN A TURBOCHARGED DIESEL ENGINE

Part of the energy supplied to a diesel engine by the fuel is converted into shaft work, and the rest leaves as heat, either by different coolers or with the exhaust gases. The different energy flows into and out of a typical turbocharged, charged-cooled diesel engine are shown in Fig. 7.7. The cycle top temperature is very high, and in a conventional system, the components making up the combustion chamber (cylinder head, cylinder liner, piston; see Fig. 7.8) can be cooled by water, oil, or gas. A typical combustion chamber heat flow distribution is as follows: to piston, 35%; to liner, 25%; to cylinder head, 25%; and to exhaust port, 15%.

In the radiator, heat from cooling by the aforementioned agents and heat caused by friction leave the engine at low temperatures. Coolant flow circulates through the cylinder head and around the liner, while the piston is usually cooled by oil jets. Part of the energy in the exhaust is used to drive the turbocharger and is returned to the system as high-temperature, high-pressure air by the turbocharger compressor. This heat is cooled away in the charge cooler and leaves the system.

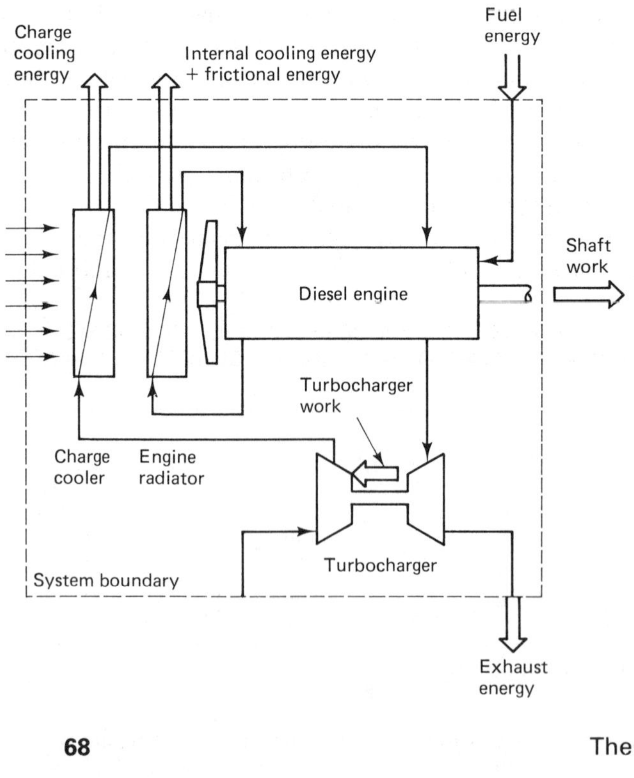

Figure 7.7 Energy flows into and out of a typical turbocharged, charged-cooled diesel engine.

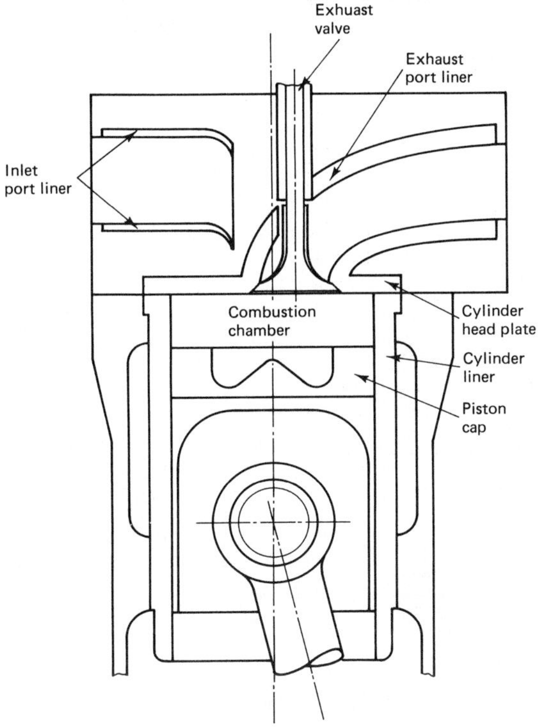

Figure 7.8 Insulated diesel engine components.

In Table 7.6, the portion labeled "internal cooling" can be achieved by methods other than air, water, or oil. In recent years, a viable alternative has been the use of suitable insulation materials for the construction of the combustion chamber components. Among the advantages afforded by this option are these:

- A large part of the low-temperature internal cooling energy will be converted into high-temperature exhaust energy that may be recovered via (a) expansion in one or more turbines or (b) the use of some kind of heat engine (such as Rankine or Stirling) after the diesel proper, a so-called bottoming cycle (see Fig. 7.9).

TABLE 7.6 ENERGY BALANCE IN A TURBOCHARGED, CHARGE-COOLED DIESEL ENGINE

Energy	Proportion of total energy (%)
Shaftwork	40
Friction energy	5
Charge cooling energy	7
Internal cooling energy	13
Exhaust energy	35

Source: B. Palm, "On the use of ceramics in diesel engines" in Ceramics in Advanced Energy Technologies, Proceedings of the European Colloquim, Petten, Holland, September 1982, edited by H. Krockel, M. Mertz, and O. Van der Biest, Dordrecht, Holland: D. Reidel Publ. Co., 1984.

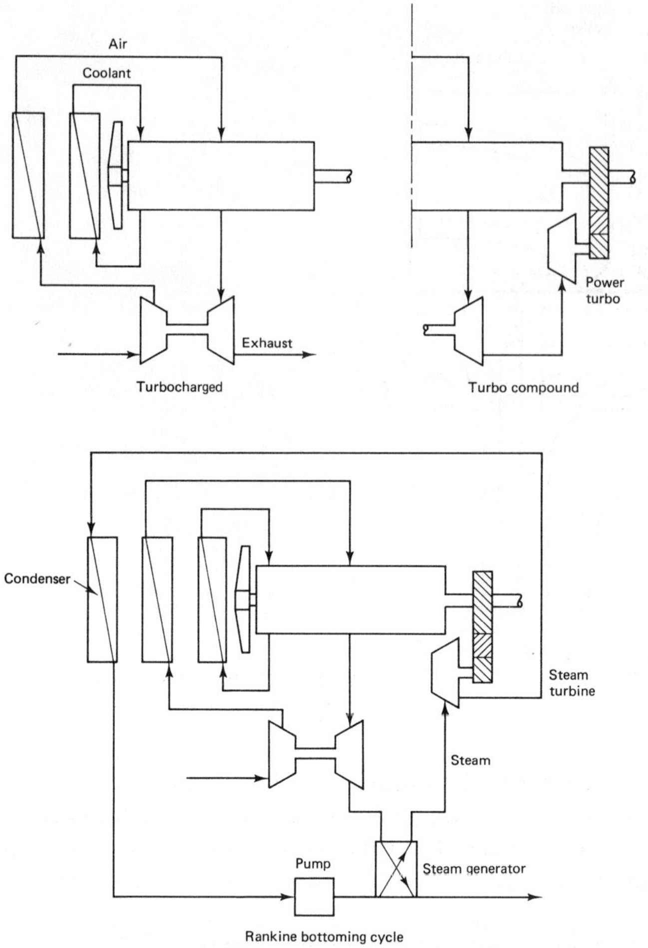

Figure 7.9 Diesel engine exhaust recovery systems.

- Shaft work or indicated efficiency increases; with use of an insulated material leading to about 70% total reduction in heat flux, there may be a marked reduction in the indicated fuel consumption.
- Brake-specific fuel consumption decreases (see Table 7.7).
- Cooling fan and water pump power consumption drop.

TABLE 7.7 BRAKE-SPECIFIC FUEL CONSUMPTION (BSFC) REDUCTION POTENTIAL OF TURBOCHARGED DIESEL ENGINES

Factor causing reduction	Reduction in bsfc (g kWh^{-1})
Improvement in thermal efficiency	1.75
Insulation of combustion chamber	3.20
Internal cycle change	4.75

Source: Same as for Table 7.6.

TABLE 7.8 EFFECT OF A 6 mm THICK CERAMIC LINING[a] ON THE OPERATING PARAMETERS IN A DIESEL ENGINE WITH EXHAUST GAS-OPERATED TURBOCHARGER (SPEED = 4500 r min^{-1}; EQUIVALENCE RATIO = 80% FULL LOAD)

Operating parameter	Effect
Exhaust gas heat	Increase of 10%
Indicated efficiency	Increase of 4%
Torque, output	Negligible effect
Air mass flow	Decrease of 5%
Specific fuel consumption	Decrease of 5%
Wall heat	Decrease of 35%

[a]Mean temperature of combustion chamber wall 530°C.

Source: P. Walzer, H. Heinrich, and M. Langer. "Ceramic components in passenger-car diesel engines," *SAMPE Quart.* *17* (3), p. 33, 1986.

Table 7.8 summarizes the influence of an insulated combustion chamber on some operating parameters in a diesel engine.

Of course, the success of the insulated combustion chamber is predicated on the availability of suitable insulation materials. In addition to the required improvement in efficiency, an insulation material must have the capability to withstand both mechanical and thermal loads in the engine and be compatible with the surrounding parts. The ideal insulation material must therefore be a simple, inexpensive ceramic that is able to withstand temperatures of up to about 1000°C and is resistant to the engine environment, with low or no porosity, low thermal conductivity, low density, low specific heat, high toughness, sufficient strength, and suitable tribological properties (especially for applications in exhaust and inlet port liners and valve seats). But perhaps the most important material requirement is excellent resistance to thermal shock. Calculations show that magnesium aluminum silicate and hot-pressed silicon carbide are excellent candidates (Table 7.9).

Table 7.9 presents information on the relevant material properties and relative resistance to thermal shock for some candidate materials for this application.

Only limited information is currently available concerning research and development on the use of ceramic materials as combustion chamber insulation in turbocharged diesel engines. In the United States, some data have originated from the Cummins/TARADCOM U.S. Army Tank project [6]. In Japan, Komatsu Ltd. has published results from tests from a completely insulated multicylinder DJ diesel engine [7], and the capability of sintered silicon nitride to function, without additional insulation, in a small NA diesel engine has been reported [8]. In Europe, work has been reported on ceramic pistons in a Gardner single-cylinder diesel engine in England [9].

TABLE 7.9 COMPARISON OF THERMAL SHOCK RESISTANCE OF CANDIDATE MATERIALS FOR THE INSULATION OF THE COMBUSTION CHAMBER COMPONENTS OF A TURBOCHARGED DIESEL ENGINE

Material[a]	Flexural Strength, σ_F (MPa)	ν	k ($Wm^{-1} K^{-1}$)	E (GPa)	α ($10^{-6} K^{-1}$)	R_2 (K)	R_1 (Wm^{-1})
Reaction-bonded Si_3N_4 (RBSN)	270	0.27	15	190	3.1	335	5025
Hot-pressed Si_3N_4 (HPSN)	783	0.27	20	303	3.0	629	12580
Hot isostatically pressed Si_3N_4 (HIPSN)	700	0.27	25	310	3.4	485	12125
Sintered Si_3N_4 (SSN)	470	0.27	17	300	3.3	347	5899
Sintered silicon carbide (SSIC)	480	0.27	60	400	4.6	190	11400
Hot-pressed silicon carbide (HPSIC)	700	0.27	90	450	4.7	242	21780
Aluminum, Al_2O_3	280	0.26	33	370	8.8	64	2112
Partically stabilized zirconia (PSZ)	500	0.24	4	170	8.7	257	1028
Alumina titanate, Al_2TiO_5	38	0.30	2	18	2.3	643	1286
Magnesium aluminum silicate	50	0.26	1	12	0.6	5139	5139

[a]Means in the 300 K–1260 K temperature range.

REFERENCES

1. T. H. Allen, C. R. Johnson and E. L. Rusert, "High-temperature emittance of coated refractory metals," paper presented at the SAMPE symposium in St. Louis, Mo., 1967; A. A. Travassoli, "Metallic materials for the space shuttle," *Aeronautical J. 76* (735), pp. 152–156, 1972.
2. W. A. Gray and R. Muller, *Engineering Calculations in Radiative Heat Transfer,* Oxford: Pergamon Press, p. 26.
3. H. S. Carlshaw and J. C. Jaeger, *Conduction in Solids*. Oxford: Clarendon Press, 1959.
4. S. Timoshenko and J. N. Goddier, *Theory of Elasticity*. New York: McGraw-Hill, 1951.
5. D. P. H. Hasselman, "Unified theory of thermal shock fracture initiation and crack propagation in brittle ceramics," *Amer. Ceramic Soc. 52* (11), pp. 600–604, 1969.
6. R. Kamo, "Cycles and performance studies for advanced diesel engines," *Proc. 5th Army Materials Technical Conf.*, Brook-Hill Publishing Co., Chesnut Hill, Mass., 1978; W. Bryzik, "Adiabatic engine program," Automotive Technology Development Contractors' coordination meeting, Dearborn, Mich., November 1980.
7. T. Yoshimitsu, et al., "Capabilities of heat insulated diesel engines," SAE 820431.
8. Y. Hamano and Y. Nakahara, "Ceramic parts for a diesel engine," Automotive Technology Development Contractors' coordination meeting, Dearborn, Mich., November 1980.
9. D. A. Parker and R. F. Smart, *An Evaluation of Silica Nitride Diesel Pistons,* Rugby, England: Associated Engineering Developments Ltd., 1980.

ELECTROMAGNETIC SHIELDING

CHAPTER 8

8.1 INTRODUCTION

There are a myriad of sources of electromagnetic or radio frequency waves, causing electromagnetic interference (EMI) in a large number of components in service. This point is illustrated by the following news flashes, which are anecdotal but by no means apocryphal. "Gasoline pump manufacturer finds CB radio zeroes remote-pump readouts"; "Airport radar signals erase taxpayers' records and computer memory bank"; "Aircraft radar operates remote door opening devices in the area"; "Microwave oven interference knocks pacemaker off drive"; "Doctor putting finishes on computer program loses all from finger sparks"; "Garage door openers in the Santa Barbara area jammed by signals to and from the U.S. President's communications plane"; "Microwave, radio, and radar signals induce EMI that cause stabilator of the U.S. Army Black Hawk helicopter to force the craft into an unrecoverable dive, resulting in five crashes"; "Computer printers in the offices of the British science magazine *New Scientist* switch themselves on for no apparent reason"; "Electronic stage lighting at the Royal Shakespeare Theater at Stratford-upon-Avon, England, stop working when nearby radio broadcast transmitters are switched on at low power"; "Sound recorders used by radio news reporters burn out as current induced in the circuit wiring by ships in the British Task Force during the Falklands War in 1983 fused the transistors"; "The metalwork of the King Fahd International Stadium functions as a giant radio-receiving aerial because of the presence, 3 km away, of medium- and shortwave radio stations broadcasting at powers in excess of 1 MW''; Television sets in homes near London's Heathrow Airport have very poor reception''; ''Cars and taxis equipped with mobile transmitters, if they come within 5 m of an office with a digital exchange, cause interference fields of strengths of 5 Vm^{-1}''; ''The Radio Astronomy Observatory at Jodrell Bank in Cheshire, England, suffers interference from a police radio transmitter sited 30 km away.''

EMI sources may be broken down into two major categories, each with a number of subgroups.

Natural Sources

- Terrestial sources, such as atmospherics (thunderstorms, etc.), lightning discharges, precipitation static, and whistlers
- Extraterrestial sources, such as cosmic noise, radio stars, and the sun

Man-made Sources

- Electric power sources, such as generators, faulty wiring or transformers, poor grounding, and transmissions
- Communication and electronics sources, such as broadcast (VHF/FM, VHF/UHF TV), communications systems (ham radios, citizen band radios, telemetry, telephone circuits, facsimile), navigational aids (aircraft beacons, instrument landing systems), radar (police speed monitors, air search, surface detection and traffic control systems), and relay communication systems (microwave relay links, satellite relay systems)
- Tools and machines, such as power tools (bandsaws, electric drills, sanders), appliances (air conditioners, fans, ovens, sewing machines, electric lawn mowers), office and business machines (calculators, cash registers, electric typewriters, computers), transporters (conveyor belts, elevators, moving sidewalks), and industrial machines (electric cranes, fork lifts, lathes, milling machines)
- Ignition systems, such as in internal combustion engines in motor vehicles, tools (auxiliary generators, portable saws), and those in aircraft, tanks, and tractors
- Industrial and consumer items, such as heaters, welders, industrial process controllers, and medical equipment (CAT, x-ray, and ultrasonic machines)

Here are a few examples of systems that require adequate EMI shielding:

1. In the Minuteman missile system, there are some 50,000 electronic and electromechanical parts in the single missile and the launch-essential ground equipment located in a launcher and the launch control facilities necessary to accomplish the launch.

2. Radio astronomy is the study of stars, galaxies, quasars, pulsars, and other extraterrestial objects as conducted at radio wavelengths. The telescopes that are used closely resemble radar and communications antennas, although the operation is generally limited to the receiver mode only. By operating in the microwave region of the electromagnetic spectrum (typically from a few megahertz to several hundred gigahertz), such telescopes are susceptible to electromagnetic interference from a wide variety of sources, ranging from periodic pulses from a digital clock for a high-speed printer to random bursts from relay switches or radio transmissions.

3. The recent introduction of power electronics in rail-traction-power solid-state dc vehicle propulsion applications, to replace older cam controllers, has raised the question of the effect of EMI on the safety and reliability of rapid transit systems. The problem was first detected in the Bay Area Rapid Transit System in San Francisco with special reference to control equipment (such as audio-frequency signaling) in automatic trains. The problem was eventually solved.

4. The increased use of microwave ovens in restaurants, homes, and cafeterias is known to have an adverse effect on the function of implanted ventricular-sensing cardiac devices ("pacemakers").

5. Personal computers, disc drives, printers, and TV monitors have to be properly shielded to prevent emission of unwanted electromagnetic signals from their electronic components, as well as to protect these components from all the stray electromagnetic waves that are normally present in the location of use.

6. Hang gliders are fitted with a special device to record the rate of climb and descent. Such devices have to be housed in shielded containers in order to reduce the negative effect that EMI (arising from microwave towers and other antennas) would have on the proper function of the device.

Since October 1, 1963, with the establishment of Federal Communications Commission (FCC) rules on the permissible levels of EMI noise emission, shielding has been a major concern of design engineers of equipment for digital computer devices. The FCC has issued regulations limiting the field strength of emitted signals from these devices having a timing signal with a clock frequency exceeding 10 kHz. The FCC has designated two classes of emitting devices, A and B, which are subject to slightly different regulations because of their different potential for affecting broadcast signals (see Table 8.1).

In Canada, the EMI issue is controlled by voluntary compliance with a standard issued by the Canadian Standards Association (CSA). The standard, C102.8-M1983, is modeled after the FCC regulations and establishes the same control limits and classes as set out in Table 8.1.

Perhaps the most stringent regulations in a major marketplace are those set by the Deutsche Bundespost. This organization, through the Verband Deutscher Elektrotechniker (VDE), sets standards for EMI emission for all electronic devices sold in West Germany.

An international committee is currently in the process of formulating guidance for EMI protection that will fall somewhere between the present FCC and VDE limits.

TABLE 8.1 FCC REGULATIONS ON ALLOWABLE EMI EMISSIONS

Class	Frequency Range (MHz)	Distance (m)	Field Strength Limit (μV m^{-1})	Applications
A	30–88	30	30	Commercial, business or industrial devices
A	88–216	30	50	
A	216–1000	30	70	
B	30–88	3	100	Personal computers, electronic games, computing devices marketed for use in a residential environment
B	88–216	3	150	
B	216–1000	3	200	

8.2 SHIELDING THROUGH THE USE OF APPROPRIATE MATERIALS

The most effective way of shielding susceptible components from EMI is through the use of certain materials in the construction of these components. Materials that conduct or absorb electromagnetic radiation should be used. The former are the more effective, which leaves the attentuation of backscattered waves as the primary application of absorbing materials.

The simplest absorbing materials are characterized by their conductivity σ and relative permittivity ε. The losses are, in this case, caused by the conductivity. More complicated materials may use hysteresis losses of permeability or permittivity or use a loss mechanism based on the damping of the precessional motion of the magnetization vectors.

Conductive Coatings

One widely used approach is to mold the components out of a polymeric material. Then a conductive layer is applied, usually on the inside. This is used in cathode ray tubes, personal computer keyboards, and related devices.

Although there are currently five methods of applying conductive coatings (paint, arc or flame spray, vacuum metallization, chemical deposition, and cathode sputtering), only paint and arc or flame spray are commercially practicable at the present time, although a recently introduced specialist chemical deposition technique has proved to be very competitive (Table 8.2). Since neither paint nor the arc or flame spray process produces a fully dense metallic layer, coating thicknesses of at least 50 μm are needed to give adequate shielding. Both coating techniques use manual spraying, so the uniformity of coating thickness is a function of operator skill. This is the main disadvantage of these coating techniques. A further drawback is the embrittlement of the film by the relatively thick low-ductility coatings. These drawbacks may be overcome by certain processes, such as vacuum metallization, magnetron sputtering, and electroless plating, which

TABLE 8.2 COMPARATIVE PERFORMANCE OF THE EMI SHIELDING EFFECTIVENESS OF VARIOUS MATERIALS

Material[a]	Shielding Effectiveness at 1 GHz (dB)	Total Cost of Shielding and Finishing on a Cathode Ray Tube (CRT) Base[b] ($ m^{-2})
Ronabond VEST, nickel-plated SMC (1.5 μm thick)	62	
Zinc-arc-sprayed SMC (75 μm thick)	61	
Nickel-painted SMC (75 μm thick)	47	
Vacuum metallized SMC	45	
Zinc arc spray		38
Silver-rich paint		52
Nickel-rich paint		27
Ronabond VEST spray top finish		23
Ronabond VEST electrostatic top finish		16

[a] VEST, vapor etch shielding technology; SMC, fire-retardant sheet molding compound (a polymer).

[b] Area of CRT base cabinet, 0.33 m^2.

Source: J. E. McCaskie, C. Tsiamis, and H. Gerhardt, "Vapor etching process for EMI/RFI shielding," *Modern Plastics 60* (3), pp. 66–68, 1983.

provide effective shielding from EMI with very thin (about 1 μm) coatings but require very expensive equipment. Table 8.3 contains a summary of the comparative merits and demerits of vacuum metallization, sputtering, zinc arc or flame spray, and paints.

"Conductive" Polymers

Another method involves the compounding of conductive constituents into a polymeric matrix, yielding a "conductive" filled polymer. Such constituents are added in the form of fibers, flakes, or powder by any of several techniques. The amount of filler added depends on the application requirements, the aspect ratio of the filler, the processing conditions needed for the composite, the conductivity of the filler, and the particular resin matrix.

These filled materials suffer from a number of disadvantages, especially poor physical properties, poor surface appearance, and high prices (see Table 8.4). In addition, these materials have poor impact strength (typically requiring about 35% loading to achieve about 40 dB attentuation level of 1 Ω cm^{-2} resistivity). This limits the use of these materials in the business machine housings market. As far as stainless steel fiber-polymeric materials are concerned, although fairly high in price, they can be used at low loadings, retaining much more of the mechanical properties of the resin. The fibers, however, have the tendency to entangle during processing and can cause problems in attaining reproducible shielding results. Some manufacturers are now offering fibers treated with coupling agents in order to improve processing and dispersion.

The conductivity of conductive polymer-based materials is primarily a function of the aspect ratio and the initial amount of the filler. Aspect ratio is defined

TABLE 8.3 SOME CHARACTERISTICS OF FOUR METHODS OF APPLYING A CONDUCTIVE COATING

Method	Cost of Capital Equipment	Advantages	Disadvantages
Vacuum metallizing	$200,000 for a 2-m metallizer	Easily controlled film thickness; excellent adhesion; excellent conductivity over long distances	Masking is usually necessary; susceptible to corrosion in humid conditions
Sputtering	$300,000 to $3,000,000 for sputtering equipment	Protected parts can be masked and shielded in "as molded" so long as they are clean; very uniform and consistent electrical properties; very clean operation, with excellent productivity, high throughput, and low scrap rate (<5%)	Uniform thickness difficult to achieve, particularly on side walls; masking is necessary; coating loses conductivity on environmental testing
Zinc arc spray or flame spray	$6000 for a basic zinc arc spray unit, but water washbooths and air supplies for the operator raise the cost to about $15,000 per booth	A metal shell inside a decorative thermoplastic housing is obtained; process is widely understood	Toxic zinc is produced during the process, constituting serious health and safety hazards; inconsistent adhesion (flaking), especially if surface preparation is incomplete or poor
Paints	No additional capital equipment needed (apart from normal spraying equipment)	Ease of application; no special equipment required; excellent long-term adhesion properties; coatings may be overlayed with color coats to achieve styling without compromising shielding, adhesion, or other physical properties.	Control of film thickness for both economic viability and shielding performance is critical

TABLE 8.4 "CONDUCTIVE" FILLED POLYMERS: COST OF RAW MATERIALS AND ADDITIVES

Filler Material	Raw Material Cost ($ kg^{-1})	Additive Cost ($ kg^{-1})
Aluminum		
Fiber	10.00	3.00
Flake	3.00	2.00
Powder	2.25	0.75
Carbon		
Carbon black	0.75	0.30
Furnace black	1.25	0.50
Graphite mat	5.00	1.75
Copper powder	6.50	2.00
Metallized glass fiber	5.50	1.50
Nickel-coated graphite fiber	60.00	6.00
Stainless steel fiber	85.00	4.50

Source: Business Communications, Norwalk, Conn.

as the ratio of the largest to the smallest dimension of the filler (for a short fiber, it is length/diameter; for a flake, it is mean diameter/thickness). Critical composition of the filler is the point on the ordinate where the plot of its resistivity as a function of its volume percentage drops sharply (Fig. 8.1). A collation of data from the literature leads to the approximation of the relationship between aspect ratio and critical composition of shield materials by the "master" curve given in Fig. 8.2. A low aspect ratio corresponds to a low critical filler composition, which in turn produces a material with low resistivity or high electrical conductivity.

Processing conditions also affect electrical conductivity and hence shielding effectiveness, as illustrated in Table 8.5.

Figure 8.1 The dependence of the resistivity on filler concentration for filled polymers.

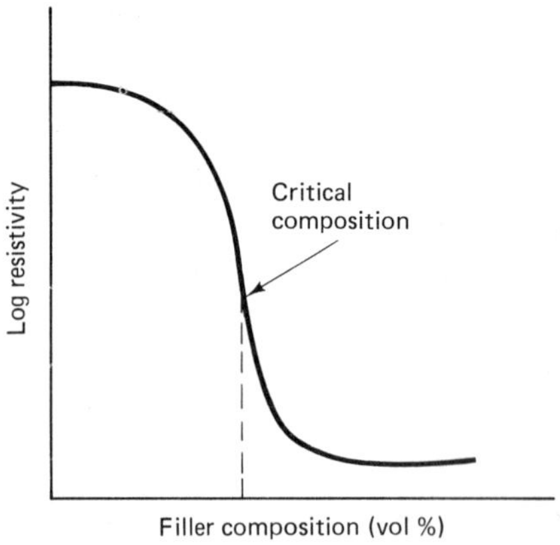

Figure 8.2 Critical filler composition versus aspect ratio for shield materials.

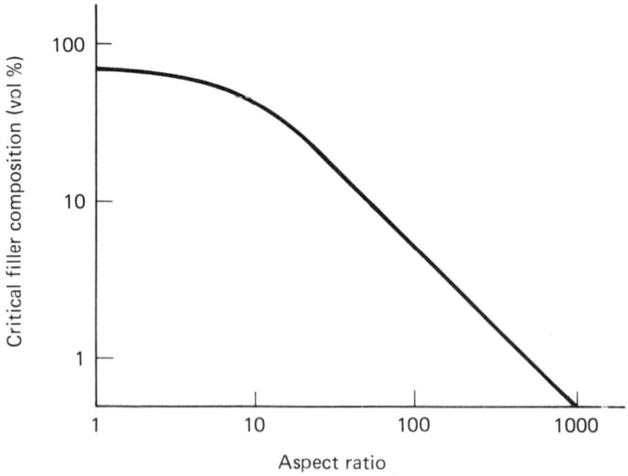

TABLE 8.5 EFFECT OF DEGREE OF MIXING, SCREW SPEED, AND THROUGHPUT RATE ON THE SHIELDING EFFECTIVENESS OF POLYCARBONATE ALLOY CONTAINING 40% ALUMINUM FLAKE

(a) Mixing Element	Degree	Shielding effectiveness at 1 GHz (dB)
Conveying	Gentle	46
Mixing blocks	Moderate	20
(b) Throughout at screw speed of 300 r min^{-1} (kg h^{-1})		Shielding effectiveness at 1 GHz (dB)
68		46
91		43
114		43

Source: J. Curry and J. Farrell, "Care and feeding of high aspect ratio EMI fillers, *Plastics Engineering 42*, p. 41, May 1986.

Filamentary Composites

A third method involves the use of filamentary composite materials. Due to their low electrical conductivity relative to metals, these materials do not have as good a shielding effectiveness against electromagnetic or radio frequency interference as metals do. In addition, the anisotropic nature of electrical conductivity in these materials must be appreciated. Thus an empirical expression for the electrical conductivity of a graphite-epoxy panel has been given as [1]

$$\sigma_L = \frac{\sigma_0(L + N/2)}{N_T}$$

where σ_L is the electrical conductivity along the desired direction, σ_0 is the electrical conductivity of a unidirectional 0° laminate, L is the number of layers in the desired direction, N is the number of layers in the $\pm 45°$ orientation from the desired direction, and N_T is the total number of layers. To increase the conductivity and hence the shielding effectiveness of a filamentary composite material, it is normal practice to protect it with a thin layer of, for example, aluminum flame spray. Then electrical conductivity of the "protected" material becomes [1]

$$\sigma_s = \frac{(\sigma_c t_c + \sigma_p t_p)}{(t_c + t_p)}$$

where σ_c and σ_p are the electrical conductivities of the filamentary composite material and the protective layer material, respectively, and t_c and t_p are their respective thicknesses.

Inherently Conducting Polymers

A fourth method, a very recent innovation, involves fabricating the component to be protected from an inherently conducting polymer, such as polypyrrole, polyphenylene sulfide, or polyacetylene, with or without dopants. The main ad-

TABLE 8.6 THE MAIN ADVANTAGES AND DISADVANTAGES OF THREE CLASSES OF EMI SHIELDING MATERIALS

Material	Advantages	Disadvantages
Conductive coatings	Excellent conductivity, hence good shielding effectiveness Inexpensive, effective use of metal Does not alter mechanical properties Does not affect aesthetics if coating is applied to interior of part	Extra processing step A chip or scratch undermines EMI protection Adhesion and corrosion must be controlled High capital cost; robot technology may be necessary Coating thickness hard to control, especially when coating complex parts
"Conductive" filled polymers	Economical one-stage processing Little modification to existing machinery Reduced cycle times Increased thermal conductivity Surface damage does not affect integrity of shield	Large loadings required to provide adequate conductivity Fillers often degrade mechanical properties Color limited to silver, brown, or black unless painted
Inherently conducting polymers	Conducting without modification	Experimental Highly unstable Expensive

vantage of this method is that the material is inherently conducting, thus eliminating the need for modification. However, such polymers are very expensive and are still experimental. They are not expected to be commercially available before the early 1990s.

Each of the four approaches outlined here has been used in a number of applications to avoid recourse to metals. This is necessary because, although it is true that components made from metals have the advantage of providing physical protection as well as excellent EMI shielding, they have the serious disadvantages of high density and low design flexibility. The main features of the three classes of materials in current use for achieving EMI shielding are summarized in Table 8.6.

8.3 RELIABILITY OF SHIELDING MATERIALS

The reliability problem in EMI shielding is very important but rather complex. No data bank of information currently exists on this subject. Very few studies have been conducted. One is that reported by Chung and Leung [2]. These re-

TABLE 8.7 CALCULATED TIME-TEMPERATURE RELATIONSHIP FOR THE DEGRADATION OF RESISTIVITY OF EMI SHIELDING MATERIALS

Time	Temperature (°C)
1 week	150
1 month	125
1 year	112
10 years	75

Source: K. T. Chung and L. M. Leung, "EMI/RFI shielding effectiveness: Theory and practice, paper presented at the 31st international SAMPE symposium, Los Angeles, April 1986, fig. 4.

searchers generated hypothetical thermal aging conditions (Table 8.7) and performed detailed experiments on two pure silver-based silicone and epoxy materials, Prima Solder EG 8020 and Prima Solder SH 8850, with thermal aging at 150°C, 175°C, and 200°C. Based on the results, these workers concluded that each of the materials tested will survive for more than 20 years as far as EMI shielding is concerned. Another study [3] reported that products containing solidified flake-conducting modifiers showed no serious changes in electrical conductivity or other properties when subjected to a variety of environmental aging tests over a period of three years.

Recently, it has been suggested [4] that the shield material loses no more than 5% thickness following an environmental test (the conditions are given in Table 8.8).

In general, it has been estimated that "conductive" filled polymers and filamentary composite materials perform better than coatings in hostile environments (as where temperature variations, chemicals, and vibration occur simultaneously).

TABLE 8.8 ENVIRONMENT CONDITIONS FOR TESTING OF EMI SHIELDING MATERIALS

Thermal cycling	Aging
1 h at 85°C 1 h at 23°C; 50% relative humidity 1 h at −29°C 1 h at 23%; 50% relative humidity Cycle 2 more times	56 d at 85°C 56 d at 35°C; 90% relative humidity

Source: J. T. Hoback, and J. J. Reilly, "Conductive coatings for EMI shielding", *J. Elastomers and Plastics 20* (1), p. 68, 1988.

8.4 EMI SHIELDING THEORY

Shielding effectiveness (SE), expressed in decibels, is a measure of the reduction of the interfering field achieved by use of a material. It is defined as

$$SE = 10 \log \left(\frac{E_i}{E_0} \right) \tag{8.1}$$

where E_i is the intensity of the electromagnetic field originally incident on the material, and E_0 is the intensity of the transmitted electromagnetic field or signal. Fig. 8.3 shows the link between the ratio E_i/E_o, SE, and shielding attenuation.

Attenuation of the field is achieved by a combination of absorption and reflection of the incident rediation; thus

$$SE = A + R + M \tag{8.2}$$

where A is the loss due to absorption, R is the loss due to reflection, and M is a correction factor that accounts for multiple reflections. The relative levels of effective EMI attenuation are as follows:

0–10 dB	Very little shielding
10–30 dB	Minimum range of meaningful shielding
30–60 dB	Average shielding; should solve many moderate interference problems
60–90 dB	Above-average shielding
90–150 dB	Excellent shielding

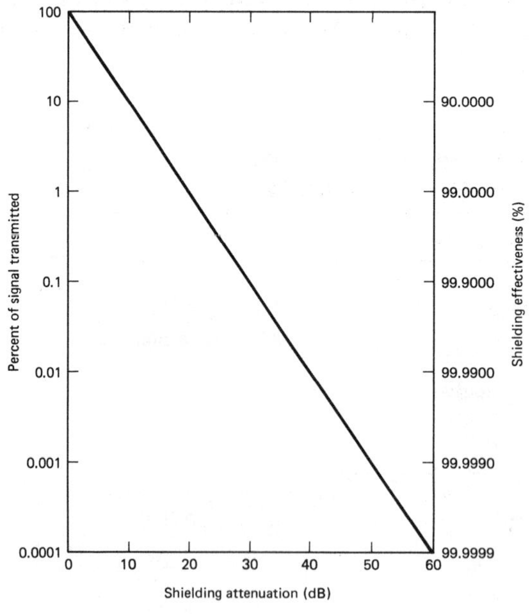

Figure 8.3 Relationship among percentage of signal transmitted through a material, shielding effectiveness of the material, and its shielding attenuation.

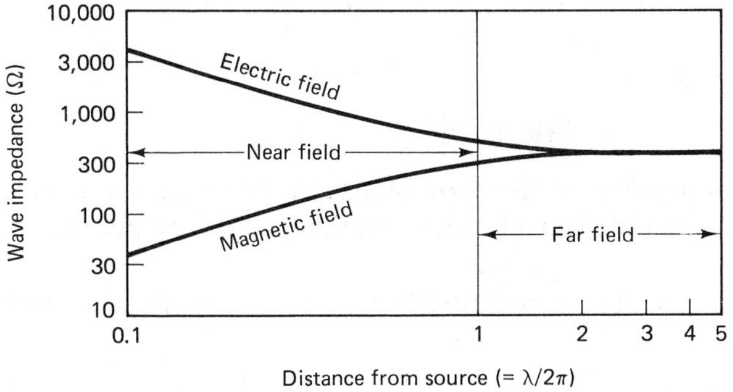

Figure 8.4 Relationship between wave impedance and distance from transmitter relative to signal frequency.

When an electromagnetic wave passes through a material, it loses energy due to its interaction with the electrons and atoms in the material. This loss is given in terms of the initial wave intensity E_i and its intensity E_t at a distance t in the material (i.e., t is the thickness of the shield on the component) as

$$E_t = E_i \exp(-t/\delta) \tag{8.3}$$

where δ is the distance required for the incident wave to be attentuated to about 37% of the initial value and is called the "skin depth" of the material. It is given by relation

$$\delta = \sqrt{\frac{1}{\pi f \mu_\mu \sigma}} \tag{8.4}$$

where f is the frequency of the electromagnetic wave (in hertz), μ_0 is the permeability of free space ($= 1.257 \times 10^{-6}\ \text{Hm}^{-1}$), and σ is the electrical conductivity of the material. It is customary, in EMI shielding theory, to express δ in terms of the relative permeability of the material μ_r (= permeability of material/μ_0) and the relative electrical conductivity of the material σ_r (= electrical conductivity of material/electrical conductivity of copper).

Electromagnetic radiation can be divided into near-field and far-field regions. In the near field, the electromagnetic signal can be predominantly an electric vector or predominantly a magnetic vector, depending on the source antenna. In the far field, plane waves exist in which the electric and magnetic vectors are in phase and orthogonal to each other. Figure 8.4 shows the relationship between wave inpedance and distance from the source (note that λ is the wavelength of the signal).

Plane Wave Radiation

Absorption loss A is given by

$$A = 8.69 \frac{t}{\delta}\ \text{dB} \tag{8.5}$$

Considering that $\mu_\mu = \mu_r \times 1.257 \times 10^{-6}$ and $\sigma = 5.8 \times 10^7\ \sigma_r$, equation (8.4) becomes

$$\delta = \frac{0.066}{\sqrt{f \mu_r \sigma_r}} \tag{8.6}$$

Put this result in equation (8.5) to give

$$A = 131.7t \sqrt{f \mu_r \sigma_r} \tag{8.7}$$

The absorption loss is independent of the type of electromagnetic wave and, for a given coated material, is a function of shield thickness and frequency of the waves.

The reflection loss R is a function of the type of electromagnetic wave. In general,

$$R = 20 \log \left(\frac{|Z_w|}{4|Z_s|} \right)$$

where Z_w is the impedance of the wave (Z_w is high for electric waves, low for magnetic waves, and equal to Z_0, the impedance of free space, for plane waves), and Z_s is the characteristic impedance of the shield. For plane waves, whose impedance falls between that of magnetic and electric waves, $Z_w = 377\ \Omega$, and for the shield, $Z_s = \sqrt{2\pi f \mu / \sigma}$. Thus

$$R = 168 - 20 \log \sqrt{\frac{f \mu_r}{\sigma_r}} \tag{8.8}$$

Thus the reflection loss, for plane waves, is independent of the shield thickness.

Multiple reflections can occur within the shielding film thickness if it is thin, and such losses are given by

$$B = 20 \log [1 - \exp(-2t/\delta)] \tag{8.9}$$

Put equation (8.9) in equation (8.6) to yield

$$B = 20 \log [1 - \exp \{-30.3t \sqrt{f \mu_r \sigma_r}\}] \tag{8.10}$$

This term is always negative; that is, it reduces shielding efficiency, and it decreases with decreasing shield thickness.

Equations (8.7), (8.8), and (8.10) are applicable to plane wave situations. This is a reasonable approximation when the electromagnetic wave source is more than 100 m away and the frequencies are in excess of about 10 MHz. For the shielding of TV and radar devices, a plane wave approximation is acceptable.

Magnetic and Electric Wave Radiation

For the case where the source of electromagnetic radiation is much closer to the component to be shielded, the following relations may be used for the shielding effectiveness

FOR MAGNETIC FIELDS

$$A = 3.338t\sqrt{f \mu_r \sigma_r} \tag{8.11}$$

$$R = 20 \log \left[\frac{0.462}{S} \sqrt{\frac{\mu_r}{\sigma_r f}} + 0.136S\sqrt{\frac{\sigma_r f}{\mu_r}} + 0.354 \right] \tag{8.12}$$

$$B = 10 \log \{1 + (C_3^2 + C_4^2)10^{-A/5} - 2 \times 10^{-A/10}[(C_3 \cos 0.23A) - (C_4 \sin 0.23A)]\} \quad (8.13)$$

FOR ELECTRIC FIELDS

$$A = 3.338t\sqrt{f\mu_r\sigma_r} \quad (8.14)$$

$$R = 353.6 + 10 \log \frac{\sigma_r}{\mu_r S^2 f^3} \quad (8.15)$$

Equation (8.13) gives the multiple reflection correction term B for electric fields as well, provided we note that in that equation,

$$C_3 = \frac{4C_1^4 + C_2^4 - 8C_1^2C_2^2}{(2C_1^2 + C_2^2 + 2C_1C_2)^2}$$

$$C_4 = \frac{8C_1^3C_2 - 4C_1C_2^3}{(2C_1^2 + C_2^2 + 2C_1C_2)^2}$$

$$C_1 = 3.69 \times 10^{-7}\sqrt{\frac{\mu_r f}{2\sigma_r}}$$

$$C_2 = 0.2 \times 10^{-6} fS \quad \text{(for magnetic fields)}$$

$$C_2 = \frac{-0.71 \times 10^{12}}{fS} \quad \text{(for electric fields)}$$

In equations (8.12) and (8.13), S is the distance (in inches) between the EMI/RFI source and the component. It should be noted that in equations (8.11) and (8.14), t is the shield thickness (in inches).

8.5 CANDIDATE MATERIALS

The best shielding is, of course, obtained with metallic materials. But especially in electronic equipment, lighter, less corrodible materials with very poor shielding effectiveness (such as polymers) are most often employed. The usual approach, therefore, is to make use of the latter materials with or without a conductive coating or filler or to use a filamentary composite material (although, in this case, the shielding is rather poor). The electrical conductivities of some candidate materials are given in Table 8.9.

For a specific application, the housing for an avionic module in an aircraft, the relative effectiveness of five candidate materials, for various fields and material thicknesses and radiation frequencies, is presented in the results* given in Tables 8.10 through 8.12 and Figure 8.5. The best method, other than using aluminum, is to cover the selected metallic substrate (whatever that may be) with a thin coat of arc-sprayed zinc paint.

*In all calculations, μ_r is put equal to 1.00.

TABLE 8.9 ELECTRICAL CONDUCTIVITIES OF EMI SHIELDING MATERIALS

Material	Electrical Conductivity, σ ($10^5\ \Omega^{-1}\ m^{-1}$)	Electrical Conductivity Relative to That of Copper,[a] σ_r
Conductive Coats		
Silver epoxy	0.2	3.3×10^{-4}
Electroless nickel	17.9	3.0×10^{-2}
Arc-sprayed zinc	15.0	2.5×10^{-2}
Nickel	0.24	4.0×10^{-4}
Ecco-Coat CC2[b]	2.5	4.2×10^{-3}
"Conductive" filled polymers[c]		
30% PAN carbon fiber in SMA copolymer	0.002	3.3×10^{-6}
40% aluminum flake in SMA copolymer	0.001	1.7×10^{-6}
5% stainless steel fiber in nylon 6,6	0.001	1.7×10^{-6}
15% nickel-coated PAN carbon fiber in nylon 6,6	0.0025	4.2×10^{-6}
EMI-X-DA-40[d]	0.0001	1.7×10^{-7}
Filamentary Composite Material		
Graphite fiber reinforced epoxy resin	0.175	2.9×10^{-4}
Inherently Conducting Polymer		
AsF_5-doped polyacetylene	1.300	2.2×10^{-3}

[a]σ for copper = $598 \times 10^5\ \Omega^{-1}\ m^{-1}$.

[b]Product is a silver and organic resin formulation.

[c]PAN, polyacrylonitrile polymer; SMA, styrene and maleic acid anhydride.

[d]Product is aluminum flake dispersed in a polycarbonate matrix.

TABLE 8.10 SHIELDING EFFECTIVENESS[a] (IN DECIBELS) OF VARIOUS MATERIALS, WITH A THICKNESS OF 25 μm, EXPOSED TO VARIOUS FIELDS

Frequency of Radiation (MHz)	Plane Waves			Magnetic Fields[b]			Electric Fields[b]		
Material	10	100	1000	10	100	1000	10	100	1000
Aluminum	103	112	159	69	96	163	9	26	82
Arc-sprayed zinc paint	74	74	78	41	61	83	—[c]	5	17
AsF_5-doped polyacetylene	52	52	53	20	40	60	—	—	4
Graphite fiber reinforced epoxy resin	35	35	35	4	22	42	—	—	—
Nickel-coated PAN carbon fiber in nylon 6,6	—	—	—	—	—	6	—	—	—

[a]Using equations presented in section 8.4.

[b]Distance of field from component = 3 in.

[c]Negligible quantity.

TABLE 8.11 CALCULATED SHIELDING EFFECTIVENESS[a] (IN DECIBELS) OF VARIOUS MATERIALS, WITH A THICKNESS OF 50 μm, EXPOSED TO VARIOUS FIELDS

Frequency of Radiation (MHz)	Plane Waves			Magnetic Fields[b]			Electric Fields[b]		
Material	10	100	1000	10	100	1000	10	100	1000
Aluminum	112	134	241	77	122	245	17	52	165
Arc-sprayed zinc paint	80	82	95	47	67	99	1	11	33
AsF_5-doped polyacetylene	58	58	60	26	46	66	—[c]	—	10
Graphite fiber reinforced epoxy resin	41	41	41	9	28	48	—	—	1
Nickel-coated PAN carbon fiber in nylon 6,6	4	4	4	—	1	12	—	—	—

[a]Using equations presented in section 8.4.

[b]Distance of field from component = 3 in.

[c]Negligible quantity.

TABLE 8.12 CALCULATED SHIELDING EFFECTIVENESS[a] (IN DECIBELS) OF VARIOUS MATERIALS, WITH A THICKNESS OF 100 μm, EXPOSED TO VARIOUS FIELDS

Frequency of Radiation (MHz)	Plane Waves			Magnetic Fields[b]			Electric Fields[b]		
Material	10	100	1000	10	100	1000	10	100	1000
Aluminum	130	191	407	93	175	410	33	104	330
Arc-sprayed zinc paint	99	108	190	53	77	132	7	21	66
AsF_5-doped polyacetylene	64	65	71	32	52	75	—[c]	6	20
Graphite fiber reinforced epoxy resin	47	47	48	14	34	54	—	—	7
Nickel-coated PAN carbon fiber in nylon 6,6	10	10	10	0	2	17	—	—	—

[a]Using equations presented in section 8.4.

[b]Distance of field from component = 3 in.

[c]Negligible quantity.

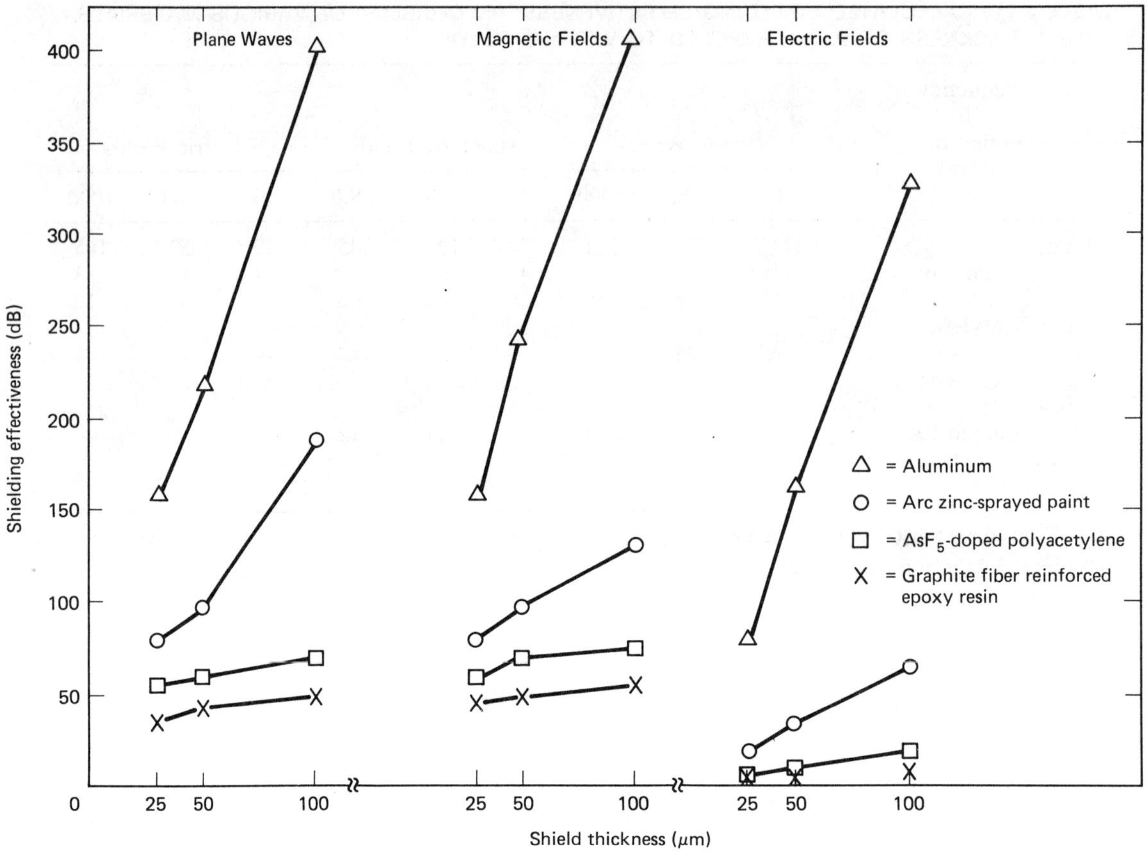

Figure 8.5 Variation in shielding effectiveness with shield thickness for three types of field at a frequency of 1 GHz.

REFERENCES

1. V. Volpe, "Estimation of electrical conductivity and electromagnetic shielding characteristics of graphite/epoxy laminate," *J. Compos. Mater. 14,* pp. 189–197, 1980.
2. K. T. Chung and L. M. Leung, "EMI/RFI shielding effectiveness: Theory and practice," paper presented at the 31st international SAMPE symposium, Los Angeles, April 1986.
3. R. M. Simm, "EMI shielding can be made of conductive plastics," *Ind. R&D 24,* pp. 104–109, 1982.
4. J. T. Hoback, and J. J. Reilly, "Conductive coatings for EMI shielding," *J. Elastomers and Plastics 20* (1), pp. 54–69, 1988.

WINDOW IN A THERMAL IMAGING DEVICE

CHAPTER 9

9.1 INTRODUCTION AND MATERIAL REQUIREMENTS

The function of an infrared transmitting window is to protect thermal imaging equipment, such as forward-looking infrared (FLIR), that operates in the 8 μm to 12 μm band, against aerodynamic loads and against rain and dust exposure. The window must perform this function without seriously degrading the nominal performance of the system.

- The window material should not inject significant distortion or loss of resolution in the optical image.
- It must transmit useful signal radiation without gross attenuation.
- It should exhibit minimal radiance in the wavelength region of system operation, even at temperatures reached in a high-speed flight environment.
- It must have high transmittance at wavelengths ranging from 0.5 μm to 12 μm, which also implies the availability of environmentally-stable broadband antireflection (AR) coatings.
- It must have low emmittance at temperatures up to 200°C.
- It must have excellent resistance to rain erosion in a high-velocity environment and, somewhat less critical, resistance to particulate impact (such as battlefield dust).

Let us examine in some detail the three most important aspects: resistance to rain erosion, drop in transmittance, and degradation of signal-to-noise ratio. The mechanisms that cause rain erosion as a result of multiple raindrop impacts are fairly complex [1] but are believed to involve two essential steps: (a) the development of a network of internal fractures that may extend to significant depths (this is why broadband AR coatings, which are usually less than 5 μm thick, cannot be expected to improve the rain erosion resistance of window materials) and are triggered by the high tensile stresses generated when a raindrop hits the surface, and (b) the removal of a finite amount of material, which may occur in conjunction with the rapid lateral outflow of liquid that escapes from the impact zone. The ability of a monolithic window material to withstand rain erosion thus relates directly to its fracture strength and surface hardness. The threshold velocity, or the velocity at which a complete ring-fracture pattern emerges, should be a function of intrinsic fracture toughness and the ability of the crystal lattice to resist microcrack propagation.

The decrease in transmittance of two candidate materials as a function of exposure time in a small-drop rainfield is shown in Table 9.1. The loss of transmittance in CVD ZnSe may be attributed to reflections at internal fracture surfaces. With CVD ZnS, there is an "incubation period" during which transmittance does not seem to be affected; subsequent to this, substantial degradation is associated with the nucleation and growth of erosion pits on the surface.

TABLE 9.1 COMPARISON OF THE RAIN EROSION RESISTANCE OF CVD[a] ZnS AND CVD ZnSe (SMALL-DROP RAINFIELD IS 0.75 mm DIAMETER WATER DROPS FALLING AT A RATE OF 1 cm h^{-1})

Exposure time (min)	Relative transmittance (%)	
	CVD ZnSe	CVD ZnS
0.0	1.00	1.00
0.7	0.97	0.97
1.3	0.90	—
2.0	0.70	—
2.8	0.62	—
3.2	0.55	—
6.0	—	0.97
12.0	—	0.73
15.0	—	0.65
18.0	—	0.55

[a]CVD, chemical vapor deposited

Source: V. Hackworth, and L. Kochner, "Behavior of infrared window materials exposed to raindrop environments at velocities of 2000 fps (610 m/s)," Report No. AFML-TR-78-184, Air Force Materials Laboratory, Wright-Patterson AFB, Ohio, 1978.

TABLE 9.2 RELATIVE FLIR—SYSTEM PERFORMANCE, AS A FUNCTION OF THE WINDOW TEMPERATURE, FOR THREE CVD MATERIALS

Window temperature (°C)	System Performance (%)		
	ZnSe window (thickness = 15 mm)	ZnS/Zn Se window (thickness = 1–12 mm)	ZnS window (thickness = 11 mm)
20	100	98	80
50	100	97	75
75	100	96	72
100	100	95	65
125	100	94	60
150	100	93	55
175	100	90	50
200	98	88	45

Source: C. A. Klein, B. DiBenedetto, and J. Pappis, "ZnS, ZnSe, and ZnS/ZnSe windows: Their impact on FLIR systems performance," *Opt. Eng. 25* (4), p. 528, 1980.

A further requirement is that the window material must not undergo substantial degradation in system performance, defined as the ratio of signal to noise with the window to that without the window, as a function of increase in window temperature. Table 9.2 shows that ZnS performs very poorly, with degradation increasing almost linearly with window temperature from about 20% at 20°C to about 50% in supersonic flight. This degradation reflects (a) a phonon-dominated absorption of target radiation and (b) the emission of noise radiation (both occurring at the long-wavelength end of the band pass). By contrast, ZnSe performs admirably, losing only 2% performance rating even at 200°C.

9.2 CANDIDATE MATERIALS

There are five types of candidate materials.

1. *Germanium* exhibits superior mechanical properties but is not suitable for many applications involving aerodynamic heating because of excess free-carrier absorption above room temperature. For example, sustained flight at altitudes less than about 1500 m and speeds of no more than Mach 0.8 reduces the transmittance of an AR-coated germanium window from 90% to 60% [2], leading to a substantial degradation in the signal-to-noise ratio.

2. The main disadvantage of *chemical vapor-deposited (CVD) zinc sulfide* (ZnS) is that even in a benign environment, structurally sound ZnS windows suffer a loss in thermal stability of about 30%, caused by intrinsic multiphonon absorption processes at wavelengths beyond 10.5 μm. ZnS has the advantages of being relatively inexpensive and behaving reasonably well in terms of rain erosion resistance. Recently, an improved form of ZnS, CLEARTRAN, was introduced on the market. This product has wider transmission limits but lower flexural strength and hardness compared to standard CVD ZnS (see Table 9.3). It is made by subjecting standard CVD ZnS to a postdeposition treatment at high temperature and pressure. The mechanism of formation is not fully understood at the moment,

TABLE 9.3 COMPARISON OF OPTICAL, MECHANICAL, AND THERMAL PROPERTIES OF STANDARD CVD ZnS, STANDARD CVD ZnSe, AND CLEARTRAN ZnS

	Standard CVD ZnS	Standard CVD ZnSe	CLEARTRAN ZnS
Transmission limits (μm)	1–14	0.5–22	0.3–14
Refractive index (at 10.6 μm)	2.19	2.40	2.19
Index of refraction inhomogeneity (ppm)	<100	<3	<20
Flexural strength[a] (MPa)	103	55	60
Modulus of elasticity (GPa)	74	67	74
Knoop hardness (50 g load)	250	130	160
Grain size (μm)	2–8	50–70	20–35
Density (kg m^{-3})	4080	5270	4090
Specific heat (kJ kg^{-1} K^{-1})	468	339	514
Coefficient of linear expansion (10^{-6} K^{-1})	7.85	7.57	6.50
Thermo-optic coefficient (10^5 K^{-1})	4.33[b]	6.10[c]	3.87[c]

[a] In a four-point loading test.

[b] At 3.39 μm.

[c] At 10.60 μm.

although it is possible that visible transmission is improved by sweeping the smaller grains with their accompanying microporosity to grain boundaries and allowing the pores to vent along these grain boundaries.

3. *Chemical vapor-deposited* (*CVD*) *zinc selenide* (ZnSe) has the advantages of excellent low absorption in the 0.5 to 12 μm range and no degradation anticipated, but it is too soft and too sensitive to raindrop or solid-particle impingement for it to be acceptable under adverse weather conditions.

Typical transmittance of CVD ZnS, CLEARTRAN ZnS, and CVD ZnSe are presented in Fig. 9.1.

4. The current practice is to use a *composite material* made up of a ZnSe substrate with a thin (about 1 mm) layer of ZnS on it; such a material combines the best of the two constituents: the good system performance of ZnSe and the good rain erosion resistance of ZnS. For the system performance of this composite material, see Table 9.2. A number of issues arise in the use of such a composite material, though. The first involves the preparation of the substrate so that no large scattering sites are produced. The second involves the stress buildup, upon cooling, as a result of the difference in the coefficient of thermal expansion of ZnS and ZnSe.

5. Some new materials have recently been introduced for use in the present application. These are *alkaline earth–rare earth sulfides* of composition AB_2S_4, where A is a divalent metal and B is a trivalent rare earth metal. Among these materials, particular attention is now focused on compounds with the cubic thorium phosphide structure, such as $SrLa_2S_4$, $MgSc_2S_4$, and $CdEr_2S_4$. The best of the genre is calcium lanthanum sulfide, $CaLa_2S_4$.

The relevant material properties of $CaLa_2S_4$ and a proprietary CVD ZnS material are presented in Table 9.4.

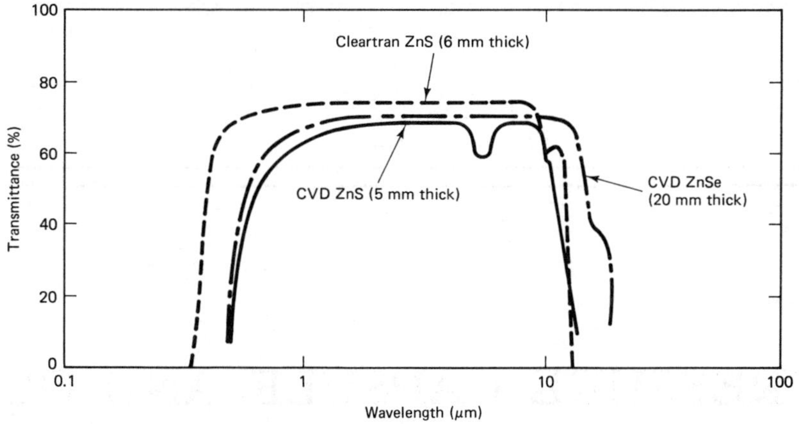

Figure 9.1 Typical transmittance data of CLEARTRAN ZnS (6mm thick), CVD ZnSe (20 mm thick), and CVD ZnS (5mm thick) (courtesy of CVD Incorporated, Woburn, Mass.).

TABLE 9.4 COMPARISON OF OPTICAL, MECHANICAL, AND THERMAL PROPERTIES OF $CaLa_2S_4$ AND ZnS

Property	$CaLa_2S_4$	RAYTRAN CVD ZnS
Optical		
Peak IR transmittance (%)	74	72
Absorption coefficient (cm^{-1})		
8.0 μm (apparent)	0.20	0.10
10.6 μm (calorimetric)	0.17	0.30
12.0 μm (apparent)	0.50	0.40
Mechanical		
Knoop hardness (kg mm^{-2})	570	500
Grain size (μm)	10–50	2–6
Flexural strength (MPa)	106	103.4
Modulus of elasticity (GPa)	96	75
Poisson's ratio	0.26	0.27
Fracture toughness (MPa$\sqrt{m}$)	0.53	0.63
Rain erosion resistance	Excellent	Acceptable
Thermal		
Melting point (°C)	1810	1830
Coefficient of linear expansion (10^{-6} K^{-1})	14.7	7.4
Thermal conductivity (W m^{-1} °C^{-1})	1.68	17.22
Thermal shock resistance (W m^{-1})	84.0	235.20

Source: W. F. Adler, "Evaluation of waterdrop impact damage in infrared transmitting materials," final report to Naval Weapons Center, USA Contract No. N60530-82-M-G164, October 1982.

REFERENCES

1. W. F. Adler, "Degradation of optical materials exposed to erosive environments" in S. Musikant (ed.), *Scattering in Optical Materials, Proc. SPIE 362,* p. 116, 1985.
2. H. Hafner and G. Cronin, "Development of GaAs for infrared windows," Report No. AFML-TR-75-49, Air Force Materials Laboratory, Wright-Patterson AFB, Ohio, 1975.

PRESSURE CAPSULE AND OUTER HULL OF A DEEP SEA–OPERATING VEHICLE

CHAPTER 10

10.1 INTRODUCTION

Humans have an almost mystical fascination with the deep sea environment. Such interest runs the gamut from the commerical (as illustrated by the recovery of sunken treasures and artifacts) and the research (as evidenced by the location of the sunken *Titanic* in the mid-Atlantic in 1987) to the military and defensive. All such activities require the use of vessels that can operate safely at substantial water depths. Over the years, both the operating depth and the weight of these vessels have been increasing (Fig. 10.1). In the United States, recent efforts to expand the military, defense, strategic, economic, commercial, and other uses of the deep sea environment have focused on improvements in the design and performance of mobile deep sea craft, the so-called deep sea–operating vehicles (DSOVs).

One type of DSOV is the deep submersible, of which the *Deep Quest* is an example (Fig. 10.2). Typically, the outer hull has an elongated lenticular shape, which is necessary for hydrodynamic streamlining. The pressure capsule is in the form of a communicating bisphere, a shape that combines the structural efficiency of the sphere and the habitability of a cylinder (Fig. 10.3). The payload is contained

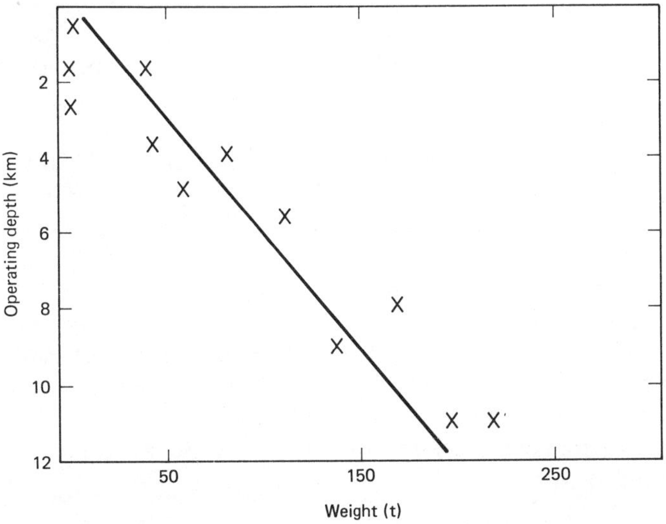

Figure 10.1 Weight versus operating depth of submarines in the United States.

Figure 10.2 The deep submergence operating vehicle *Deep Quest*. (Courtesy of Lockheed Missiles & Space Co.)

Figure 10.3 Bispherical arrangement of the pressure vessel of the *Deep Quest*.

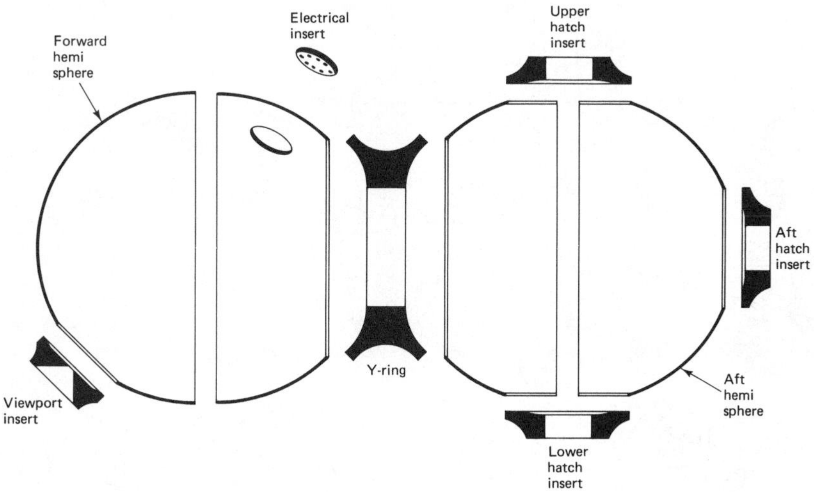

in a module that is readily removable and fitted for various types of missions, such as search and rescue, oceanographic surveys, recovery of objects, and coring of the seabed.

Another type of vessel is the deep submergence rescue vehicle (DSRV), made up of a pressure capsule within an outer hull. The pressure capsule has a trispherical shape [1] (Fig. 10.4). This consists of three equal-sized spheres axially aligned and joined to approximate a cylinder. This combines the structural advantages of a sphere and the habitability of a cylinder. Typically, the outer hull and cylindrical midbody have diameters of 15 m and 2.5 m, respectively. The vessel is designed to be a ''flying submarine,'' capable of a rescue mission anywhere in the world within two days of a call being received in the United States. An example of this type of vehicle is the U.S. Navy's DSRV-1 (see Fig. 10.5), devised as part of the Navy's submarine rescue capability in response to the loss of the submarine USS *Thresher* in 1963.

Figure 10.4 Principal features of a deep submergence rescue vehicle (DSRV), showing the trispherical arrangement of the pressure capsule [1]. (reprinted with permission from John Wiley, Inc.).

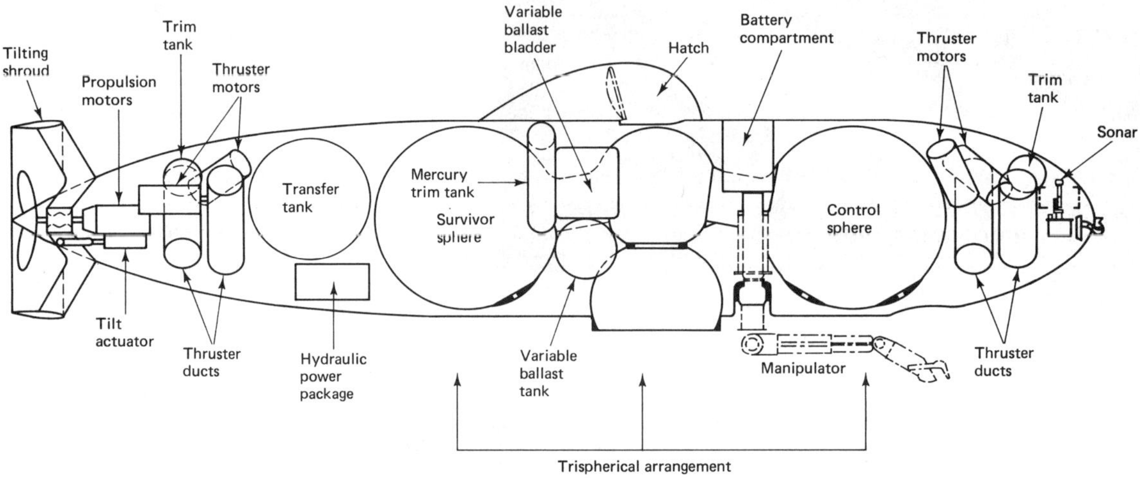

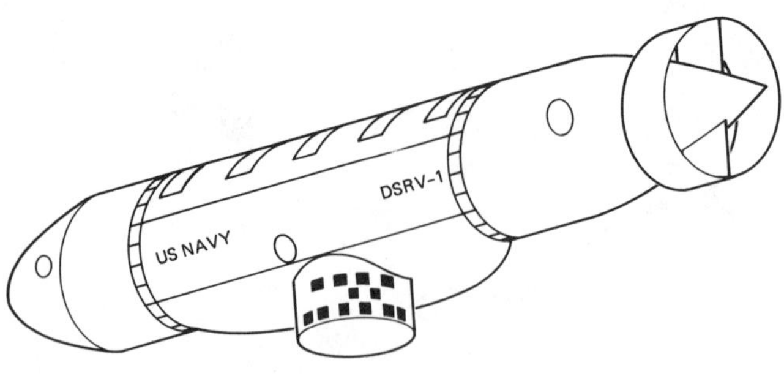

Figure 10.5 The DSRV-1. (Diagram Courtesy of the U.S. Navy)

10.2 CHOICE OF MATERIALS

Both the deep submersible and the deep submergence vehicles require critical designs to obtain the optimum mix of operating depth and payload. One way of achieving such a mix is through proper selection of materials for the key structural components, the pressure capsule and the outer hull.

The main requirements in a candidate material for the pressure capsule (other than the viewports) are as follows:

- High specific yield strength under both dry and long-term water exposure conditions
- High specific compressive strength under both dry and long-term water exposure conditions (at about 6 km, a DSOV is subjected to a compressive pressure of about 60 MPa)

In Table 10.1, five candidate materials are compared on the basis of these criteria.

The main functions of the outer hull are as follows:

- To provide the streamlined hydrodynamic shape necessary to minimize power consumption when at sea or in port
- To house and serve as a foundation for the many tanks, components, plumbing, electrical conductors, and buoyancy items necessary to fulfill the required functional role of a mobile hydrocraft

Four relations may be used to evaluate the performance of the outer hull under external pressure.

TABLE 10.1 PROPERTIES AND PERFORMANCE INDICES OF SOME CANDIDATE MATERIALS FOR THE PRESSURE CAPSULE OF A DEEP SEA–OPERATING VEHICLE

Material	Yield Strength, σ_y (MPa)	ρ (kg m^{-3})	Compressive Strength, σ_c (MPa)	E (GPa)	σ_y/ρ (10^3 m)	σ_c/ρ (10^3 m)
Pressure vessel steel[a]	965	7800	1034	215	12.61	13.51
Fiberglass[b]	200	1993	780	41	10.22	39.90
Glass-ceramic[c]	69	2500	345	86	2.81	14.07
Al alloy[d]	400	2796	280	70	14.58	10.21
Ti alloy[e]	830	4540	827	116	18.64	19.16

[a]HY140, a quenched and tempered Ni-Cr-Mo-V steel.

[b]Unidirectional E glass fiber reinforced epoxy resin (fiber volume fraction = 0.60); all property values are in the fiber direction.

[c]Pyroceram 9608.

[d]1100-0

[e]Ti-6Al-4V

- $$P_c = \frac{CE}{(1 - \nu^2)^{0.5}} \left(\frac{t}{R_0}\right)^2 \tag{10.1}$$

where P_c is the critical pressure for buckling or elastic unstability, t is the hull wall thickness, R_0 is the external diameter of the hull, and C is a constant that depends on the shape of the hull (for example, $C = 1.15$ when the hull is a spherical shell). Thus for a given hull, t and R_0 are fixed, P_c is directly proportional to the material parameter $E/(1 - \nu^2)^{0.5}$.

- $$P_s = C' \left(\frac{t}{R_0}\right) \sigma_y$$

where P_s is the limiting pressure for compressive failure, σ_y is the yield strength of the hull material, and C' is a constant (for example, $C' = -2$ for a spherical shell). Thus for a given hull, P_s is directly proportional to σ_y.

- $$\frac{W}{S} = C'' \left(\frac{\rho}{\rho_{sw}}\right) \left(\frac{t}{R_0}\right)$$

where W/S is the weight-to-displacement ratio of the hull, ρ is the density of the hull material, ρ_{sw} is the density of the seawater, and C'' is a constant (for example, $C'' \approx 3$ for a spherical shell). Thus for a given hull, W/S is directly proportional to ρ.

- The compressibility (the change in volume per unit change in pressure) K of the pressure hull is given by

$$K = C'''\sigma_y \frac{1 - \nu}{EP_y}$$

where P_y is the pressure at which the stress in the shell of the hull equals the yield strength of the hull material—in fact, P_y is approximately equal to the collapse depth; and C''' is a constant (for a thin spherical shell, $C''' = 3$). Thus for a given hull and collapse death, K is dependent on the material properties parameter $\sigma_y(1 - \nu)/E$.

Table 10.2 compares four candidate hull materials on the basis of these criteria. In addition, a suitable outer hull material should also have these qualities:

- High specific strength
- High specific modulus
- High retention of strength and modulus with exposure to seawater
- Excellent resistance to seawater corrosion
- Ease of maintainability
- Low raw material and fabrication costs
- Excellent resistance to fatigue loading imposed by wave-slapping
- Good resistance to attack or fouling by biological organisms
- Good resistance to ''service abuse,'' such as scuffing at dockside or against a surface support ship

TABLE 10.2 PROPERTIES AND PERFORMANCE INDICES OF SOME CANDIDATE MATERIALS FOR THE OUTER HULL OF A DEEP SEA–OPERATING VEHICLE

Property	Unidirectional E Glass Fiber Reinforced Epoxy Resin[a]	Unidirectional Boron Fiber Reinforced Epoxy Resin[a]	Unidirectional Kevlar Fiber Reinforced Epoxy Resin[a]	Aluminum Alloy 1100-0
Tensile strength[b] (MPa)	1103	1300	1670	380
Compressive strength[b] (MPa)	780	2500	276	280
Shear strength (MPa)	70	69	69	160
Yield strength, σ_y (MPa)	200	1100	1240	400
Endurance strength at 10^6 cycles (MPa)	260	1070	1000	275
Tensile modulus,[b] E (GPa)	41	207	86	70
Compressive modulus[b] (GPa)	35	207	86	70
Shear modulus (GPa)	2	7	2	28
Density, ρ (kg m^{-3})	1993	1990	1366	2796
Poisson's ratio, ν	0.25	0.30	0.30	0.35
$E/(1 - \nu^2)^{0.5}$ (GPa)	42	217	90	75
$10^3\sigma_y(1 - \nu)/E$	3.66	3.72	10.09	3.71

[a]Fiber volume fraction = 0.60.

[b]In the fiber direction, except for the Al alloy, which has isotropic properties.

TABLE 10.3 SPECIFIC STRENGTH AND MODULUS OF SOME CANDIDATE MATERIALS FOR THE OUTER HULL OF A DEEP SEA–OPERATING VEHICLE

Property	Unidirectional E Glass Fiber Reinforced Epoxy Resin[a]	Unidirectional Boron Fiber Reinforced Epoxy Resin[a]	Unidirectional Kevlar Fiber Reinforced Epoxy Resin[a]	Aluminum Alloy 1100-0
Specific tensile strength (10^3 m)	56.42	66.59	124.62	13.85
Specific compressive strength (10^3 m)	39.90	128.06	20.60	10.21
Specific shear strength (10^3 m)	3.58	3.53	5.15	5.83
Specific endurance strength (10^3 m)	13.30	54.81	74.62	10.03
Specific tensile modulus[b] (10^6 m)	2.10	10.60	6.42	2.55
Specific compressive modulus[b] (10^6 m)	1.79	10.60	6.42	2.55
Specific shear modulus (10^6 m)	0.10	0.36	0.15	1.02

[a] Fiber volume fraction = 0.60.

[b] In the fiber direction, except for the Al alloy, which has isotropic properties.

TABLE 10.4 EFFECT OF ENVIRONMENTAL CONDITIONS ON SOME MECHANICAL PROPERTIES OF COMPOSITE MATERIALS

Condition	Interlaminar Shear Strength of Kevlar 49/Epoxy Resin (MPa)	Interlaminar Shear Strength of S glass/Epoxy Resin (MPa)	Fracture Toughness of $\pm 10°/90°$ Graphite/Kevlar 49 Epoxy Resin Laminate (MPa $\sqrt{m}$)	Fracture toughness of $\pm 10°/0°/\pm 45°/90°$ Graphite Epoxy Resin Laminate (MPa $\sqrt{m}$)
As received	40.6	79.2	—	—
In salt water for $3\frac{1}{2}$ years	40.6	75.8	—	—
As received (93°C)	—	—	7.49	10.61
Three load cycles to 75% ultimate, 48 h in salt water	—	—	8.56	9.74

Sources: "Characteristics and uses of Kevlar 49 aramid high modulus organic fiber," Bulletin K-5, Wilmington, Del.: E.I. du Pont de Nemours, 1981; H. Bau and S. W. Beckwith, "Fracture mechanics evaluation of filament wound case materials subjected to operational environments," *J. Spacecraft and Rockets 21* (3), pp. 281–286, 1984.

The values of the specific strength and specific stiffness for four candidate materials are given in Table 10.3. Note that the material property data in Tables 10.2 and 10.3 are for dry conditions. It is not expected, however, that these properties would be severely impaired by prolonged exposure to seawater. This is a speculative or projected conclusion based on published data on the effect of environmental conditions (seawater exposure is the relevant condition for the present application) on various mechanical properties of composite materials (Table 10.4).

The results in Tables 10.2 and 10.3 indicate that the best materials are the boron and Kevlar fiber reinforced epoxy resins.

REFERENCE

1. J. F. Brahtz (ed.), *Ocean Engineering: Goals, Environment, Technology,* New York: Wiley, 1968, p. 389.

SHAPE MEMORY EFFECT APPLICATIONS

CHAPTER 11

11.1 INTRODUCTION

The shape memory effect (SME) is the ability of certain alloys to ''remember'' the dimensions that a basic shape (wire, rod, sheet, table) possessed at a previous point in its thermomechanical history and to work to recover those dimensions upon heating above and/or cooling below a characteristic temperature, termed the transformation temperature or temperature range T_K.

The SME was first observed in a gold-cadmium alloy in 1932 [1], but it was not until 1951 that the phenomenon was characterized as being the result of a martensitic transformation in that alloy system. Since 1951, the effect has been observed in many alloy systems, such as Cu-Zn, Cu-Zn-Al, Cu-Zn-Ga, Cu-Zn-Sn, Cu-Zn-Si, and Fe-Pt. Their generally low ductility (making them very prone to fracture upon shape recovery) kept them from becoming commercially important until the discovery of the SME in the 50 wt% nickel, 50 wt% titanium alloy (NITINOL)* by Wiliam Buehler at the Naval Ordnance Laboratory in 1965 [2].

*The nickel-titanium alloy system (the acronym stands for Nickel-Titanium Naval Ordnance Laboratory).

TABLE 11.1 TYPICAL PROPERTIES OF TWO NICKEL-TITANIUM (NITINOL) SHAPE MEMORY ALLOYS

Material Property	NITINOL for Bimetal Substitution	NITINOL for Connection Applications
Tensile strength at elongation of 0.2% (MPa)	60	240
Fracture tension (MPa)	840	580
Fracture expansion (%)	60	20
Endurance strength at 10 cycles (MPa)	—	350
Electrical conductivity (Ω^{-1} m^{-1})	2×10^6	2×10^6
Specific heat (MJ m^{-3} K^{-1})	3	3
Latent heat (MJ m^{-3})	100–150	100–150
Unit cost,[a] 1986 (\$ kg^{-1})	440	400

[a]Varies with the size, shape, and form of the alloy.

Source: K. N. Melton and O. Mercier, "Alloys with two-way shape memory effect," *Mech. Eng. 102* (3), pp. 42–43, 1980.

This alloy has many attractive mechanical, electrical, and thermal properties (Table 11.1) as well as excellent corrosion resistance and room-temperature damping characteristics.

In addition to the NITINOL alloys and their proprietary offshoots, such as TINEL (see Tables 11.2), other SME alloys of engineering interest are the general-performance Ni-Ti, Cu-Zn-Al and the Cu-Al-Ni alloys (see Table 11.3).

The magnitude of T_K is a critical function of alloy composition, so chemistry control is one of the major factors in the fabrication of these alloys. Although all of the alloys can easily be hot-worked (by extrusion, rolling, or forging), they can

TABLE 11.2 MATERIAL PROPERTIES OF TINEL SHAPE MEMORY ALLOY

Transformation		**Electrical**	
Transition temperature[a] (°C)	−200 to 100	Resistivity at 300 K ($\mu\Omega$ cm)	
Latent heat of transformation		Austenite	100
(martensite to austenite) (J g^{-1})	167.2	Martensite	80
Percentage shape memory (%)	8.0	Relative magnetic permeability	<1.002
Physical		Magnetic susceptibility	3×10^6
Melting point (°C)	1300	**Mechanical**	
Density (kg m^{-3})	6500	Modulus of elasticity at 300 K (GPa)	96
Thermal conductivity (W m^{-1} °C^{-1})		Poisson's ratio	0.33
At 25°C	18.00	Ultimate tensile strength (MPa)	862
At −196°C	8.60	Yield strength (MPa)	414
Coefficient of thermal expansion (10^{-6} °C^{-1})		Elongation at failure (%)	15 to 20
Austenite	11.0		
Martensite	6.6		

[a]Depends on composition.

Source: R. W. Benson, R. F. Flot, and C. L. Sandberg, "The use of shape memory effect alloys as an engineering material," paper presented at the 15th national SAMPE technical conference, Cincinnati, October 1983.

TABLE 11.3 PROPERTIES OF THREE SHAPE MEMORY ALLOYS IN THE HEAT-TREATED CONDITION

Property	General-performance Ni-Ti	Cu-Zn-Al	Cu-Al-Ni
	Alloy		
Melting point (°C)	1240–1310	950–3020	100–1050
Thermal dilation coefficient (10^{-6} °C^{-1})			
Austentitic form	10		
Martensitic form	6.6	16–18	16–18
Specific heat (J kg °C^{-1})	470–620	390	400–480
Transformation heat (kJ kg^{-1})	3.2	7–9	7–9
Modulus of elasticity (GPa)	98	70–100	80–100
Yield strength (MPa)			
Austentitic form	200–800		
Martensitic form	150–300	150–300	150–300
Tensile strength, martensitic form (MPa)	800–1100	700–800	1000–1200
Fracture strain, martensitic form (%)	40–50	10–15	8–10
Fatigue limit (MPa)	350	270	350
Grain size (μm)	50–100	50–100	25–60
Transformation temperatures (°C)	−100 to +120	−200 to +120	−200 to +170
Maximum one-way shape memory (% strain)	8	5	6

be cold-worked to only a limited extent since they tend to have very high values of the strain-hardening exponent (reductions between anneals of 20% to 30% are not unusual). Joining of SME alloys normally involves mechanical means since welding or brazing requires temperatures that destroy the shape memory effect.

11.2 METALLURGICAL ASPECTS

Basically, SME involves the appearance and disappearance of martensite in an alloy when heated and/or cooled through T_K. In SME, martensite takes up two possible structures: an ordered body-centered cubic phase or a complex crystal structure that can be twinned and faulted. In all SME alloys, at or around T_K, the ordered phase transforms to self-accommodating martensite plates characterized by six plate groups, each group consisting of four variants, such that the average shape deformation in a particular plate group is zero.

The general scheme of the SME transformation process is shown in Fig. 11.1.

The SME may manifest itself as a one-way or a two-way form. In one-way shape memory, following deformation of the alloy part, spontaneous recovery of the original shape occurs when heated above T_K. This process can be repeated, but the deformation must also be repeated each time; see Fig. 11.2(a). In two-way shape memory, the part spontaneously changes from one shape to another upon heating or cooling through T_K; see Fig. 11.2(b). To develop this behavior,

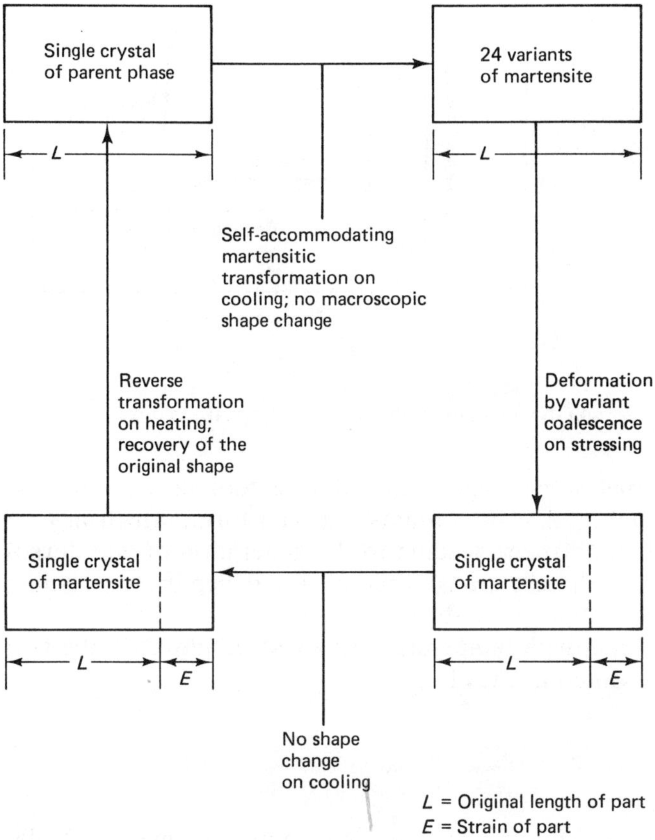

Figure 11.1 Schematic representation of the macromechanisms involved in the shape memory effect.

Figure 11.2 Shape memory alloy effect: (a) one-way; (b) two-way.

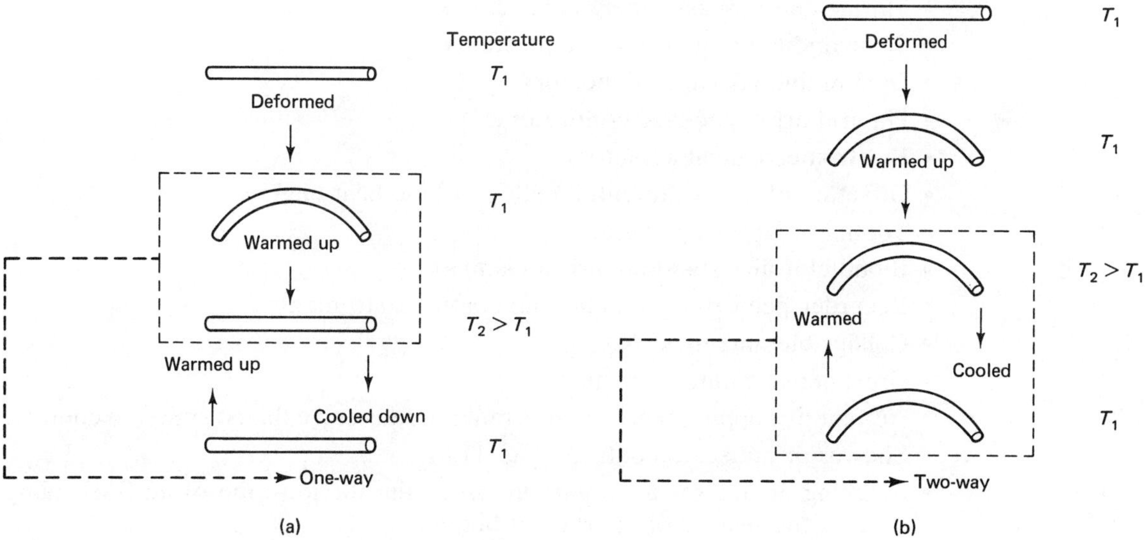

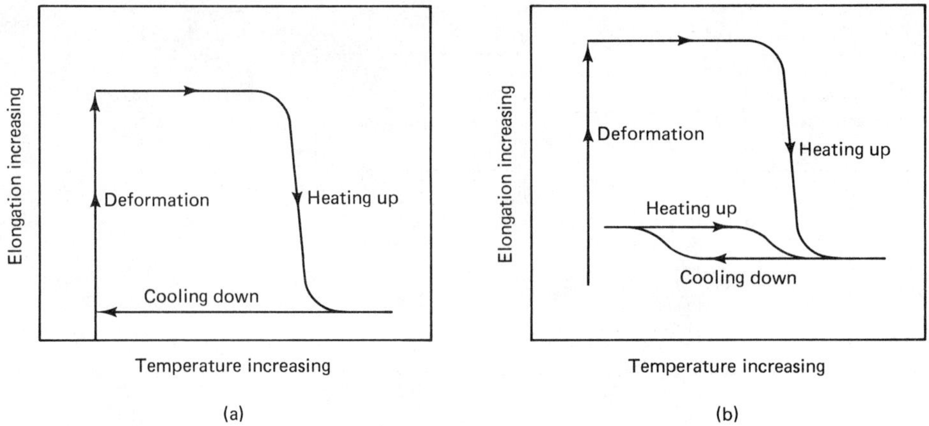

Figure 11.3 Strain as a function of temperature: (a) one-way effect; (b) two-way effect.

the part must have an external stress applied to it during cooling to form martensite, thus limiting the number of plate variants that can form, which in turn reduces the self-accommodation aspect, leading to the generation of an internal stress. This stress initiates the phase transformation upon heating or cooling through T_K giving rise to the spontaneous shape change.

The variation of elongation with temperature of an SME alloy, for the two types of SME effects, is presented in Fig. 11.3.

11.3 APPLICATIONS

A wide variety of applications using the SME have recently become available, and more are being developed. Here is a list of some of them:

- Aircraft hydraulic coupling systems
- Marine and process piping connections
- Semiconductor gas tubing connections
- Dual in-line package connectors
- Pin grid array package connectors
- Electromechanical actuators
- Offset crank and differential pulley turbine heat engines
- Tooth-straightening braces
- Blood clot filters and aneurism clamps
- Recorder pen drives in industrial control instruments
- Collapsible antennas
- Greenhouse window openers
- Automotive applications, such as radiator fans and exhaust emission controls
- Shower or antiscald controls (Fig. 11.4)
- Latching device for a satellite in which the torsion tube of an SME alloy triggers the release of instrument booms

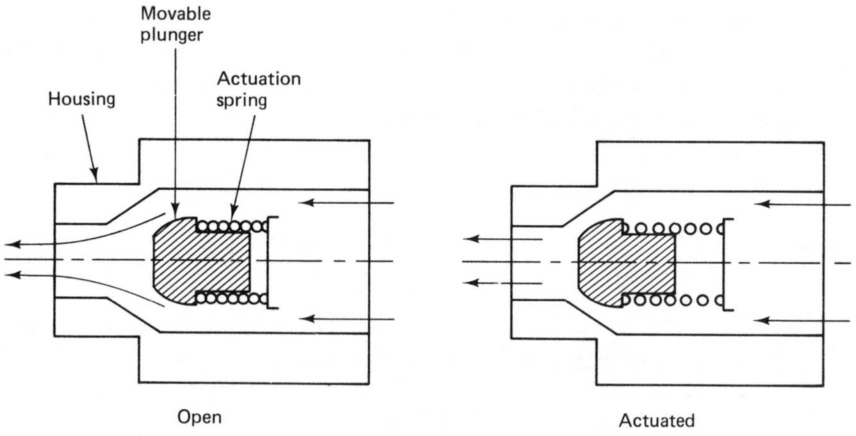

Figure 11.4 An antiscald control valve made of a shape memory alloy.

SME alloys are particularly useful as thermostatic devices, such as fire safety units. Such a unit takes the form of a perimeter signal consisting of a prestretched wire in the martensitic condition mounted in such a way that it forms a continuous loop around the area to be protected. A fire at any point—say, near a wall—will cause a local shape recovery that will in turn produce tension in the wire and cause the activation of a switch to sound an alarm or to set off a sprinkler system.

SME alloys and applications became commercially viable in 1986 when Japan's Wacoal Corporation began marketing its memory wire bras in the USA. Shortly thereafter, Beta Phase Inc., of Menlo Park, Calif., a producer of the SME alloy Memorite, and Universal Optical Inc., of East Providence, R.I., unveiled wire eyeglass sidepieces that returned to their original form when dipped in water at about 55°C. Innovation Technology, Inc., of Beltsville, Md., used NITINOL to convert heat into motion in an engine that runs on low-cost heat sources, such as the sun or industrial waste fluids. Heavily involved in SME alloy applications is Raychem Corp., of Menlo Park, Calif., which produces the Vease linear actuator, the Cryocon semipermanent mate-break connecting device, the CryoTact zero insertion force socket, the Cryofit pipe and tube joining system, and Tinel-Lock heat-recoverable rings for cable shielding termination. One of Raychem's products is a prototype device that uses the SME alloy to lift the flexure arm in a hard disc drive. The device promises improved reliability by reducing the probability of a head "crash" and subsequent loss of data, as well as eliminating modes that produce debris that interfere with the head function. Memory Metals, Inc., of Stamford, Conn., has a range of products that use the Memrytec SME alloy, including electrical and optical connectors, heating and ventilating air-conditioning controls, oil-drilling devices, and automotive components, such as a heat-controlled fuel-air meter. Hitachi of Japan has developed a three-fingered robot "hand," with each joint activated by a bundle of 12 NITINOL alloy wires, each only 200 μm in diameter. The heating and cooling are effected by passing electric currents along these wires.

In early 1988, Toki Corporation of Japan announced a new type of Ni-Ti SME alloy with a uniformly oriented crystalline structure. Its shape transformation temperature is about 100°C, but its shape memory characteristics are marked up to about 300°C. Two potential applications of this alloy (named Bio Metal Fiber) are (a) a pneumatic control needle valve that can reciprocate a miniature-sized piston more than 40,000 times on the power of a small battery and (b) a safety fuse breaker.

REFERENCES

1. L. C. Chang and T. A. Read, *Trans. AIME 47,* p. 191, 1951.
2. W. J. Buehler and R. C. Wiley, "Nickel-based alloys," U.S. Patent No. 3,174,851, March 23, 1965.

OPTICAL COMPUTER COMPONENTS

CHAPTER 12

12.1 INTRODUCTION

Optical computing may be defined as any process that manipulates signal-carrying light. Thus defined, its applications cover a wide continuum: image processing using spatial filtering of coherent light, signal processing using acoustoptic devices, pattern recognition using matched filters, and optical computational devices that perform matrix-vector multiplication, Fourier transformations, and polynomial evaluations. We shall concentrate here on the use of photons in computational devices, the so-called optical computers.

Electrons and photons can carry information from one place to another. In this sense, they are similar. There are, however, key differences between them. Let us outline some of these.

1. Because electrons can interact easily with one another (through their electric charges and magnetic fields), undesired cross talk and electromagnetic interference occur in electron-based devices. Photons, by contrast, have no charge and hence do not interact easily. Beams of light can pass through each other with little or no effect, and light-carrying fibers can be next to each other and suffer no cross talk or electromagnetic interference.

2. The speed at which an electronic signal can be transmitted from one component to the next, in an integrated circuit, is limited by the product of the resistance and the capacitance of the wire. This limit is practically unaffected by the wire diameter. Further, the bandwidth of the electrical signal is limited by inductance at higher frequencies. The bandwidth of light traveling through a fiber, however, is not limited by resistance, capacitance, or inductance.

3. Photons need not be confined to a waveguide—the optical equivalent of a wire—but can travel freely through space.

4. The potential for exploiting the inherent parallelism of light to obtain connectivity and wider bandwidth is enormous. The setting up of decision planes can act on thousands of pixels of information in parallel instead of individual bits of data in fast sequence, as in the case of most electronic computers.*

12.2 RESEARCH AND DEVELOPMENT WORK

In the early 1960s, workers at AT&T Bell Laboratories in the United States invented the room-temperature laser diode, a technology that has since spawned fiber-optic communications. Since the mid-1970s, multiple quantum-well materials have been developed with the aid of increasingly sophisticated crystal growth techniques, such as molecular beam epitaxy and metal organic chemical vapor deposition, making it possible to have materials with much stronger nonlinear properties than bulk materials or such properties being obtained at room rather than at cryogenic temperatures. Such technological advances, together with a growing perception that standard digital electronic computers may be fast reaching their limits in many application areas, have provided the impetus for the burgeoning theoretical interest and research and development efforts in optical computers.

Note that the goals of such interest and efforts among the various groups working in this field are by no means identical. Three categories of goals can be discerned.

The objective of groups with "modest" goals is (a) to make electronic computers work faster, (b) to devise a special-purpose machine that can better handle classes of specialist problems that are not easily amenable to solutions with the aid of electronic devices, or (c) to build a machine that is more resistant to damage by electromagnetic interference or nuclear electromagnetic pulses. These goals may be achieved through the use of "hybrid" systems: an electronic computer with optical interconnecting components.

The objective of groups with "reasonable" goals is to develop an optical computer whose speed of computation will outstrip that of the fastest electronic supercomputers.

*Some examples of parallel electronic computers are the Concurrent Computer (Beaverbrook, Oreg.), the Connection Machine (Thinking Machines, Cambridge, Mass.), the NCUBE/Ten (NCUBE, Beaverbrook, Oreg.), and the T Series (Floating Points Systems, Portland, Oreg.). All of these computers use the hypercube or binary *n*-cube architecture. In March 1987, Fifth Generation Computer Corp. (New York) introduced the 32-bit DADCO, based on an architecture developed at Columbia University.

The intention of groups with "ambitious" goals is to exploit the physical differences between photons and electrons to develop a whole new generation of computer architectures that reason and behave more like the human brain.

These goals are being pursued actively in a growing number of centers in the United States, Japan, and Europe.

United States

Research and development activities in optical computing are being conducted throughout the country by organizations such as these:

- Optical Circuitry Cooperative, a joint effort of the University of Arizona, Tucson (through its Optical Sciences Center), and 12 major companies, such as Boeing, DuPont, and TRW.
- Center for Photonic Technology at the University of Southern California
- Guided Wave Optoelectronic Manufacturing Technology Development Program at Battelle Memorial Laboratories, Columbus, Ohio
- California Institute of Technology, Pasadena (associative memories, photorefractive materials)
- Naval Research Laboratory, Washington, D.C. (analog signal processing, matrix-vector associative memories)
- Stanford University, Stanford, Calif. (optical interconnects)
- Center for Applied Optics at the University of Alabama, Huntsville (architectures, image processing, computing components)
- Center of Excellence in Optical Data Processing, a consortium of some 15 laboratories, based at Carnegie-Mellon University, Pittsburgh

In fiscal 1986, U.S. federal government funding of optical computing work totalled about $20 million, most of the money being channeled through defense-related agencies, such as the Defense Advanced Research Projects Agency (about $11 million), the Air Force Office of Scientific Research (about $4 million), the Innovative Science and Technology Office of the Strategic Defense Initiative Office (about $3 million), and the Army Research Office (about $700,000). In addition, it is estimated that in 1986, major U.S. companies, such as Boeing, AT&T Bell Laboratories, BDM, GTE, Lockheed, Honeywell, Unisys, and Texas Instruments, spent about $3 million on optical computing work.

Japan

In 1984, a research body called the Optical Computer Group was established under the auspices of the Japan Society of Applied Physics. The group coordinates the efforts of over 100 researchers in universities, government laboratories, and private-sector companies.

In June 1986, thirteen of Japan's electronic giants (including NEC, Matshushita, Hitachi, and Fujitsu) announced a 10-year cooperative program, with funding from the Ministry of International Trade and Industry, to consolidate efforts in developing optoelectronic chips.

Europe

Among the groups with active optical computer programs are these:

- European Joint Optical Bistability Project, involving eight universities and institutions in Great Britain, Belgium, West Germany, Italy, and France. Its appropriation, in 1984, was about $2 million.
- In France, the University of Paris at Orsay (liquid crystal components) and Thompson-CSF (optical processing).
- In West Germany, Erlangen University in Bavaria (parallel logic based on symbolic substitution, optical crossbar switching, spatial filtering), the University of Duisburg (electrooptic multipliers, broadcast buses for communication between modules in a computer), and the Fraunhofer Institute in Karlsruhe (optical networking).

The Soviet Union reportedly has the world's largest research and development program in optical computing. Work seems to be concentrated at three main centers—the A.F. Ioffe Physio-Technical Institute in Leningrad, the Institute of Radio Engineering in Novosibirsk, and the P.N. Lebeder Physics Institute in Moscow—in the fields of materials, devices, architectures, algorithms, and applications.

12.3 THE OPTICAL COMPUTER

The principal components of an optical computer (Fig. 12.1) are these:

- A central processing unit (CPU), which performs the essential implementable functions
- A data management processor that orchestrates the flow of data and sequence of operations (usually considered part of the CPU in electronic computers)
- Several types of memory elements for both short-term and long-term data storage
- Buffering and format devices, which organize input data fields spatially
- Input devices, which convert the data input types into a form suitable for subsequent processing
- Output devices, which convert the processed results to detectable and interpretable forms
- Detectors, which produce externally addressable results

In Fig. 12.1, the feedback interconnects are shown explicitly as separate components in anticipation of interactive architectures and algorithms that are now being investigated. The layout of Fig. 12.1 thus represents a distillation of current research and development directions.

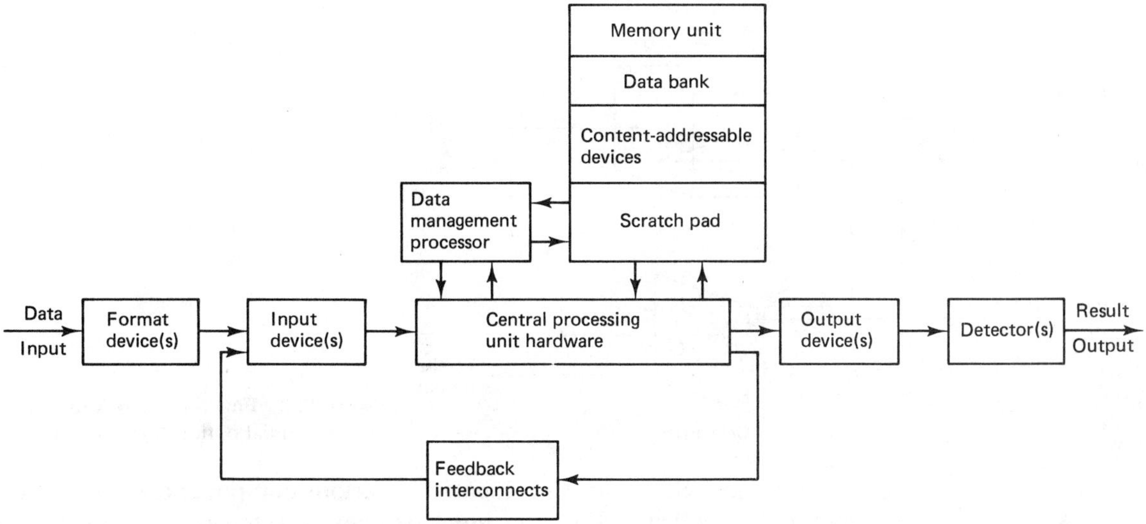

Figure 12.1 Schematic diagram of the elements of a generalized optical computer.

12.4 THE SPATIAL LIGHT MODULATOR

Perhaps the heart of the optical computer is the spatial light modulator (SLM), whose primary function is to detect an incident light intensity distribution that represents the data field to be processed, use the detected intensity to modify a given material property, and encode the amplitude and/or phase of the readout illumination with material state-dependent information.

The SLM thus functions as a two-dimensional array of independent logic gates. Implementing the logic system requires a device to provide the gate function and an interconnect element (Fig. 12.2). The main functional requirements of an optical logic gate are, therefore, as follows:

- Switching energies comparable to those of electronic logic gates
- High gain, so that each device can switch at least two (if not more) succeeding devices in order to achieve fan-out
- High contrast, in order to show a large change between a logic zero and a logic unity
- External access, so that other parts of the optical circuit can address elements in cascade in order to achieve fan-in
- Low power consumption at high speed, thereby minimizing heat dissipation; power on the order of a few milliwatts or less for operating frequencies of 1 GHz or more is desirable
- Quick restoration of logic level after each switching action, so that errors resulting from imperfect devices and signals do not accumulate throughout

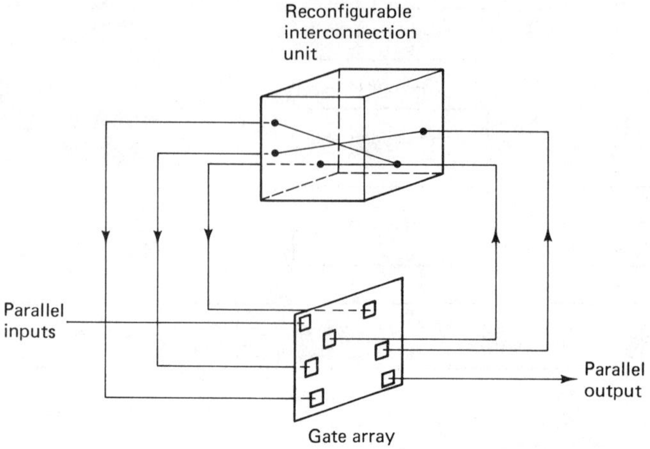

Figure 12.2 Functional block diagram of a sequential optical logic system.

the computation; SLMs are thus located in various components of the optical computer (Fig. 12.1), such as in input devices, scratch pad memory, programmable Fourier plane filtering and correlation, and programmable detector masking

12.5 THE ELECTROOPTIC SPATIAL LIGHT MODULATOR

In the past few years, the list of the different types of SLMs has expanded to include liquid crystal light valves and deformable mirror, magnetooptic, photochromic, electroabsorptive, photodichroic, and acoustooptic types. The main effort, though, appears to be focused on the electrooptic SLM, which is made up of a layer of photoconductive electrooptic material sandwiched between dielectric blocking layers overcoated with transparent counterelectrodes (Fig. 12.3).

The dielectric blocking layers provide the voltage division necessary to generate space-variant potential distributions across the electrooptic material. Po-

Figure 12.3 (a) An electrooptic spatial light modulator; (b) one in which the photoconductive voltage division and the electrooptic modulation functions are physically separated.

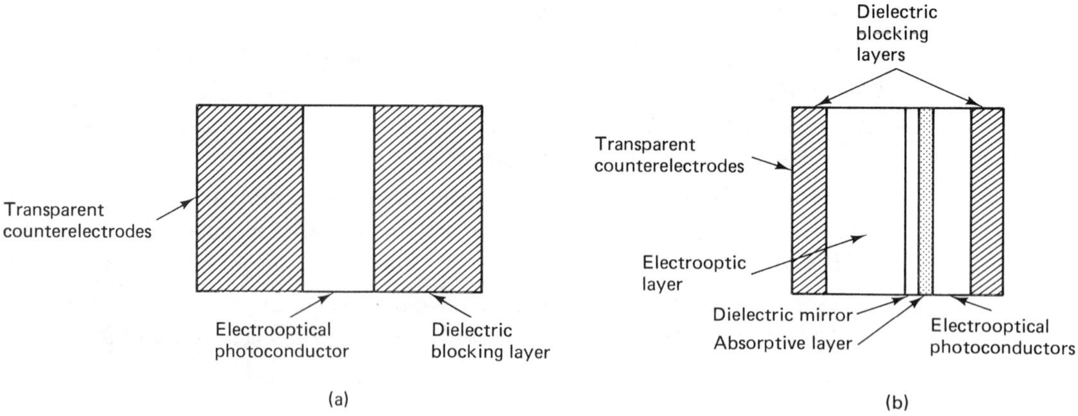

larized readout light is then encoded by the resultant refractive index modulation in the electrooptic material. If the index modulation is anisotropic in the plane of the electrooptic crystal, proper choice of the readout polarization will induce a space-invariant birefringence (differential refractive index modulation for each of two orthogonal polarization components), which can be converted to pure amplitude modulation by means of a suitably-oriented polarizing filter.

The process of incoherent-to-coherent conversion in an electrooptic spatial light modulator (ESLM) is summarized in Fig. 12.4. An input intensity distribution $I(x, y)$ gives rise to a net charge separation $P(x, y, z)$, which induces a change in the space-charge field within the electrooptic material $E(x, y, z)$. The longitudinal integral of $E(x, y, z)$ corresponds to a space-variant potential difference across the blocking layer $V(x, y)$, which results in an output amplitude that is proportional to $I(x, y)$.

The relevant requirements for a material for the electrooptic crystal in an ESLM are as follows:

- Cubic space group symmetry, for the enhancement of readout contrast ratio
- Minimum uniaxial anisotropy, in order to limit the induced birefringent polarization effects for all readout propagation directions not parallel to the optic axis
- A large nonlinear electrooptic coefficient, thereby reducing the magnitude of the required applied voltage and relaxing the constraints of thickness and dielectric breakdown strength of the dielectric material in the blocking layer
- High dark sensitivity (together with a large dielectric constant of the blocking layer material), which yields long storage times for applications requiring lengthy interrogations of the input data or multiple interrogations
- Large product of mobility and lifetime for the contributing charge carrier, resulting in high device sensitivity (defined as change in potential difference per unit change in the concentration of optically excited charge carriers)
- High photoconductive quantum efficiency, which leads to high sensitivity because that efficiency determines the fraction of absorbed photons that generates photoexcited charge carriers
- Good spectral dispersion of the absorption, which affects the available readout gain as well as the number of multiple interrogations, at a given signal-to-noise ratio, that can be tolerated without appreciable degradation of the image
- Low degree of optical activity—that is, the use of crystallographic point

Figure 12.4 Schematic diagram illustrating the physical relationship between input intensity and output amplitude and the linear electrooptic effect for the Pockels Readout Optical Modulator.

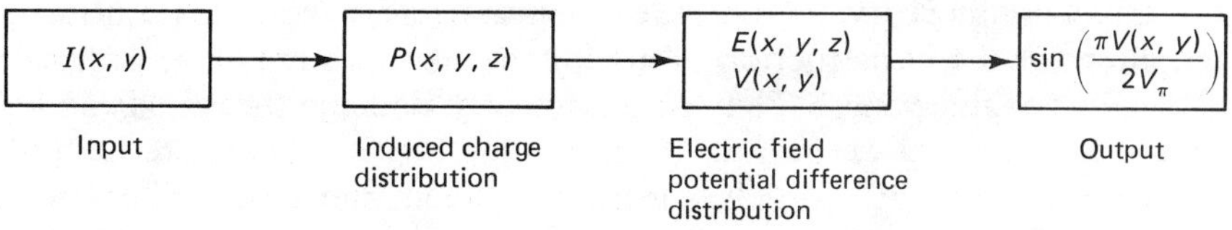

groups that practically excludes the possibility of natural optical activity, thereby significantly reducing the complications arising from nonlinearly polarized eigen states and accompanying signal dependent phase distortion

- Easy crystal growth to produce an excellent optical-quality material with the minimum of residual strain birefringence, rotation-induced striae, resistivity inhomogeneities, and other flaws

12.6 MATERIAL SELECTION

Three categories of materials possess properties that make most of the requirements achievable.

Single-Crystal Photorefractive Materials

These are materials in which the free carrier can be moved when the material is exposed to light; the concentration or depletion of charges induces electric fields and hence a change in the refractive index of the material. Examples of such materials are lithium niobate, $LiNbO_3$; barium titanate, $BaTiO_3$; bismuth germanium oxide, $Bi_{12}GeO_{20}$; bismuth silicon oxide, $Bi_{12}SiO_{20}$; strontium barium niobate, $Sr_{0.5}Ba_{0.5}Nb_2O_6$; and lithium tantalate, $LiTaO_3$.

Nonlinear Materials

These are materials that (a) will yield an output whose intensity is not linearly related to the input (at a certain input, the material may go from transparent to opaque or vice versa) and (b) are susceptible to large changes in their refractive indices in response to changes in the intensity of incident light. Examples include superlattice structures fabricated from ultrathin periodic layers of III–V semiconductors—such as gallium arsenide, GaAs; and aluminum gallium arsenide, $Al_{1-x}Ga_xAs$—and multiple quantum-well envelope state transition (QWEST) materials. A QWEST material is obtained by sandwiching a thin layer of material of smaller band gap (such as GaAs) between layers of a larger-band-gap material (such as AlGaAs), thereby quantum-mechanically confining the electrons to the smaller-band-gap material. Layer thicknesses can range from a few atomic layers to 500 nm.

Organic Materials

Certain organic materials look promising. Incorporating long carbon-chain molecules, such materials can be synthesized into crystals. They have demonstrated nonlinear effects that are several orders of magnitude weaker than, say, GaAs, but theoretical modeling shows that these materials have the potential for very strong nonlinearity. Active research into these compounds continues because they allow for the construction of an almost infinite number of materials, using "molecular engineering." They also have high damage thresholds and short response times to optical excitation. Probably the best of this genre is *N*-(4,nitrophenyl)-(*L*)-prolinol (NPP), which is grown between transparent silica plates.

Of these three categories of candidate materials, the single-crystal electrooptic or photorefractive materials are currently recognized to hold the greatest promise.

12.7 THE PHOTOREFRACTIVE EFFECT

The change in the refractive index of a material n due to an applied electric field E is expressed as

$$\Delta \frac{1}{n^2} = rE + gE^2 \tag{12.1}$$

The first term in equation (12.1), which has a linear dependence on the applied electric field, is known as the Pockels effect. The second term, which has a quadratic dependence on the field, is known as the Kerr electrooptic effect. The constants r and g are known as Pockels and Kerr coefficients, respectively. The Pockels effect depends on the polarity of the field; the Kerr effect does not.

The electrooptic effect can best be represented as a deformation of the optical indicatrix, which is an ellipsoid of revolution, and is expressed (Fig. 12.5) as

$$\frac{x^2}{n_o^2} + \frac{y^2}{n_o^2} + \frac{z^2}{n_e^2} = 1 \tag{12.2}$$

If the major and minor axes are lined up with the x, y, and z axes, the ellipsoid may be expressed as

$$\frac{x^2}{n_x^2} + \frac{y^2}{n_y^2} + \frac{z^2}{n_z^2} = 1 \tag{12.3}$$

Once the ellipsoid is deformed, however, and these axes move away from the x, y, and z axes, the cross terms (xy, yz, zx) become necessary to represent the new ellipsoid. The general expression for such an ellipsoid is

$$a_{11}x^2 + a_{22}y^2 + a_{33}z^2 + 2a_{23}yz + 2a_{31}zx + 2a_{12}xy = 1 \tag{12.4}$$

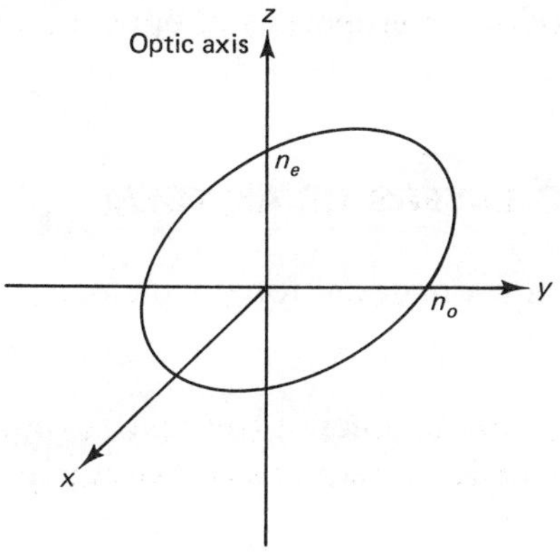

Figure 12.5 The optical indicatrix of a uniaxial crystal; n_o is the ordinary refractive index, and n_e the extraordinary refractive index.

The values of the coefficients a_{11}, a_{22}, a_{33}, a_{23}, a_{31}, and a_{12} change according to the applied electric fields. They may be represented as

$$\begin{bmatrix} a_{11} - 1/n_x^2 \\ a_{22} - 1/n_y^2 \\ a_{33} - 1/n_z^2 \\ a_{23} \\ a_{31} \\ a_{12} \end{bmatrix} = \begin{bmatrix} r_{11} & r_{12} & r_{13} \\ r_{21} & r_{22} & r_{23} \\ r_{31} & r_{32} & r_{33} \\ r_{41} & r_{42} & r_{43} \\ r_{51} & r_{52} & r_{53} \\ r_{61} & r_{62} & r_{63} \end{bmatrix} \begin{bmatrix} E_x \\ E_y \\ E_z \end{bmatrix} \tag{12.5}$$

where r_{11}, r_{12}, r_{13}, . . . are the Pockels constants in the respective crystallographic directions; E_x, E_y, and E_z are the components of the applied electric field in the x, y, and z directions, respectively; and n_x, n_y, and n_z are the refractive indices of the material, in the absence of the electric field, in the x, y, and z directions, respectively. Note that when the field is absent, equation (12.4) reduces to equation (12.3).

Equation (12.5) looks rather unwieldy and complex, but many of the 18 elements of the matrix are zero, and several of the nonzero elements have the same value. The reduction of these terms is a function of the symmetry of the crystal. Thus lithium niobate has only four different values for its eight nonzero elements, and gallium arsenide has only one value for its three nonzero elements.

Some properties of three photorefractive materials are given in Table 12.1.

12.8 LITHIUM NIOBATE

At the present level of knowledge, the best photorefractive material is lithium niobate. It can be easily fabricated into optical switches (it may be grown in single-crystal form, measuring 75 mm in diameter by 610 mm in length, from the melt by the Czochralski method, yielding very homogeneous and high-purity crystals). It is relatively insensitive to the plane of polarization of the incident light wave and has a high Curie temperature. Incidentally, this last-mentioned property is the main advantage of lithium niobate vis-à-vis lithium tantalate. The latter has a lower Curie temperature, thus bringing the material to a high temperature for the titanium in-diffusion step in the fabrication of devices. This necessitates a subsequent repolarization step, adding to both complexity and cost.

A comparison of the relevant properties of optical switch materials is given in Table 12.2.

12.9 THE DIELECTRIC BLOCKING LAYERS OF AN ESLM

The relevant requirements for a material for the dielectric blocking layers in an ESLM are as follows:

- High volume resistivity (of the order of 10^{16} Ωm), to prevent charge transport across the layers even under conditions of extremely high electric fields (of the order of 10^{11} V m^{-1})

TABLE 12.1 ELECTROOPTIC CONSTANTS AND REFRACTIVE INDICES OF THREE PHOTOREFRACTIVE MATERIALS

Material	Crystal Symmetry[a]	Pockels Coefficient, r (10^{-12} m V^{-1})	Refractive Index
$LiNbO_3$	3m	$r_{13} = 8.6$	$n_x = n_y = 2.286$
		$r_{33} = 30.8$	$n_z = 2.200$
		$r_{22} = 3.4$	
		$r_{51} = 28.0$	
$LiTaO_3$	3m	$r_{13} = 7.9$	$n_x = n_y = 2.176$
		$r_{33} = 35.8$	$n_z = 2.180$
		$r_{22} = 1.00$	
		$r_{51} = 20.0$	
GaAs	$\bar{4}$ 3m	$r_{41} = 1.2$	$n_x = n_y = n_z = 3.42$

[a]3m electrooptic matrix:

$$\begin{bmatrix} 0 & -r_{22} & r_{13} \\ 0 & r_{22} & r_{13} \\ 0 & 0 & r_{33} \\ 0 & r_{51} & 0 \\ r_{51} & 0 & 0 \\ -r_{22} & 0 & 0 \end{bmatrix}$$

$\bar{4}$ 3m electrooptic matrix:

$$\begin{bmatrix} 0 & 0 & 0 \\ 0 & 0 & 0 \\ 0 & 0 & 0 \\ r_{41} & 0 & 0 \\ 0 & r_{41} & 0 \\ 0 & 0 & r_{41} \end{bmatrix}$$

Sources: I. P. Kaminow and E. H. Turner, "Linear electrooptic materials," in R. J. Pressley (ed.), *Handbook of Lasers with Selected Data on Optical Technology*. Cleveland, Ohio: Chemical Rubber Co., 1971, pp. 447–458; T. Okada (ed.), *Crystal Optics*. Tokyo: Morikita Publishing Co., 1975; J. Koyama and H. Nishihara, *Optoelectronics*. Tokyo: Corona-Sha, 1978, pp. 298–300.

TABLE 12.2 SOME ROOM-TEMPERATURE PERFORMANCE RATINGS OF FOUR DEVICES TO BE USED AS AN OPTICAL SWITCH

Device	Suitable Materials	Switching Energy	Switching Speed[a]	Switching Power (mW)
Self-electrooptic effect device	GaAs, AlGaAs, QWESTs	100 pJ	On: 30 ns Rec: 30 ns	2
Bistable interference filters and	GaAs	3 pJ	On: 50 ps Rec: 50 ps–50 ns	10
etalons	ZnS	10 μJ	On: 100 ps Rec: 100 μs–10 ms	10
Optical logic gate etalon	GaAs	100 pJ	On: 1 ps Rec: 40 ps	8
QWEST device[b]	GaAs, QWESTs	50 fJ	5 ps	13

[a]On (switch-on): time to get an answer; Rec (recovery): time required for device to relax before the next operation.

[b]Values of all ratings are author's estimates, based on information available in mid 1988.

- High dielectric constant (of the order of 20 to 500), for proper impedance matching between the layers and the electrooptic crystal
- High dielectric breakdown strengths (of the order of 10^9 V m^{-1}), to withstand the high electric fields that would be applied across the electrooptic crystal
- Excellent optical transparency in the write and read spectral windows
- Low residual scattering arising from inhomogeneities in the refractive index (these contribute to increased noise-to-signal ratios in the Fourier plane of the processor)
- Very low concentrations of mobile ions that can migrate under high electric field conditions to produce significant history-dependent bias voltages
- Ease of deposition (a) with negligible interfacial strain on the electrooptical crystal layer, (b) with excellent thickness uniformity on substrates of under 20–50 mm in size, and (c) at low substrate temperatures, to avoid increase in residual birefringent strain

The most promising blocking layer material is currently parylene polydichloroparaxylene (PPDPX), which is already being used extensively in ESLMs. It is a vapor-deposited polymer with conformal coating properties, good optical transparency in the visible range, adequate dielectric breakdown strength, and high volume resistivity. Its disadvantages are mainly (a) a low dielectric constant ($\approx$3) and (b) high sensitivity to humidity and ultraviolet radiation, which can cause dielectric breakdown at fields substantially less than the specified value.

12.10 FINAL COMMENTS

Although a prototype digital optical computer has been built* and performed well in tests (150 to 200 demonstrations in 1984), computer professionals continue to express a number of reservations as to the viability of a full-scale general-purpose optical computer. Here are some of these reservations.

1. It is necessary to replace the von Neumann architecture, on which electronic computers are based, with other structures having greater inherent parallelism, designed to match the parallelism of optics (see Fig. 12.6).

2. The information must be kept from degrading during all the processes and transmissions from one point to another. It is widely recognized now that optical computing is more suitable to areas, such as signal processing, that involve a relatively short series of fixed processes rather than to general-purpose computations.

3. Optical devices inevitably involve precision surfaces and precision distances. This is a serious drawback.

4. Optical logic gates require a certain number of outputs to have the right input light level coming from a fixed output power. Again, this is a disadvantage

*SAOBIC (Systolic Acoustooptic Binary Computer), built by GuilTech Research Company, Sunnyvale, California.

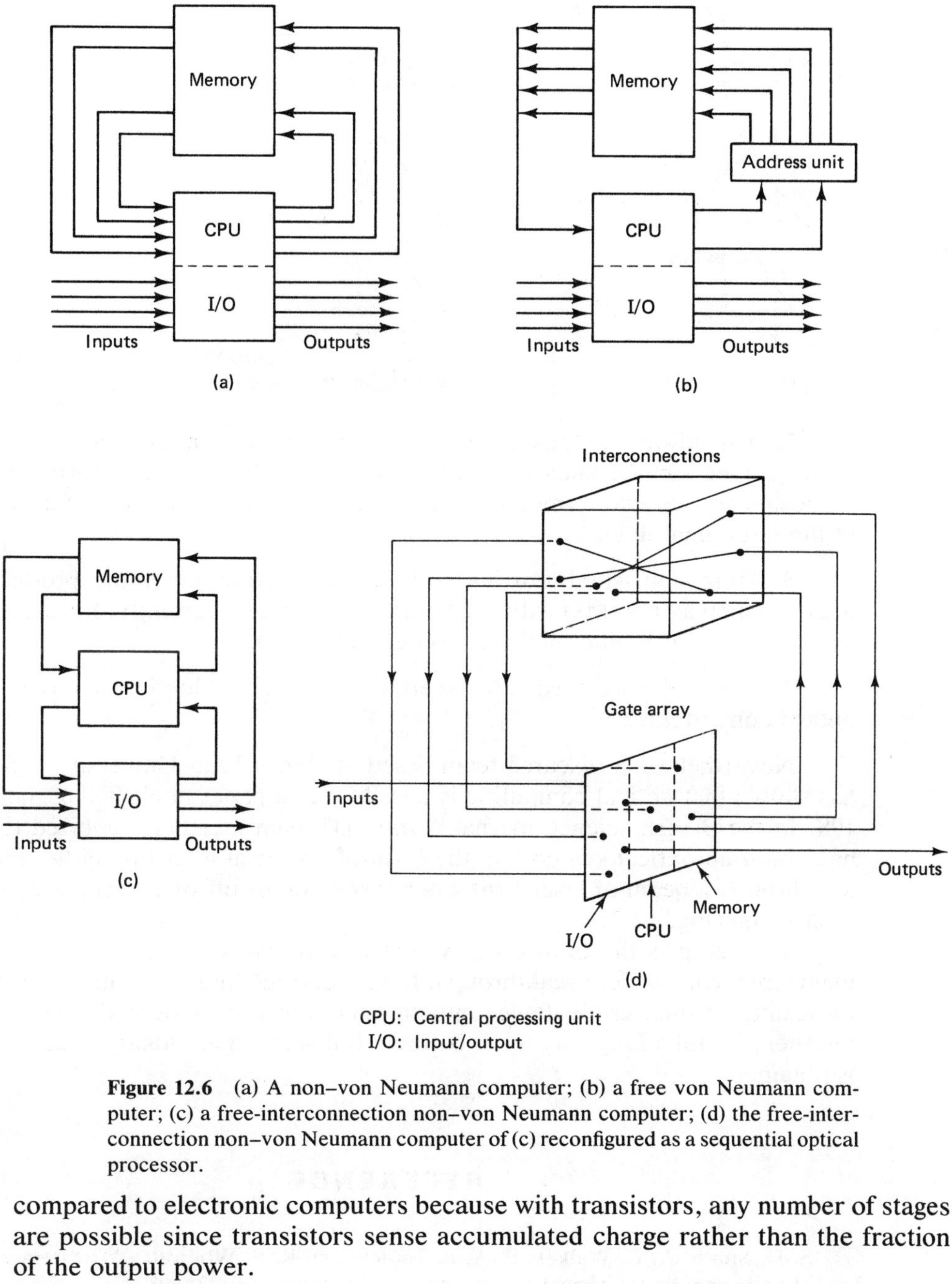

Figure 12.6 (a) A non–von Neumann computer; (b) a free von Neumann computer; (c) a free-interconnection non–von Neumann computer; (d) the free-interconnection non–von Neumann computer of (c) reconfigured as a sequential optical processor.

compared to electronic computers because with transistors, any number of stages are possible since transistors sense accumulated charge rather than the fraction of the output power.

5. Optical devices are inevitably limited by the wavelength of light, thus imposing a size limitation. By contrast, the possibilities for miniaturization of transistors in electronic computers are much greater, as is evidenced in the growth of very large scale integrated (VLSI) circuits.

6. The possibilities of using the output from an optical logic gate to drive subsequent stages is very limited.

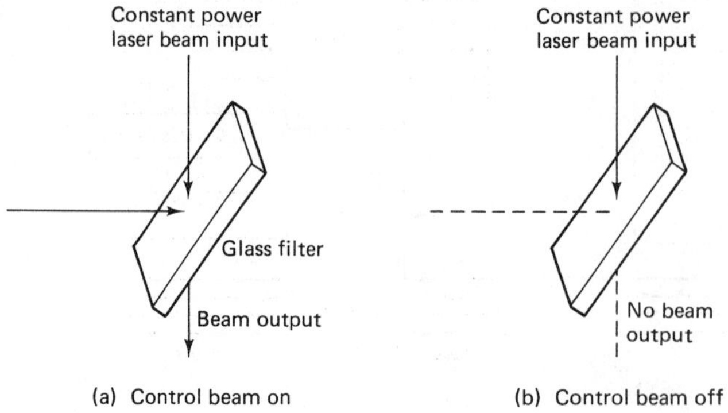

Figure 12.7 Optical switching with a laser light beam [1].

7. A good optical logic gate, especially a three-terminal device that would be a genuine optical analog to the transistor, is lacking. Currently, the main proposals are for two-terminal devices, which do not have the circuit flexibility of three-terminal devices.

8. More consistent starting materials are required for the production of devices (such as ESLMs) with predictable properties. Currently, for example, no standard form of lithium niobate has been obtained.

9. More tests are needed to ascertain the expected lifetime and reliability of optical computers.

Nevertheless, a research team based at Heriot-Watt University in Scotland has shown that optical computing is possible. In a paper published at the start of 1987 in the British science journal *Nature* [1] team members reported that they have built an optical switch,* in the form of a special filter that either blocks or lets through a beam of laser light when turned on or off by a second, controlling laser beam (Fig. 12.7).

The setup is the optical equivalent of a transistor, and the results by this team represent a major breakthrough. It was reported that the team is now working on scaling up their crude device into an operational machine with 1000 switches, but there is still a long way to go before a full-scale operational optical computer is obtained.

REFERENCE

1. S. D. Smith, A. C. Walker, F. A. P. Tooley, and R. S. Wherrett, "The demonstration of restoring digital optical logic," *Nature 325* (6099), p. 27, 1987.

*In late 1988, it was suggested by a number of researchers that an alternative approach is to do away with switching in optical computer altogether, using analog rather than digital processing with lenses and holograms instead of electronic devices.

SOUNDING BOARD OF AN ORCHESTRAL HARP

CHAPTER 13

13.1 INTRODUCTION

The harp is a chordophone (stringed musical instrument) consisting of a set of parallel strings of different lengths mounted on an approximately triangular frame, stretched between a resonator and a neck that are joined together at one end. The modern or orchestral double-action harp (Fig. 13.1), credited to Sébastien Érard, dates from 1810 and has the largest range in the orchestra: about $6\frac{1}{2}$ octaves (3 below and $3\frac{1}{2}$ above middle C). The chromatic flexibility offered by this harp makes it fully adapted to the needs of freely modulating Western music and of a clear and telling, though not especially brilliant or expressive, tone. It is extraordinarily beautiful in the orchestra when composed for and has the virtue of being decidedly loud!

One problem has dogged manufacturers of harps: The sounding board is not strong enough to take the string load without creeping. When the strings are tightened in tuning, the board is subjected to about 2 t of load. The resulting elastic deformation leads to further tightening to keep the strings in tune. The strain progressively becomes so great that permanent plastic deformation begins to occur with time; that is, the board creeps. Eventually, the sounding board can separate from the sounding box, rendering the instrument unplayable.

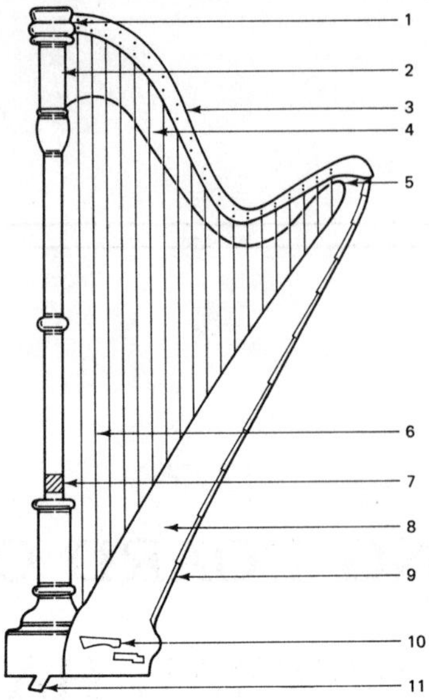

Figure 13.1 Principal parts of a modern orchestral harp: (1) neck; (2) metal plate; (3) tuning pegs; (4) bridge; (5) rotating discs; (6) strings; (7) pillar containing pedal rods; (8) sounding board; (9) eyelets; (10) pedals; (11) feet.

One way of solving this problem is through a redesign of the sounding board, but any changes (such as stiffening the board with ribs or redistributing the load to different points) can seriously alter the instrument's tone and musical quality. A change in the material used in its construction is, presumably, the only practical solution.

The sounding board of an orchestral harp has traditionally been made from eastern spruce, a softwood, although some hardwoods, such as sugar maple, beech, and white oak, have been used. This choice of material is both deliberate and sensible since it is well established that the resonance of wood plays a vital role in the quality of sound that emanates from an instrument made of it.

There is one disadvantage, however, attached to the use of wood in this application. Its properties vary so much (as a function of relative humidity, for example) that considerable skill is required of the artisan in selecting the wood and determining the thickness required to produce consistently fine instruments. Few people in the trade possess this ability.

Thus a substitute for wood that is durable, is stable with environmental conditions, and has fairly predictable properties is desirable. Filamentary composite materials are, arguably, the most suitable candidates.

13.2 THE FLEXURAL EQUATION OF MOTION

The sounding board may be approximated as a flat, rectangular plate under the combined action of a lateral uniformly distributed load per unit area p and self-mass unit area m. It has been shown [1] for such a plate made of an orthotropic material that the midsurface at any point $A(x_a, y_a)$ has a deflection w (Fig. 13.2), whose governing differential equation is

$$D_x \frac{\partial^4 w}{\partial x^4} + 2H \frac{\partial^4 w}{\partial x^2\, \partial y^2} + D_y \frac{\partial^4 w}{\partial y^4} = p + m \frac{\partial^2 w}{\partial v^2} \qquad (13.1)$$

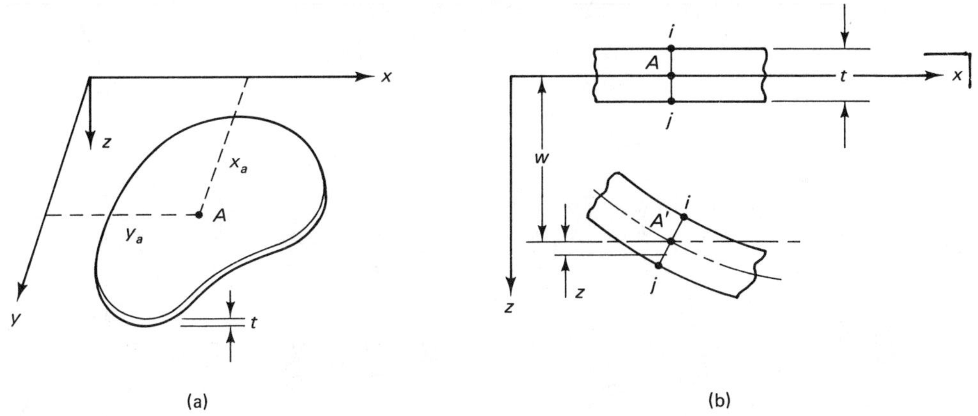

Figure 13.2 Deflection of the midsurface of a loaded flat, rectangular plate.

where D_x and D_y are flexural rigidities, H is the warping stiffness of the plate, and υ is the time of vibration. Note that $D_x = t^3E_x/12$; $D_y = t^2E_y/12$; $H = D_{xy} + 2G_{xy}$ (where E_x and E_y are the moduli of elasticity of the material, G_{xy} is the shear modulus of the material, D_{xy} is the flexural rigidity of the plate, and t is the thickness of the plate).

The wavelengths at the operational frequencies of a sounding board are sufficiently long that the inhomogeneities in the sounding board material may be neglected. Equation (13.1) does not take into account internal damping losses (which may be quite high in wood, for example). One acceptable modification to equation (13.1) involves putting the warping stiffness equal to zero.

13.3 NATURAL FREQUENCIES OF THE SOUNDING BOARD

Approximating the sounding board as a rectangular plate simply supported along two opposing edges and vibrating from side to side (Fig. 13.3) leads to the natural frequencies being given by

$$f_n(\mathrm{s}^{-1}) = n^2 \frac{\pi}{2L^2} \sqrt{\left(\frac{EI}{\rho bt}\right)} \qquad (13.2)$$

where I is the moment of inertia of the cross section of the plate and ρ is the density of the plate material.

Figure 13.3 Idealized vibration modes of the sounding board of a harp.

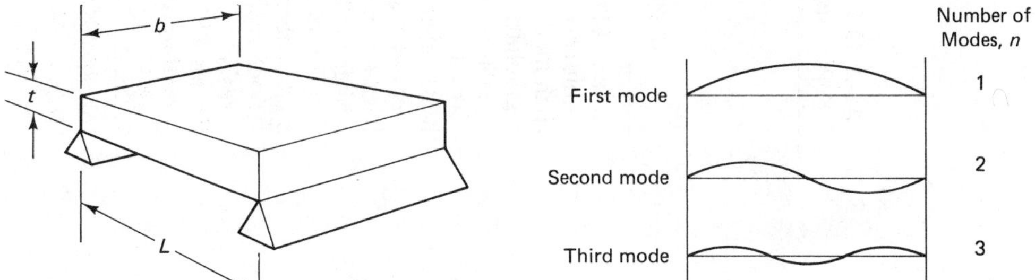

TABLE 13.1 COMPARISON OF RELEVANT PROPERTIES OF FILAMENTARY COMPOSITE MATERIALS

Material[a]	Longitudinal Modulus of Elasticity, E_x (GPa)	Transverse Modulus of Elasticity, E_y (GPa)	Density, ρ (kg m^{-3})	Areal Density, $m(t = 2.5\text{ mm})$ (kg m^{-2})	D_x/D_y	D_x/m[b] (m^4 s^{-2})
Spruce (12% moisture content)	11	0.86	400	1.00	12.80	14.32
High-modulus graphite fiber reinforced epoxy resin	207	6.00	1412	3.53	34.50	76.35
Boron fiber reinforced epoxy resin	207	21.00	1990	4.98	9.86	54.12
High-strength graphite fiber reinforced epoxy resin	141	8.00	1570	3.93	17.63	46.71
S glass fiber reinforced epoxy resin	56	17.00	1997	5.00	3.29	14.58
Kevlar 49 fiber reinforced epoxy resin	86	7.00	1366	3.42	12.29	32.74

[a] All composite materials have fiber volume fraction of 0.60.

[b] $D_x/m = t^3E_x/12m$.

Since the sounding board is to be made from either a natural filamentary composite material (such as spruce wood) or a manufactured one (such as carbon fiber reinforced epoxy resin), there will be along-grain or along-fiber and across-grain or across-fiber vibrations, having frequencies given, respectively, by

$$f_{nal}(\mathrm{s}^{-1}) = n^2 \frac{\pi}{2L_{al}^2} \sqrt{\left(\frac{E_{al}I_{al}}{\rho b_{al}t_{al}}\right)} \qquad (13.3)$$

and

$$f_{nac}(\mathrm{s}^{-1}) = n^2 \frac{\pi}{2L_{ac}^2} \sqrt{\left(\frac{E_{ac}I_{ac}}{\rho b_{ac}t_{ac}}\right)} \qquad (13.4)$$

13.4 CONDITIONS FOR SIMILARITY

Leading on from equation (13.1), the conditions for similarity between a sounding board made of spruce wood and one made of another orthotropic material thus become these:

1. The ratio of the longitudinal (along-grain or along-fiber) bending stiffness to the transverse (across-grain or across-fiber) bending stiffness, D_x/D_y, should be low, ideally no more than about 13 (which is the value for spruce at 12% moisture content; see Table 13.1).

2. The ratio of the longitudinal bending stiffness to the areal density must be very high, ideally no less than about 14 $\mathrm{m}^{-4}\ \mathrm{s}^{-2}$ (which is the value for spruce; see Table 13.1).

3. Areal density is to be as low as possible, ideally be about 1 kg m^{-2}, which is the value for spruce (Table 13.1). In other words, the density should be about 400 kg m^{-3}.

4. The natural frequencies both along and across the fibers in the competitive orthotropic material should match those in spruce.

5. The logarithmic decrement (which is a measure of the damping or decay of flexural vibrations) along the fibers, δ_x, and across the fibers, δ_y, are functions of frequency. It is seen from Table 13.2 that the "average" transverse damping

TABLE 13.2 THE VARIATION OF DAMPING WITH FREQUENCY FOR SPRUCE WOOD AND A GRAPHITE-EPOXY SANDWICH

	Graphite-epoxy sandwich		Spruce wood	
Frequency (Hz)	Transverse logarithmic decrement	Longitudinal logarithmic decrement	Transverse logarithmic decrement	Longitudinal logarithmic decrement
300	0.070	0.010	0.050–0.102	0.010–0.022
1000	0.072	0.020	0.052–0.110	0.015–0.028
2000	0.074	0.040	0.062–0.130	0.018–0.030
5700	0.010	0.055		0.030–0.038

Source: D. W. Haines and N. Chang, "Making better musical instruments from graphite composite," *Mech. Eng. 98* (3), pp. 26, 27, 1976.

ability of spruce is about 50% better than that of a graphite-epoxy sandwich material, but longitudinal vibrations in such a sandwich are damped about 30% better than in spruce.

13.5 SELECTION OF MATERIAL

Consideration of the first three conditions in section 13.4, in the light of the values in Table 13.1, leads to the conclusion that epoxy resin reinforced with fibers of high-modulus graphite, boron, or Kevlar 49 would be a suitable alternative to spruce for the sounding board of an orchestral harp.

In fact, if the board is made from high-modulus graphite fiber reinforced epoxy resin (HMGFRP), the required ideal ratio of D_x to D_y of 12.8 can be obtained by using a volume fraction of fibers of 13%, as shown in Table 13.3. For such a material, $E_x = 49.41$ GPa and $E_y = 3.44$ GPa, and its density $\rho = 1244$ kg m^{-3}.*

For a given vibration mode, with sounding boards made of spruce wood and HMGFRP having the same length and width (both along and across the grains or fibers), and with the spruce board having a thickness of 10 mm, we find that the corresponding thickness of the HMGFRP board would be (a) 8.3 mm [when the natural frequencies of the longitudinal free vibrations are matched; see Equation (13.3)], and (b) 8.8 mm [when the natural frequencies of the transverse free vibrations are matched; see Equation (13.4)]. This means that the weight of the spruce sounding board is only about 40% that of the HMGFRP.

TABLE 13.3 MODULUS ANISTROPY FOR HMGFRP

Fiber Volume Fraction	Modulus Anisotropy[a]
0.0	1.00
0.2	21.33
0.4	34.99
0.6	41.99
0.7	42.99
0.8	42.33
1.0	36.00

[a]Modulus anisotropy =

$$\frac{D_x}{D_y} = \frac{E_x}{E_y} = \frac{E_f V_f + E_m(1 - V_f)}{[V_f/E'_f + (1 - V_f)/E_m]^{-1}}$$

where E_f and E'_f are the moduli of elasticity of the fiber along and across the fiber axis, respectively; E_m is the modulus of elasticity of the matrix material; and V_f is the volume fraction of the fiber. For HMGFRP, $E_f = 360$ GPa, $E'_f = 10$ GPa, and $E_m = 3$ GPa.

*Using a simple rule of mixtures, $\rho = \rho_f V_f + \rho_m(1 - V_f)$, where ρ_f and ρ_m are the densities of the fiber and matrix, respectively. In this case, $\rho_f = 1870$ kg m^{-3} and $\rho_m = 1150$ kg m^{-3}.

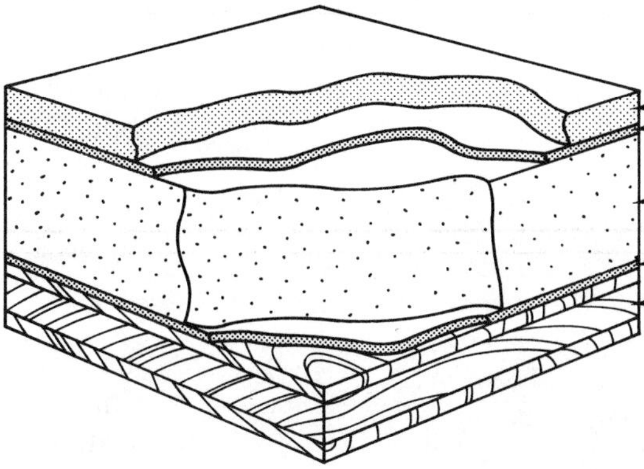

Figure 13.4 General arrangement of a sandwich construction.

In light of these results, as well as condition 3 in section 13.4, which is a corollary of these results, the HMGFRP sounding board should be made of a sandwich, consisting of a low-density isotropic core (ideally, high-quality paper, with a density of 200 kg m^{-3}), with very thin sheets of HMGFRP bonded onto the core at each face using high-quality adhesive (Fig. 13.4). For weight matching between such a sandwich and spruce wood, we can write a "weight balance" equation:

$$(400 \times 9.81 \times t_{sp}) = [1244 \times 9.81(t_c - t_p)] + (200 \times 9.81 \times t_p)$$

or

$$t_{sp} = 3.11t_c - 2.61t_p \tag{13.5}$$

where the subscripts *sp*, *c*, and *p* stand for spruce wood, HMGFRP, and paper, respectively. Since the weights are to match, natural frequency matching requires, from equation (3.3) or modified forms thereof, that

$$E_{spal}I_{sp} = E_{cal}I_c$$

or (13.6)

$$\frac{E_{spal}}{E_{cal}} = \frac{t_c^3 - t_p^3}{t_{sp}^3}$$

or

$$t_{sp}^3 = 4.55(t_c^3 - t_p^3) \tag{13.7}$$

From equations (13.5) and (13.7), we get $t_p = 0.61t_c$ and $t_{sp} = 1.52t_c$. Thus if the spruce wood sounding board is 10 mm thick, the replacement sandwich must consist of a paper core 4 mm thick and two facings of HMGFRP each 3.3 mm thick.

Several experimental musical instruments using this type of sandwich material for the sounding board or top plate have been constructed and are reportedly giving excellent acoustical performance and aesthetic satisfaction.

REFERENCE

1. A. C. Urgural, *Stress in Plates and Shells*. New York: McGraw-Hill, 1981.

A VAULTING POLE

CHAPTER 14

14.1 INTRODUCTION

Pole vaulting is, arguably, one of the most elegant track and field events. It is always a graceful athletic performance, encompassing the beauty of controlled trajectory motion (Fig. 14.1).

Figure 14.1 Trajectory of an athlete performing a pole vault.

From the viewpoint of material selection, the objective is to choose a material that has a very high value of specific endurance strength and whose use would result in the lowest values of (a) the weight of the vaulting pole, for a given pole "spring stiffness,"* and (b) the total cost of the pole (raw material plus fabrication costs).

14.2 STRESS ANALYSIS

In use, a vaulting pole is a cantilever beam, one end being fixed on the field (momentarily, before "takeoff" of the athlete) and a force (essentially the weight of the athlete) being applied at an angle of about 30° to the pole at some distance from the other end. (Usually the point of application of this force is 85% to 95% of the length of the pole, measured from the "fixed" end; see Fig. 14.2.)† In addition, a uniformly distributed wind load acts on the pole.

The stress analysis of the pole may thus be reasonably approximated by the slope and deflection calculations for a cantilever beam (a) with a point force W not at the free end and (b) carrying a uniformly distributed load w per unit length of pole over the entire pole span.

Case 1

Referring to Fig. 14.3, consider a section XX at a distance x from the fixed end. The bending moment at section XX, M_x, is equal to $-W(L_1 - x)$.

Figure 14.2 Stages in pole vaulting: (a) approach; (b-e) climb; and the clearance (f).

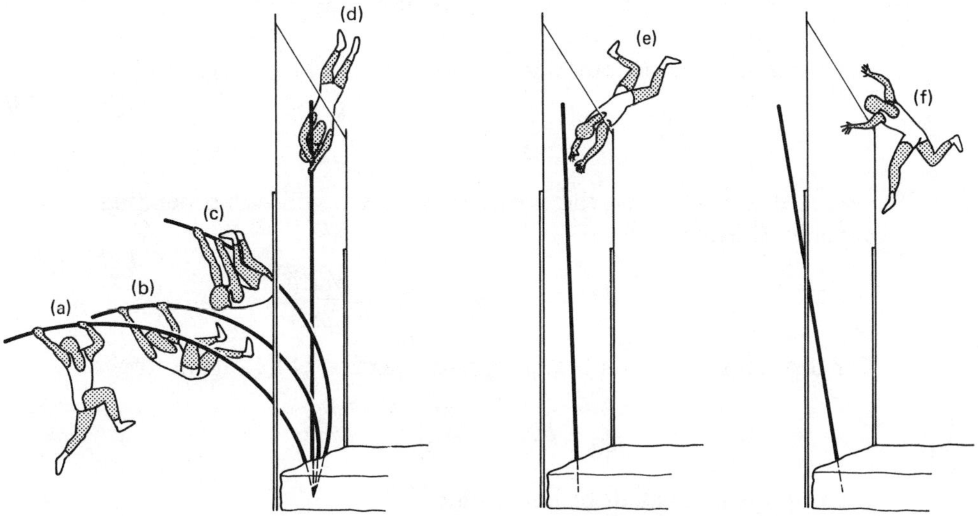

*Defined as applied load on the pole per unit deflection of the pole.

†Both the angle of grip and the position of grip on the pole depend on the style or technique of the athlete.

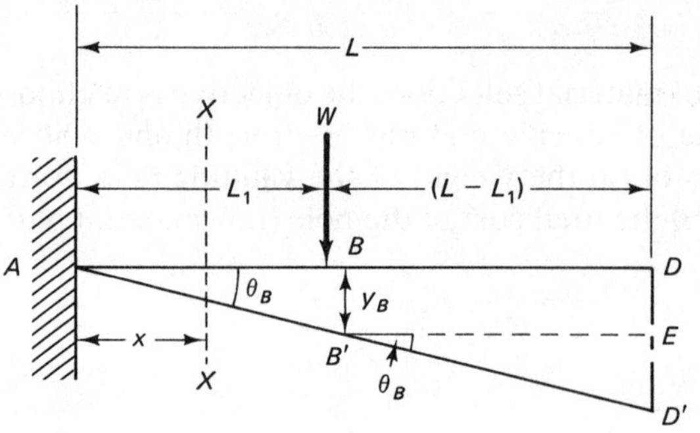

Figure 14.3 A point load W not at the free end of a cantilever.

The flexural equation of a beam is given by

$$EI\,\frac{d^2y}{dx^2} = -M$$

where I is the moment of inertia of the cross section of the beam, M is the bending moment at any section, and y is the deflection at any section, distance x from the built-in end. Thus, in this case,

$$EI\,\frac{d^2y}{dx^2} = -M_x = W(L_1 - x)$$

Integrating for slope, $\theta = dy/dx$, we have

$$EI\,\frac{dy}{dx} = WL_1x\,\frac{-Wx^2}{2} + C_1$$

where C_1 is a constant of integration. At point A, $x = 0$ and $dy/dx = 0$, $\therefore C_1 = 0$. Hence

$$EI\,\frac{dy}{dx} = WL_1x - \frac{Wx^2}{2} \tag{14.1}$$

For slope at point B, put $x = L_1$ in equation (14.1) to give

$$\theta_B = \frac{1}{EI}\left(WL_1L_1 - \frac{WL_1^2}{2}\right) = \frac{WL_1^2}{2EI} \tag{14.2}$$

As there is no load on the portion BD, there will be no bending moment in that portion. Hence

$$\theta_D = \theta_B = \frac{WL_1^2}{2EI}$$

For deflection y at point B, integrate equation (14.1) again to give

$$EIy = \frac{WL_1x^2}{2} - \frac{Wx^3}{6} + C_2$$

where C_2 is a constant of integration.

At point A, where $x = 0$ and $y = 0$, $C_3 = 0$. Hence

$$EIy = \frac{WL_1x^2}{2} - \frac{Wx^3}{6}$$

For deflection at point B, put $x = L_1$:

$$y_B = \frac{1}{EI}\left(WL_1\frac{L_1^2}{2} - \frac{WL_1^3}{6}\right) = \frac{WL_1^3}{3EI}$$

But $y_B = DE$ (Fig. 14.3), and $D'E = EB' \tan\theta_B = \text{BD} \tan\theta_B = BD\theta_B$ (since θ_B is small, $\tan\theta_B \approx \theta_B$). Thus

$$D'E = L - L_1 \frac{WL_1^2}{2EI}$$

But

$$y_D = DD' = DE + ED' = \frac{WL_1^3}{3EI} + \frac{WL_1^2}{2EI}(L - L_1) \tag{14.3}$$

Case 2

Referring to Fig. 14.4, consider a section XX at a distance x from the fixed end A. The bending moment $M_x = \dfrac{-w(L - x)^2}{2}$; therefore,

$$EI\frac{d^2y}{dx^2} = \frac{w(L - x)^2}{2}$$

Integrating both sides gives

$$EI\frac{dy}{dx} = \frac{w}{2}\left(L^2x - Lx^2 + \frac{x^3}{3}\right) + C_3$$

where C_3 is a constant of integration. At point A, the slope, θ, is zero, thus $dy/dx = 0$ when $x = 0$, and we have $C_3 = 0$. Thus

$$EI\frac{dy}{dx} = \frac{w}{2}\left(L^2x - Lx^2 + \frac{x^3}{3}\right) \tag{14.4}$$

For the slope at point D, put $x = L$ into equation (14.4) to give

$$\theta_D = \frac{wL^3}{6EI}$$

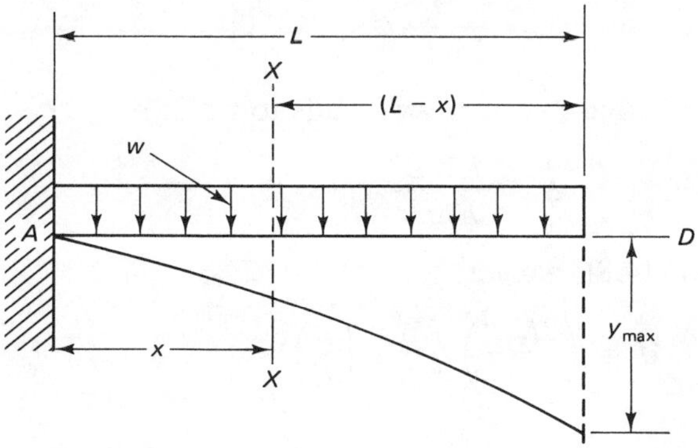

Figure 14.4 A cantilever subjected to wind load w per unit length.

Integrating equation (14.4) to give the deflection y we have

$$EIy = \frac{w}{2}\left(\frac{L^2x^2}{2} - \frac{Lx^3}{3} + \frac{x^4}{12}\right) + C_4 \tag{14.5}$$

where C_4 is a constant of integration. At point $A(x = 0)$, $y = 0$; therefore, $C_4 = 0$. Hence the deflection at point D, the maximum deflection y_{max} is given by putting $x = L$ in equation (14.5) to obtain

$$y_{max} = \frac{wL^4}{8EI}$$

The total maximum deflection of the pole vault—the sum of y_D in case 1, with $(L - L_1) = 0.10L$, and y_{max} in case 2—is given by

$$\Delta = \frac{W(0.729L^3)}{3EI} + \frac{W(0.810L^2)(0.10L)}{2EI} + \frac{wL^4}{8EI} = 0.2835\,\frac{WL^3}{EI} + \frac{0.125wL^4}{EI}$$

To simplify the mechanics of materials aspect of this application, we consider the general case of the deflection of a cantilever beam subjected to a load W applied vertically at some distance x from the fixed end of the beam. Then the deflection of the vaulting pole would be obtained by putting $W = W' \sin \alpha$ (where W' is the weight of the athlete and α is the angle of grip with respect to the pole; the later parameter is usually about 30°). But putting $W = W' \sin 30°$ and expressing w in terms of W, the maximum elastic deflection is given by

$$\Delta = \frac{KWL^3}{EI} \tag{14.6}$$

where K is a numerical constant.

14.3 MATERIAL SELECTION METHODOLOGY

Consider a vaulting pole with a thin tubular cross section, in which the length L and the bore $2r$ are fixed by the designer but the wall thickness t may be varied.

The mass of the pole, $M = 2\pi rtL\rho$, where ρ is the density of the pole material. Thus the variable cross section dimension t becomes

$$t = \frac{M}{2\pi rL\rho} \tag{14.7}$$

From equation (14.6), and noting that $I = \pi r^3 t$ leads to

$$\Delta = \frac{KWL^3}{E\pi r^3 t} \tag{14.8}$$

From equations (14.7) and (14.8), we get

$$M = \left(\frac{2KL^4}{r^2}\right)\left(\frac{W}{\Delta}\right)\left(\frac{\rho}{E}\right)$$

Since L and r are fixed by the designer, for a given pole "spring stiffness" (W/Δ) the mass of the pole is directly proportional to the material parameter, ρ/E. The raw material cost of the pole is subsequently directly proportional to the material parameter, $p\rho/E$, where p is the raw material cost per unit mass.* The third material parameter to consider is the specific endurance strength, defined as the endurance strength (FL) after 10^6 cycles divided by the density.

A comparison of four candidate materials on the basis of these three material parameters is given in Table 14.1.† On consideration of performance and cost, the best choice is the glass fiber reinforced epoxy resin. Most pole vaulters, in fact, now use a pole made of this material. For example, before 5000 cheering spectators at Paris's Jean Bouin Stadium on July 13, 1985, Sergei Bubka of the Soviet Union set a pole vaulting world record of 6 m using an AMF Pacer vaulting pole (made by AMF Pacer, Carson City, Nev., using a unidirectional prepreg glass fiber material supplied by Fiberite, Winona, Minn.).

TABLE 14.1 COMPARISON OF CANDIDATE MATERIALS FOR A VAULTING POLE

Material	ρ (kg m^{-3})	E (GPa)	p (\$ t^{-1})	FL after 10^6 Cycles (MPa)	ρ/E (10^{-7} m^{-1})	$p\rho/E$ (10^{-4} \$ t^{-1} m^{-1})	FL/ρ (10^3 m)
Low-alloy steel	7800	200	210	320	3.83	0.80	4.18
2014 Al alloy	2796	73	2500	200	3.76	9.38	7.29
S glass fiber reinforced epoxy resin[a]	1997	56	5000	300	3.50	17.50	15.31
Carbon fiber reinforced epoxy resin[a]	1661	138	130,000	530	1.18	153.40	32.53

[a] Volume fraction of fiber = 0.60.

*Fabrication costs are not considered because data on them, for different poles, are very difficult to obtain from pole manufacturers.

†In the early days of pole vaulting, bamboo was a common pole material (for example, Earle E. Meadows of the USA won the gold medal at the 1936 Olympic Games in Berlin with a cleared jump of 4.35 m using a bamboo pole). Since about 1950, however, bamboo has not been used for this application.

COMPRESSED-AIR TANK OF A SUPERSONIC WIND TUNNEL

CHAPTER 15

15.1 INTRODUCTION

The application, made from a bank of large thin-walled cylindrical pressure vessels, must satisfy the following criteria:

- The maximum stress in the vessel must be below the yield strength of the material used.
- The vessel must not fail by fast fracture.
- The vessel must not fail by fatigue.

15.2 STRESS ANALYSIS OF A THIN-WALLED CYLINDRICAL PRESSURE VESSEL

Such a vessel is subjected to internal pressure p only, which is taken to be uniformly distributed over the internal surfaces, giving rise to both circumferential stress σ_H and longitudinal stress σ_L (Fig. 15.1). In arriving at the expression for σ_H and σ_L, the following assumptions are made:

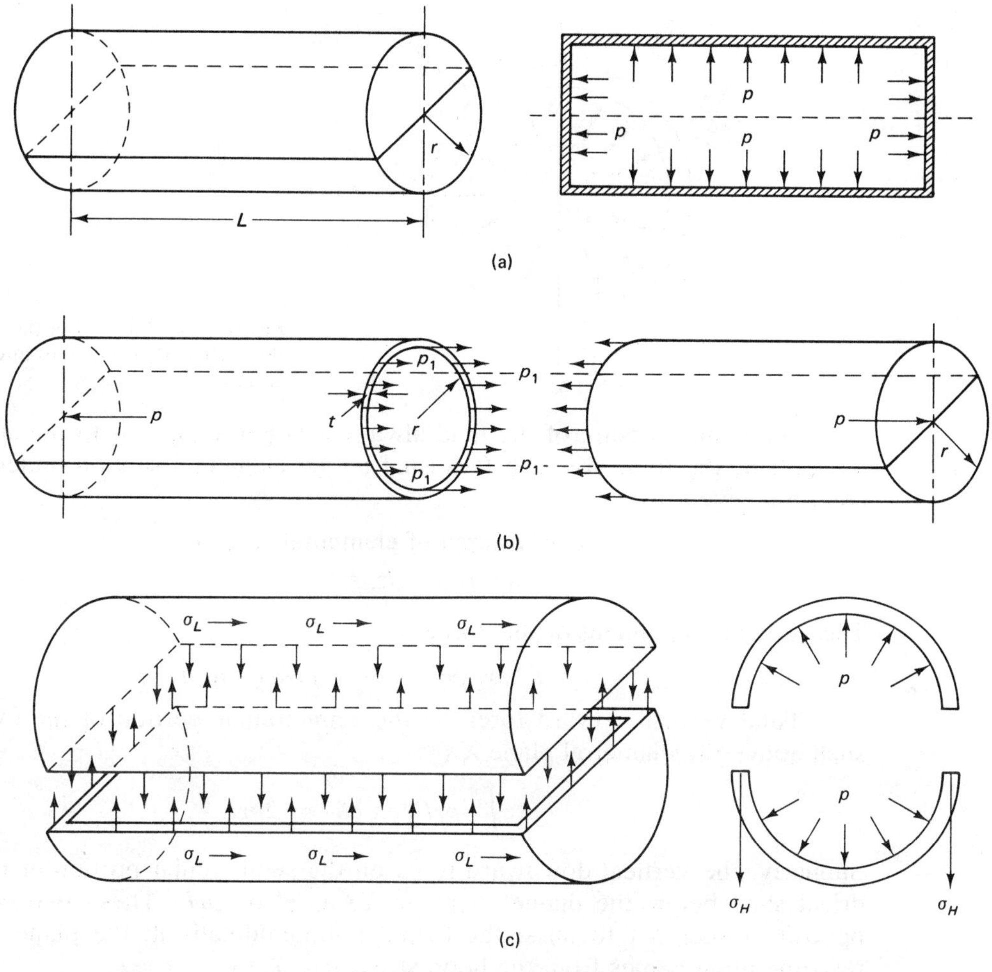

Figure 15.1 Stresses in a thin cylinder subjected to internal pressure p: (a) cylindrical shell under internal fluid pressure; (b) longitudinal stress development; (c) hoop stress development.

- That the radial stresses in the cylinder wall are negligible
- That there are no longitudinal stays in the cylinder
- That the stresses are uniformly distributed throughout the section wall

Hoop Stress, σ_H

Consider an elemental portion of the cylinder wall, of width δx, subtending an angle $\delta\theta$ at the center and at an angle θ with XX (Fig. 15.2). The internal radius of the cylinder is r and (away from the ends) the length L is under consideration (Fig. 15.1).

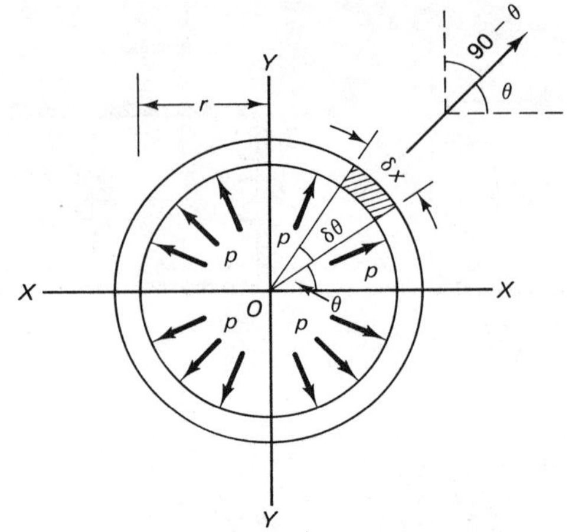

Figure 15.2 Element of the wall of a thin cylinder subjected to internal pressure *p*.

Since the pressure of the fluid always acts perpendicular to the surface of its contact, the total pressure *P* normal to the elemental section under consideration is given by

$$P = p\text{ (area of elemental section)}$$
$$= p\delta xL = pr\delta\theta L$$

The vertical component of this force

$$= P\cos(90 - \theta) = pr\delta\theta L\sin\theta$$

Total vertical upward force on the semicircular portion of the cylindrical shell above the diametral plane *XX*

$$= \int_0^{\pi} prL\sin\theta\delta\theta = 2prL$$

Similarly, the vertical downward force on the semicircular portion of the cylindrical shell below the diametral plane *XX* is also $2prL$. These two equal and opposite forces act to burst the cylinder longitudinally at the plane *XX*. The resisting force comes from the hoop stress σ_H. Thus

$$2prL = 2\sigma_H tL$$

making

$$\sigma_H = \frac{pr}{t}$$

Longitudinal Stress, σ_L

The total force at the end of the cylindrical shell is equal to $p\pi r^2$. The longitudinal stress developed in the shell (Fig. 15.1) is then obtained from the force balance equation:

$$p\pi r^2 = 2\pi rt\sigma_L$$

making

$$\sigma_L = \frac{pr}{2t}$$

The maximum stress in the vessel is thus $\sigma_H = pr/t$.

15.3 MATERIAL SELECTION METHODOLOGY

1. To avoid yield of the tank,

$$\sigma_{\max} = \frac{pr}{t} < \sigma_y \tag{15.1}$$

where σ_y is the yield strength of the vessel material. The higher the value of σ_y, the higher the hoop stress that can be tolerated in the vessel.

2. The plane-strain or opening mode (Fig. 15.3) stress intensity factor, for the case of a component containing an edge crack in tension, is given by

$$K_I = 1.12\sigma\sqrt{\pi a} \tag{15.2}$$

where σ is the applied stress and a is the length of the edge crack in the vessel.

To avoid failure by fast fracture,

$$1.12\sigma_{\max}\sqrt{\pi a_{cr}} < K_{IC} \tag{15.3}$$

where $\sigma_{\max}$ is the maximum applied stress, a_{cr} is the critical flaw size, and K_{IC} is the fracture toughness of the vessel material.

From equation (15.3), $a_{cr} = [K_{IC}/(1.12\sigma_{\max})]^2/\pi$. That is, a_{cr} is directly proportional to $(K_{IC}/\sigma_y)^2$, since $\sigma_{\max}$ is a fraction of σ_y. Thus the larger the value of a_{cr}, the more attractive is the material, since cracks can be easily detected without the use of sophisticated equipment.

3. Assume that the fatigue crack growth law in the vessel material is describable by the Paris-Erdogan Law,

$$da/dN = B(\Delta K_I)^n$$

where da/dN is the fatigue crack growth rate, ΔK_I is the range of stress intensity factor, and B and n are material constants. The number of cycles to failure N_f of the component is given by

$$N_f = \frac{2}{(n-2)B\alpha^n(\Delta\sigma)^n\pi^{n/2}}\left(a_i^{(1-\frac{n}{2})} - a_{cr}^{(1-\frac{n}{2})}\right) \tag{15.4}$$

when $n \neq 2$. In the present case, $\alpha = 1.12$ [see equation (15.2)]. So for the crack to grow from an initial size a_i to a critical size a_{cr}, the rate of growth is dependent on K_{IC}.

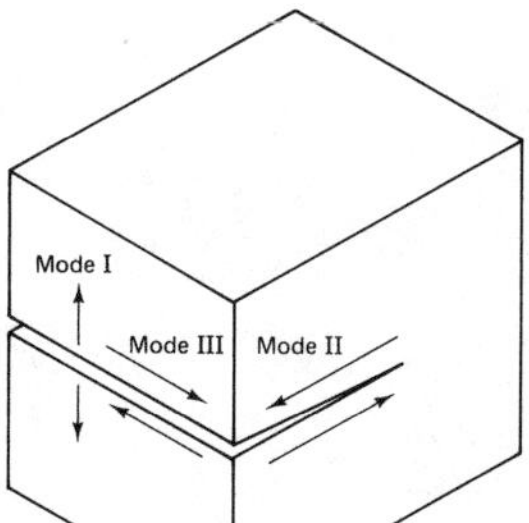

Figure 15.3 Displacements associated with the three planar modes of fracture: opening mode (I), edge-sliding mode (II), and tearing mode (III).

A comparison of three candidate materials on the basis of all three criteria is presented in Table 15.1. On the basis of criterion 1, the best material is the maraging steel, but from the viewpoints of criteria 2 and 3, the titanium alloy is obviously superior.

TABLE 15.1 COMPARISON OF THREE MATERIALS FOR A COMPRESSED-AIR TANK

Material	σ_y (MPa)	K_{IC} (MPa$\sqrt{m}$)	B	n	a_{cr}[a] (mm)	N_f[b]
18% Ni maraging steel	1340	100	1.32×10^{-10}	2.32	5.65	1955
Al-Zn-Mg alloy	490	25	6.34×10^{-11}	3.14	2.64	2205
Ti-6Al-4V	830	120	9.60×10^{-13}	3.80	21.21	4513

[a]In equation (15.3), $\sigma_{max} = 0.5\,\sigma_y$.

[b]In equation (15.4), $a_i = 1$ mm and $\Delta\sigma = 0.5\sigma_y$; that is, stress varies from zero to $\sigma_{max} = 0.5\sigma_y$.

SKATEBOARD WHEELS

CHAPTER 16

16.1 INTRODUCTION

From the first push forward on a home driveway to intricate freestyle routines in a purpose-built park, and at all levels in between, skateboarding (Fig. 16.1) is a very exhilarating and fun-filled recreational activity. It is an all-around athletic exercise that synthesizes many of the best features of several sports, such as

Figure 16.1 A skateboarder in action.

surfing, skiing, dance, gymnastics, and martial arts, combining strength, grace, and concentration. It is one sports activity in which the potential for individual, distinctive, and innovative styles and moves is very large. Although skateboarding is no longer as popular as it was during its heyday in the 1970s, there are still a large number of devotees.

16.2 MAIN PARTS OF A SKATEBOARD

There are three main parts of a skateboard: the deck or flexboard, the wheels, and the truck (Fig. 16.2).

The deck comes in a variety of lengths and shapes (Fig. 16.3). Lengths range from short (0.61 m) "hotdogging" models to 1.52 m speedsters designed for flat-out downhill racing, the average length being 0.77 m. As far as shapes are concerned, some decks have kicktails engineered on them to provide better leverage and traction. Others have rockers installed to lower the center of gravity of the skateboard. In addition, there are cambered skateboards for slalom and downhill racing. Skateboards produced in the early 1980s employ a number of advanced themes in flex, camber, and torsion in deck designs. The materials commonly

Figure 16.2 Main parts of a skateboard.

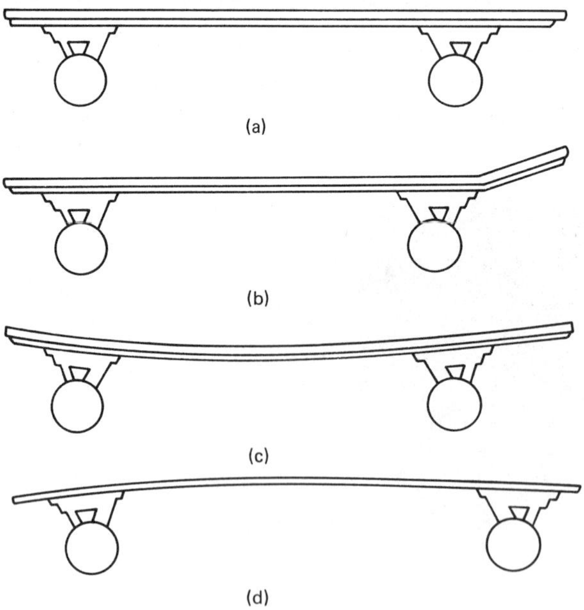

Figure 16.3 Common skateboard decks (side view): (a) basic flat board; (b) kicktail board for rear-foot leverage; (c) rocker board for lower center of gravity; (d) cambered board for slalom and downhill racing.

Figure 16.4 Wheels of a skateboard.

used for decks are wood (ash, teak, oak), polymers (polypropylene, polycarbonate), composites (glass fiber reinforced epoxy resin), wood and fiberglass combinations, and aluminum and its alloys.

The explosive growth of modern skateboarding would probably not have been possible without major advances in both the design of the wheels (Fig. 16.4) and the materials used for their manufacture. Wheels come in a wide range of sizes (from width of 29 mm by diameter of 38 mm to width of 58 mm by diameter of 67 mm). The wheels contain either loose ball bearings or double-sealed precision bearings. The former have the disadvantages of requiring regular lubrication and the possibility of the wheels being slowed by sand, road grit, and other particles.

Connecting the deck to the wheels is a steering and suspension system called the truck. Again, this comes in different designs. All trucks consist of "shock absorbers" of elastomeric or polymeric bushings or cushions or metallic coiled springs. Some trucks are adjustable in length and can be tightened or loosened to accommodate the weight of the rider. Various parts of a truck (such as the baseplate and the axle) are made of cast iron, forged iron, heat-treated aluminum alloys, or steel. Figure 16.5 features a schematic representation of a truck and wheel assembly showing two types of wheels and some parts of the truck.

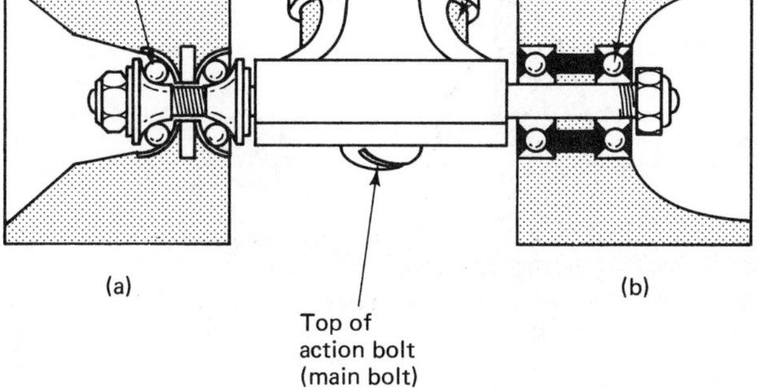

Figure 16.5 Truck and wheel assembly of a skateboard: (a) standard wheel with open loose bearings; (b) flex wheel with double-seated precision bearings.

16.3 MATERIAL SELECTION CONSIDERATIONS

To perform satisfactorily, the materials for the wheel (and bearing assembly) must have the following characteristics:

- Lowest possible rolling friction resistance, in order to increase the speed of the skateboard (this is the primary requirement)
- High frictional resistance to lateral movement, especially important for slalom and freestyle motions
- Good shock absorption qualities, in order to make riding on rough surfaces not too uncomfortable
- High resistance to abrasive wear
- High resistance to degradation by corrosives such as lubricating oil

The contact stress on a perfectly elastic wheel in motion on a rigid surface and carrying a static load is shown in Fig. 16.6(a). Upon rolling, one part of the "flat" (formed at the wheel-surface interface) suffers an increased strain, while the strain on other parts is reduced. For "real" materials, however, the stress in the section of the flat where the strain is constantly decreasing lags behind that in the other section of the flat, resulting in the stress distribution shown in Fig. 16.6(b).

Figure 16.6 Distribution of contact stress between the wheels of a skateboard and the road surface: (a) wheels made of a perfectly elastic material; (b) wheels made of a "real" material; (c) wheels with no resilience.

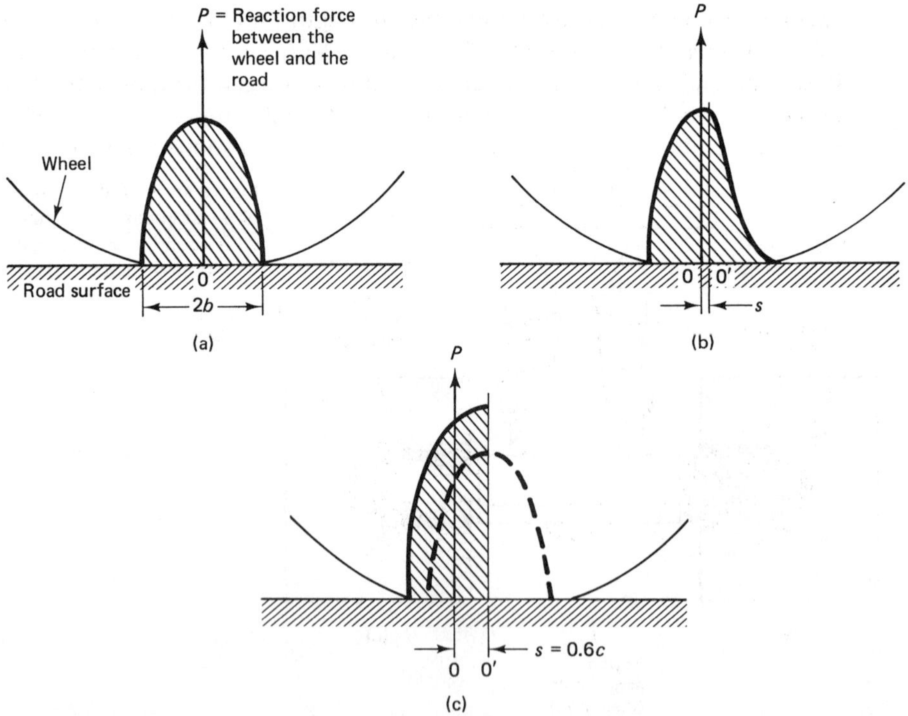

The final outcome is that the center of pressure 0′ moves forward of the center of the wheel 0 by a small amount s, giving rise to a bending moment Ps that resists the rolling. The rolling resisting force R is the equivalent force acting at the center of the wheel in the direction opposite to the bending moment; that is,

$$R = \frac{Ps}{D_w/2}$$

where D_w is the diameter of the wheel. The coefficient of friction μ_{rs} is thus

$$\mu_{rs} = \frac{R}{P} = \frac{2s}{D_w} \tag{16.1}$$

Some knowledge of the value of s is needed in order to estimate μ_{rs} from equation (16.1). For a perfectly elastic wheel material, $s = 0$; for a "worst case" material, with the stress distribution given in Fig. 16.6(c), $s = \alpha c$, where c is the half-width of the flat and α is a numerical constant, usually about 0.6. For many "real" materials, therefore, we can assume that as a first approximation,

$$s = \alpha(1 - A)c \tag{16.2}$$

where A is the percentage rebound and is equal to the square of the coefficient of restitution e of the wheel material. Now

$$c = \beta\sqrt{\frac{D_w P}{E_{ef} h}} \tag{16.3}$$

where β is a numerical constant, usually about 0.8, h is the width of the wheel and E_{ef} is the effective modulus of the wheel and surface materials. E_{ef} is given by

$$E_{ef} = \left[\frac{1 - \nu_w^2}{E_w} + \frac{1 - \nu_{rs}^2}{E_{rs}}\right]^{-1} \tag{16.4}$$

where E_w and ν_w are the modulus of elasticity and Poisson's ratio of the wheel material, respectively, and E_{rs} and ν_{rs} are the corresponding values for the road surface material.

From equations (16.1), (16.2), and (16.3), we obtain

$$\mu_{rs} = \frac{2}{D_w}\left[\alpha(1 - A)\beta\sqrt{\frac{PD_w}{hE_{ef}}}\right] \approx \frac{(1 - A)}{\sqrt{E_{ef}}}\sqrt{\frac{P}{hD_w}} \tag{16.5}$$

An examination of equation (16.5) shows that, for a given wheel with specified values of wheel plan area ($h \times D_w$) and wheel load P, the primary design requirement of low friction may be obtained by using a material with high values for modulus of elasticity (for a given road surface material) and coefficient of restitution.

On rougher surfaces (such as tarmac or concrete roadways), an important additional factor affects friction, leading to a reconsideration of the desirable properties for a wheel material. Upon rolling over a rough surface, the wheel load will be supported on the road asperities where the contact pressure is high. The

TABLE 16.1 MECHANICAL AND FRICTION CHARACTERISTICS OF CANDIDATE MATERIALS FOR THE WHEELS OF A SKATEBOARD[a]

Material	E (MPa)	ν	E_{ef}[b] (MPa)	e	μ_{rs}[c]	μ'_{rs}[d]
Polyurethane (polyester type)	6.38	0.40	7.60	0.80	0.051	0.077
Polyurethane (polyether type)	25.50	0.40	30.36	0.70	0.036	0.054
Glass fiber reinforced polyimide	31,028	0.40	14,265	0.30	0.0030	0.0045

[a]Wheel plan area = 2300 mm^2; wheel load = 350 N.

[b]Road material elastic characteristics are E_{rs} = 23 GPa and ν_{rs} = 0.11.

[c]Coefficient of friction on a smooth road surface.

[d]Coefficient of friction on a rough road surface, assuming that $\mu'_{rs} = 1.5\mu_{rs}$.

resulting penetration must be countered by the elastic deflection of the wheel if energy loss and friction are to be minimized. This requires a wheel material with large elastic strain and high elastic modulus, with the former requirement becoming more important as road roughness increases. The presentation outlined earlier for the friction coefficient for the wheel is applicable to the wheel on a rough surface provided the value for μ_{rs} is increased by a suitable factor, usually 50%.

On the basis of the data in Table 16.1, the best material, from the viewpoint of both μ_{rs} and μ'_{rs}, is glass fiber reinforced polyimide. However, because of its high cost (\$6.50 kg^{-1} in 1988), low shock absorption qualities, and relatively low wear resistance, the materials currently favored for the construction of the wheels of skateboards are hot-cured polyether-based polyurethanes.

ARMOR PLATE FOR A SECURITY VAN

CHAPTER 17

17.1 INTRODUCTION AND THEORY

The three most common modes of failure experienced by metal armor plate are ductile hole formation, dishing, and plugging (Fig. 17.1). In all these modes, considerable plasticity is exhibited by the target material, thus making the work done in plastic flow a reasonable guide to the energy required to pierce an armor plate.

The kinetic energy of a projectile that pierces the armor plate (in any of the failure modes) is equal to the work done in expanding a hole in the armor plate from zero to the projectile diameter and is given by [1]

$$\frac{mU_m^2}{2} = \frac{\pi D^2 \sigma_y t}{2} \tag{17.1}$$

where m is the mass of the projectile, U_m is the minimum velocity of the projectile, D is the diameter of the projectile, t is the thickness of the armor plate, and σ_y is an appropriate strength parameter of the armor plate material, such as tensile or compressive yield strength at some high value of true strain, such as 100%. (This is so because in the piercing of the armor plate, the strains are quite large.) Note that equation (17.1) accounts for the most significant mode of energy consumption (plastic flow) and neglects secondary effects such as friction and inertia.

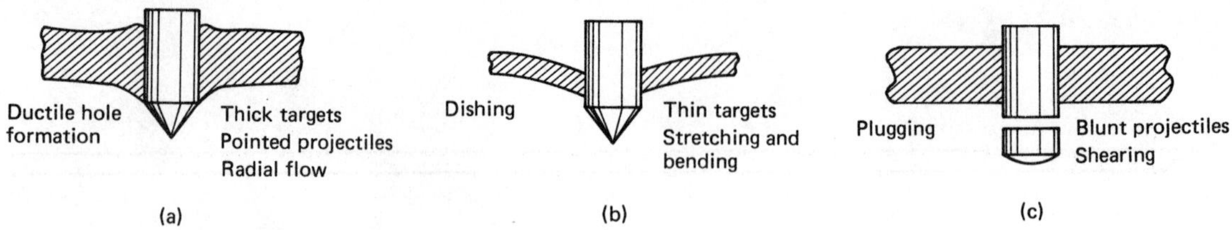

Figure 17.1 Schematic representation of the three principal failure modes associated with the penetration of homogeneous metal armor by armor-piercing projectiles.

Making t the subject of equation (17.1) leads to

$$t = \left(\frac{m}{\pi}\right)\left(\frac{U_m}{D}\right)^2\left(\frac{1}{\sigma_y}\right) \tag{17.2}$$

The mass of the armor plate material per unit area $= \rho t = (m/\pi)(U_m/D)^2(\rho/\sigma_y)$, where ρ is the density of the armor plate material. The cost to prevent penetration of a unit area = unit cost of target material, p, times the mass per unit area $= p\rho t = (m/\pi)(U_m/D)^2 p(\rho/\sigma_y)$.

17.2 EVALUATION OF CANDIDATE MATERIALS

Thus, to defend an armor plate material from a projectile of fixed mass and velocity, the candidate armor plate materials must be evaluated on the basis of the minimum value of (a) the thickness factor T, which is proportional to $(1/\sigma_y)$; (b) the mass factor M, which is proportional to (ρ/σ_y); and (c) the cost factor C, which is proportional to $(p\rho/\sigma_y)$. Such evaluations for four candidate materials are presented in Table 17.1.

TABLE 17.1 PROPERTIES OF FOUR CANDIDATE ARMOR PLATE MATERIALS

Material	Tensile Yield Strength (MPa)	Density (kg m^{-3})	Unit Cost (\$ kg^{-1})	T (MPa)$^{-1}$	M (10^{-5} m^{-1})	C (10^{-5} \$ kg^{-1} m^{-1})
Hot-rolled 1020 mild steel	976	7830	0.30	1.03×10^{-3}	7.87	2.36
P8 Hadfield's steel (hot-rolled)	1690	7870	0.40	0.59×10^{-3}	4.57	1.83
Titanium (commercial purity; annealed)	1167	4540	6.00	0.86×10^{-3}	3.82	22.92
Titanium 318 alloy (forged and annealed)	1685	4420	7.20	0.59×10^{-3}	2.57	18.50

17.3 CONCLUSION

Hot-rolled mild steel is clearly an attractive candidate material, but its use here incurs a substantial weight penalty. The solution is to increase σ_y by suitable heat treatment.

In terms of thickness and weight, P8 Hadfield's steel appears to offer considerable advantages, but it is extremely difficult to machine, which fact raises its fabrication cost.

Commercial-purity titanium does not offer substantial weight savings and has a high cost disadvantage. The titanium alloy performs extremely well on both thickness and weight bases although its cost is almost prohibitive for this application.

On the whole, therefore, the choice is between the Hadfield's steel and the titanium 318 alloy as the best material for this application. Police riot shields, made from certain titanium alloys, are already in use in some countries.

REFERENCE

1. R. L. Woodward, "The penetration of metal targets which fail by adiabatic shear plugging," *Int. J. Mech. Sci. 20,* pp. 599–607, 1978.

HEAT STORAGE MATERIAL FOR A LOW-TEMPERATURE LATENT-HEAT (SOLAR) THERMAL ENERGY SYSTEM

CHAPTER 18

18.1 INTRODUCTION

To increase the sun's usefulness as an energy source requires the development of an efficient heat storage system. Latent heat storage is particularly attractive for a number of reasons associated with its ability to provide a high-energy storage density and to store heat at a constant temperature corresponding to the phase transition of the heat storage substance.

Latent systems generally store heat in the form of either the latent heat of fusion in appropriate materials (those that melt or freeze at a requisite temperature) or the heat of crystallization, as the substance is transformed from one solid phase to another.

A temperature range of 0°C to 120°C to cover a variety of low-temperature, fairly-constant-load solar thermal applications, such as domestic water heating, direct or heat-pump-assisted space heating, greenhouse heating, and solar cooling, is considered. Any material that melts with an accompanying high heat of fusion in this range may be used as a latent heat storage material if it meets certain criteria. These requirements for a particular class of latent heat storage materials are given in section 18.2.

A latent heat thermal storage system has three components:

- A heat storage material that undergoes a solid-to-liquid phase transition within the desired operating temperature range
- A containment vessel for the storage material
- A heat exchanger for transferring the heat from the heat source to the heat storage material and from the heat storage material to the heat sink

18.2 MATERIAL SELECTION CONSIDERATIONS

Desirable properties of a phase-change heat storage material are as follows:

- A melting point suitably matched to the service temperature range of the thermal system
- High storage capacity; that is, high latent heat of fusion per unit mass and per volume, thereby reducing the amount of material needed to store a given amount of energy
- High density in both solid and liquid phases, thereby reducing the volume of the container needed to hold the material
- High thermal conductivity in both solid and liquid phases, reducing the temperature gradients required for charging and discharging the storage material
- Congruent melting; that is, the ease with which the material melts completely leading to identical compositions for the liquid and solid phases, thus eliminating the problem of segregation arising from a difference in density between the solid and liquid phases
- Small volumetric changes during phase transition, thereby enabling the use of a containment vessel and a heat exchanger having simple cross sections and geometries
- Minimal supercooling during freezing; the melt should crystallize at its thermodynamic freezing point, aided, if necessary, by a nucleating agent added to the material
- Good chemical stability
- No chemical decomposition tendency, thereby assuring a long system lifetime
- Noncorrosiveness to containment vessel materials such as stainless steel, mild steel, tin-plated mild steel, copper, and aluminum and its alloys
- Nontoxic, nonflammable, and nonexplosive properties
- Ready availability in large quantities at a reasonable price
- Reliability; the phase change process should be completely cyclable without undue degradation
- Low vapor pressure at operational temperatures

18.3 CANDIDATE MATERIALS

These may be grouped into the categories shown in Fig. 18.1.

Paraffins

These materials have a waxy consistency at room temperature and are basically straight-chain hydrocarbons with only a small amount of branching near the end of the chain. Paraffins contain one major constituent, alkanes, with several compositions C_nH_{2n+2}. The *n*-alkane content in paraffin waxes is of the order of 75% to 100%. Pure paraffins consist only of alkanes; for example, paraffin octadecane, $C_{18}H_{38}$.

There are two solid-phase allotropes of paraffins: (a) the high-temperature form, which exists at a temperature that is marginally higher than the melting point of the substance and is soft, with an acicular crystal structure, and (b) the low-temperature form, which exists at a temperature below the melting point of the substance and is hard and brittle, with a disc-like crystal structure.

Paraffins are a candidate material primarily because of their ready availability in a large temperature range and their high heat of fusion. They also freeze without supercooling but, except for technical-grade paraffins, are probably too expensive for use as phase-change materials. They have low thermal conductivity, show large volume change upon change of phase, and are flammable.

Nonparaffin Organics

The prime examples in this class are fatty acids, the general composition being $CH_3(CH_2)_{2n}COOH$. They have (a) heats of fusion comparable to those of paraffins, (b) reproducible melting and freezing behavior, and (c) minimal supercooling tendencies upon freezing. They are, however, costly (150% of the cost of paraffins). In addition, they have low thermal conductivity, have varying levels of toxicity, and are unstable at elevated temperatures.

Figure 18.1 Taxonomy of phase-change heat storage materials.

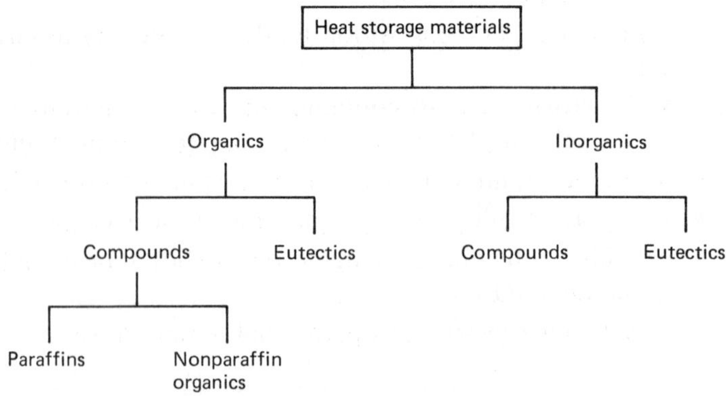

Eutectics of Organic and Inorganic Compounds

Eutectics have the attraction of fixed melting and freezing points. Their potential as phase-change storage materials has only recently been appreciated, thus only limited data on their properties are available.

Inorganic Compounds

One group of these is salt hydrates, with general composition $M \cdot nH_2O$, where M is an inorganic compound. Their main attribute is their high volumetric latent storage density; their major drawbacks are (a) that most of them melt incongruently,* leading to a phenomenon called decomposition (this is the segregation and precipitation, at the bottom of the containment vessel, of the solid phase following melting of the hydrate), and (b) their poor nucleating properties, resulting in supercooling of the liquid salt hydrate prior to freezing.

Some properties of these candidate materials are given in Table 18.1.

18.4 OPERATIONAL SYSTEM CONSIDERATIONS

Corrosion

Only limited corrosion data is available [3], but the general conclusions are these:

- Stainless steel is compatible with most PCMs.

Figure 18.2 Phase diagrams for calcium chloride hexahydrate: (a) incongruent melting; (b) congruent melting [1]. (reprinted with permission from Pergamon Press plc.)

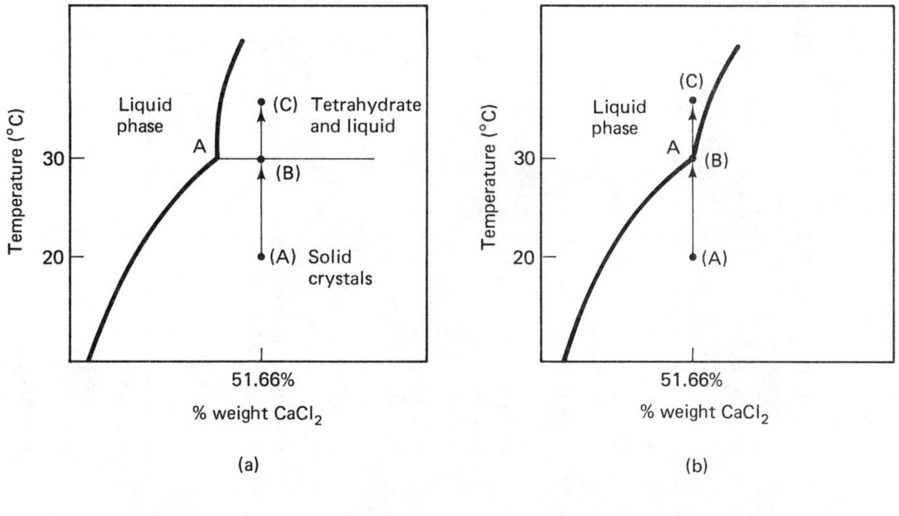

*There have been reports of some success with congruent melting; for example, this is accomplished in $CaCl_2 \cdot 4H_2O$, calcium chloride hexahydrate, by adding small quantities of strontium chloride hexahydrate; see Fig. 18.2.[1].

TABLE 18.1 SOME PROPERTIES OF CANDIDATE HEAT STORAGE MATERIALS

Material	Freezing/Melting Range (°C)	Heat of Fusion (kJ kg^{-1})	Density (kg m^{-3})	Specific Heat (kJ kg^{-1} K^{-1})	Thermal Conductivity (W m^{-1} K^{-1})
Paraffins					
Octadecane	28	244	774[a]	2.16[b]	0.150 (solid phase)
Sunoco P116 paraffin wax	47	209	786	2.89 (solid phase)	0.500 (solid phase)
Fatty Acids					
Lauric acid	42–44	178	870[c]	1.60	0.147[c]
Stearic acid	70	203	941[d]	2.35[e]	0.172[a]
Salt Hydrates					
$Na_2SO_4 \cdot 10H_2O$	32.4	254	1485 (solid phase)	1.93 (solid phase)	0.544
$Na_2HPO_4 \cdot 12H_2O$	35	281	1520 (solid phase)	1.70 (solid phase)	0.514[f]
Eutectic Compounds					
$Mg(NO_3)_2 \cdot NH_4Br$	50	130	1633 (solid phase)	2.01 (solid phase)	0.556 (solid phase)
Urea$\cdot NH_4Br$	76	151	1551 (solid phase)	1.59 (solid phase)	0.669 (solid phase)

[a] At 70°C.
[b] At 100°C.
[c] At 50°C.
[d] At 40°C.
[e] At 125°C.
[f] At 32°C.

- Mild steel is fairly resistant to most PCMs excepting $Zn(NO_3)_2 \cdot 6H_2O$ (an inorganic salt hydrate) and the $Mg(NO_3)_2$—$MgCl_2$ eutectic.
- Copper is also fairly resistant to all PCMs excepting $Na_2S_2O_3 \cdot 5H_2O$ (sodium thiosulfate 5-hydrate).
- Aluminum and its alloys generally show poor resistance to salt hydrates except for $Na_2S_2O_3 \cdot 5H_2O$.
- Most polymeric materials are generally corrosion-resistant to a large number of inorganic salt hydrates and eutectic compounds.

Thermal Cycling

One of the most severe conditions that PCMs must endure during service is thermal cycling, that is, repeated melting and freezing. For example, where only one-day heat storage is required, the PCM is subjected to over 7000 melting/freezing cycles during a system life of 20 years. The effect that thermal cycling has on the characteristics of PCMs is thus a major consideration in PCM material selection. Ideally, laboratory tests consisting of thousands of cycles are needed in order to ascertain the long-term stability of these materials. The measurements should include the variation of (a) temperature within the material with time, (b) the heat transfer rate during charging and discharging of the storage substance with time, and (c) the calculated stored thermal energy with the number of thermal cycles.

Very few detailed studies of these factors have been reported. A few reports exist of limited or short-term thermal cycling of some paraffins and fatty acids [4] and of two hydrates, $Na_2HPO_4 \cdot 12H_2O$ and $Na_2SO_4 \cdot 10H_2O$ [5]. The paraffins and lauric acid reportedly showed no degradation over 120 cycles [4], whereas subjecting $Na_2HPO_4 \cdot 12H_2O$ to 90 thermal cycles led to congruent melting of the material when the formation of the heptahydrate was prevented by use of a suitable nucleation agent [5]. The temperature profiles of Glauber's salt, $Na_2SO_4 \cdot 10H_2O$, plus 3 wt % of borax with and without a thickener during thermal cycling are given in Fig. 18.3.

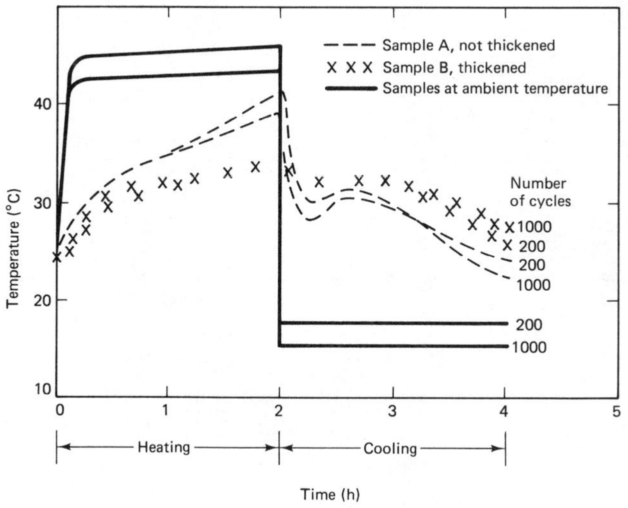

Figure 18.3 Temperature profiles measured during thermal cycling tests with two samples of Glauber's salt plus 3 wt % borax. Sample B contains 8% thickener as suspension medium [2]. (reprinted with permission from Pergamon Press plc.)

Recently, Marks [6] reported on the effect of thermal cycling on the heat stored in Glauber's salt, with and without additions of borax and attapulgite clay. The results reveal a steep decline in the thermal capacity of the pure salt over 40 thermal cycles, from 238 to 63 kJ kg^{-1}, but a slower drop in the thermal capacity of the thickened salt, from 238 to 125 kJ kg^{-1} in 140 thermal cycles.

REFERENCES

1. B. Carlsson, H. Styme, and G. Wettermark, "An incongruent heat of fusion system—$CaCl_2 \cdot 6H_2O$—made congruent through modification of the chemical composition of the system," *Solar Energy 23,* p. 347, 1979.
2. A. Abhat, (Low-temperature latent heat storage energy; heat storage materials) *Solar Energy,* vol. 30, p. 328, 1983.
3. N. Yoneda and S. Takanashi, "Eutectic mixtures for solar heat storage," *Solar Energy 21,* pp. 61–63, 1978; D. Heine and A. Abhat, "Investigation of physical and chemical properties of phase change materials for space heating/cooling applications," in F. de Winter and M. Cox (eds.), *SUN: Mankind's Future Energy Source,* vol. 1. Elmsford, N.Y.: Pergamon Press, 1978, pp. 500–506; J. Schroder "R&D systems for thermal energy storage in the temperature range from −25°C to 150°C," *Proc. Seminar New Ways to Save Energy* (Dordrecht, West Germany: Reidel Publishing Co., 1980, pp. 495–504); D. Heine, "The chemical compatibility of construction materials with latent heat storage materials," *Proc. Int. Conf. on Energy Storage,* Brighton, England, 1981, pp. 185–192.
4. A. Abhat, et al., "Development of a nodular heat exchanger with an integrated latent heat store" [in German], Report No. BMFT FB-T 81-050. Bonn: German Ministry for Science and Technology, 1981.
5. M. Telkes, "Thermal storage for solar heating and cooling," *Proc. Workshop on Solar Energy Storage Subsystems for the Heating and Cooling of Buildings,* Charlottesville, Va., 1975, pp. 17–23.
6. S. Marks, "An investigation of Glauber's salt with respect to thermal cycling," *Solar Energy 25,* pp. 255–258, 1980.

COATINGS ON COMPONENTS IN A LIQUID-METAL-COOLED NUCLEAR REACTOR

CHAPTER 19

19.1 INTRODUCTION

Liquid sodium at elevated temperatures, as in a nuclear reactor (Fig. 19.1), is a strongly reducing medium, tending to remove oxide films from most metal surfaces. This results in a number of undesirable features, such as increased galling, high friction, and self-welding of metal surfaces. Arguably, the most acceptable way of controlling this tendency is to use appropriate surface coatings.

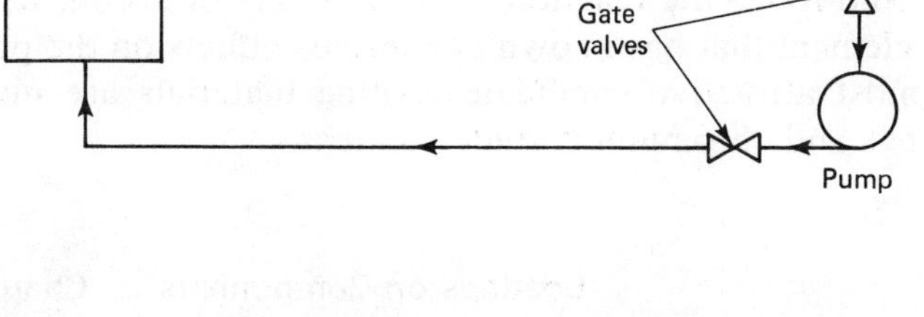

Figure 19.1 Liquid-metal cooling system for a nuclear reactor.

These coatings, however, have to operate under very severe conditions. Not only do they have to survive the liquid metal environment, but they must also be capable of withstanding the neutron irradiation flux present in all liquid-metal-cooled nuclear reactors. Thus, these coatings may be exposed to liquid sodium corrosion at temperatures up to 650°C, thermal cycling between 200°C and 650°C, and neutron irradiation dosages that may exceed 6×10^{26} neutrons m^{-2}.

19.2 DESIRABLE MATERIAL PROPERTIES

Let us examine the key requirements for a suitable coating material.

1. Corrosion resistance. The coating must have sufficient resistance to the operational corrosive medium so that, over the life of the component on which it is placed, the component's intended functions are not compromised. In particular, the performance of the coating must not be undermined by selective leaching of critical constituents after long-term exposure to sodium.

2. Resistance to wear and damage. The coating is usually required to have adequate wear resistance such that coating thickness loss (due to corrosion and wear) over the life of the component be less than 50% of original coating thickness.

3. Low friction.

4. Good ability to prevent self-welding. Self-welding may be defined as the potential for material to bond to itself at high temperatures. Sodium encourages the occurrence of this phenomenon through its ability to wrench oxide films from metal surfaces. Coatings must be capable of resisting self-welding tendencies by, say, reducing diffusing rates (as in refractory metal coatings) or promoting film formations that reduce the potential for metal-to-metal contact.

5. Excellent spalling resistance. A coating must have a low tendency to spall. Spalling is the cracking of a coating due to stresses developed in it during temperature changes.

6. Resistance to the various effects of exposure to fast neutrons.

19.3 CANDIDATE MATERIALS

The coatings must contain no elements that can change, in the neutron environment, to elements or highly radioactive isotopes that cause problems. The severest of these problems is the increase in cost of handling activated components when they are to be removed from service in the reactor for either replacement or maintenance. Thus the coatings must not contain cobalt, which transforms to the highly radioactive cobalt 60 through the reaction $^{59}Co(n, \gamma)^{60}Co$, or boron, which transforms to helium, an element that has known deleterious effects on the properties of coatings. The most attractive candidate coating materials are nickel aluminide diffusion coatings and chromium carbide coatings.

Nickel Aluminide Diffusion Coatings

Nickel aluminide diffusion coatings have provided some of the best tribological performance to date. They are made by the pack cementation process [1]. In this process, also known as aluminizing, nickel-base alloy parts, such as alloy 718, are placed in a retort with a powdered compound containing aluminum. The retort is heated in the presence of a halide gas capable of reacting with and permitting the movement of aluminum as a metallic halide to the part surface, where the halide breaks down to deposit aluminum. The equilibrium reaction for this process is

$$2AlCl_3 \rightleftharpoons 2Al + 3Cl_2$$

The diffusion of aluminum into the substrate permits it to combine with the nickel in the alloy, producing the intermetallic compound NiAl on the surface.

The principal limitation on the use of aluminide diffusion coatings in liquid-metal systems is the high processing temperature (1000°C or higher) that is necessary for the formation. Such temperatures encourage thermal distortion of substrates, affecting tolerances in components.

Some of the benefits bestowed on Inconel alloy 718 by aluminizing are shown in Table 19.1.

Chromium Carbide Coatings

Chromium carbide coatings may be applied by either of two processes. The *detonation gun process* [2] involves mixing oxygen gas and acetylene gas with the powdered coating material and then detonating the mixture in the chamber of a gun. The ensuing explosion provides sufficient heat to the powder to nearly melt it and throws the hot particles out of the barrel of the gun at supersonic velocities.

TABLE 19.1 EFFECT OF ALUMINIZING ON THE CORROSION, WEAR, AND FRICTION BEHAVIOR OF INCONEL ALLOY 718 IN SODIUM

Material Property	Inconel Alloy 718	Aluminized Alloy 718
Corrosion rate after 2000 h in sodium at 625°C (μm y^{-1})	6	0.75
Maximum wear depth for 70 MPa rubbing for 20 min at 520°C, self-mated (μm)	53	Practically zero
Friction coefficient at 625°C		
Against A286 steel	0.60–0.85	0.15–0.60
Against Stellite 6	0.30–0.40	0.20–0.35

Sources: R. N. Johnson, et al., *Proc. Int. Conf. Liquid Metal Technology in Energy Production;* USERDA Conf. Publ. No. 760504-PL. Washington, D.C.: U.S. Energy Research and Development Administration, 1976, p. 122; R. N. Johnson and D. G. Farwick, *Proc. 2d Int. Conf. Liquid Metal Technology in Energy Production,* USDOE Conf. Publ. No. 800401-P1. Washington, D.C.: U.S. Department of Energy, 1980, pp. 3-1 to 3-11.

The impingement of these high-temperature, high-kinetic-energy particles on the substrate produces a dense, strong, well-bonded coating. The process is repeated at about 250 to 500 shots per minute to build up the coating to the desired thickness (about 0.1 mm for most applications).

The limitations associated with detonation gun coatings include (a) critical dependence on substrate preparation, (b) rigid process and quality controls, and (c) the need to ensure that contact stresses are below the yield strength of the substrate material.

The *electrospark discharge* (*ESD*) *process* for applying chromium carbide coatings is essentially a microwelding process that uses short-duration, high-current electrical pulses to deposit an electrode material on a metal substrate. One attribute of this process is that the coatings are fused to a metal surface with such a low total heat input that the bulk of the substrate material remains at or near ambient temperature. This minimizes thermal distortions or changes in the microstructure of the substrate material.

Because the pulses are so brief, exceptionally rapid solidification occurs, resulting in an extremely fine-grained coating structure that approximates (and with some materials actually is) an amorphous material or a metallic glass. It has been suggested that some of the excellent wear and corrosion properties of the ESD coatings derive from this fine structure.

The limitations of the ESD process include (a) a maximum coating thickness currently limited to about 0.3 mm, (b) stress relief cracking in thicker coatings, (c) minimum as-deposited roughness, currently 2.5 μm, (d) the need for both substrate and coating materials to be electrically conductive, and (e) a maximum effective coating rate currently limited to about 0.09 $m^2\ h^{-1}$ per applicator head.

REFERENCES

1. R. L. Watchell, in B. N. Chapman and J. E. Anderson (eds.), *Science and Technology of Surface Coatings,* Orlando, Fla.: Academic Press, 1974, p. 105; C. Duret and R. Pichoir, in E. Lang (ed.), *Coatings for High Temperature Applications.* New York: Applied Science Publishers, 1983, p. 33.
2. E. P. Cashon, *Tribology Int. 8* (3), pp. 111–115, 1975.

THE RAILGUN: AN ELECTROMAGNETIC LAUNCHER

CHAPTER 20

20.1 INTRODUCTION

The concept of electromagnetic launching or propulsion uses an acceleration force (the Lorentz force) generated by the interaction between a high current pulse (of the order of kiloamps to megaamps) with its associated strong magnetic field flowing in parallel conductors of the launcher, and a movable current-carrying armature (which may be a solid conductor or a plasma) attached to the projectile (typically 10 g in mass). The latter, if free to move, accelerates at very high speeds (typically 2 km s^{-1}). The railgun is an example of an electromagnetic launcher. The barrel of the gun is made up of two narrow conducting rails (each typically 3 m long) extending the full length of the barrel and separated by insulating rails to form the bore (typically 15 mm). A schematic representation of the railgun and its power source is shown in Fig. 20.1, while a typical crossection is shown in Fig. 20.2. Other cross sections under study are shown in Fig. 20.3.

One of the major engineering barriers to successful development of this system is the fabrication of barrels and bore components that will survive a large number of shots, under rapid-fire burst service conditions, without compromising performance.

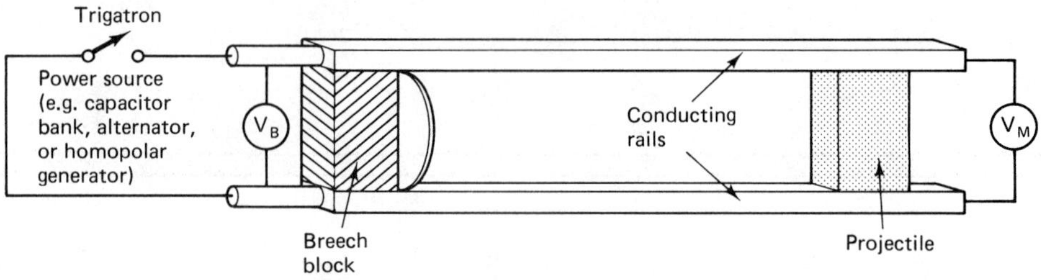

Figure 20.1 A simple schematic representation of a railgun and its associated power source. (V_B, breach voltage; V_M, muzzle voltage)

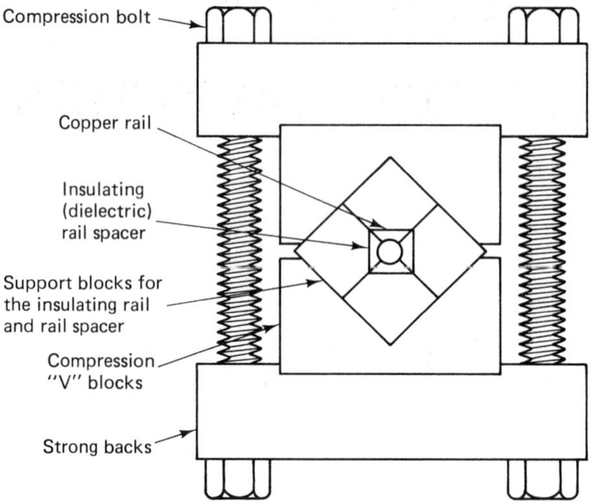

Figure 20.2 The cross section of a railgun.

Figure 20.3 Other cross sections of railguns under study.

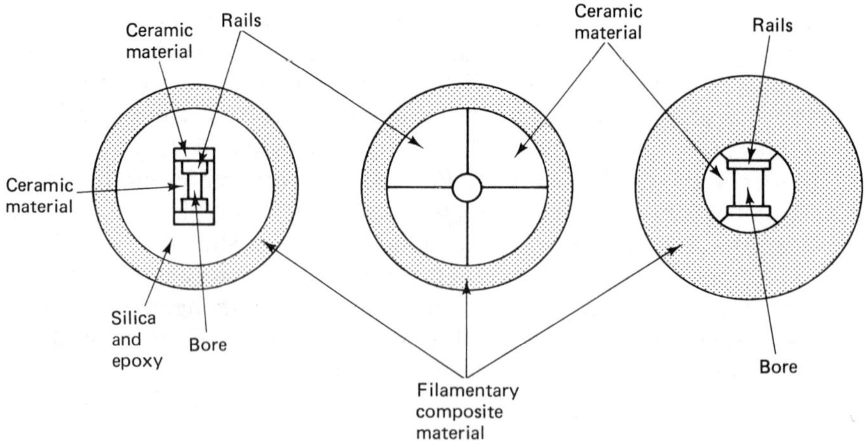

20.2 INSULATING RAILS OR BORE

Ablation melting and erosion of materials are the most important material selection considerations here. Material loss is a result of thermal energy from the plasma armature and, to a certain extent, the mechanical interactions of the projectile with the bore. Ablation erosion undermines railgun performance by effects such as adding mass to the plasma, causing secondary arc restrike, and changing bore tolerances. To reduce material loss (especially in rapid-fire launches), the material at the plasma interface must have a high melting point, high hardness, and good abrasion resistance. The resistance of a material to arc melting and erosion may be obtained from the approximate transient heat flux to induce melting in a pulse time from the expression [1]

$$\text{AMER} = \frac{\Delta T}{2}\left(\frac{\pi \rho C_p k}{t}\right)^{0.5}$$

where AMER is the arc melting and erosion resistance or the heat flux to initiate melt, t is the pulse time, ΔT is the difference between the melting temperature of the material T_M and 298 K, and ρ, C_p and k are the density, specific heat, and thermal conductivity of the material, respectively.

The thermal-structural performance of insulator spacer materials will depend, to a large extent, on their resistance to thermal shock effects, which may produce surface cracking as the insulators are thermally pulsed by the plasma arc. For the given conditions, the thermal shock resistance of the materials is reflected in the value of R, which is defined as the ratio of the flexural bending strength σ_f or modulus of rupture (MOR) of the material to the operational thermal stress. That is,

$$R' = \frac{\sigma_F^*(\rho C_p k)^{0.5}}{E\alpha/(1 - \nu)}$$

where α, E, and ν are the coefficient of linear expansion, modulus of elasticity, and Poisson's ratio, respectively, of the insulating material.

It is extremely important that the insulator spacer material maintains its integrity under dynamic structural loading conditions. Thus (a) the compressive strength of the material must be greater than both the maximum plasma pressure (of the order of 200 to 400 MPa) and the bore preload stress (of the order of 150 to 600 MPa), (b) the tensile or bending strength depends on the bore pressure, support stiffness, and insulator space geometry, and (c) dynamic stress wave interaction effects must be considered. A reasonable measure of the performance of an insulator material, under these conditions, is its fracture toughness, K_{IC}.

There are several classes of potential insulator materials. Table 20.1 presents the values of the relevant parameters for one example from each of these classes.

The values of the various indices of performances of these candidate materials, given in Table 20.2, indicate that the promising material, sintered SiC, belongs to the advanced ceramics category. Recent railgun environment screening tests, performed to correlate actual with predicted performance of candidate ma-

*Or MOR.

TABLE 20.1 ROOM-TEMPERATURE PROPERTIES OF RAILGUN BORE CANDIDATE MATERIALS

Material	σ_f (MPa)	E (GPa)	K_{IC} (MPa m^{-1})	ρ (kg m^{-3})	α (10^{-6} K^{-1})	k (W m^{-1} K^{-1})	C_p (kJ kg^{-1} K^{-1})	T_M (K)	ν
Lexan 141 polycarbonate (organic polymer)	93	2.4	—	1200	80.0	0.30	1.20	663	0.37
Glass fiber reinforced polymer (60% volume fraction)	470	15.9	—	1950	10.0	0.68	1.57	463	0.35
MM-1100 inorganic glass composite	83	73.0	—	2800	9.4	5.90	0.46	898	0.33
99.9% Al_2O_3 (conventional ceramic)	552	386.0	2.5	3960	6.4	39.00	0.88	2273	0.26
Sintered SiC (advanced ceramic)	462	406.0	4.6	3100	4.02	126.00	0.67	2773	0.27
Al_2O_3-SiC (reinforced ceramic)	641	393.0	8.0	3740	6.00	33.00	0.83	2313	0.30

TABLE 20.2 INDICES OF PERFORMANCE OF RAILGUN BORE CANDIDATE MATERIALS

Material	AMER[a] (MW m^{-2})	10^{-6} R' (W m^{-2} $s^{0.5}$)	K_{IC} (MPa$\sqrt{m}$)
Lexan 141 polycarbonate	21.3	0.20	1.8
Glass fiber reinforced polymer	21.10	2.78	50.0
MM-1100 inorganic glass composite	146.6	0.22	20.0
99.9% Al_2O_3	2040.6	1.92	2.5
Sintered SiC	3548.6	3.34	4.6
Al_2O_3-SiC	1807.5	1.93	8.0

[a]Pulse time t = 100 μs.

terials, at the U.S. Army Ballistics Laboratory [2,3] on some advanced ceramics (sintered Si_3N_4, hot-pressed Si_3N_4, partially stabilized zirconia, PSZ, and transformation-toughened alumina, TTA) have yielded mixed results. Although the thermal shock resistance index of performance indicates that these advanced ceramics should have excellent resistance to thermal shock (R′, in W m^{-2} $s^{0.5}$, is 2.12×10^6 for sintered Si_3N_4, 4.23×10^6 for hot-pressed Si_3N_4, 1.00×10^6 for PSZ, and 5.10×10^6 for TTA), it was found that they suffered varying degrees of thermal microcracking [3]. It was also reported, though, that all these materials showed insignificant weight losses and showed excellent resistance to gross surface melting and ablation [3]. Work in progress at various research laboratories is directed toward a full characterization of these materials for this application.

20.3 CONDUCTING RAILS

To maximize heat conduction in the gun, a conductive rail material with a high thermal conductivity is required. To minimize arc damage (typically, the rails have to withstand the passage of 1 MA arcs traveling at 50 km s^{-1} in a 5 m rail launcher), melting, and material loss through evaporation, a material with a high melting temperature, high density, and high specific heat is required. A structure that meets these requirements most practically would have a core made of high thermal conductivity material and a refractory metal surface.

The most refractory metals are rhenium, osmium, and tungsten; the materials with the highest thermal conductivities are silver and copper (Table 20.3). From considerations of cost, both osmium and silver may be dismissed.

TABLE 20.3 PROPERTIES OF CANDIDATE CONDUCTING RAIL MATERIALS

Material	k (W m^{-1} K^{-1})	T_M (K)	ρ (kg m^{-3})	C_p (kJ kg^{-1} K^{-1})
Copper	397	1356	8,930	0.385
Osmium	90	3300	22,480	5.825
Rhenium	76	3450	20,500	0.146
Silver	419	1234	10,500	0.235
Tungsten	167	3650	19,320	0.142

Although electrical considerations would seem to favor copper or its alloys as a core material, tests [4] have shown that these materials cannot withstand the arc conditions within the railgun core without severe damage resulting. The solution to the problem has been to use chemically vapor-deposited refractory metal coatings on copper substrates. The most promising of these coatings currently being investigated are based on tungsten and tungsten/rhenium. Typically, the deposition temperature for tungsten from tungsten fluoride is between 450°C and 700°C, while for the deposition of rhenium from rhenium fluoride a temperature between about 1000°C and 1100°C is required. For this reason coatings based on tungsten are preferred.

20.4 CONCLUSION

The emergence of a significant effort in the United States to develop electromagnetic launch technology is largely the result of focused interest by the Department of Defense since about 1975. The Defense Advanced Research Projects Agency (DARPA) is the major sponsor of university and industrial efforts, with such funds being funneled through various U.S. defense organizations, such as the U.S. Army Ballistics Research Laboratory (Aberdeen Proving Ground, Md.), the Defense Nuclear Agency (Washington, D.C.), the Strategic Defense Initiative Office (Washington, D.C.), and the U.S. Air Force Armament Division (Eglin AFB, Fla.).

The major recipient of this funding, as far as universities are concerned, is currently the Center for Electromechanics at the University of Texas at Austin. A number of private-sector companies are actively involved, too. These include SPARTA, Inc., Components and Materials Technology Division (Del Mar, Calif.); General Electric's Electromagnetic Launcher Operation (Ballston, N.Y.); GA Technologies (San Diego, Calif.); and General Dynamics, Valley Systems Division (Pomona, Calif). Although there are a number of railgun test facilities now in existence,* most of the results of tests to evaluate the performance of various materials under actual railgun conditions are either not available because of the classified nature of the work or incomplete because testing is still being conducted.

The indications, though, are that copper coated with tungsten, rhenium, or an alloy of both is the most promising material for the conducting rails and that advanced ceramics will be used for the base insulation.

REFERENCES

1. W. M. Rohsenhow and J. P. Harnett, *Handbook of Heat Transfer,* New York: McGraw-Hill, 1973.

*Examples are CHEMCATE, a high-energy railgun user test facility at Maxwell Labs, Inc., San Diego, Calif.; a Distributed Energy Source railgun laboratory at LTV Aerospace and Defense Co., Dallas, Texas; the railgun test laboratory at the Center for Electromechanics at the University of Texas at Austin; and the United States Air Force Armament Laboratory, Okaloosa Island, near Eglin Air Force Base, Fla.

2. K. A. Jamison, H. S. Burden, J. D. Powell, and M. Marquez-Reines, "Non-steady phenomena in railgun plasma armatures," *IEEE Trans. Magnetics MAG-22* (6), pp. 1546–1551, 1986.
3. S. N. Rosenwasser and R. D. Stevenson, "Selection evaluation of insulator materials for high-performance railgun bore," *IEEE Trans. Magnetics MAG-22* (6), pp. 1722–1729, 1986.
4. A. J. Bedford, "Rail damage in a small-caliber railgun" and "Rail damage and armature parameters for differential railgun materials," *IEEE Trans. Magnetics MAG-20* (2), pp. 348–351, 352–355, 1984.

PIPING FOR COAL SLURRIES IN A COAL LIQUEFACTION PLANT

CHAPTER 21

21.1 INTRODUCTION

One of the constraints to the development of coal liquefaction technologies (Fig. 21.1) is the erosive wear of such components as pumps, valves, and piping used in the processing and conveyance of flowing coal slurries. The problem is notoriously acute in valves used to feed the process steam from the high-pressure side of the valve to the low-pressure side. In these valves, which are designed to maintain the pressure drop, the slurry stream velocity is quite high. This speed is further enhanced by the expansion of the dissolved gases in the slurry stream. The consequence of this is erosive wear in various valve components. Nominal wear life as short as 14 days has been reported for letdown valves in some liquefaction plants [1]. The problem is no less severe for the piping.

The selection of piping material in this application is thus an example of choice of material for resisting erosion by solid particles.

21.2 THE SOLID-PARTICLE EROSIVE PROCESS

Figure 21.2 illustrates a solid particle impinging on a surface. Experimentally, it has been found that erosive wear rate depends on the impingement angle α, the

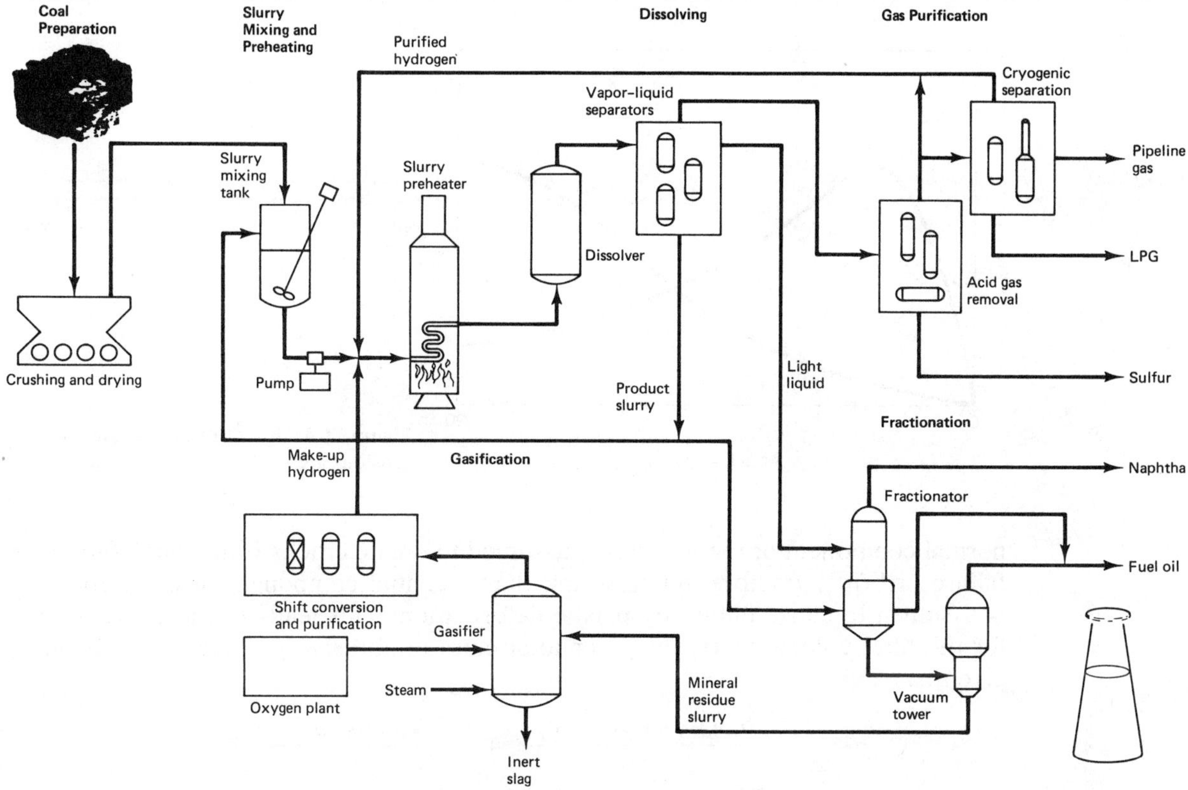

Figure 21.1 Process flow diagram for the solvent-refined coal (SRC II) coal liquefaction process. (Courtesy of the U.S. Department of Energy)

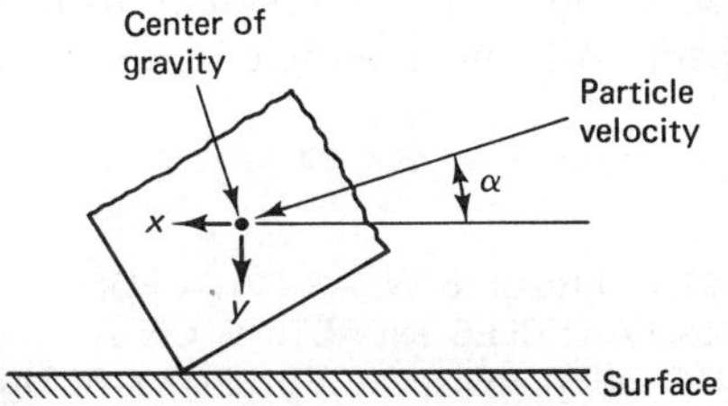

Figure 21.2 An abrasive grain striking a surface.

particle velocity V_o, and the size and density of the particles, as well as the properties of the target material. It has been found that the maximum erosive wear of ductile materials occurs at α equal to about 20°. The maximum erosive wear of brittle materials occurs near $\alpha = 90°$, and this is related to the efficiency of crack formation around the impact crater (Fig. 21.3).

Actually, most materials exhibit behavior that is neither completely ductile nor completely brittle, and quantitative procedures have been developed to evaluate the relative amounts of brittle and ductile erosive wear that occur for a given material-particle system.

A simple postulate for solid particle erosive wear has been derived and is based on experimental observations [2]. The damage done by the tangential and normal components of the stream velocity are assumed to be uncoupled. The

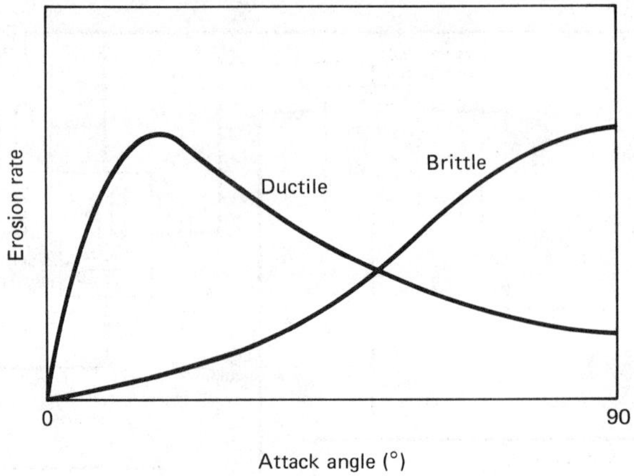

Figure 21.3 Effect of attack angle on erosion rate.

normal component of the velocity is assumed to create cracks from which fatigue failure or brittle fracture initiates, and the tangential component of the velocity is assumed to cause failure by plastic deformation. This leads to the expression for the total volume worn away per total mass of abrasive particles (v/m) being given by

$$\frac{v}{m} = \frac{0.5(V_o \sin \alpha - V_L)^2}{U_f} + \frac{0.5(V_o^2 \cos^2 \alpha - V_p^2)}{U_s}$$

where U_f is the amount of energy needed to erode a unit volume of material by the fatigue mechanism, U_s is the specific energy needed in removing a unit volume of the surface material by shear deformation, V_L is the limiting velocity at which no fatigue failure takes place, and V_P is the residual horizontal component of the particle velocity after impact. All four coefficients may be obtained for a set of experimental results (Table 21.1).

TABLE 21.1 EROSIVE WEAR DATA FOR ALUMINA PARTICLES INPACTING ON A GRAPHITE FIBER REINFORCED EPOXY RESIN COMPOSITE MATERIAL

Impingement Angle, α (°)	Erosion Weight Loss (g)[a]
10	50×10^{-4}
20	75×10^{-4}
25	170×10^{-4}
40	130×10^{-4}

[a]Per gram of alumina impacted at a velocity of 91.4 m s^{-1}.

Source: N. P. Suh and A. P. L. Turner, *Elements of the Mechanical Behavior of Solids*. Washington, D.C.: Scripta Book Co., 1975, p. 577.

The dependence of erosive wear rate, $\dot{W}$ on particle velocity V has been observed for many materials and found to be of the form

$$\dot{W} = f(V^{\beta})$$

In most cases $\beta = 2$ or 3 (Table 21.2), but detailed studies have shown some dependence on the attack angle (Table 21.3), temperature of erosion (Table 21.4),

TABLE 21.2 INFLUENCE OF IMPACT VELOCITY OF EROSIVE PARTICLES ON THE EROSION OF MATERIALS

Impact velocity ($m\ s^{-1}$)	Volumetric Erosion (cm^3kg^{-1}) of: Steel and nickel-based alloy	Aluminum alloy	Glass-nylon composite
50	0.02	0.06	0.15
150	0.12	0.60	1.50
300	1.00	3.00	10.0

Source: J. E. Goodwin, W. Sage, and G. P. Tilly, *Proc. Inst. Mech. Eng. 184,* p. 288, 1969.

TABLE 21.3 INFLUENCE OF ATTACK ANGLE AND VELOCITY OF EROSIVE PARTICLES ON THE EROSION OF COPPER

Velocity ($m\ s^{-1}$)	Mass Loss Per Mass of Abrasive ($mg\ g^{-1}$) at: 20° attack angle	90° attack angle
20	0.020	0.009
50	0.020	0.100
100	1.000	0.750

Source: L. K. Ives, and A. W. Ruff. *Proc. ASTM Symp. Erosion Prevention Useful Appl.* STP 664, 1978.

TABLE 21.4 INFLUENCE OF VELOCITY OF EROSIVE PARTICLES AND TEMPERATURE OF TYPE 310 STAINLESS STEEL ON THE EROSION OF THE STEEL

Velocity ($m\ s^{-1}$)	Mass Loss Per Mass of Abrasive Particle ($mg\ g^{-1}$) at: 25°C	975°C
10	0.002	0.015
30	0.020	0.100
60	0.150	0.500

Source: L. K. Ives, *U. Engr. Mat. Trans. ASME 99,* p. 130, 1977.

TABLE 21.5 INFUENCE OF EROSIVE PARTICLE SIZE ON THE EROSION OF 11% CHROMIUM STEEL (VELOCITY OF ABRASIVE PARTICLES = 244 m s^{-1})

Average Abrasive Particle Diameter (μm)	Erosion Loss (mg g^{-1})
20	1.0
60	3.0
100	4.5
200	4.5

Source: J. E. Goodwin, W. Sage, and G. P. Tilly, *Proc. Inst. Mech. Eng. 1984,* p. 290, 1969.

and particle size (Table 21.5). Other possible influential parameters, such as particle concentration in the erosive stream, particle hardness, particle strength, and frangibility, have been examined in only a few instances [3].

21.3 THEORIES OF EROSIVE WEAR BY SOLID PARTICLES

Ductile Materials

The earliest consideration of ductile processes of erosive wear was done by Finnie [4], who considered a macromachining mechanism as a model. He treated the problem by assigning a plastic response character to the material through a flow stress, σ_f. The trajectory of a particle cutting and removing material was calculated, and the eroded volume v was determined as

$$v = \frac{mV_0^2}{\sigma_f C t} g(\alpha)$$

where m is the mass of the impacting particles, V_o is its velocity, C is the ratio of the vertical force to the horizontal force on the particle, t is the depth of cut, and $g(\alpha)$ is a function describing the effect of the attack angle α. The model predicts that the wear rate is proportional to the square of the velocity of the solid particles and leads to a complex angular dependence that predicts a maximum erosion rate at 20°. The wear rate is also postulated to be inversely dependent on the dynamic flow strength of the target material, which is equated to its hardness. The model has been very successful in predicting the maximum erosive wear rates observed in metals at small angles and in explaining the inverse hardness dependence of erosive wear rate. In refinements of the original analysis [5] the interaction force between the particle and the target surface was modified. This has led to a velocity exponent of about 2.5, which is close to that obtained from experiments.

Brittle Materials

The Sheldon-Finnie model [6] assumes that erosive wear occurs as a result of Hertzian contact stresses during impact that cause cracks to grow from preexisting flaws on the target surface. The loads at which crack propagation occurs is related to the distribution of surface flaws through the Weibull statistics. The erosive wear rate $\dot{W}$ (weight lost from material per weight of impacting particles) is dependent on particle size r and V_o thus:

$$\dot{W} = k_1 r^a V_o^b \tag{21.1}$$

where a and b are related to the Weibull constant; thus

$$a = 3\,\frac{m - 0.67}{m - 2} \qquad \text{for round particles}$$

$$a = 3.6\,\frac{m - 0.67}{m - 2} \qquad \text{for angular particles}$$

$$b = 2.4\,\frac{m - 0.67}{m - 2} \qquad \text{for all types of particles}$$

The constant k_1 in equation (21.1) is given by

$$k_1 = E^{0.8(m+1)/(m-2)} \rho^{1.2(m-0.67)/(m-2)} \sigma_o^{-\frac{2}{m-2}}$$

where E is the modulus of elasticity of the target material, ρ is the density of the particle material, and σ_o is another Weibull constant.

Although the Sheldon theory is reasonable, its physical basis must be questioned because it assumes Hertzian crack formation, whereas it is now known that lateral crack formation is the main cause of material removal during erosion.

The theory by Evans et al. [7] explains solid-particle erosive wear better because it countenances lateral crack formation.

The theory leads to an expression for the volume lost from the target material per volume of impacting particles S being given by

$$S = C' V_o^{19/6} r^{11/3} \rho^{19/12} K_{Ic}^{4/3} H^{-1/4} \tag{21.2}$$

where C' is a constant, K_{IC} is the fracture toughness of the target material, and H its hardness.

The elastic-plastic theory used by Weiderhorn and Lawn [8] can also be applied to the development of a model of erosive wear. Assuming that the lateral crack size is proportional to the maximum particle penetration, the following expression for the erosive wear rate S was obtained.

$$S = C'' V_o^{22/9} r^{11/13} \rho^{11/9} K_{IC}^{-4/3} H^{1/9} \tag{21.3}$$

where C'' is a constant.

The velocity and particle size exponents of equations (21.1), (21.2), and (21.3), representing the three models, are compared with experimental data in Table 21.6. It is interesting to note from these results that all the theories or models perform reasonably well given the fact that the elastic model uses the Weibull constants and the modulus of elasticity and hardness of the target material to predict target material erosion loss, while the relevant material parameters in the elastic-plastic theories are the fracture toughness and hardness of the target material.

TABLE 21.6 COMPARISON OF VELOCITY AND PARTICLE SIZE EXPONENTS, $\dot{W}$, $S = f(r^a V_o^b)$

	a	b
Elastic theory [eq. (21.1)]	~4.0–4.2	~2.6–3.0
Elastic-plastic theory [eq. (21.2)]	3.7	3.2
Elastic-plastic theory [eq. (21.3)]	3.7	2.4
Experimental results	3–4	2–4

It should be added here that other correlations between erosion rate and the physical properties of the target material have been suggested. These include (a) the "thermal pressure" parameter [9], which is equal to $\gamma K \Delta T$, where γ is the coefficient of linear expansion of the target material, K is its bulk modulus, and ΔT is the difference between the exposure temperature and the melting point of the target material; it was reported [9] that the erosion rate is inversely proportional to this parameter; and (b) the parameter $\rho C_p \Delta T$ [10], where C_p is the specific heat of the target material. A summary of the general factors affecting erosive wear of isotropic materials is presented in Table 21.7.

Recent experimental results for the erosive wear of certain candidate ceramic materials in coal slurries (Table 21.8) seem to indicate that the wear process can be attributed to either a ductile mechanism (hardness-controlled, Fig. 21.4) or an

TABLE 21.7 FACTORS THAT AFFECT THE EROSIVE WEAR OF ISOTROPIC MATERIALS

Factors	Type of Isotropic Material	Effect on Erosive Wear[a]
Hardness of eroded surface	Ductile	W inversely proportional to Vickers hardness
	Brittle	No correlation
Stress on eroded surface	Ductile	Little effect on W
	Brittle	Some effect on W
Size of eroding particles	Ductile	W decreases for particle diameters $\leq$ 100 μm; W is not affected by particle size when diameter $>$ 100 μm
	Brittle	Ductile behavior when particle diameter $>$ 10 μm
Shape of eroding particles	Ductile/brittle	Angular particles produce more erosive wear
Hardness of eroding particles	Ductile/brittle	Harder particles produce more erosive wear
Angle of impingement of eroding particles	Ductile	W_{max} at ~20°
	Brittle	W_{max} at ~90°
Eroding particle velocity, V_o	Ductile	$W \alpha V_o^{2-3}$
	Brittle	$W \alpha V_o^{3-5}$

[a] W is the ratio of the mass of the target material removed to the mass of the eroding particles.

TABLE 21.8 WEAR RATE OF CANDIDATE PIPING MATERIALS BY COAL SLURRIES

Material	H^a (GPa)	K_{IC} (MPa$\sqrt{m}$)	$1/H$ (GPa^{-1})	$K_{IC}^{4/3}H^{1/4}$ (MPa$\sqrt{m}$)$^{4/3}$ (GPa)$^{1/4}$	Mean Erosive Wear Rate[b] (μm h^{-1})
Cemented WC	16.3	10.50	0.06	45.95	36
Cemented WC	17.5	9.50	0.06	41.12	34
Cemented WC	17.5	9.50	0.06	41.12	29
SiC	32.5	3.75	0.03	13.90	10
SiC	26.1	3.90	0.04	13.87	20
SiC	19.0	4.00	0.05	13.25	200
Si_3N_4	14.0	3.75	0.07	11.26	241
Al_2O_3	18.2	3.10	0.05	9.33	152

[a]Vickers hardness, with indenter load of 2.5 kg.

[b]Test temperature = 343°C; slurry jet velocity = 100 m s^{-1}; angle of attack of slurry steam = 50°.

Source: D. K. Shetty, et. al., "Erosive wear of advanced ceramics in coal-slurry streams," *Corrosion 38* (9), pp. 500–509, 1982.

elastic-plastic fracture mechanism (both hardness- and fracture toughness–controlled, Fig. 21.4). The deviations of the experimental data points from the correlation lines are significant in each case. On the whole, though, these materials showed "superior coal-slurry erosion wear resistance" [11] and are therefore prime candidates for the application considered in this chapter.

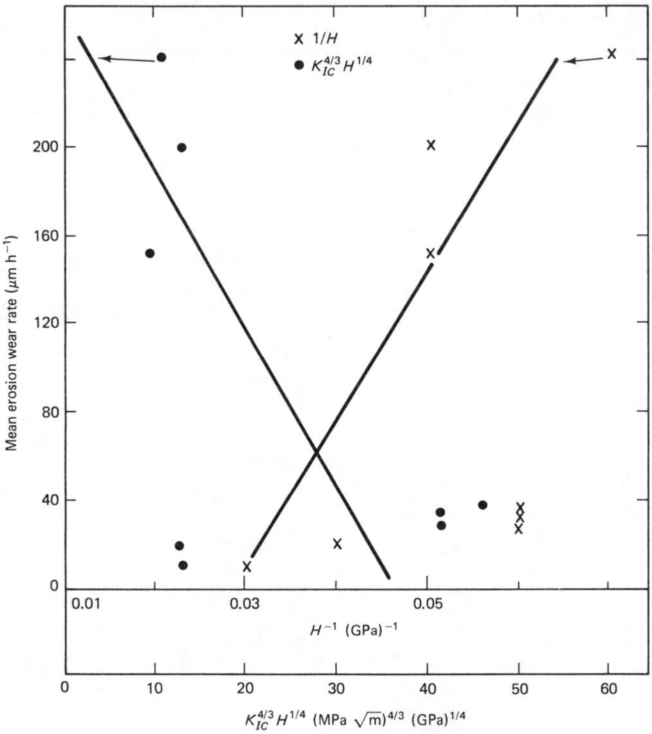

Figure 21.4 Dependence of the erosion rate of selected ceramic materials on two predictive parameters based on hardness H and fracture toughness K_{IC} of these materials. Vickers hardness with an indenter load of 2.5 kg; test temperature = 343°C; slurry velocity = 100 m s^{-1}; angle of attack of slurry = 50° [11].

In this context, ARCO Chemical Company, of Greer, S.C., in 1987 reported the successful choice of a ceramic composite called ARtuff SA 15, comprising of pure aluminum oxide reinforced with silicon-carbide whiskers, in fabricating parts of high-pressure pumps to be used in the pumping of coal slurries.

REFERENCES

1. J. J. Mueller, et al., "Application of advanced materials and fabrication technology to let down valves for coal liquefaction systems," Electrical Power Research Institute final report AF-305, January 1977.
2. J. H. Nielson and A. Gilchrist, "Erosion by a stream of solid particles," *Wear 11,* pp. 111–122, 123–143, 1968.
3. H. Unemois and I. Kleis, *Wear 31,* pp. 359–371, 1975.
4. I. Finnie, *Proc. U.S. Nat. Cong. Appl. Mech.,* p. 527, 1958.
5. I. Finnie and D. H. MacFadden, "On the velocity dependence of the erosion of ductile metals by solid particles at low angles of incidence," *Wear 48,* pp. 181–190, 1978.
6. G. L. Sheldon and I. Finnie, *Trans. ASME J. Eng. Ind. 88,* pp. 393–400, 1966.
7. A. G. Evans, M. E. Gulden, and M. Rosenblatt, "Impact damage in brittle materials in the elastic plastic range," *Proc. Roy. Soc. A361,* pp. 343–365, 1978.
8. S. M. Weiderhorn and B. R. Lawn, *J. Amer. Ceramic Soc. 62,* pp. 66–70, 1979.
9. P. Ascarelli, Army Materials and Mechanics Research Center Report No. TR71-47, 1971.
10. I. M. Hutchings, *Wear 35,* pp. 371–374, 1975.
11. D. K. Shetty, et al., "Erosive wear of advanced ceramics in coal-slurry streams," *Corrosion 38* (9), pp. 500–509, 1982.

ANCHORAGE STEM OF AN ARTIFICIAL HUMAN HIP REPLACEMENT

CHAPTER 22

22.1 INTRODUCTION

One of the most rewarding areas of research for the modern materials scientist or engineer is that of biomaterials,* especially the study of materials used in total joint anthroplasty (repair to or replacement of human joints). Implantable materials for use as bone replacements or as internal fixation devices represent a significant advance in modern orthopedic surgical practice. Such replacement or supplementation becomes necessary either for the relief of pain and incapacity due to degeneration of the musculoskeletal system, as in osteoarthrosis, or for reconstructive surgery, following an automobile accident, for example.

Applications of these components include (a) prostheses, which are total synovial joint replacements made necessary when the normal weight-bearing joint fails; (b) orthoses such as intermedullary rods, intertrochanteric nails, plates, pins, and screws used to provide internal fixation during fracture healing; and (c) small localized replacements for missing bone segments. Total joint replacement components predominate; some 300,000 total joint anthroplasties are now performed

*A biomaterial has been defined as "a systemically, pharmacologically inert substance designed for implantation within or incorporation with living tissue" (Clemson Advisory Board for Biomaterials at the 6th Annual Biomaterials Symposium, Clemson University, Clemson, S.C., April 1974).

annually throughout the world, with hip replacement devices accounting for 90% of this work. One analysis of data from the National Center for Health Statistics in the United States [1] showed that 159,000 hip anthroplasty procedures were conducted in 1983, having increased from 130,000 procedures in 1979.

Joints operate to permit relative movement of bones under the action of applied loads, with the minimum of energy loss through friction and practically no wear of the articulating surfaces. All of this is made possible through a complex, composite skeletal structure consisting of two materials:

- Bone, a hard, elastic matrix of calcified collagen whose main function is the transmission of loads imposed by muscle forces and the bearing of weight through the joint surfaces
- Cartilage, which coats the articulating surface within each joint, lubricated with synovial fluids to produce a coefficient of friction of the order of 0.003 to 0.015 under a wide range of combinations of joint load and sliding speed

For the composition and mechanical properties of a typical human bone, see Table 22.1.

A total joint replacement consists of the following elements:

- The anchorage stem, placed within the central core or medullary canal of long bones, and an articulating head

TABLE 22.1 COMPOSITION AND PROPERTIES OF HUMAN BONE[a]

Composition (vol %)	
Calcium hydroxyapatite	40–45
Collagen	~40
Fluids	~20
Properties	
Elastic modulus (GPa)	
Parallel to axis of long bones	17.10
Normal to axis of long bones	11.50
Shear modulus (GPa)	
Parallel to axis of long bones	3.30
Ultimate strength (MPa)	
Parallel to axis of long bones	
• Tension	133
• Compression	193
Normal to axis of long bones	
• Tension	51
• Compression	133
Poisson's ratio	
Parallel to axis of long bones	0.46
Normal to axis of long bones	0.58
Density (kg m^{-3})	1600–1700
Elongation at fracture (%)	3–4

[a]Composition varies with compactness of bone; a Haversian cortical bone provides typical data.

- Bearing surfaces to provide the necessary lubrication and wear resistance between the two solid counterfaces (the femoral and acetabular components); an almost universal design of the bearing in joint replacement components employs a metal anchorage stem and articulating head and a nonmetallic counterface to simulate natural articulation. Among the nonmetallic materials in use today are ultra-high-molecular-weight polyethylene (a polymer), polycrystalline alumina (a ceramic), vapor-deposited carbon (a turbostratic graphite), and Delrin 150 (a polyoxymethylene homopolymer)
- A fixation agent, to provide mechanical continuity of the replacement component and the surrounding bone; today, most artificial joints are held within the bone with a two-component mixture usually called acrylic bone cement*

Recently, some new cement formulations, including butyl methacrylate, zinc polycarboxylate and silicone elastomer, and carbon fiber reinforced methylmethacrylate, have been proposed [2]. Attempts to use fixation stems, as a revolutionary new alternative to bone cement and for stimulating direct mechanical continuity between prosthetic surfaces and body tissue [3], have not yet proved successful [4]. The use of a soft, porous coating of polytetrafluorethylene graphite on femoral components, which attempts to stabilize the prosthesis through the growth of fibrous tissues into the interstices of the porous surface layer, shows promise [5]. Other very recent innovative fixation techniques are precoating the prostheses with PMMA bone cement [6] or a thin PMMA film [7] and the use of electrical stimulation [8].

The various parts of a total hip replacement are shown in Fig. 22.1.

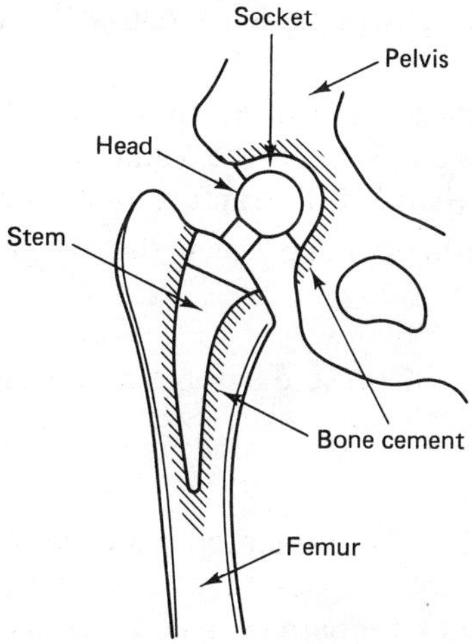

Figure 22.1 Schematic diagram of implanted total hip replacement.

*The complete formulation is ±40 g powder (polymethyl methacrylate, PMMA, powder, ±1% benzoyl peroxide catalyst), and ±20 g liquid (methylmethacrylate monomer; ethylene dimethylacrylate for cross linking; dimethylparatoluidine promoter; and hydroquinone stabilizer). The product is polymerized in situ during surgery.

22.2 DESIRABLE MATERIAL PROPERTIES

The design of joint replacements necessitates consideration of the composite mechanics of the artificial component, the host bone, and the normal function of the healthy joint. The artificial component must be fabricated from materials that fulfill a number of basic criteria, notably:

- Tissue-material interface tolerance (biocompatibility)
- Nontoxicity
- Extreme resistance to environmental degradation (such as corrosion) in conditions that include crevice exposure and fretting at interfaces between dissimilar materials
- Resistance to abrasion, in combination with low sliding friction in the case of bearing surfaces
- Low cost and high reproducibility of material properties

In addition, the material must meet certain minimum specifications in terms of mechanical properties

- Minimum yield strength of 500 MPa, to prevent permanent deformation of the stem within the bone
- Fatigue strength exceeding 400 MPa (10^7 cycles, fully reversed bending) in the body fluid environment, to ensure that fatigue failure does not occur in the case of shafts that anchor bearing surfaces to long bones and sustain large forces in bending and torsion
- Ultimate tensile strength of at least 650 MPa, to protect the component against overload
- Elongation to fracture of at least 8% to ensure that, should the operational stress in the component exceed the yield strength of the material, plastic deformation, rather than brittle fracture, will occur
- High specific strength, to keep down the weight of the component while providing the desired strength

Let us examine some of these desirable material properties in some detail.

Biocompatibility

Biocompatibility means that the implanted material and its possible degradation products are tolerated by the surrounding biological tissue; that is, the function of the surrounding tissue is unimpaired. Tolerance is determined by studying the tissue reaction to the material by suitable clinical and/or biological testing procedures and making a decision as to whether the recorded value is acceptable in terms of the spectrum of possible responses (Fig. 22.2). Clearly, it is imperative that the materials used have as benign an effect as possible—one that is as close to the left of Fig. 22.2 as possible.

No effect → Loss of specific tissue function → Allergy → Tissue necrosis → Neoplasia

Figure 22.2 Biocompatibility continuum.

Nontoxicity

Nontoxicity of a wide assortment of materials in direct contact with body tissues or fluids can be determined from readily available data.

Corrosion Resistance

Corrosion resistance is an extremely important requirement; indeed, the corrosion of metals when used as surgical implants is a major design concern. These metals will be used under conditions and in environments that are conducive to fast corrosion: large cyclic stresses and highly corrosive fluids. In addition, the conditions are dynamic; that is, the environment provides a corrosive medium for the metals, and the metal and its corrosion products in turn affect the surrounding body tissues and fluids. Thus the implant material may be altered sufficiently for its physicochemicomechanical properties to be seriously affected. Also, toxic substances that before were isolated from the tissue-implant interface may now be released, generating allergenic and/or carcinogenic reactions, as well as causing or contributing to deficient body tissue function. It is therefore important that implant materials show as little corrosion tendency as possible. A stock of data is available on the in vitro and in vivo corrosion of a large number of materials [9].

Low Wear and Friction

Tribological excellence means low wear of the stem and low sliding friction of the counterfaces. Wear can lead to loss of material from the device into the contiguous tissues, with subsequent tolerance problems (such as necrosis of large areas of muscle and osseous tissue), as well as a change in shape, and may thus result in dysfunction of the prosthesis. Friction within the joints has to be kept to the minimum practicable in order to prevent loosening of components through repetitive transmission of frictional twisting moment across the component-bone interface.

Fatigue Strength

Adequate fatigue life is necessary in order to provide acceptable long-term performance because, for example, it has been estimated that the stress on the head of the femur during a step amounts to several megapascals, with the number of such impacts being 1 to 4 million per year [10].

Mechanical Compatibility

Compatibility with bone in terms of mechanical properties (see Table 22.1) is not only a prime consideration where long usage of a device is required but is also of paramount concern where bone replacement is involved. Serious mismatches between the prosthesis and the repository (the bone) can lead to a gross deterioration of the interface between the two systems or even complete failure of the prosthesis.

22.3 PROPER DESIGN OF THE PROSTHESIS

In addition to these material property requirements, the prosthesis must be properly designed and engineered for the particular application. Even if an "ideal" material were used, improper structural design could lead to engineering failure of the device. It is therefore important that optimum design principles be adhered to. This point is illustrated in Fig. 22.3. Part (a) shows a 135° nail plate as used in the fixation of intertrochanteric fractures of the femur. This plate is subject to considerable impact stresses and tends to absorb a large fraction of the load-bearing stress because of its low angular configuration. Part (b) shows a 150° nail plate, in which the nail is parallel to the stress lines in the proximal part of the femur. This revised component is more stable, facilitating the impaction of the fragments and hence stimulating more osteogenic activity—quicker production and growth of the cells necessary for bone re-formation.

22.4 CANDIDATE MATERIALS

Candidates are limited to a few stainless steels, cobalt-chromium base alloys, and titanium alloys.

Stainless Steels

Originally, grades of 302, 304, a molybdenum-bearing version of 18/8, and 316 were used, but problems were often encountered due to unacceptable perform-

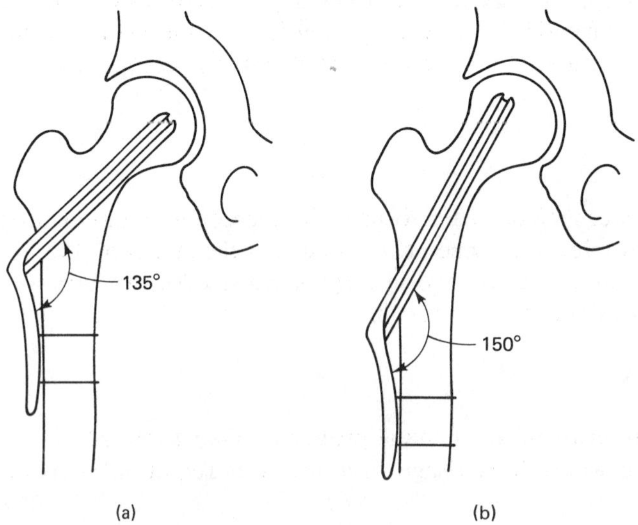

Figure 22.3 Nail plates: (a) conventional 135° nail plate; (b) 150° nail plate.

Figure 22.4 Microstructure of 316L stainless steel. (Cold-worked, 50X magnification; courtesy of Richards Medical Co.)

ance, especially in the case of sensitized components made of 316 stainless steel. Also, often encountered was the associated problem of the formation of irregular, often duplex grain structures through recrystallization during hot forging.

Since the mid-1980s, there has been substantial improvement in the reliability of stainless steel joint replacements, this having been achieved through the adoption of stricter controls of alloy chemistry and microstructure. For example, the sulfur content of orthopedic stainless steels have been progressively reduced, from in excess of 0.01 wt % in 1975 to typically 0.0020 wt % today, through the use of a vacuum-melting steel production process. Nowadays, the corrosion resistance index [(wt % Cr) + 3.3(wt % Mo)] of an orthopedic stainless steel is about 23–24, thus improving the material's resistance to pitting corrosion.

Currently, the most widely recommended stainless steel for these applications is 316L in its annealed or cold-work state. The main disadvantage of this material is its tendency for crevice corrosion. It has a high active pitting potential in body fluids (+150 mV, versus the standard calomel electrode, SCE) and does not repassivate easily under the acidic conditions that characterize the zone between the stems of joint replacement components and the supporting shell of the bone cement. In addition, 316L alloys have low fatigue strengths under crevice corrosion conditions, thus limiting their use to joint replacement components for older, lighter, and less active persons. Some of these shortcomings of conventional 316L alloys have been overcome with the recent introduction of a new nitrogen-strengthened manganese-bearing austenitic stainless steel, 22Cr-9Ni-4Mn alloy.

The microstructure of a cold-worked 316L stainless steel is shown in Fig. 22.4.

Cobalt-Base Alloys

Alloys with nominal composition Co-27Cr-5Mo-0.3C were first introduced to orthopedics in the mid 1930s and have been used extensively for the manufacture of joint replacement components ever since. The main advantages of these alloys are their high corrosion resistance and excellent biocompatibility. However, a number of disadvantages are associated with these alloys. In the cast condition, cobalt-base alloys show inconsistent mechanical properties, notably in fatigue resistance (due mainly to the difficulty of casting components with fine, uniform interdendritic micropore-free grains).

Two methods for improving the reliability of components made of these alloys have been particularly successful. One is the homogenization of castings through annealing at about 1200°C (a temperature that is marginally less than the eutectic temperature of the interdendritic mixture). Materials so treated have shown large increases in toughness and corrosion resistance and modest improvement in endurance strength. The second successful method is the use of multistage heat treatment, such as annealing at about 820°C (to precipitate and grow embrittling phases), followed by annealing at about 1200°C (to dissolve all these precipitates) and quenching (to generate a high stacking fault density). Such treatments have produced alloys with ultimate tensile strength exceeding 1150 MPa and endurance strength exceeding 450 MPa.

In recent years, the trend has been toward the development of processes other than casting that yield cobalt-based materials with improved performance. Two of these approaches are alloy chemistry adjustment and powder metallurgy. Alloy chemistry adjustment (to improve forgeability) involves reduction of the carbon content (to about 0.1 wt %) and modest changes in the chromium and molybdenum concentrations. Typically, the forged product not only has an endurance strength two to three times that of the cast product but also displays an endurance ratio of about 55%, which is explained by the favorable dislocation structure arising from the processing. The use of powder metallurgy for the fabrication of joint components involves molding, by way of hot isostatic pressing under high pressure (about 100 MPa) and high temperature (about 1100°C), of the fine alloy powder (maximum particle size about 250 μm) to yield components with an equiaxed austenitic structure and a dispersal of intergranular spheroidal carbides with a very fine grain size (typical grain diameter of about 6 μm). The ultimate tensile and endurance strengths of this fine-grained material are typically 1200 MPa and 700 MPa, respectively. The typical microstructure of a cast and sintered Co-Cr-Mo alloy is shown in Fig. 22.5.

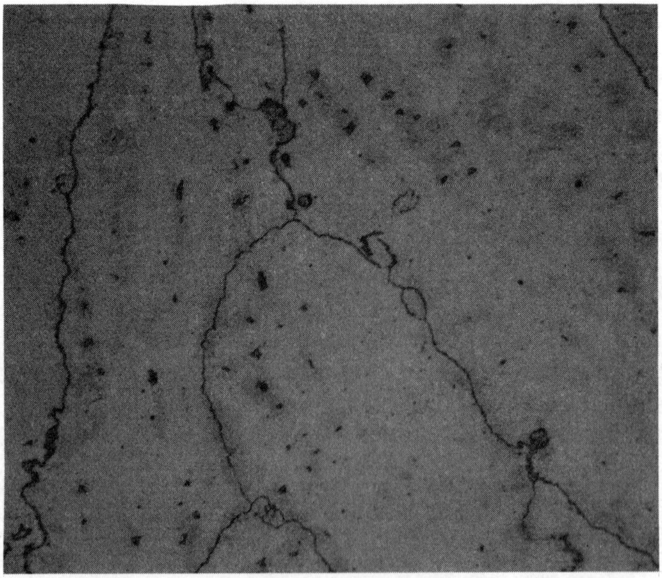

Figure 22.5 Microstructure of a Co-Cr-Mo alloy. (Cast and sintered, 50X magnification; courtesy of Richards Medical Co.)

Figure 22.6 Microstructure of Co-Ni-Cr-Mo (MP 35N) alloy. (100X magnification; courtesy of Richards Medical Co.)

Wrought Co-Ni-Cr-Mo Alloys

Alloys with a suitable combination of strain-hardening properties (and hence endurance strength) and corrosion resistance have the composition Co + Ni + Fe (65–75 wt %) and Cr + Mo (25–35 wt %). One alloy of special interest in this category is MP35N, which has a mixed Co-Ni base and may be strain-hardened, through reductions of about 50% at 550°C, to produce a material with an ultimate tensile strength in excess of 1800 MPa. The inhomogeneous microstructure of the cold-worked alloy does, however, lead to the formation of fatigue cracks at fairly low stress levels, making the endurance strength about 480 MPa. In recent years, the fabrication of total joint replacements using MP35N has employed hot forging at temperatures greater than 600°C. By carefully controlling the rate of deformation during hot working, the best possible dislocation microstructures could be obtained, thereby increasing the critical stress necessary for martensitic transformation in the ductile zones of newly formed cracks and hence raising the endurance strength to over 500 MPa.

The typical microstructure of MP35N alloy is shown in Fig. 22.6.

Titanium Alloys

Numerous clinical tests have shown titanium and a number of its alloys to be the most biocompatible and corrosion-resistant of implantable metallic materials [11]. Because commercially pure titanium has inadequate mechanical properties, the application of this metal in total joint replacement components has been in the form of the duplex alloy Ti-6Al-4V in the hot-forged and annealed state. Titanium alloy castings show mechanical properties that are comparable to wrought products (ultimate tensile strength of 900 MPa and elongation at failure of 10%).

One drawback associated with the use of high-strength Ti-6Al-4V alloys is their notch sensitivity, which results in reduced endurance strength. Such reduction is strongly affected by the surface condition of the alloy, so even forged and cast components of the highest-quality porous-coated titanium alloy have endurance strengths of only 200 MPa. Although this impaired performance is not an important consideration in the choice of material for conventional anchorage stems of joint replacement components, it limits its use for direct skeletal attachment of joint replacements to bone.

It has recently been shown that implantation of nitrogen ions significantly improves the resistance of Ti-6Al-4V alloy to wear-accelerated corrosion [12]. A graded-energy technique was employed (the alloy was implanted first with ions at 60 keV with a dosage of 1.3×10^{21} ions m^{-2}, then at 35 keV with a dosage of 3.2×10^{20} ions m^{-2}, and finally at 17.5 keV with a dosage of 2.0×10^{20} ions m^{-2}). It was shown that this technique produced a more uniform implanted-element concentration through the near-surface volume than would have been achieved by a single-energy implantation procedure. An average concentration of 20 at % nitrogen through a surface layer thickness of about 100 nm was achieved [12]. The typical microstructure of Ti-6Al-4V alloy in the wrought condition is shown in Fig. 22.7.

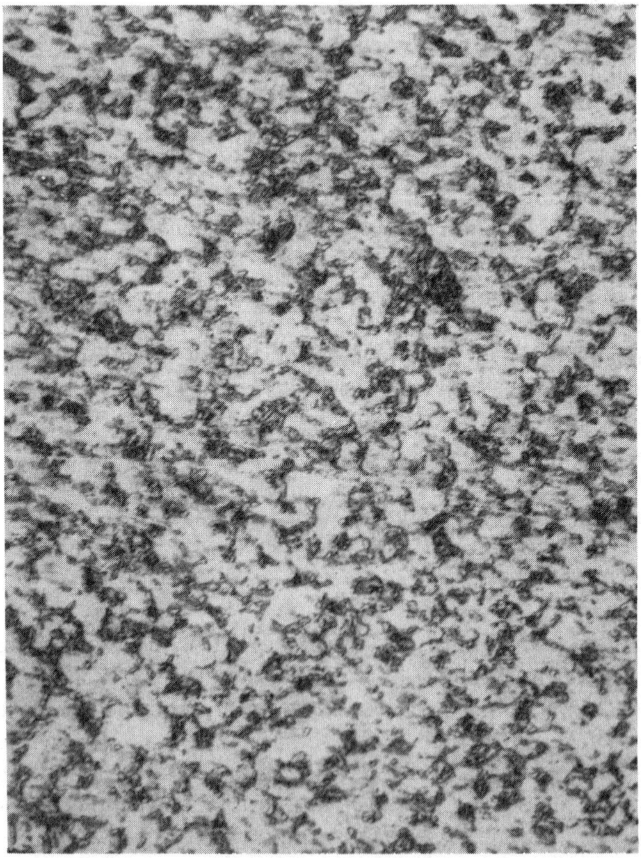

Figure 22.7 Microstructure of Ti 6Al 4V alloy. (ELI grade, wrought; 640X magnification; courtesy of Richards Medical Co.)

Comparison of Candidate Metallic Materials

The typical chemical compositions of four candidate materials are given in Table 22.2, and typical values of some of their mechanical properties are given in Table 22.3. The characterizing corrosion (electrochemical) parameters of these materials are presented in Table 22.4.

One of the most important considerations in the final choice among these materials is the effect of the implantation of metal anchorage stems on "stress protection" of the bone [13,14]. This is the phenomenon whereby normal physiological stress levels within the bone are reduced due to the relative stiffness of

TABLE 22.2 CHEMICAL COMPOSITION OF METALLIC CANDIDATE MATERIALS FOR TOTAL HIP REPLACEMENT COMPONENTS (TYPICAL VALUES IN WT %)

Alloying Elements	Fe-Cr-Ni-Mo (316L)	Co-Cr-Mo (Cast, Powder Metallurgy, Wrought)	Co-Ni-Cr-Mo (MP35N)	Ti-Al-V (Cast, Wrought)
Al	—	—	—	6.0
C	0.02	0.27	0.01	0.02
Co	—	Balance	—	—
Cr	17.5	28.0	20.0	0.2
Cu	0.2	—	—	—
Fe	Balance	0.5	0.2	0.2
H	—	—	—	0.005
Mn	1.8	0.5	0.05	—
Mo	2.6	6.1	9.5	—
N	—	0.1	—	0.015
Ni	13.5	0.3	35.0	—
O	—	0.015	—	0.15
P	0.02	0.010	0.003	—
S	0.006	0.015	0.003	—
Si	0.65	0.75	0.05	0.05
Ti	0.02	—	0.83	Balance
V	—	—	—	4.3

TABLE 22.3 MECHANICAL PROPERTIES OF ALLOYS FOR TOTAL HIP REPLACEMENT COMPONENTS (TYPICAL VALUES)

Alloy	0.2% Yield Strength (MPa)	Ultimate Tensile Strength (MPa)	Elongation at Fracture (%)	E (GPa)	Fatigue or Endurance[a] Strength (10^7 cycles) (MPa)
316L (cold-worked)	700	875	12	196	383
Co-Cr-Mo (hot-forged)	890	1250	8	230	552
MP35N (hot-forged)	1000	1200	13	230	500
Ti-Al-V (hot-forged/shot-peened)	950	1075	13	120	580

[a]Fully reversed bending (rotating cantilever).

TABLE 22.4 ELECTROCHEMICAL PARAMETERS OF CANDIDATE MATERIALS FOR TOTAL HIP REPLACEMENT COMPONENTS IN SIMULATED BODY FLUIDS AT 38°C

Material	Breakdown Potential (mV, SCE)[a]	Repassivation Potential (mV, SCE)	Crevice Breakdown Potential (mV, SCE)	Corrosion Rate (μm y^{-1})
316L (annealed cold-	0 to +250	−170 to −130	+10 to +70	—
worked)	−100 to +50	—	—	25–50
Co-Cr-Mo (cast)	+360 to +630	+400 to +580	+400	81–230
MP35N (annealed)	+500 to +710	+500 to +600	+200	30–50
Ti-6Al-4V (annealed)	+800 to +1000	—	—	17–102

[a]SCE, saturated calomel electrode.

Sources: H. J. Mueller and E. H. Greener, "Polarization studies of surgical materials in Ringer's solution," *J. Biomed. Mater. Res. 4,* pp. 29–41, 1970; F. W. Bultitude and J. R. Morris, "The corrosion of surgical implants," Technical Report No. CoRo/44/83/29, United Kingdom Atomic Energy Authority (UKAEA), Aldermaston, England, 1969; D. F. Gibbons, "Biomedical materials," in B. G. Fleming and B. N. Feinberg (eds.), *Handbook of Engineering in Medicine and Biology*. Cleveland, Ohio: CRC Press, 1976, p. 275.

the metal implant, which reduction, in the case of the upper end of the femur, can be as much as 90% [15]. One way to reduce the degree of stress protection of bone is to ensure that the replacement components have a rigidity that is similar to that of natural bone. In this respect, the use of Ti-6Al-4V alloy makes possible the design of components of normal flexibility, given the constraints imposed by the size and geometry of anchorage stems and by the human musculoskeletal system.

Ceramic Materials

These are being used, especially in Europe, for the manufacture of spherical heads of hip prostheses, bone segment replacements for traumatic or genetic defects, and bone wedges to correct gait, among other devices. The main ceramic material in current use is polycrystalline alumina, a material fabricated by molding and sintering powders containing 99.7 wt % Al_2O_3 and 0.25 wt % MgO to yield low residual void levels (<0.2%) and extremely fine grain size (max. 7 μm). The final product has low fracture toughness, high modulus of elasticity, and high ultimate tensile strength. The most attractive features of this material, however, are its tribological properties. Rubbing against polymeric counterfaces, alumina has friction coefficients that are much lower than in metal-polymer combinations. To sustain such low friction, though, adequate lubrication is necessary. The wear rates for alumina-polymer counterfaces are much lower than for metal-polymer ones.

There are two main drawbacks to the use of ceramic joint replacement components. One is the loss of strength in these materials following prolonged exposure to body fluids. In alumina, for example, this degradation is the result of preferential attack by these fluids of grain boundary segregates, particularly calcium and silicon ions. This usually means that ceramics to be used in the manufacture of joint replacement components must be both silicon- and calcium-free. The other drawback is that as a result of the low fracture toughness, there

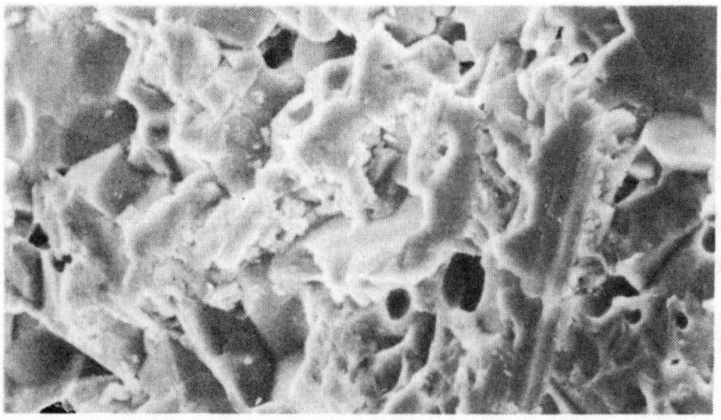

Figure 22.8 Microstructure of alumina, Al_2O_3. (1000X magnification; courtesy of Dr. S. D. Cook, Tulane Medical School)

is a possibility of slow crack growth and propagation occurring at modest levels of cyclic stress. Also, the endurance strength of most alumina materials is very low—currently the best value is about 70 MPa. The typical microstructure of alumina is shown in Fig. 22.8.

Carbons

Interest has focused on two turbostratic carbons: low-temperature isotropic (LTI) and vapor-deposited carbons. The former is formed by pyrolysis of a gaseous hydrocarbon, leading to deposition of a thick carbon layer on a refractory substrate, such as graphite. The microstructure is made up of small graphitic crystallites (3–4 nm) arranged in a random manner with extensive cross linking. This produces a bulk solid of high strength, low density, and isotropic mechanical properties. The strain energy to fracture of the material is comparable to that of human bone and about six times that of dense alumina ceramics. The major attribute of pyrolytic carbons is the absence of fatigue during fluctuating loading [16]. LTI carbon currently finds use in joint replacement for the manufacture of smaller joint prostheses (fingers, toes, etc.).

Vapor-deposited carbon, i.e. ultra-low-temperature isotropic carbon, has excellent biocompatibility and corrosion resistance [17]. The typical microstructure of LTI pyrolitic carbon is shown in Fig. 22.9.

Polymers and Polymer-based Filamentary Composites

Polymeric materials have poor strength characteristics for load bearing; thus their use in load-bearing joint replacement components such as stems is limited, although they do have potential for specialized use in components under compressive loading.

Figure 22.9 Microstructure of low-temperature isotropic (LTI) carbon. (1000X magnification; courtesy of Dr. S. D. Cook, Tulane Medical School)

The mechanism of failure of hip anthroplasties is still a matter of controversy, but one favored theory is that resorption of the bone proximal to the implant can lead to cement failure and failure of the replacement. For example, a finite element modeling of the prosthesis-cement-bone system [18] has shown that reduced stem modulus could lead to increased calcar longitudinal and hoop stresses in the bone (and hence reduced stresses in the cement) and therefore reduced bone resorption. Polymer composites have begun to be introduced into hip anthroplasty to address the failure problem. Several researchers have reported the use of carbon and glass fiber reinforced polymers (epoxy, polysulfone, PMMA, etc.) [19]. Filamentary composite materials based on a synthetic aramid fiber (such as Kevlar 49) have been shown [20] to be biocompatible and have found use in smaller bone replacements, such as tendon prostheses [21].

The attractive feature of these composite materials is that the orientation and fiber content can be varied in the implant to provide the mechanical property orientation necessary for good function. The fiber can be distributed in the matrix material to provide strength in only those locations and directions where it is needed. The implant must be designed so that fatigue failure does not occur and the matrix material is not attacked by the physiological environment.

22.5 CONCLUSION

The modern total hip replacement system, based on metallic implants, was introduced by Sir John Charnley about 1965. Today, a large number of companies produce these components, among them the Total System (Zimmer, Warsaw, Ind.), which is based on a proprietary Ti-6Al-4V alloy, Tivanium; the Charnley Total System (Chas. F Thackeray, USA, Inc., Woburn, Mass.), which uses a proprietary Fe-Cr-Ni-Mn-Mo alloy; the APR Universal System (Intermedics Orthopedics, Austin, Texas), which uses a titanium alloy coated with cancellous titanium; and the Dimension/PC Hip System (Kirschner Medical, Timonium, Md.), which uses a forged and thermally processed Ti-6Al-4V alloy.

REFERENCES

1. I. M. Rutkow, "Orthopaedic operations in the United States, 1979 through 1983," *J. Bone and Joint Surgery 68A* (5), pp. 716–719, 1986.
2. J. R. DeWijn, "Reduction of maximum temperature in the polymerization of cold and heat curing acrylic resins," *J. Biomed. Mater. Res. 8,* pp. 421–434, 1974; R. M. Pillar, R. Blackwell, I. Macnab, and H. U. Cameron, "Carbon-fiber-reinforced bone cement in orthopedic surgery," *J. Biomed. Mater. Res. 10,* pp. 893–906, 1976.
3. G. Lord, et al., *Rev. Cir. Orthop. 64,* p. 5, 1978; C. A. Homsby, et al., *J. W. Clin. Orthop. 89,* p. 220, 1972.
4. H. U. Cameron, *Clin. Orthop. 165,* p. 188, 1982.
5. C. A. Homsby, et al., *J. Bone and Joint Surgery 52A,* p. 604, 1970; C. A. Homsby and M. S. Anderson, *Proc. Conf. Mater. for Use in Med. and Biol.,* p. 85, 1974.

6. W. Barb, J. B. Park, G. H. Kenner, and A. F. von Recum, "Intramedullary fixation of artificial hip joints with bone cement: Impregnated implants. I: Interfacial strengths; II. Density and histological study," *J. Biomed. Mater. Res. 16.*, pp. 447–458, 459–489, 1982.
7. S. Raab, A. M. Amed, and S. W. Provan, "Thin film PMMA precoating for improved implant bone-cement fixation," *J. Biomed. Mater. Res. 16,* pp. 679–704, 1982.
8. S. M. Collela, et al., "Fixation of porous titanium implants in cortical bone enhanced by electrical stimulation," *J. Biomed. Mater. Res. 15,* pp. 37–46, 1981.
9. S. A. Brown and K. Merritt, "Electrochemical corrosion in saline and serum," *J. Biomed. Mater. Res. 14,* pp. 173–175, 1980; V. J. Colangelo and N. D. Greene, "Corrosion measurements in vivo," *J. Biomed. Mater. Res. 1,* pp. 405–414, 1967.
10. C. Bechtol, A. B. Fergusson, and P. G. Laing, *Metals and Engineering in Bone and Joint Surgery,* chap. 5, pp. 92–106. Baltimore: Williams and Wilkinson, 1959.
11. D. F. Williams (ed.), *Bicompatibility of Clinical Implant Materials,* vol. 1, p. 9. Boca Raton, Fla: CRC Press, 1981.
12. R. A. Buchanan, E. D. Rigney, and J. M. Williams, "Ion implantation of surgical Ti-6Al-4V for improved resistance to wear accelerated corrosion," *J. Biomed. Mater. Res. 21,* pp. 355–366, 1987.
13. B. Weightman, in M. Sehaldach and D. Hohmann (eds.), *Advances in Artificial Hip and Knee Joint Technology,* p. 138. New York: Springer-Verlag, 1976.
14. R. R. Tarr, et al., in B. R. Simon, P. C. Johnson, and J. G. Gross (eds.), *Finite Elements in Biomechanics,* p. 345. England: Wiley, 1982.
15. R. D. Crowninshield et al., *Proc. 9th Open Scientific Meeting of the Hip Society,* p. 111, 1981.
16. F. J. Schoen, *Carbon 11,* p. 4313, 1973.
17. A. D. Haubold, H. S. Shim, and J. C. Bokros, in D. F. Williams (ed.), *Biocompatibility of Clinical Implant Materials,* vol. 2, p. 3. Boca Raton, Fla.: CRC Press, 1981.
18. R. D. Crowninshield et al., "An analysis of femoral component stem design in total hip anthroplasty," *J. Bone and Joint Surgery 62A,* pp. 68–78, 1980.
19. P. G. Rose, et al., "New aspects of the design and application of carbon-fiber-reinforced carbon for prosthetic devices," and U. Soltesz and H. Richter, "Investigation of mechanical behavior of fiber-reinforced materials for endoprosthetic devices," papers presented at the First World Biomaterials Congress, Baden Austria, April 1980.
20. J. D. Henderson, R. H. Mullarky, and D. E. Ryan, "Tissue biocompatibility of Kevlar aramid fiber and polymethylmethacrylate composites in rabbits," *J. Biomed. Mater. Res. 21,* pp. 59–64, 1987.
21. J. M. Hunter and S. H. Jaeger, "Tendon implants: Primary and secondary usage," *Orthop. Clin. N.A. 8,* pp. 473–489, 1977.

FLAT SPIRAL SPRING

CHAPTER 23

23.1 INTRODUCTION

The objective is to use the spring to store energy and to recover the same either slowly or rapidly, depending on the function of the spring. Thus the flat spiral spring finds uses in applications such as clocks, watches, and gramophones, as are found in antique shops or in private bric-à-brac collections.

23.2 STRESS ANALYSIS

A flat spiral spring consists of a uniform thin metallic strip wound into a spiral in one plane, the outer end being fixed to a pin P and the inner end being attached to a winding sprindle S (Fig. 23.1). To wind the spring, a torque T is applied to the spindle. Let H and V be the reactions at P along and normal to the line joining the axis of the spindle to the center of the pin P.

For an elemental section EF, of length δs, the bending moment $= (Vx - Hy)$. Under the action of the applied torque, the radius of the circle, of which EF is the arc, changes from R_i to R_f; consequently, the change

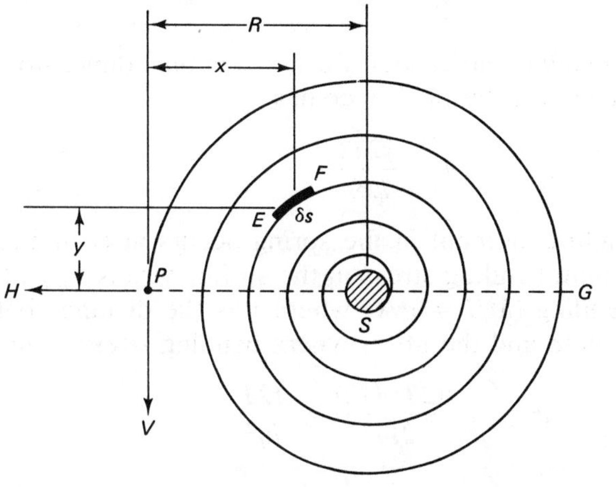

Figure 23.1 A flat spiral spring.

in the angle subtended by the arc EF at the center, $\delta\theta$, may be obtained from

$$\delta\theta = \frac{\delta s}{R_f} - \frac{\delta s}{R_i} \tag{23.1}$$

From the simple theory of bending, $M/I = E/R$, where M is the applied bending moment, I is the moment of inertia of the cross section, and R is the radius curvature. Thus we can write, in this case,

$$\frac{T}{EI} = \frac{1}{R_f} - \frac{1}{R_i} \tag{23.2}$$

From equations (23.1) and (23.2),

$$\delta\theta = \frac{T\delta s}{EI} = \frac{Vx - Hy}{EI}\,\delta s$$

Total change in the angle between the tangents at the ends of the spring is thus

$$\begin{aligned} &= \int \frac{Vx - Hy}{EI}\,ds \\ &= \frac{V}{EI}\int x\,ds - \frac{H}{EI}\int y\,ds \end{aligned} \tag{23.3}$$

Now $\int x\,ds = LR$, where L is the length of the spring (note that $\int x\,ds$ is the moment of the whole spring length about the Y direction). Also $\int y\,ds = 0$, since $\int y\,ds$ is the moment of the whole spring length about the axis joining the centers of S and P. Hence equation (23.3) becomes

$$\theta = \frac{VLR}{EI}$$

Now, VR is the applied torque T; thus

$$\theta = \frac{TL}{EI}$$

For the spring made from a wire strip having cross-sectional dimensions of width b and thickness t, the angular rotation θ becomes

$$\theta = \frac{12TL}{Ebt^3} \tag{23.4}$$

The maximum bending moment in the spring occurs at G and is given by $2RV = 2T$. If the maximum bending stress in the spring wire is σ_{max}, then, from the simple theory of bending ($M/I = \sigma/y$, where y is the distance between the centroid of the cross section and the fibers where bending stress σ acts),

$$\sigma_{max} = \frac{(2Tt)(12)}{2bt^3} = \frac{12T}{bt^2} \tag{23.5}$$

23.3 MATERIAL SELECTION METHODOLOGY

For the spring not to take up a permanent set, the yield strength of the spring material σ_y must be greater than σ_{max}; that is,

$$\sigma_y > \frac{12T}{bt^2} \tag{23.6}$$

Since $\theta = (12TL)/(Ebt^3)$,

$$T = \frac{Ebt^3\theta}{12L} \tag{23.7}$$

Hence, from equations (23.6) and (23.7),

$$\frac{\sigma_y}{E} > \frac{t\theta}{L}$$

Thus, for a given spring cross section ($b \times L$) and angular twist θ, the material parameter σ_y/E must have as high a value as possible in order to avoid general yield.

The results in Table 23.1 show that the best material is the titanium alloy. This alloy (B120VCA) was the first high-strength β-titanium alloy found suitable for spring application. It became available in the 1950s and can be fabricated into wire and subsequently heat-treated to ultimate tensile strength in excess of 1300 MPa.

TABLE 23.1 COMPARISON OF CANDIDATE MATERIALS FOR A FLAT SPIRAL SPRING

Material	σ_y(MPa)	E (GPa)	$10^3(\sigma_y/E)$
Cold-rolled brass	638	120	5.32
Spring steel	1300	200	6.50
Nimonic	614	200	3.07
Ti-13V-11Cr-3Al	800	110	7.27

This alloy is being used in the McDonnell Douglas DC-10 aircraft in applications such as nose and main landing gear down-lock springs and elevator and aileron control springs. This alloy is difficult to process and is therefore expensive. Recently, new β-type alloys that have fewer processing and fabrication problems have been introduced. One such alloy, Beta-C (Ti-3Al-8V-6Cr-4Mo-4Zr), is finding increasing use as a spring material in aerospace and oil tools and in high-performance vehicles.

RECEIVER OF A HIGH-TEMPERATURE SOLAR ENERGY THERMAL SYSTEM

CHAPTER 24

24.1 INTRODUCTION

An awareness of the profligate consumption of fossil fuels coupled with the dislocations of the 1973–74 and 1980–81 oil crises provided the impetus for massive injections of capital into research and development projects associated with the exploration for and exploitation of alternative or renewable energy resources, especially solar energy. Regrettably, with the de facto demise of the cartelizing role of the Organization of Petroleum Exporting Countries (OPEC) and the attendant dramatic drop in the price of oil,* the incentives for further research and development work have nosedived in recent years in the industrialized world.† This is particularly unfortunate considering the pattern of fossil fuel consumption and alternative energy resource potential in this group of countries (Table 24.1).

It is necessary to add a couple of caveats to the foregoing observations. First, there are pockets of sustained activity in the field of alternative energy resources, such as at the Solar Energy Research Institute in Golden, Colo., and

*The price of Saudi Arabian light crude oil dropped from $34 per barrel in June 1982 to $15 per barrel in June 1988.

†The abandonment of solar energy R&D efforts in the United States was reflected in the cancellation of the federal government's solar energy tax credits in December 1986.

TABLE 24.1 PETROLEUM CONSUMPTION AND SOLAR ENERGY POTENTIAL IN VARIOUS INDUSTRIALIZED COUNTRIES

Country	Value of Petroleum Consumption, 1986 ($\$ \times 10^9$)	Solar Energy Potential ($kWh\ m^{-2}\ d^{-1}$)[a]
United States	37.64	3.93 (Madison, Wisconsin)
Japan	24.80	3.05 (Tokyo)
Australia	1.04	4.95 (Canberra)
Great Britain	10.00	2.50 (Kew, England)
Sweden	3.14	2.81 (Stockholm)

[a]Global irradiation on a horizontal surface.

ARCO Solar, an industrial concern with corporate headquarters in Woodland Hills, Calif.

Also, much work is being conducted in various developing countries, as at the Center for Energy Studies, India Institute of Technology, Hauz Khas, New Delhi, India, and the Industrial Development Center for Arab States in Cairo, Egypt.

Solar energy thermal systems (SETS) are perhaps the best known and most widely used of the class of systems that employ alternative energy resources. SETS fall into two broad categories: low-temperature types and high-temperature types. Examples of the former include water heating, small-scale space heating, and cooling systems. These systems typically (a) use flat-plate collectors, (b) use water or air as the heat transfer fluid, (c) have a basic storage medium, such water in insulated tanks, and (d) employ a process temperature in the 45°C to 100°C range. Examples of high-temperature SETS include steam production plants and direct process heating systems. Typically, these systems (a) use concentrating collectors and receivers of the nontracking or active-tracking type, (b) use steam for the process application, (c) employ a process temperature in the range 90°C to 300°C, and (c) use some form of energy storage, such as organic oils (up to 350°C) and rock beds.

In SETS, the collectors are used to concentrate the incident global solar irradiation, and the receiver transfers this energy to a working fluid that is used to provide process heat. This heat may then be used for either industrial applications or driving a suitable heat engine to produce electricity. (Some of the operational solar industrial process heat and electricity projects in the world are presented in Table 24.2.)

The receivers in SETS are the key components that determine the operating temperature at which power conversion or thermal processes can take place. The achievement of high operating temperatures (greater than 800°C) makes possible higher-efficiency engines and an increased number of useful industrial processes.

24.2 RECEIVER DESIGN REQUIREMENTS

Since the receiver is located between the concentrator and the power conversion or thermal process subsystem, its design must meet the requirements imposed by

TABLE 24.2 SOLAR INDUSTRIAL PROCESS HEAT AND ELECTRIC POWER PROJECTS

Company/Location	Process Application	Process Temperature (°C)	Collector Type[a]
Industrial Process Projects			
Campbell Soup Co., Sacramento, Calif., USA	Can washing	82–91	PT, P
West Point–Pepperell, Inc., Fairfax, Ala., USA	Fabric drying	160	PT
Ore-Ida Foods Co., Ontario, Oreg., USA	Steam production	214	PT
CARCESA, Merida, Spain	Sterilization of milk	162	PT
Maryborough Hospital, Maryborough, Victoria, Australia	Hospital hot water production	50	P
Electric Power Projects			
	Electric Power (kW)		
Coolidge, Ark., USA	200		
Willard, N.M., USA	25		
Gila Bend, Ariz., USA	50		
Daggett, Calif., USA	13,800		

[a]PT, parabolic trough; P, flat plate.

these components or subsystems. Thus the receiver design requirements are inextricably woven with the specific application and system configuration (the temperature and pressure needed by the thermal process or power conversion system, the heat transfer characteristics of the working fluid, the mechanical configuration of the system, and the optical properties of the concentrator).

The following, therefore, are the desirable properties of the receiver material:

- High tensile strength
- High fracture toughness
- Excellent resistance to thermal shock induced by step changes in temperature
- Good fatigue strength
- Good creep strength
- Stability at elevated temperatures
- Excellent resistance to both corrosion and oxidation
- Easy fabricability
- Commercial availability at reasonable cost
- Capacity to withstand the flux nonuniformities created by the concentrator
- Capacity to withstand the stress induced by nonsymmetric heating and thermal gradients
- Very high maximum use temperature (especially important in receivers that operate at temperatures in excess of 800°C)

24.3 CANDIDATE MATERIALS

For high-temperature receivers, material temperature limits are a major design constraint. Stainless steels are limited to about 700°C, but superalloys can be used up to about 1000°C. However, these superalloys contain a number of strategic alloying elements whose continued availability is now being questioned (especially in light of the political instability of the major supplier of these elements, South Africa). Therefore, ceramic materials are now being seriously considered. Table 24.3 lists three of these ceramic materials and their relevant properties.

In fact, a number of ceramic receivers designed for operation in the 1000°C to 1400°C temperature range are currently at different stages of research and development (Table 24.4).

TABLE 24.3 CERAMIC MATERIALS FOR USE IN HIGH-TEMPERATURE SOLAR THERMAL RECEIVERS

Ceramic Material	Maximum Use Temperature (°C)	σ_f (MPa)	ν	E (GPa)	α (10^6 K^{-1})	Thermal Shock Resistance,[a] R_2 (K)
Fused silica	950	100	0.17	72	0.55	2096
Lithium aluminosilicate	1000	136	0.27	70	1.00	1418
Hot-pressed silicon carbide	1300	700	0.27	450	4.70	242

[a]See Chapter 7 for details.

TABLE 24.4 CERAMIC SOLAR THERMAL RECEIVERS UNDER DEVELOPMENT

Organization	Receiver Size (kW Thermal Power)	Design Operational Temperature (°C)	Working Fluid	Ceramic Material
Jet Propulsion Laboratory, Pasadena, Calif.	75	1370	Air (helium, nitrogen)	Sintered β-SiC/mullite
Sandia National Laboratories, Albuquerque, N.M.	250	1060	Air	Siliconized SiC
Department of Energy, Washington, D.C., and Jet Propulsion Laboratory, Pasadena, Calif.	Variable	1040	SO_3, SO_2	Cordierite

ACOUSTOOPTIC DEVICES

CHAPTER 25

25.1 INTRODUCTION

Although the physical effects involved in acoustooptic interactions are relatively simple, no entirely satisfactory microscopic theory of the photoelastic effect has yet emerged. Our simple phenomenological treatment will suffice for understanding the observed acoustooptic behavior in both isotropic and anisotropic materials. The treatment will culminate in an expression for the figure of merit of a material to be used in acoustooptic devices. This will serve as an aid in designing and optimizing a variety of these devices, several of which are now available from a number of commercial sources. Among these devices are tunable filters (used in tuning of dye lasers, radiometry, correlators, laser detectors, multispectral imaging components, Q-switches, etc.). Under development are thin-film guided-wave optical devices, which have the potential of revolutionizing optics in much the same way that thin-film integration dramatically changed the field of electronics.

25.2 ACOUSTOOPTIC INTERACTIONS: A SIMPLE PRIMER

The basic mechanism of all acoustooptic interactions is the change induced in the refractive index of an optic medium due to the presence of an acoustic wave, which is defined as a traveling pressure disturbance that produces regions of compression and rarefaction in the material. The refractive index n is related to the density ρ, for the case of an ideal gas, through the Lorentz-Lorenz equation

$$\frac{n^2 - 1}{n^2 + 2} = (\text{constant})(\rho) \tag{25.1}$$

The elastooptic coefficient is obtained by differentiating equation (25.1) to give

$$\rho \frac{\partial n}{\partial \rho} = \frac{(n^2 - 1)(n^2 + 2)}{6n} \tag{25.2}$$

with the derivative being taken under isentropic conditions. The fundamental quantity given by equation (25.2), also called the photoelastic constant p, can be related to the applied pressure P, with the result

$$p = \frac{1}{\beta} \frac{dn}{dP} \tag{25.3}$$

where β is the compressibility or bulk modulus of the material.

The relation in equation (25.3) follows from the usual definition of the photoelastic constant

$$\Delta \left(\frac{1}{\varepsilon}\right) = \Delta \left(\frac{1}{n^2}\right) = pe \tag{25.4}$$

where ε is the dielectric constant ($\varepsilon = n^2$) and e is the strain amplitude produced by the acoustic wave. From equation (25.4), it is seen that the change in the refractive index Δn produced by the strain is

$$\Delta n = \frac{-n^3 pe}{2} \tag{25.5}$$

where, for an acoustic wave of frequency Ω at time t, the strain amplitude produced by the acoustic wave e is given by

$$e = e_0 \exp (i\Omega t)$$

Although the relation defining the photoelastic interaction, equation (25.5), has been written as a scalar in which p is independent of the directional properties of the material, it must be appreciated that even for an isotropic material, longitudinal and transverse acoustic waves cause photoelastic interaction to take up different parametric values. A complete description of the interaction, especially for anisotropic materials, requires a tensor relation among the dielectric properties, the elastic strain, and the photoelastic coefficient of the form

$$\Delta \left(\frac{1}{n^2}\right)_{ij} = \sum_{kl} P_{ijkl} e_{kl} \tag{25.6}$$

where $(1/n^2)_{ij}$ is a component of the optic index ellipsoid, e_{kl} are the Cartesian strain components, and P_{ijkl} are the components of the photoelastic tensor.

The most useful photoelastic effect is the ability of acoustic waves to diffract a light beam as a result of either a moving phase grating or collisions between light and sound in which energy and momentum are conserved.

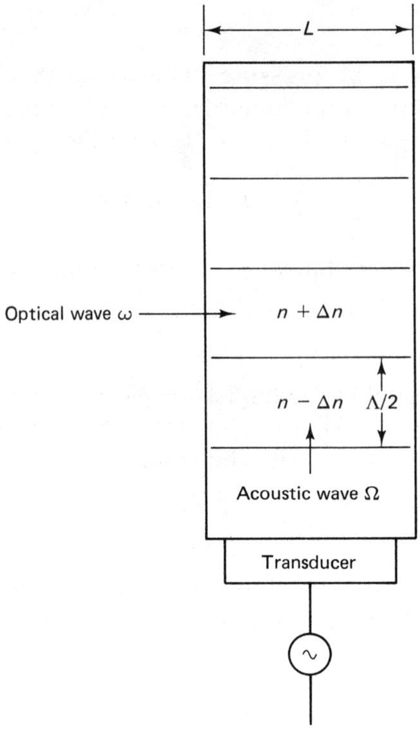

Figure 25.1 Interaction of optical and acoustical waves in a delay line.

In the simplest case of plane acoustic waves interacting with plane light waves (Fig. 25.1), the light wave, of frequency ω and wavelength λ, is incident from the left into a delay line with an acoustic wave of frequency Ω and wavelength Λ. If the refractive index of the delay medium is $n + \Delta n$ in the presence of the acoustic wave, then the phase of the optic wave will be changed by an amount $\Delta\phi$, given by

$$\Delta\phi = \frac{2\pi L \,\Delta n}{\lambda} \tag{25.7}$$

if the delay line has a width L.

If the electric field incident on the delay line is represented by

$$E = E_0 \exp(i\omega t) \tag{25.8}$$

the field of the phase-modulated emerging light will be

$$E = E_0 \exp[i(\omega t + \Delta\phi)] \tag{25.9}$$

It is a well-known radio-frequency engineering concept that the spectrum of a phase-modulated carrier of frequency ω consists of components separated by multiples of the modulation frequency Ω as shown in Fig. 25.2. [1].

Referring to the spectrum of a phase-modulated wave (Fig. 25.2), the ratio of the intensity in the first order to that in the zeroth order is obtained as

$$\frac{I_1}{I_0} = \left(\frac{J_1 \,\Delta\phi}{J_0 \,\Delta\phi}\right)^2 \tag{25.10}$$

where J_0 and J_1 denote the zeroth and first-order Bessel functions.

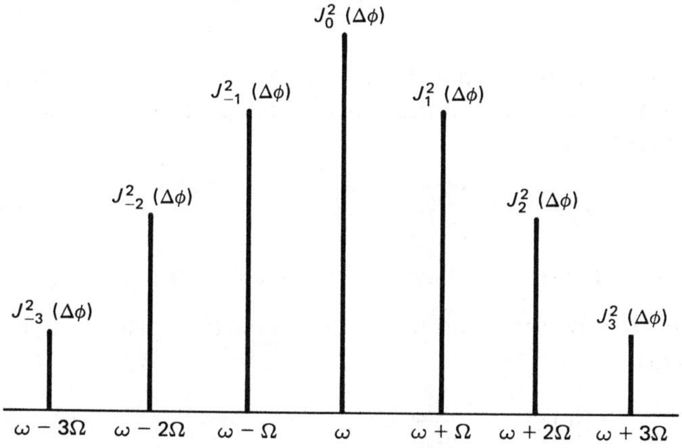

Figure 25.2 Intensity of diffracted orders due to Raman-Nath interaction [1].

For acoustooptically diffracted light, the acoustic power flow is given by

$$P_a = \frac{CUe^2}{2} \tag{25.11}$$

where C is the elastic stiffness constant, which is in turn related to β and U, the acoustic wave velocity or the velocity of sound in the delay medium ($U = 2\pi\Omega\Lambda$) through the equation

$$C = \frac{1}{\beta} = \rho U^2$$

Hence the acoustic power density

$$P_d = eP_a = \frac{\rho U^3 e^3}{2} \tag{25.12}$$

From equations (25.5) and (25.7), the phase modulation depth in terms of the acoustic power density may be expressed thus:

$$\Delta\Phi = \frac{2\pi L\ \Delta n}{\lambda} = -\pi \frac{L}{\lambda} n^3 p \left(\frac{2P_d}{U^3}\right)^{0.5} \tag{25.13}$$

For small modulation index, the zeroth- and first-order Bessel functions may be approximated by

$$\begin{aligned} J_0\ \Delta\phi &\approx \cos \Delta\phi \approx 1 - \Delta\phi \\ J_1\ \Delta\phi &\approx \sin \Delta\phi \approx \Delta\phi \end{aligned} \tag{25.14}$$

Hence, from equations (25.10) and (25.13),

$$\frac{I_1}{I_0} \approx (\Delta\phi)^2 = \frac{\pi^2}{2}\left(\frac{L}{\lambda}\right)^2 \left(\frac{n^6p^2}{\rho U^3}\right) P_d \tag{25.15}$$

Noting that

$$P_a = P_d LH \tag{25.16}$$

where H is the height of the transducer, then

$$\frac{I_1}{I_0} = \frac{\pi^2}{2}\left(\frac{L}{H}\right)\left(\frac{n^6p^2}{U^3}\right)\left(\frac{P_a}{\lambda^2}\right) \tag{25.17}$$

Since L, H, P_a, and λ are all dependent on external factors, the figure of merit (the collection of parameters that depends on the intrinsic properties of the delay line material), denoted by FOM_1, is

$$\text{FOM}_1 = \frac{n^6 p^2}{\rho U^3} \tag{25.18}$$

Two other figures of merit have been used in the choice of materials.

1. When the efficiency-bandwidth product is the important criterion, the relevant figure of merit, FOM_2, is

$$\text{FOM}_2 = \frac{n^7 p^2}{\rho U} \tag{25.19}$$

2. Another figure of merit was introduced by Dixon [2] in connection with the design of wideband acoustooptic devices. It is important to keep the transducer height H as small as possible (this reduces the power requirement of the transducer). Then the appropriate figure of merit, FOM_3, is

$$\text{FOM}_3 = \frac{n^7 p^2}{\rho U^2} \tag{25.20}$$

25.3 CANDIDATE MATERIALS

The selection of a material for any particular device will be dictated by the type of operation of the device. In general, though, it is desirable to select a material with a large value of the relevant figure of merit (materials with high refractive index, high elastooptic coefficient, low density, and low acoustic wave velocity). In addition, the following properties are desirable:

- Low high-frequency acoustic loss. This is a useful characteristic if the material is to be employed in wideband applications. A microscopic treatment of ultrasonic attenuation by Woodruff and Ehrenreich [3] yields the expression

$$\alpha = \frac{\gamma^2 \Omega^2 k T}{\rho U^5} \tag{25.21}$$

  where α is the ultrasonic attenuation, γ is the Grunneisen constant, k is the thermal conductivity of the material, and T is the absolute temperature of operation of the acoustooptic device. Since $\alpha \approx U^{-5}$, the requirement of low acoustic wave velocity (required for a high figure of merit) is in conflict with the requirement of a low ultrasonic attenuation.
- High optic quality, so that not only absorption but also scattering and large-scale inhomogeneities are small.
- Good chemical stability, obviating the need for protective enclosures to maintain integrity.
- Good chemical properties, so that the device can be cut and polished and handled easily.

- Availability in the form of large single crystals of high optical quality.
- High optical damage threshold.
- Low temperature coefficient of velocity, in order to avoid drift of scan.

Two materials have been shown to be suitable for use in acoustooptic devices: fused quartz, because of its excellent optical quality and low cost for large sizes, and lithium niobate, because of its exceptionally low acoustic loss at microwave frequencies. For infrared applications, germanium has proved suitable, as has arsenic trisulfide glass, where bandwidth requirements are not critical. Among the new materials, very good acoustooptic performance has been obtained with TeO_2, $PbMoO_4$, and several chalcogenide crystals (such as Tl_3AsS_4 and Tl_3PSe_4).

A compilation of the relevant properties and acoustooptic parameters for three classes of materials, representing amorphous materials (fused quartz), crystals for the visible range ($LiNbO_3$), and infrared crystals (germanium) are given in Table 25.1. Using figure of merit and acoustic attenuation as criteria of performance, a trade-off between these two is necessary, and the final selection of the best material is dependent on the system demands.

TABLE 25.1 PROPERTIES OF THREE CANDIDATE ACOUSTOOPTIC DEVICE MATERIALS

Material	Refractive Index, n	Acoustooptic Coefficient, p	Density, ρ ($g\ cm^{-3}$)	Acoustic Wave Velocity, U (10^5 cm s^{-1})
Fused quartz	1.46	0.273	2.20	5.96
$LiNbO_3$	2.20	0.285	4.64	6.57
Germanium	4.00	0.430	5.40	5.50

Material	FOM_1 ($10^{-18}\ s^3\ g^{-1}$)	FOM_2 ($10^{-7}\ cm^2\ s\ g^{-1}$)	FOM_3 (10^{-12} cm $s^2\ g^{-1}$)	$\rho^{-1}U^{-5}$[a] ($10^{-30}\ cm^{-2}\ s^5\ g^{-1}$)
Fused quartz	1.55	8.04	1.35	6.04
$LiNbO_3$	7.00	66.50	10.10	1.76
Germanium	843.00	10,200.00	1855.00	3.68

[a]With α as defined in equation (25.21), ultrasonic attenuation is directly proportional to $\rho^{-1}U^{-5}$.

REFERENCES

1. C. F. Raman and N. S. N. Nath, *Proc. Indian Acad. Sci. I 2*, p. 406, 1935.
2. R. W. Dixon, "Photoelastic properties of selected materials and their relevance for applications to acoustic light modulation and scanners," *J. Appl. Phys. 38*, pp. 5149–5153, 1962.
3. T. O. Woodruff and H. Ehrenreich, *Phys. Rev. 123*, p. 1553, 1961.

REFLECTOR RIBS OF THE SPACE-BORNE PARABOLIC ANTENNAS OF A COMMUNICATIONS SATELLITE

CHAPTER 26

26.1 INTRODUCTION

Space-borne antennas are indispensable parts of communications satellites, which are now a very popular means of communication (recording and transmission of data; broadcast of radio, television, and telephone signals, etc.). Their economic competitiveness in the field of signal broadcast, though, is being seriously threatened by fiber-optic systems. Although the capital costs of satellites are much less than those of fiber-optic cable systems (Table 26.1), use-related expenditures (running costs) are higher for the former than for the latter (Table 26.2).

The trend in the field of satellite communications is toward the creation of multinational consortia, such as the International Telecommunications Satellite Organization (Intelsat); the European Spacebus Consortium (comprised of Aérospatiale of France and Messerschmitt-Boelkow-Blohm of West Germany); Communications Satellite Corporation (Comsat); and the Pan-American Satellite Corporation. In the United States, since deregulation of the communications satellite business, most such satellites are made by private-sector companies, most notably Hughes Aircraft's Space and Communications Group, TRW, Ford Aerospace and Communications, Contel Corporation, and the recently formed Astro Space Division of General Electric (created in the merger of GE's Space Systems Division and RCA's Astro Electronics Division).

TABLE 26.1 HARDWARE COST TRENDS FOR OPTIC-FIBER CABLES AND COMMUNICATIONS SATELLITES FOR THE NORTH ATLANTIC REGION

	Cost ($ × 10^6) using	
Year	Fiber-optic Cables	Communications Satellites
1985	100	50
1990	130	75
1995	190	60
2000	180	60
2005	145	60

Source: J. C. Lowndes, "Hughes rates cost of satellites lower than optical fiber hardware," *Aviation Week and Space Technology 125,* p. 76, December 1, 1986.

TABLE 26.2 COST TO CARRIERS OF COMMUNICATIONS SATELLITES AND FIBER OPTICS SYSTEMS

	Annual charge per half circuit ($ × 10^3) using		
Distance (km)	Satellite	Satellite (Zero Earth Station Cost)	Two-fiber Pair Cable (280 Mbits $pair^{-1}$)
1000	7.6	2.5	0.5
2000	7.6	2.5	0.9
3000	7.6	2.5	1.2
4000	7.6	2.5	1.5
5000	7.6	2.5	2.0
6000	7.6	2.5	2.5

Source: J. C. Lowndes, "Pricing strategy is key to satellite market share", *Aviation Week and Space Technology 125,* p. 53, November 24, 1986.

Perhaps the best-known communications satellites are the Intelsat and Landsat series. Intelsat 5, the latest in that series, was launched using the European Space Agency's Arianne 2 booster in May 1988. The Landsat satellites provide data on vegetation, mineral resources, water pollution, and information used to update land-use maps. The latest in the series, Landsat 5, was used to record volcanic heat patterns in the eruption in March 1986 of the 12-km-high Augustine volcano in Cook Inlet, 290 km southeast of Anchorage, Alaska.

The successful operation of many of these advanced satellites (Figs. 26.1 and 26.2) depends on developing larger, higher-gain antennas. Two generic types of these antennas exist: the focusing antenna and the planar array.

The focusing-surface type is made up of a radiating element or family of elements located at the focal point of the focusing surface. The shape is generally parabolic or spherical. The diameter of the surface is a function of the signal-to-

Figure 26.1 Antenna system of a Hughes HS 376 communications satellite. (Courtesy of Hughes Aircraft Co.)

Figure 26.2 The European Space Agency's *Ulysses* solar observation spacecraft. (Courtesy of *Aviation Week & Space Technology*)

noise ratio that is required at the radio frequency of interest. One particular type of the focusing-surface antenna configuration is called the wrap-rib parabolic reflector (Fig. 26.3). This consists of a number of radial ribs or beams that are cantilevered from a central hub structure. Each of the ribs is attached to the hub by hinges. This radial spoke system provides the mounting for the antenna structure. This design is based on an approximation of the structure to a paraboloid of revolution.

Two rib design concepts have been applied to the wrap-rib reflector: the open or C-shaped rib cross section and the closed or lenticular cross section (Fig. 26.4).

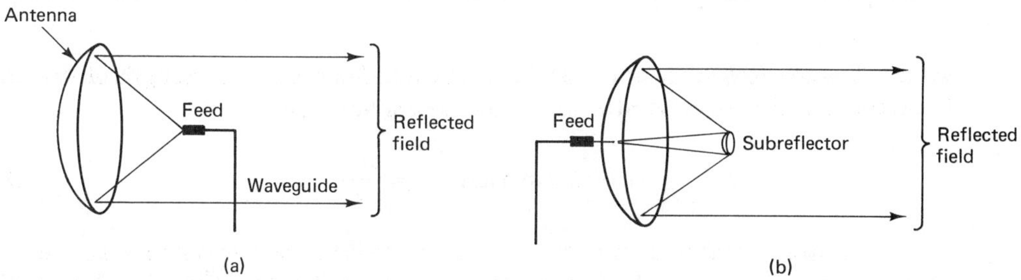

Figure 26.3 The parabolic antenna of a communications satellite: (a) feed waveguide located in front of the antenna, and (b) fed from behind using a subreflector.

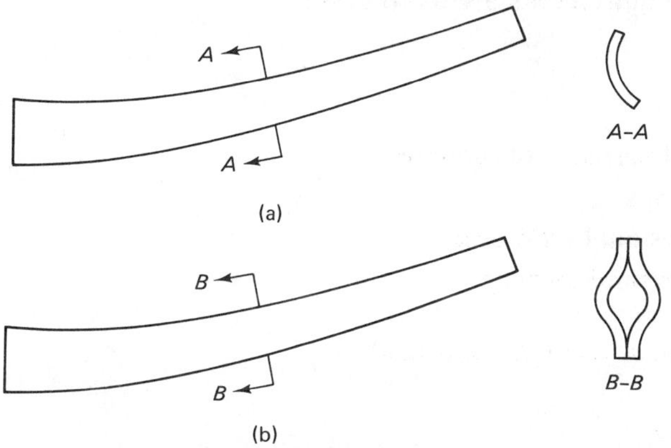

Figure 26.4 Antenna rib design configurations: (a) C-shaped cross section; (b) lenticular cross section.

The choice between these two is dictated by the diameter of the reflector. The C section can be employed up to about 50 m in diameter, while the lenticular section can be useful from ~15 m up to 100-m.

The design requirements are that the ribs provide a dimensionally stable structure in the environment of space for an indefinite length of time.

26.2 DESIGN REQUIREMENTS AND MATERIAL CHOICE

The gain of an undistorted parabolic antenna, expressed in decibels, is given by the simple relationship

$$G = 20 \log \left(\frac{\pi \eta^{0.5} D f}{0.30} \right) \tag{26.1}$$

where η is the antenna efficiency factor (the product of all contributing efficiencies), D is the diameter (in meters), and f is the radio frequency (in GHz).

Equation (26.1) does not take into account the loss of performance resulting from distortions of the reflector due to thermal and dynamic losses. These losses may be approximated by the well-known Ruze equation:

$$G = 683 \left(\frac{\Delta}{\lambda} \right)^2 \tag{26.2}$$

where Δ is the RMS distortion of the surface and λ is the wavelength of the radio frequency, which is related to the radio frequency thus:

$$\lambda \text{ (in meters)} = \frac{0.3}{f} \tag{26.3}$$

It is apparent that Δ must be small compared to the radio frequency in order to minimize the losses. However, the short wavelengths associated with high radio frequencies make it extremely difficult to limit the distortions to acceptable values, and the only alternative is to employ better materials. The design requirements for the antenna ribs are dictated by this need for low reflector distortions.

Hence the material requirements are as follows:

- High strain to yield
- Good creep resistance
- Good resistance to absorption of moisture
- Low thermal distortions
- High bending and torsional stiffnesses
- Good resistance to thermal cycling

Let us examine some of these requirements further.

High Yield Strain

The radial and camber bending stresses in the wrapped reflector are biaxial in nature and, for an isotropic material, are given by

$$\sigma_{br} = \frac{Et}{2(1 - \nu^2)} \left(\frac{1}{R_r} + \frac{\nu}{R_c} \right) \tag{26.4a}$$

and

$$\sigma_{bc} = \frac{Et}{2(1 - \nu^2)} \left(\frac{1}{R_c} + \frac{\nu}{R_r} \right) \tag{26.4b}$$

where E is the modulus of elasticity of the reflector material, ν is its Poisson's ratio, t is the material thickness, and R_r and R_c are the wrap radius of the rib and the camber radius of the cross section, respectively.

For an anisotropic material, these stresses are

$$\sigma_{br} = \frac{t}{2(1 - \nu_{rc}\nu_{cr})} \left(\frac{E_{br}}{R_r} + \frac{\nu_{rc}E_{bc}}{R_c} \right) \tag{26.5a}$$

and

$$\sigma_{bc} = \frac{t}{2(1 - \nu_{rc}\nu_{cr})} \left(\frac{E_{bc}}{R_c} + \frac{\nu_{cr}E_{br}}{R_r} \right) \tag{26.5b}$$

where E_{br} and E_{bc} are the flexural modulus of the antenna material in the direction of R_r and R_c, respectively, and ν_{rc} and ν_{cr} are the Poisson's ratio along the same axes.

As a first approximation, the bending stresses may be determined by simplifying equations (26.5a) and (26.5b) to give

$$\sigma_{br} \approx \frac{E_{br}t}{2R_t} \quad \text{and} \quad \sigma_{bc} \approx \frac{E_{bc}t}{2R_c}$$
$$t/2R \approx e \approx \sigma_b/E \tag{26.6}$$

Equation (26.6) demonstrates the fact that the higher the allowable strain e, the more compact and lighter the overall system can become, dramatizing the need for a material with a high yield strain.

Thermal Loads

Thermal gradients that arise in the ribs (due to shading of a portion of the rib either by an adjacent rib or by other spacecraft or satellite structures) produce deflections normal to the reflector and thus play a major role in the performance of the reflector. The ribs also experience longitudinal deflections due to these thermal gradients. The thermal bending deflections are dependent on the thermal gradients across the ribs and the coefficient of linear expansion α of the rib material. Since the thermal gradients are, in turn, a function of the thermal conductivity k, the resistance of the rib to thermal distortion may be evaluated with the aid of the coefficient of thermal distortion (CTD), given by

$$\text{CTD} = \frac{\alpha \text{ in the longitudinal direction}}{k \text{ in the transverse direction}}$$

An acceptable material must have as low a value of CTD as possible.

Flexural and Torsional Stiffnesses

The performance of a reflector is constrained by the orbital deflections that arise from the orbital loads (from satellite maneuvers, thermal contraction of the reflective surface material, etc.). The loads arising from thermal effects are usually applied orthogonal to the longitudinal axis, and the resulting deflections depend on the rib flexural stiffness, EI. The flexure of the rib results from its transverse motions, and because the rib is curved, such motions give rise to torsional stresses and deflections. It follows that materials with high values of specific tensile stiffness, E/ρ and specific torsional stiffness G/ρ are desirable. High specific tensile stiffness also allows for reduced rib weight and stowed volume.

The values in Table 26.3 show that the best materials are the metal-matrix composites.*

*See Chapter C7 in Part 3.

TABLE 26.3 COMPARISON OF PROPERTIES OF CANDIDATE MATERIALS FOR THE REFLECTOR RIBS OF A SPACE-BORNE PARABOLIC ANTENNA[a]

Material	E_L (GPa)	G_{LT} (GPa)	ρ (kg m^{-3})	α_L (10^{-7} °C^{-1})
Aluminum	71	27	2700	240.0
Graphite fiber				
Epoxy resin filamentary composite[b]	96	4	1550	9.0
Aluminum metal composite [c]	365	26	2380	9.0
Magnesium metal composite[c]	352	17	1827	1.8

Material	k_T (W m^{-1} °C^{-1})	CTD (10^{-9} m W^{-1})	E_L/ρ (10^6 m)	G_{LT}/ρ (10^6 m)
Aluminum	2.40	10,000	2.68	1.02
Graphite fiber				
Epoxy resin filamentary composite[b]	23.40	38.50	6.31	0.26
Aluminum metal composite [c]	38.94	23.11	15.63	1.11
Magnesium metal composite[c]	31.15	5.78	19.64	0.95

[a]The subscripts L and T refer to the longitudinal axis and transverse axis, respectively.

[b]Volume fraction of graphite fibers = 0.62.

[c]Volume fraction of graphite fibers = 0.48.

OPTICAL STORAGE DEVICES

CHAPTER 27

27.1 INTRODUCTION

The list of organizations that require sophisticated and compact facilities for the acquisition and retention of enormous quantities of data and information in general is growing—from subscribers to space information networks (comprising of a fleet of geosynchronously—located communications satellites and real-time data capture, processing, and archiving subsystems) to public libraries, and from research laboratories to the United States Congress. This demand for such facilities is now being met with the aid of optical storage devices. These devices are one of the advanced data recording and storage methods that complement (and should soon supplant) tape, which has the main disadvantage of high cost, ease of degradation of data quality over time, and low storage volume.

The alternative methods of data storage are charge-coupled devices, silicon dioxide materials in conventional recording and high-density fast-access disc configurations, laser holography, and optical digital discs. We shall concentrate on optical digital discs.

Figure 27.1 is a block diagram of a typical optical recording and playback system. For recording, the output of the laser is fed into a modulator that modulates the intensity of the light in response to an input electrical signal. The modulated laser beam is enlarged by the recording optics, is totally reflected by

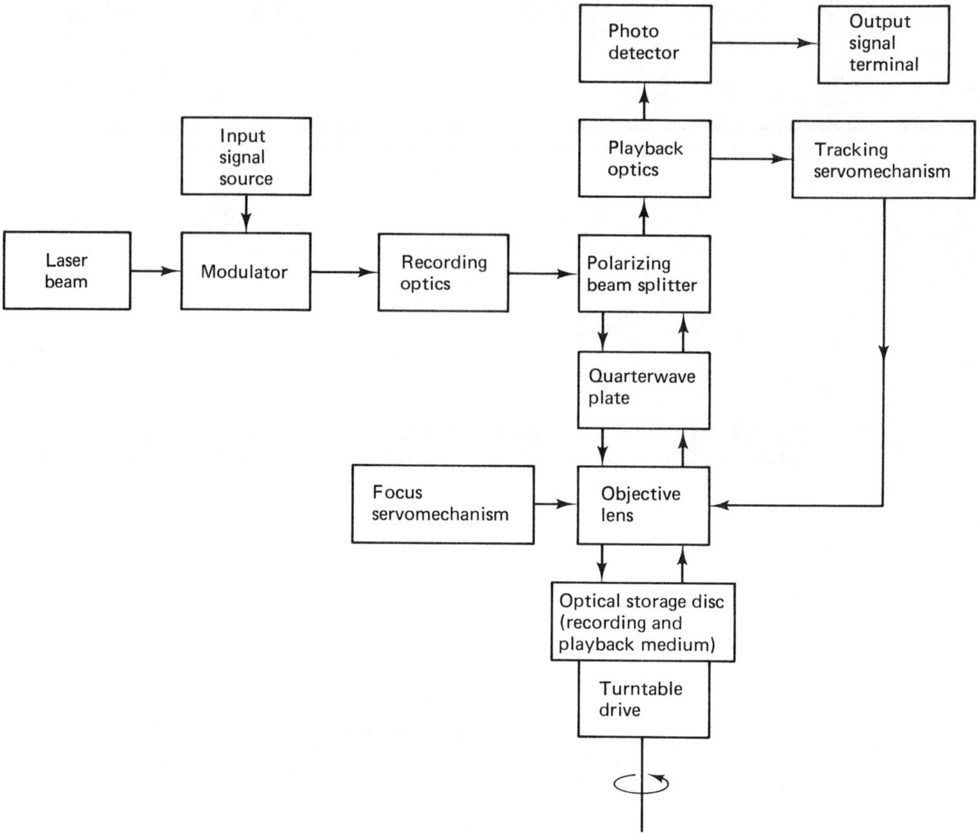

Figure 27.1 Schematic representation of an optical disc recording and playback system.

a polarizing beam splitter, and then passes through a quarter-wave plate into the objective lens. That lens focuses the laser beam onto a small area (mean diameter = 0.001 mm) on the disc so that the intensity modulation of the beam leads to information being stored as a series of "spots" along a spiral track in the surface of the disc. For readout, an unmodulated, less intense laser beam follows a path identical to that taken by the recording beam to get to the disc. That light is reflected from the recorded spots and passes back, via the objective lens, to the quarter-wave plate. The light then passes through the polarizing beam splitter and is directed by the playback optics to a photodetector, which converts the reflected light beam to an electrical signal. That signal is proportional to the original input signal to the modulator. A servomechanism tracks the light through the playback optics to ensure that the laser beam is on the recorded track during playback.

The advantages of an optical disc recording system include these:

- Very high information density (10^{11} to 10^{12} bits per disc)
- Very high data rates (up to 50 Mbits s^{-1} in a single channel recorder)

TABLE 27.1 STORAGE COSTS OF VARIOUS RECORDING SYSTEMS (IN CENT BIT^{-1})

Recording System	1980	1985	1990 (est.)
Optical disc	6.87×10^{-8}	3.33×10^{-8}	10^{-8}
Random-access memory	1.00	0.03	0.015
Magnetic tape			
Longitudinal	8.3×10^{-5}	3.33×10^{-5}	1.45×10^{-5}
Rotary	3.3×10^{-4}	10^{-5}	1.25×10^{-6}
Bubble	0.05	0.01	0.003

- Rapid random access to the data (much less than 1 s)
- Low cost (projected to be about 10^{-8} cent bit^{-1}; see Table 27.1)
- Great archival data storage capacity (10 years or more)
- Fast queuing (start/stop, fast forward, rewind, search, etc.)

The development of an optical storage system has been slow because of the difficulty in finding a suitable recording material. This is so because there are two major fundamental but conflicting demands on materials to be used for this application: high sensitivity at an appreciable wavelength and long-term stability of recorded data (excellent archivability). Other important properties of a proper recording material are these:

- An ability to support spots at least as small as the diffraction-limited laser beam, in order to pack as much information as possible onto the disc
- A high signal-to-noise ratio
- A very low raw bit error rate
- Low cost
- Easy availability
- Low toxicity

27.2 CANDIDATE MATERIALS

Six types of materials are suitable.

Thin Metal Films

These films are based on tellurium or alloys of Group 6A elements,*, in which the focused laser beam can easily produce small holes because of the high melting point (450°C) and low thermal conductivity of the metal.

One of the main advantages of tellurium-based storage media is their fine resolution. The low thermal conductivity means that heat deposited by the laser

*Although rhodium and bismuth have been investigated, thin metal film optical storage media are now almost exclusively based on tellurium.

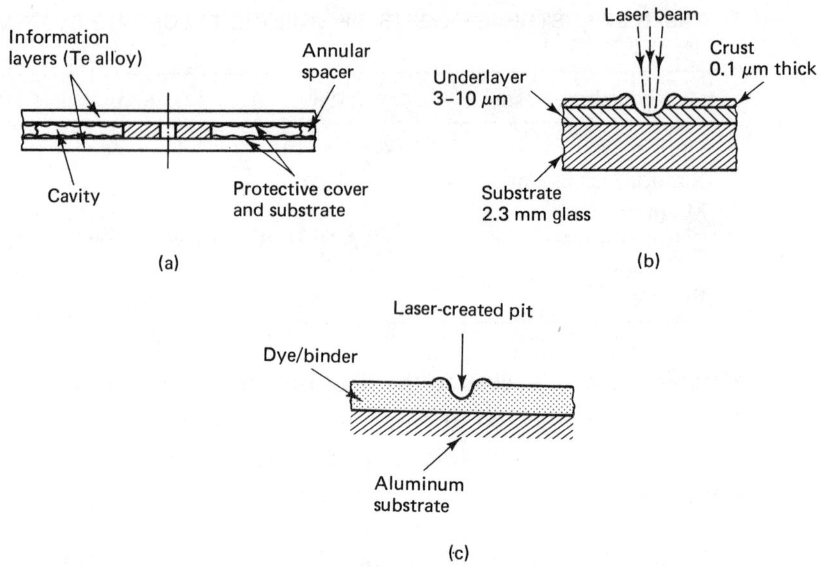

Figure 27.2 Representative archival optical storage media: (a) Philips's Air Sandwich; (b) Drexon recording system; (c) Kodak Phase-Shift System.

stays in a small area (at least for the 100 ns or so that the laser pulse is on). Holes as small as 0.4 μm in diameter can be ablated with an argon laser beam. Because the ablated pits can be packed closely, data storage density on tellurium is very high, on the order of 300 Gbits m^{-2}.

One of the disadvantages of tellurium is its proneness to oxidation due to electrolytic corrosion. But its stability can be improved by alloying it with selenium (although this reduces the infrared absorptive qualities of the resulting material) and antimony (this adversely affects the sensitivity of the resulting film). Another disadvantage is its toxicity. Although elemental tellurium is not poisonous, many of its salts, like the common alloying elements (such as selenium and antimony), are. Therefore, disposal of millions of optical storage discs could constitute a very serious environmental hazard. The three largest tellurium disc research programs are currently based at Philips Laboratories in Briarcliff Manor, N.Y.; the Philips Research Laboratories in Eindhoven, The Netherlands [see Fig. 27.2(a)]; and the RCA Research Laboratories in Princeton, N.J.

Metal-impregnated Plastics

This type is currently available commercially, marketed under the trade name Drexon [Fig. 27.2(b)]. It consists of two layers of an organic colloidal matrix, solvent-coated onto a substrate. A thin upper crust is impregnated with filamentary and spherical particles of silver metal. The thicker underlayer lacks the silver. Adjusting the relative concentration of filamentary and spherical particles alters the reflectivity of the material in the range 30% to 60%.

When the laser beam impinges on the top layer of the Drexon disc, the silver-impregnated material behaves much as tellurium does: It melts, vaporizes, and exposes the nonsilvered underlayer, which reflects less of the read laser light. Because the melting point of the composite material is less than 200°C, a hole 0.8 μm in diameter can be recorded by only 3 mW in a 100-ns pulse.

Organic Dye-Polymer Binder Systems

These consist of a dye-polymer medium in which the laser-created pit has a reflectivity almost identical to that of the unpitted surface. The basis of a recording system using this type of material is that the pit shifts the phase of the reflected light so as to interfere destructively with the incoming beam and thereby reduce the reflected intensity.

Several dye-substrate systems have attracted study for this application, including these:

- Triphenylmethane dye (~16 wt %) on nitrocellulose
- Anthroquinone or diazomethine on polyethylene terephthalate
- 4% oil red O or oil black BT on polymethylmethacrylate (PMMA), polyoxymethylene, or poly-d-MeSty
- Hydroxyl squarylium dye on PMMA (see Fig. 27.3)

The dye and a polymeric binder are solvent-coated onto a substrate to produce an amorphous film of high optical quality. The thickness of the dye-polymer layer is chosen to minimize reflection and maximize absorption at the wavelength of the write laser.

Two companies currently have large research and development investments in a recording system using these materials. In the Eastman Kodak process, the dye-polymer has a hole-forming threshold of about 1 kJ m^{-2}, similar to that of tellurium-based materials [Fig. 27.2(c)]. In the IBM process, the dye-polymer combination is hydroxyl squarylium/PMMA. It has been reported that for 100-ns pulses, a trilayer (40 nm of dye on top of 100 nm of transparent silium dioxide on top of 30 nm of aluminum) on a PMMA substrate has a sensitivity of about 6 mW of laser power. However, for 50-ns pulses, which would be necessary to write at 20 Mbits s^{-1}, the dye reportedly needs 60% more laser power than a tellurium-alloy disc.

Photographic and Electrophotographic Media

The best of such media is silver halide film, which is attractive for several reasons, particularly long-term stability (excellent archivability), low cost, and ready availability in large quantities. However, two main disadvantages are associated with the use of this medium. First, the exposed film must be processed chemically

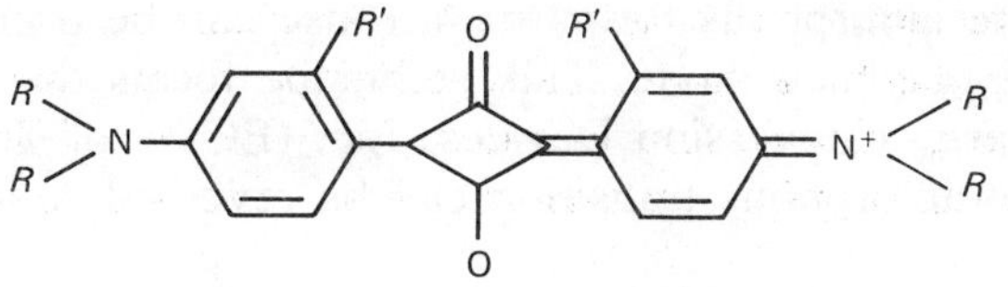

Figure 27.3 Basic structure of squarylium dyes. For hydroxyl squarylium, $R = CH_3$ and $R' = OH$.

before it can be read. This prevents the use of the common DRAW (direct read after write) error-correction schemes. Second, the cost is not as competitive as first projected. For low storage density, a film with a resolution of 5 μm is clearly cheap, but for film with a resolution of 1 μm or better (which is needed for high-capacity memories), the film costs as much as a tellurium-alloy disc.

Bit storage on electrophotographic media is similar to recording on an electrostatic copier. A scanning laser beam writes a bit pattern onto a photoconducting plate, and the application of a chemical toner transfers the pattern onto a polyester substrate that has been fused to the plate. Electrophotographic media have recording thresholds on the order of 100 to 300 kJ m^{-2}. Data density is limited by the resolution of the toner, which is currently about 0.7 μm.

Bubble Media

In this application, the laser beam creates tiny bubbles whose reflectivity differs from that of the flat surface. The main advantage here is the life of the recording system; because the top surface of the medium need not be vaporized, it can be made from a high-melting-point material.

Two companies that are currently experimenting with bubble media for optical digital discs are 3M of Minneapolis and Thomason-CSF of France. 3M uses a bubble medium consisting of a high-melting-point-material layer, an intermediate layer chosen for its outgassing properties, and a bottom reflective surface. When the laser beam impinges on this medium, the vapor produced by the middle layer pushes the top surface upward to create a bubble, typically 0.4 to 0.8 μm in diameter.

Thomson-CSF Central Research Laboratories in Orsay, France, has used a film of gold or platinum on top of a polymer layer, on top of a plastic substrate. A signal-to-noise ratio of 60 dB with a helium-neon laser power of 10 mW at 10 MHz has been reported.

Erasables

Corning Glass Works in Corning, N.Y., is developing erasables using photochromics. The write beam from a helium-neon laser imprints a polarization preference on the previously isotropic medium. Since the material is unchanged by the infrared wavelength, a polarized GaAs diode laser can read the data without altering it. To erase the data, the laser, with its polarization rotated from the original angle, writes over the old data. The main disadvantages of this system are sensitivity to light—only a few minutes in a well-lit room can bleach the data away—and relative insensitivity for a given contrast ratio.

Another erasable recording technique is to induce a crystal-to-amorphous transition in a tellurium alloy. Because the crystalline form is both more reflective and more opaque than the amorphous form, such a disc can be used in either reflective or transmissive playback mode. This technique forms the basis of a system developed by Energy Conversion Devices, Inc. (ECD), of Troy, Mich. Because the crystalline-to-amorphous transition can be reversed simply by applying energy, data recording is completely erasable. Researchers at Philips are also working on an erasable disc, using this principle, that is coated with a mixture of gallium and indium antimonides.

Magnetooptics is a third technique being investigated. A laser beam heats the surface of an amorphous magnetic film above its Curie temperature. Subsequent cooling in the presence of an external magnetic field alters the magnetization direction of the heated spot in a predictable way. Data may be erased by simply reversing the magnetic field direction. The packing density in this system is poor by optical disc standards because each magnetized spot interacts with its neighbors, preventing tight packing. One of the companies leading research and development efforts in this field is KDD Research and Development Laboratories of Tokyo. KDD has reported data density of about 50 Gbits m^{-2} on a glass disc coated with films of terbium-iron or gadolinium-terbium-iron.

BLANKET OF A FUSION REACTOR

CHAPTER 28

28.1 INTRODUCTION

The attainment of economic and safe fusion power is arguably one of the major challenges in the fields of energy engineering and physics. In recent years, there have been a number of successful experiments in Europe, the United States and Japan whose cumulative impact has been to induce confidence that the ultimate goals—of a demonstration reactor and subsequent commercial-scale operation—are achievable by the early part of the 21st century.

Quite simply, the principle of a fusion reactor is the confinement of a very hot plasma away from the physical walls of the reactor for a sufficiently long time to permit the ion to undergo a fusion reaction (Fig. 28.1). Such confinement is achieved in one of two ways, namely magnetic and inertial. One form of magnetic confinement, namely toroidal (tokamak) has attracted the most study to date, and is likely to form the basis of a commercial unit. (Figs. 28.2 and 3). In the tokamak, confinement of the plasma is in a helical magnetic field created by the superposition of a strong externally-generated toroidal field produced by super-conducting coils and the poloidal field generated by the plasma current.

The largest currently operating tokamak in the world is the Joint European Torus (JET) [Fig. 28.4] situated in Abingdon, Oxford, England, which was brought into operation on June 25, 1983. The objective of this project is to obtain and

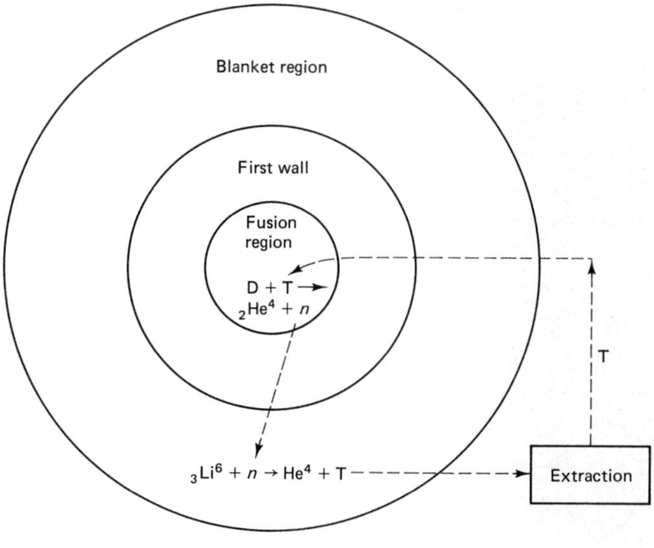

Figure 28.1 Simple representation of the principle of operation of a fusion reactor.

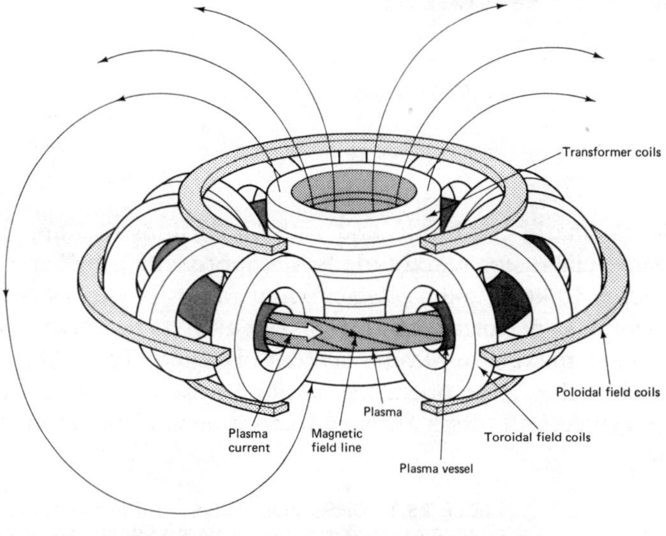

Figure 28.2 The basic tokamak configuration.

Figure 28.3 Schematic representation of a tokamak fusion power plant.

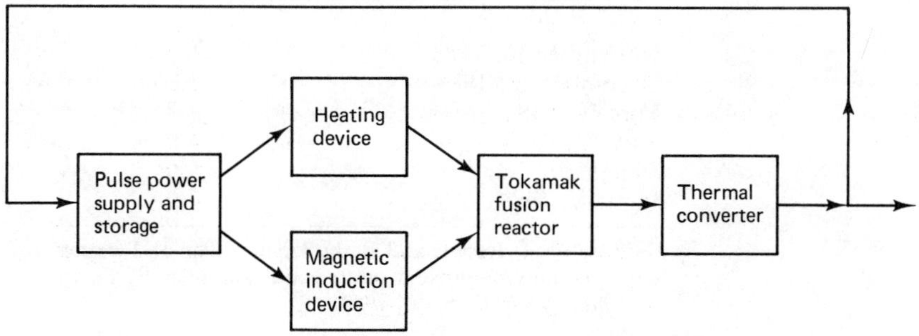

1 Magnetic circuit (core and eight return sections)
2 Mechanical structure
3 Toroidal field magnet of 32 D-shaped coils
4 Outer poloidal field coil
5 Double-walled vacuum vessel
6 Inner poloidal field coils (primary or magnetising windings)

Figure 28.4 Major components of the JET (Courtesy of JET joint undertaking, Oxford, England)

study plasma in conditions and with dimensions that approach those needed for a commercial fusion reactor (densities approaching $10^{20}\ m^{-3}$, plasma temperatures exceeding 10 keV, and a plasma minor radius in excess of 1 m). An indication of the improved performance of this system in its first two years of operation is the increase in maximum plasma current obtained (Table 28.1).

The other large operating tokamaks are the Tokamak Fusion Test Reactor at the Princeton Plasma Physics Laboratory in Princeton, N.J., and the JT-60 in Japan.

TABLE 28.1 CHRONOLOGICAL IMPROVEMENT OF PERFORMANCE OF THE JOINT EUROPEAN TORUS (JET)

Date	Maximum Plasma Current Obtained (MA)
June 25, 1983	0.1
November 10, 1983	1.2
December 16, 1983	2.2
May 10, 1984	3.2
May 23, 1985	4.0
June 5, 1985	5.0
January, 1989	7.0

Source: P. H. Rebut, and B. E. Keen, "The JET experiment: evolution, present status, and prospects," *Fusion Technology 11* (1), p. 30, 1987.

28.2 FUSION REACTIONS

Although there are a number of fuel cycles that are useful in a fusion reactor, the one most likely to be exploited first involves the reduction of the hydrogen isotopes deuterium (D) and tritium (T) to produce a 3.5 MeV helium ions and 14.1 MeV neutrons, that is

$$_1D^2 + {_1T^3} \longrightarrow {_2He^4} + n + 17.6 \text{ MeV}$$

Although deuterium is readily available, tritium is not found in nature and is bred in the reactor by the interaction of neutrons with a suitable material, such as lithium; that is

$$_3Li^6 + \text{``slow'' } n \longrightarrow T + He^4 + 4.8 \text{ MeV}$$

$$_3Li^7 + \text{``fast'' } n \longrightarrow T + He^4 - 2.5 \text{ MeV} + n' \text{ (``slow'')}$$

28.3 DESIGN ASPECTS AND COMPOSITION OF BLANKET SYSTEM

The minimum conditions to be met in order to have a self-sustaining fusion reactor are defined in terms of the containment parameter, $N\tau$ (where N is the plasma density and τ is the confinement time) and is given by the so-called Lawson criterion. For the D-T reaction, this criterion reduces to

$$N\tau = 3NkT$$

where k is the Boltzmann's constant and T is the absolute temperature of the plasma.

The blanket system consists of (a) breeder of tritium; suitable candidates include Li, 17 Li-83Pb, Li_2O, Li_8ZrO_6, Li AlO_2, and Li_2BeF_4, (b) a coolant (suitable candidates include water, Li, He, nitrate salts, and 17 Li-83Pb), (c) a neutron multiplier (suitable candidates are beryllium and lead), and (d) a structural material.

In the present case study, we focus on the selection of the structural material.

28.4 DESIGN CONDITIONS

Operating conditions must be considered when selecting materials for the blanket. These conditions are very hostile:

- Magnetic field of 4 to 12 T
- Initial vacuum of 10^{-7} Pa
- Temperature range of 150°C to 700°C
- Minimum coolant pressure of 5 MPa
- Heat flux of 0.2 to 0.8 MW m^{-2}
- Heat pulses of 10^4 to 10^5 per year
- Lithium corrosion

- High concentrations of tritium and deuterium
- Presence of lithium-breeding material
- Presence of coolant (lithium, water, helium, etc.)
- High neutron fluxes, on the order of 1 to 4 MW m^{-2} of 14.1-MeV neutrons and 3×10^{18} to 12×10^{18} nm^{-2} s^{-1} total neutrons

28.5 CANDIDATE MATERIALS

One major consideration is the thermal cycling that the blanket structural materials are subjected to in the reactor. Thus, materials used must have high thermal shock resistance; otherwise, high cyclic thermal stresses leading to fast crack growth could result. The appropriate shock resistance index in this case is R_1 (see Chapter 7). R_1 is given by

$$R_1 = \sigma_F k \frac{1 - \nu}{E\alpha}$$

In this case σ_F is put equal to σ_y, the yield strength of the material. The values of the relevant properties for various candidate materials and the respective values of R_1 are presented in Table 28.2.

Another major consideration is the amount of radiation damage incurred by the blanket material. It is now recognized that such damage is of two types: displacement damage (DP) and gas production (H, He) damage.

The values of R_1 and radiation damage for seven candidate materials are given in Table 28.3. It appears from these results (R_1 and radiation damage) that the V-20Ti and Nb-Zr alloys are the best materials. Four caveats must be attached to this suggestion, though. First, these alloys are very expensive, have poor compatibility with coolants and tritium, have a limited available property database, and are composed of metals that are not easy to obtain in the United States. Second, final selection of materials must be made following the generation of data related to the effects of radiation on their physical and mechanical properties. Apart from the stainless steels, such data are inadequate. Third, very small amounts

TABLE 28.2 PROPERTIES OF CANDIDATE MATERIALS FOR THE BLANKET OF A FUSION REACTOR AT 500°C

Material	Yield Strength, σ_y (MPa)	Thermal Conductivity, k (W m^{-1} K^{-1})	Poisson's Ratio, ν	Young's Modulus, E (GPa)	Coefficient of Linear Expansion, α (10^{-6} K^{-1})	R_1 (kWm^{-1})
Annealed 316 steel	205	12.8	0.26	193	17.8	0.56
20% cold-worked 316 steel	380	12.8	0.26	190	17.8	1.06
Ferritic steel, HT9	275	22.8	0.26	200	17.8	1.30
Ti-6Al-2Sn-4Zr-2Mo	830	7.1	0.33	114	7.7	4.50
Iconel 718 alloy	931	12.0	0.34	200	16.0	2.30
V-20Ti	700	28.0	0.36	105	9.0	13.27
Nb-1Zr	138	41.9	0.38	70	7.5	6.83

TABLE 28.3 RADIATION DAMAGE OF CANDIDATE MATERIALS FOR THE BLANKET OF A FUSION REACTOR

	Radiation Damage In a 10-MW y m^{-2} Lifetime		
Material	DP	At ppm H	At ppm He
Annealed 316 steel	110	5320	1470
20% cold-worked 316 steel	110	5320	1470
Ferritic steel, HT9	110	4500	1100
Ti-6Al-2Sn-4Zr-2Mo	N.A.	N.A.	N.A.
Inconel 718 alloy	120	7800	2400
V-20Ti	110	2430	580
Nb-1Zr	70	1050	290

N.A., Data not available.

Source: R. Nygen, "Materials for magnetic fusion," *Proc. IEEE 69,* pp. 1056–1068, August 1981.

of carbon, nitrogen, and oxygen picked up by these refractory alloys induce embrittlement in them. Fourth, the achievement of the purity levels in helium required to ensure the integrity of structures made of these alloys is so high (<1 part per trillion) as to restrict their use severely.

By contrast, niobium- and vanadium-based alloys are good candidate materials for use with liquid lithium (provided that carbon and nitrogen impurity levels are low). This is so because of the difficulty of the lithium-solubility of certain alloying elements.

All in all, then, blanket structures may be made of either vanadium- and niobium-based alloys or stainless steels. In fact, these materials have been recommended in studies of so-called best systems fusion designs carried out at the Argonne National Laboratory [1,2].

Although in all probability the structural material for the blanket will be similar to one of the alloy classes in Table 28.2, the composition and microstructure will be optimized for the fusion application, especially in the light of the degrading effect arising from the presence of helium in grain boundaries. These effects are reflected in reduced ductility and rupture strength. Helium may be kept off grain boundaries by, for example, (a) using a lower operating temperature (less than about 40% of the melting point of the material), (b) cold work, and (c) controlled precipitation of minor elements; for example small amounts (<0.25 wt %) of titanium in 316 stainless steel result in finely divided titanium carbide particles that serve as preferential helium bubble nucleation sites and thus keep helium away from the grain boundaries.

REFERENCES

1. D. L. Smith, et al., "Fusion Reactor Blanket/Shield Design Study," Argonne National Laboratory, Report ANL/FPP-79-1, Argonne, Ill., 1979.
2. D. L. Smith, et al., "Blanket Comparison and Selection Study," Argonne National Laboratory, Report ANL/FPP/84-1, Argonne, Ill., 1985.

HEATER TUBING ASSEMBLY OF AN AUTOMOTIVE STIRLING ENGINE

CHAPTER 29

29.1 INTRODUCTION: DESIGN REQUIREMENTS AND CANDIDATE MATERIALS

In a Stirling engine (Fig. 29.1) a fixed amount of gas is contained in a working volume comprising at least one region that is held at a high temperature and another at a lower temperature. Some of the gas is continuously transferred between these regions using a piston called a displacer piston. The gas pressure fluctuation results in the movement of another piston called the power piston,

Figure 29.1 The basic parts of a Stirling engine.

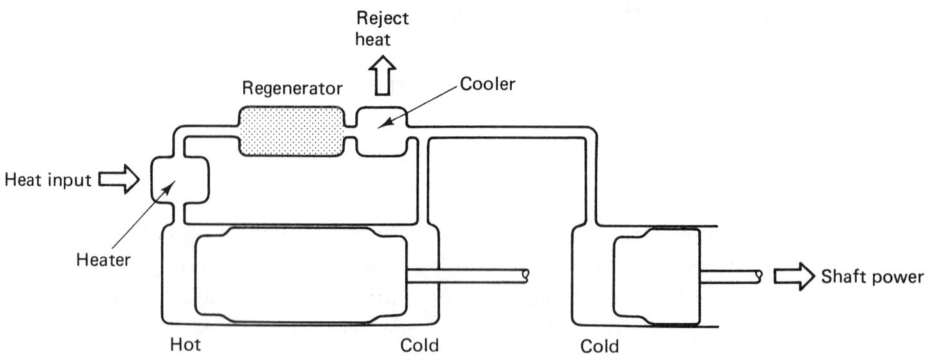

leading to mechanical work being done. In almost all real engines the pistons move simultaneously and almost sinusoidally. A regenerator, placed between the hold and cold ends of the system, is really a heat exchanger and heat store through which the gas flows during the displacer's motion. Most modern engines operate at high speed (typically 3000 r min^{-1}), resulting in near-adiabatic gas behavior in the cylinders. Thus an external heater or cooler, with a large surface area, is fitted in order to add the heat absorbed during the expansion phase and to reject the heat generated during the compression phase (the difference between these two heat loads is the mechanical work available from the power piston). The Stirling engine has five main unique features, these being:

- A variety of heat sources (providing the heat input) can be used; these range from the usual liquid fossil fuel (gasoline) to the experimental hydrogen gas and from the innovative ^{90}Sr isotope to the cheap biomass waste
- Smooth pressure variations and continuous combustion which result in a quiet engine with a torque distribution that is essentially invariant with engine speed
- A suitable choice of working fluid, say hydrogen or helium gas, having low density, low viscosity and good heat transfer properties, leads to a substantial reduction in flow losses and a major increase in heat transfer rate. This is because the working fluid is retained in the engine
- Most of the rejected heat is transferred by the coolant and thus easily recoverable for further use
- The cycle is thermodynamically efficient because heat is added only to gas that is already at a high temperature and removed only from gas that is already at a low temperature

Here is a list of some of the major research and development efforts in the application of Stirling engines:

- Implantable power source for artificial human hearts (the University of Washington in the United States)
- Power packages for underwater and marine operation (MAN-MWM in West Germany, United Stirling in Sweden, and Société ECA in France)
- Power converter for a liquid metal-cooled spaceborne fast nuclear reactor*
- Ground power units for the United States Army (Mechanical Technology, Inc., [MTI], Sunpower, Stirling Power System, and the University of Washington, all in the United States)
- Drivers in heat pumping systems (MTI in the United States, Kawasaki Co. in Japan, the Technical University of Denmark, and the Technical Research Center of Finland)
- Power sources for remote stations such as navigation beacons, lighthouses, telemetry stations, and various applications in the rural areas of developing countries (Ho Mach Systems Ltd. in England, and Sunpower in the United States)

*See Question 32, in Part 2.

But by far the largest single research and development programs are to do with automotive engine applications. We focus on this aspect in this case study. Following the passage of the Automotive Propulsion Research and Development Act of 1978, in the United States, the government initiated the Automotive Program which was directed by the Department of Energy (USDOE). Engines based on the Stirling cycle were quickly identified as cleaner, more efficient alternatives to internal combustion engines (based on the Otto and Diesel cycles) for automotive use. The program which is now managed on behalf of the USDOE by the NASA—Lewis Research Center in Cleveland, Ohio, has been at the forefront of research and development work on automotive Stirling engines. Due to federal research budget cuts, the current status of the program is under a cloud, and there has been some major redefinition of program goals. For example, the emphasis now is on science and technology rather than demonstrations. In addition to this program, other major research and development players are United Stirling in Sweden, and MTS and AM General Corporation in the United States (the Stirling engine program at Ford Motor Co. was phased out in 1978, while both General Motors in the United States and FFV Industrial Products in Sweden have sizeable research programs in Stirling engines in general but have not shown interest in automotive Stirling engines in particular).

Otto and Diesel cycle engines are clearly the leaders in today's automotive prime-mover market, with their preeminence likely to continue in the near term. Years of dedicated research and development efforts have produced engines which are unquestionably effective and reliable. But the automotive Stirling engine can present a plausible challenge to these "old faithfuls" because of a number of advantages of the former relative to the latter:

- Higher values of available energy density and indicated thermal efficiency* with comparable values for power density (Table 29.1). The indicated thermal efficiency is primarily a measure of the engine's performance because it excludes losses in external components of the system
- Fuel economy: at least 30% better than the latest internal combustion engines fitted in comparable production vehicles travelling the Environmental Pro-

TABLE 29.1 SOME CHARACTERISTICS OF OTTO, DIESEL, AND STIRLING ENGINES

Engine	Available Energy Density (kWh kg^{-1})	Indicated Thermal Efficiency (%)	Power Density (kW kg^{-1})
Otto	2.5 (gasoline)	27	0.38
Diesel (compression-ignition)	2.8 (diesel fuel)	32	0.30
Stirling[a]	3.0 (diesel fuel)	35	0.35

[a]Based on performance data for experimental Stirling engines available in 1986.

*Defined as the ratio of the indicated power to the sum of the heat input to the engine head and any other power put into the engine to make it run.

TABLE 29.2 EXHAUST EMISSIONS FROM TWO EXPERIMENTAL AUTOMOTIVE STIRLING ENGINES (IN G PER VEHICLE KM)

Pollutant	Maximum Permissible Level Set by the EPA	Stirling Engine and Test Automobile: MTI's MOD-1 in an AMC Sprite[a]	Stirling Engine and Test Automobile: United Stirling's P-40 in an Opel Sedan[b]
No_x	0.25	0.31	0.25
Hydrocarbons	0.26	0.14	0.08
Carbon monoxide	2.13	1.20	0.27
Particulates	0.13	—	—

[a]Source: W. H. Haverdink, et al, "Assessment of an Experimental Stirling-Engine-Powered Automobile," General Motors Report GMR-4817, August 1984.

[b]Source: Anon, "Assessment of the State of Technology of Automotive Stirling Engines," Mechanical Technology Incorporated, Albany, N.Y., Report number NASA CR-159631, September 1979.

tection Agency (EPA) combined city/highway driving cycle (for example, the figures for the 1988 Ford Tempo car were 17/24 kml^{-1})

- Low pollution; the United States legal automotive exhaust emission levels are easily achievable (Table 29.2) because of the well-controlled combustion characteristics that can be obtained in the Stirling engine
- Use of a wide array of different fuels without loss of efficiency, increase in exhaust emissions, or modifications to the engine
- Potentially longer service life and reliability partly because of (a) the separation between the combustion products and the moving parts and (b) a simpler electrical system
- Low noise: a well dynamically-balanced kinematic Stirling engine can run very quietly producing an average of 60 dB of noise (Table 29.3), the same level as conversational speech.

The two production Stirling engines are the SPIKE (a free piston machine designed in 1980 at Sunpower, Athens, Ohio) and the 4-95 or P-40 (a kinetic machine, developed by United Stirling AB of Sweden) (Fig. 29.2). For automotive application, the working fluid of the P-40 is hydrogen and the engine includes a number of auxiliary components: oil pump, combustion air blower, cooling water pump, alternator, hydrogen compressor, fuel atomizing air compressor, and a hydrogen pump. The preheater and combustion air blower are the most striking features of this version (Fig. 29.3).

In this case study, we concentrate on material selection considerations for the heater head and heater head tubing in the P-40 engine (Fig. 29.4). The heater tube assemblies connect the regenerators (totalling eight in all) to the cylinder head. The tubes (outside diameter 4.5 mm and wall thickness 0.75 mm, and totalling 72) rise vertically from manifolds attached to the regenerator heads (there are eight manifolds) for about 0.1 m and then turn upward and downward along an involute. The heater tubing assembly (the head and the tubes) is arguably one

TABLE 29.3 NOISE FROM STIRLING AND DIESEL-POWERED AUTOMOTIVE VEHICLES

Type of Noise	Noise Pressure Level (in dB) in a Stirling Engine	Diesel Engine
Drive-by noise		
—left side	67	79
—right side	68	80
Stationary vehicle noise at 7 mm from the car		
—front	78	79
—rear	68	72
—left side	73	77
—right side	68	74
Interior noise (microphone near driver's right ear)		
—at 70 kmh^{-1}	64	71
—at 80 kmh^{-1}	65	73
—at 90 kmh^{-1}	67	74
—at 100 kmh^{-1}	73	77

Source: Anon, "Assessment of the State of Technology of Automotive Stirling Engines," Mechanical Technology Incorporated, Albany, N.Y., Report Number NASA CR-159631, September 1979.

Figure 29.2 The main components of the P-40 Stirling engine (courtesy of United Stirling AB).

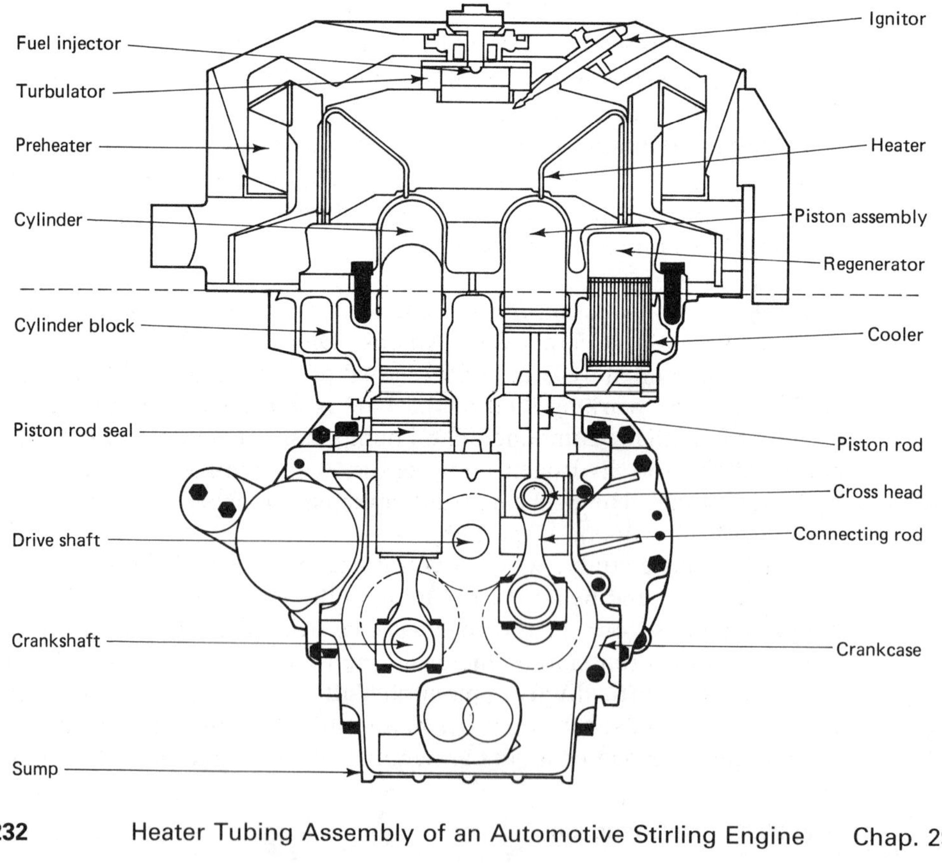

Figure 29.3 A P-40 Stirling engine adapted for automotive use (Courtesy of United Stirling AB).

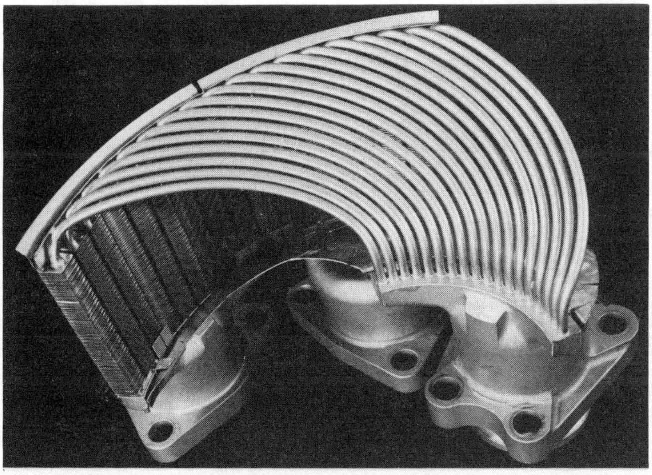

Figure 29.4 One quadrant of the heater head, heater tubes, cylinder, regenerator housing and fins on outer tube row of the P-40 Stirling engine (Courtesy of United Stirling AB).

of the most important engine components. Selection of proper materials thus has far-reaching engineering and economic implications. Two important considerations in material selection are:

- Containment of hydrogen at high operating temperatures and pressures (typically 750°C and 15 MPa, respectively)
- Ability to resist the long-term degrading effects of hydrogen on mechanical properties

On the basis of these requirements the most suitable candidate materials are iron-nickel-based alloys. The compositions of four of these are given in Table

TABLE 29.4 CHEMICAL COMPOSITION OF HIGH-TEMPERATURE ALLOYS (WT %)

Element	19-9DL	CRM-6D	A-286	XF-818 Cast Alloy
C	0.31	1.11	0.05	0.23
Si	0.80	0.48	0.16	0.38
Mn	0.83	3.30	0.11	0.17
P	0.023	0.016	0.015	0.009
S	0.01	0.020	0.008	0.015
Fe	66.28	63.46	55.76	53.28
Ni	9.28	5.45	25.05	18.80
Cr	18.86	22.67	14.16	18.21
Mo	1.48	1.14	1.72	7.30
W	1.30	1.05	—	—
Cb + Ta	0.42	1.20	—	0.42
Ti	0.20	—	2.35	—
Co	—	—	—	—
Other	0.13Cu	—	0.16Al 0.004B	1.1B

29.4. Each of these materials has been evaluated against the background of the major requirements.

Values for permeation of hydrogen through the material have been obtained at temperatures ranging from 650°C to 815°C, leading to the values for the preexponential coefficient A_p and activation energy Q_p in the Arrhenius equation

$$\text{Rate of permeation} = A_p \exp\left(-\frac{Q_p}{\text{RT}}\right)$$

These are given in Table 29.5.

Diffusion values for the same conditions yielded the values of the preexponential term A_d and the activation energy Q_d in the equation

$$\text{Rate of diffusion} = A_d \exp\left(-\frac{Q_d}{\text{RT}}\right)$$

There are also presented in Table 29.5.

Results are also presented in the table for the values of the preexponential term A_s and the activation energy Q_s for the solubility of hydrogen in the candidate materials. It is seen in Table 29.5, that alloy 19-9DL is the best material.

TABLE 29.5 VALUES OF THE PREEXPONENTIAL COEFFICIENT AND ACTIVATION ENERGY FOR PERMEATION, DIFFUSION, AND SOLUBILITY OF HYDROGEN IN CANDIDATE MATERIALS

Material	A_p ($\text{cm}^{-1}\ \text{s}^{-1}\ \text{MPa}^{-0.5}$)	Q_p (kJ mole^{-1})	A_d ($\text{cm}^2\ \text{s}^{-1}$)	Q_d (kJ mole^{-1})	A_s ($\text{MPa}^{-0.5}$)	Q_s (kJ mole^{-1})
19-9DL	0.140	80	0.032	65	4.200	15.0
CRM-6D	0.920	69	0.015	54	6.100	15.0
A-286	0.021	60	0.033	62	0.600	−1.7
XF-818	0.011	52	0.013	42	0.850	2.5

Source: S. Bhattacharya, E. J. Vesely, and V. L. Hill, "High pressure/high temperature hydrogen permeability in candidate Stirling engine alloys," *J. Mater. Energy Sys. 3*, pp. 12–21, April 1982.

APERTURE PLATE OF A SOLAR THERMAL DISH COLLECTOR

CHAPTER 30

30.1 INTRODUCTION

Point-focus is one type of active tracking collector used in a solar thermal system. Global irradiation is focused by a concentrator (which could take the form of spherical lens, circular Fresnel lens, Fresnel mirror, or paraboloid mirror) onto a receiver, where the radiation is absorbed and its energy transferred, with minimum losses, to heat a working fluid. Figure 30.1 shows schematic representations of reflecting concentrators. The hot fluid emanating from the receiver may be used directly (as in a solar industrial process heat system) or indirectly (in which case the hot fluid is fed to a heat engine that converts the thermal energy into mechanical energy used to drive a generator to produce electrical power).

A plant employing such a solar system typically has one or more collectors, each comprising of a concentrator (5 to 15 m in diameter) with a receiver (usually a cavity type) positioned near its focus (focal ratio ranging from 0.5 to 1.0).

In operation, the concentrator is oriented directly toward the sun and through tracking devices (which rotate the concentrator about two axes) follows its motion as faithfully as possible. The concentrated radiation enters the receiver through an aperture. Should a malfunction of the system occur, the concentrator may come to rest. Then, as rotation of the sun relative to the earth proceeds, the spot of concentrated radiation slowly moves away from the aperture of the receiver

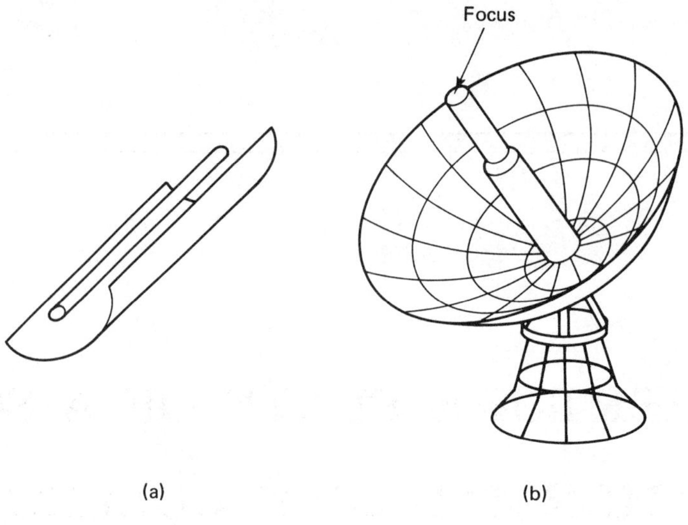

Figure 30.1 Reflecting solar concentrators; item (a) 2-dimensional type; item (b) 3-dimensional type.

and across various sections of the system (edge of concentrator, receiver faceplate, aperture plate, etc.). When this occurs, those sections are subject to high dosages of heat (temperatures in excess of 3000°C are not uncommon), and severe damage may result. (The ''sunspot'' at the focus of the concentrator is about 0.2 m in diameter, with the flux density at or close to the center of that spot being on the order of 1 to 15 MW m^{-2}).

30.2 MATERIAL SELECTION CONSIDERATIONS

The dominant consideration in the selection of materials for the aperture plate is, clearly, the ability of the material to withstand the effects of ''sunspot walk-off'' without the aid of active cooling. It should be noted that, since the speed of the ''sunspot'' is the same as the earth's rotational speed, it takes about 15 min for the spot to traverse a distance equal to its diameter.

During normal operational modes in many solar receiver power plant designs, a ''sun circle'' moves across and off the aperture plate every time the concentrator is rotated to direct it at or away from the sun. These are called the acquisition and deacquisition positions. Although the active times in each of the positions are mere seconds, over the course of a day the aperture plate could be subjected to a large number of acquisition-deacquisition cycles. Therefore, another consideration in the selection of an aperture plate material is its ability to withstand a large number of these cycles.

A third requirement is the ability of the material to withstand lip heating, which arises from the small proportion of the solar radiation that fails to pass through the aperture but instead hits the edge of the aperture plate. This ''radiation overspill'' is the result of a number of factors—the diffuse nature of global irradiation, tracking errors, warpage of the collector during operation, among others.

30.3 CANDIDATE MATERIALS

Ten materials are good candidates for dish collector aperture plates:

- Graphite, with or without heat-resistant coatings
- Silicon carbide
- Silica
- Silicates (such as mullite, cordierite, and processed kaolin)
- Alumina
- Zirconia
- Steel
- Polytetrafluoroethylene (PTFE)
- Aluminum, with or without heat-resistant coatings
- Copper, with or without heat-resistant coatings

The performance of seven of these materials in relevant tests may be gleaned from data summarized in Tables 30.1 and 30.2. These results show that the best material is graphite grade G90, which is an extruded material that is impregnated a number of times with coal-tar pitch and then regraphitized to reduce its porosity and increase its bulk density. It is premium-grade graphite and hence quite expensive (1988 price, \$50 kg^{-1}), making the raw material cost of an aperture plate (typically about 15 kg in mass) nearly \$800.

A commercial grade of extruded graphite, type CS, also performed reasonably well in all the tests. Since the unit price of this material is about an order of magnitude less than that of grade G90, it should also be considered a viable candidate. When using a grade CS in the design of an aperture plate, it is necessary to pay close attention to the fact that its properties are markedly anisotropic (with respect to the direction of extrusion). Suitable strengthening procedures could be introduced to compensate for this effect.

TABLE 30.1 PERFORMANCE OF CANDIDATE MATERIALS IN "WALK-OFF" TESTS[a]

Material	Thickness (mm)	Time (s)	Failure Mode
Graphite, type CS	14–50	10–840	Cracked halfway across but did not fall apart
Graphite, type G90	24–26	1800	Survived
High-purity slip-cast silica	18–21	90–240	Slumped
Mullite	32–38	1–4	Melted
ZrO_2, cast and sintered	29	20	Melted
Copper	26	60–180	Melted
PTFE	38	120	Melted

[a]Tests conducted in the Test Bed Concentrator at the Jet Propulsion Laboratory's Parabolic Dish Test Site, Edwards AFB, Calif. Actual insolation = 720 W m^{-2}; peak flux density = 7 MW m^{-2}.

Source: L. D. Jaffe, "Solar tests of aperture plate materials for solar thermal dish collectors," DOE/JPL-1060-62 (JPL Publication No. 83-68). Pasadena: Jet Propulsion Laboratory, California Institute of Technology, 1983.

TABLE 30.2 PERFORMANCE OF GRAPHITE MATERIALS IN "WALK-OFF" OXIDATION, ACQUISITION-DEACQUISITION, AND LIP HEATING TESTS

Material	Oxidation during "Walk-off": Loss of Thickness, after 15 min, Due to Oxidation (mm)	Acquisition/Deacquisition Test: Loss of Thickness after 700–2000 Cycles[a] (mm)	Lip Heating Test: Oxidation of Material[b]
Graphite, type CS	0.2–8.0[c]	5.0–7.0	Negligible
Graphite, type G90	0.2–8.0[c]	—	Negligible

[a] Equivalent to 1–2 y service; this loss of thickness, which is equivalent to 15% to 20% of the sample's weight, is acceptable.

[b] This test approximately represented the conditions for an organic Rankine receiver with an aperture diameter of about 0.36 m. Flux at the lip = 10 kW m^{-2}; lip temperature = 107°C.

[c] Equivalent to a loss of mass of 2% to 22%; this loss rate is acceptable for walk-off protection because walk-off is expected to be an infrequent event, and the test was more severe than anticipated service situations.

Source: L. D. Jaffe, "Solar tests of aperture plate materials for solar thermal dish collectors," DOE/JPL-1060-62 (JPL Publication No. 83-68). Pasadena: Jet Propulsion Laboratory, California Institute of Technology, 1983.

Part II

Questions and Sample Answers

QUESTION 1
Components of a Military Helicopter

While the civilian helicopter market has been languishing, there has been a growing appreciation of rotary-wing aircraft on the military side as the "battlefield tanks" of the future. As a result of the growing array of threats to military helicopters (such as ground-based air defense systems and the emergence of new air-to-air Soviet combat helicopters) and the age of the U.S. Army helicopter fleet, the development of a new generation of U.S. military rotorcraft has become more urgent.

Research and development in this field has been focused on new rotary-wing concepts (tilt-wing, X-wing, high-speed conventional rotors, etc.), electronics, aerodynamics, propulsion, and the use of advanced materials in various components (in order to reduce payload weight and improve performance).

Write a brief account of the use of advanced materials in some specified components of a military helicopter. Make reference to applications of these materials in the latest helicopter models, both in Western Europe and in the United States.

There has indeed been a flurry of research, development, and production activities in the field of military helicopters with the objective of producing more sophisticated, maneuverable, lighter, faster, and versatile craft (Table Q1.1) for the armed forces of Western Europe and the United States (Figs. Q1.1, Q1.2, and Q1.3 and Table Q1.2). The use of advanced materials, in various helicopter components and structures, will contribute to the achievement of the goals of these rotorcraft programs. Use of these materials will reduce helicopter life cycle costs by 10% to 15% by 1995 and 25% to 30% by the year 2000.

Applications of advanced materials in military helicopters include the following.

TABLE Q1.1 AVERAGE HOVER DISC LOADINGS AND MAXIMUM SPEEDS OF FOUR ROTOR SYSTEMS

System	Hover Disc Loading (kPa)	Maximum Speed (km h^{-1})
Helicopters	0.3	230
Tilt rotors	2.0	520
X-wing	0.5	940
Direct lift	50.0	970

Figure Q1.1 An advanced version of the AH-64A Apache military helicopter. (Courtesy of McDonnell Douglas)

Figure Q1.2 The Bell/Boeing V-22 Osprey military helicopter. (Photo courtesy of Bell Helicopter Textron, Inc.)

Figure Q1.3 The computer-aided design image of the Tonal helicopter. (Courtesy of *Aviation Week and Space Technology;* details correct as of November 1988)

TABLE Q1.2 SOME NEAR-TERM ADVANCED MILITARY HELICOPTERS, UNITED STATES AND WESTERN EUROPE, 1987–1997

Country	Company	Model	Status in 1988
USA	Bell Helicopter, Textron	Updated version of the AH-1W Super Cobra attack helicopter	U.S. Marine Corps has begun receiving these craft
USA	McDonnell Douglas	Advanced version of the AH-64 Apache	Flight demonstrations planned
USA	Bell Helicopter, McDonnell Douglas, in competition with Boeing/Sikorsky	LHX light scout, attack, and utility helicopter	Design review stage
Italy	Agusta of Italy	A129 Mangusta attack helicopter	First production craft delivered to the Italian army in 1988
Italy, Britain, Spain, Netherlands	Agusta of Italy, Westland of Britain, CASA of Spain, and Fokker of the Netherlands	Tonal attack helicopter	Feasibility and cost-definition study under way
Netherlands, Britain, Italy, France, West Germany	Fokker of the Netherlands, Westland of Britain, Agusta of Italy, Aérospatiale of France, and MBB of West Germany	NH-90	Production begun, for planned 1995 introduction
France, West Germany	Eurocopter GmbH (a consortium)	PAH-2, HAP, HAC	Introduction planned for 1996

Gearbox Housing

For this application, materials with excellent high-temperature corrosion resistance are required. Thus candidates include some metals (cast aluminum, cast magnesium, and investment-cast titanium alloys), filamentary composite materials, metal matrix composite materials, and intermetallic compounds. Two examples of metallic alloys are a cast aluminum alloy, C354, and a cast magnesium alloy, WE54. Examples of filamentary composite materials are the graphite fiber reinforced bismaleimide material used in the input housing of the intermediate gearbox units on the Sikorsky UH-60 helicopters, and polyimide-based materials whose use in fabricating gearbox housings can permit their operational temperatures to be as high as 350°C. Examples of metal matrix composite materials are silicon carbide fiber and silicon-titanium-carbon fiber reinforced aluminum metal matrix composites. Suitable intermetallic compounds are iron and chromium aluminides. It is expected that once these materials are fully developed and characterized (see Chapter C4), their use in gearbox housings would result in an 83% reduction of the wall thickness (compared to a metal housing).

Driveshaft

The main requirement in a driveshaft is excellent fatigue resistance. Polymer-based composite materials, containing three-dimensionally braided fibers, are an attractive alternative to metals for this application. Other candidates belong to a new class of materials, called ultrastructure materials. These materials are the result of some manipulation of microstructures at the molecular level to create new polymers and metals. Thus these new materials offer the designer the possibility of tailoring a material to a particular application. One such material that is attracting considerable research attention is a self-reinforcing polymer. This polymer has molecules with rigid segments acting as reinforcements and more flexible segments that act as the matrix.

Landing Gear

The main requirement in a suitable material for this application is high specific stiffness. Metal matrix composite materials are thus one class of advanced materials that are suitable. They are preferred to both polymer-based composite materials and high-performance metallic alloys. For example, the specific stiffness of boron fiber reinforced 6061 aluminum alloy* is 10.3×10^6 m longitudinally, compared to 6.4×10^6 m for Kevlar 49 fiber reinforced epoxy resin and 2.6×10^6 m for Ti-6Al-4V alloy.

Rotor

The main requirement here is a material with high fatigue or endurance strength. Thus suitable materials are polymer-based composite materials in which the fibers (glass, graphite, carbon, or Kevlar 49) are either woven three-dimensionally to prevent lamination or braided so that the composite laminate may carry interlaminar shear. The rotor head in the Sikorsky S-76 helicopter is made of the former type of composite material.

Airframe

The main requirement here is high specific compressive strength. It is expected that by 1992, thermoplastic polymer-based composite materials (see Chapter C6) would be widely used for this application. One major attractive feature of these materials for this application is the fact that heat and pressure can be applied following failure of the part to reflow the resin, thereby recovering partial or full design performance. This is the chosen material for the fuselage of the Boeing Vertol CH-47 cargo helicopter.

Engine Parts

Materials used in these applications must have, among other things, good creep resistance. This brings forward two particularly strong candidates for such ap-

*Fiber volume fraction = 0.60.

plications. The first is intermetallic compounds. Their use could increase the operational temperature in the hot sections of helicopter turbine engines from 1040°C to 1210°C. This, in turn, results in decreased engine-specific fuel consumption. Even higher temperatures are possible with the use of the second class of materials, advanced ceramics, in both their unreinforced and reinforced forms.

QUESTION 2
Driveshaft of a Medium-sized Car

The use of filamentary composite materials in automotive vehicle components is on the increase.

(a) *Demonstrate the plausibility of this statement.*

(b) *State the attractive features of these materials and discuss the relevant considerations for the choice of materials for one automotive vehicle component, the driveshaft of a medium-sized car.*

The widespread adoption of composite materials for automotive vehicle components may be gleaned from the data in Table Q2.1, as well as in the number of components now in use or being considered for use in the near term. These include driveshafts, side rails, doors and door beams, cross members, suspension arms, brackets, leaf-spring suspension members, bumper energy absorbers, and road wheels. Some recent examples are in certain body panels of the BMW Z1 sports car (Fig. Q2.1)—the hood, the trunk lid, and the cover for the convertible-top storage well, which are made of glass-epoxy mat sandwiches cored with polyurethane foam.

In the United States, Chrysler (through its Liberty II Composite Vehicle Materials Engineering Group and Genesis Project) and General Motors (through its GM80 and GM98 projects) had been in the forefront of determining the feasibility, and eventual manufacturing, of relatively high volume vehicles from polymers and composite materials. All of these projects were mothballed in 1986 and 1987, but these disappointments on the high-volume side were compensated for

TABLE Q2.1 CHANGES IN THE SHARES OF WEIGHT OF MATERIAL USAGE IN A TYPICAL COMPACT U.S. PASSENGER CAR, 1976–1996

	Share (%)		
Material	1976	1986	1996
Iron and steel	76	71	65
Plastics and composite materials	4	7	12
Aluminum	3	4	5
Glass	2	3	3
Others	15	15	15

Sources: Society of the Plastics Industry, Inc., Washington, D.C.; Springborn Laboratories, Enfield, Conn.; author's own estimates.

Figure Q2.1 The BMW Z1 roadster. (Courtesy of General Electric Plastics)

by gains on the low-volume side of the market, notably GM's plastic-bodied GM200 minivan project, Chrysler's collaboration with the Budd Company on the hybrid steel-composite CIV H-body Dodge Lancer project (Fig. Q2.2), and Ford Motor Co's Low Investment (LIV), a concept car based on a European Ford Fiesta XR2 chassis.

One of the main attributes of filamentary composites, as far as their use in automobiles is concerned, is the weight reduction (and hence energy saving and fuel consumption improvement) they yield. This point is illustrated in the data given in Tables Q2.2 and Q2.3. These materials have other attractive features for use in automotive components as well:

- Decreased noise and increased vibration damping
- Elimination of the problem of corrosion associated with the use of metals
- Ease of fabrication: a one-piece filamentary composite driveshaft instead of a two-piece metallic driveshaft, for example

Figure Q2.2 Upper structure of the Composite Intensive Vehicle developed by Chrysler Corp. and Budd Co. (Courtesy of Budd Co.)

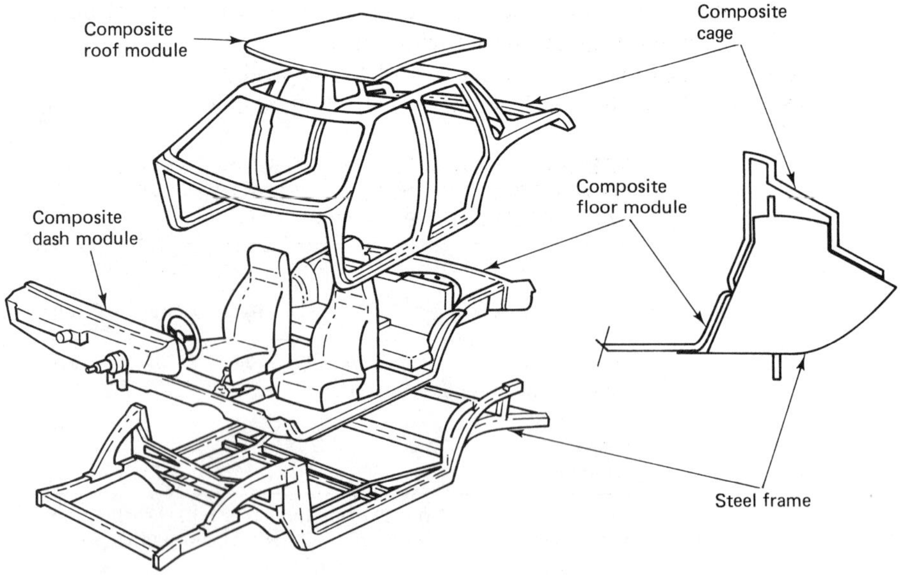

TABLE Q2.2 COMPONENTS AND WEIGHT SAVINGS FROM THE USE OF COMPOSITE MATERIALS[a] IN THE FORD MOTOR COMPANY'S EXPERIMENTAL COMPOSITE CAR, LWV[b]

Component	Weight Savings (kg)
Hood	11.35
Hinge, upper left front	0.81
Hinge, lower left front	0.86
Door guard beam	0.66
Suspension arm, upper front	0.99
Suspension arm, lower front	0.74
Transmission support	0.82
Driveshaft	2.45
Air conditioning, lateral brace	2.83
Air conditioning, compressor bracket	1.95

[a]Graphite fiber reinforced epoxy resin.

[b]Lightweight vehicle; car has the same size as a 1978 Ford Granada. Total weight of the LWV is 1143 kg, or 657 kg less than that of the standard Ford Granada. Thus a smaller engine (2.8-l V6), instead of the 5.8-l V8 engine of the standard Granada, can be used without affecting performance (0–100 km h^{-1} in 12 s); fuel economy increases from 7.23 km l^{-1} to 9.78 km l^{-1}.

Source: J. F. Young and K. S. Shane (eds.), *Materials and Processes,* Part A: Materials, 3d ed. New York: Marcel Dekker, 1985, p. 792.

TABLE Q2.3 COMPONENTS AND WEIGHT SAVINGS ACCRUING FROM THE USE OF COMPOSITE MATERIALS IN THE CHRYSLER CORPORATION EXPERIMENTAL COMPOSITE CAR, THE POLY CAR OR PXL

Components Made of Glass Fiber Reinforced Epoxy Resin	Weight Savings (kg)	Components Made of Carbon/Graphite Fiber Reinforced Epoxy Resin	Weight Savings (kg)
Road wheels	23	Rear leaf spring	14
Bumper energy absorbers	9	Driveshaft	4
Door inner panels	6	Door hinges	3
Grille opening panel	1	Engine accessory drive bracket	2
Transmission support crossmember	1	Engine valve pushrod	1

Source: J. M. Margoles, "Automobile developments in the 1980's: Composites versus metals," *SAMPE J. 16* (4), p. 682, 1980.

As far as the driveshaft is concerned, the main design criteria are in five areas:

- Geometric: length, diameter
- Mechanical: critical rotational speed; torsional moment (both maximum torque and load history) that the shaft would be subjected to during its service life; torsional stiffness of the shaft material; longitudinal and torsional damping characteristics of the shaft material

- Thermal: maximum and minimum temperature of the shaft under service conditions
- Environmental: resistance of the shaft material to degradation due to exposure to humidity, engine oils, etc.
- Impact (resistance of the shaft material to serious damage due to impact from tools, road objects, etc.)

Data on some of these criteria are summarized in Table Q2.4.

The first critical frequency N_{cr} of a simply supported thin rotating tube (which approximates a driveshaft) is given by

$$N_{cr} = Cd_m \left(\frac{E}{\rho}\right)^{0.5} L^{-2} \text{ rev min}^{-1} \qquad \text{(Q2.1)}$$

where C is constant, d_m is the mean diameter of the shaft, L is its length, E is the modulus of elasticity of the shaft material, and ρ is its density.

Equation (Q2.1) can be rearranged to give (a) the critical length of the shaft L_{cr} as being proportional to $(E/\rho)^{0.25}$ for given values of d_m and N_{cr} and (b) the minimum required mean diameter of the shaft as being proportional to $(E/\rho)^{-0.5}$ for given values of N_{cr} and L_{cr}. The specific tensile modulus (E/ρ) of the material is clearly an important parameter in this application. In view of the fact that the main mechanical stresses in the driveshaft will arise from torsional loading, the specific shear strength and the specific shear modulus are also important material properties.

Torsional buckling determines the minimum wall thickness of the driveshaft for a given diameter. In general, the torsional buckling stress τ_{scr} is proportional to a collection of material properties [1]; that is,

$$\tau_{scr} = \frac{KE_{xx}^{3/8}E_{yy}^{5/8}}{(1 - \nu_{xy}\nu_{yx})^{5/8}} \qquad \text{(Q2.2)}$$

where K is a constant. The critical or minimum wall thickness t_{cr} then becomes

$$t_{cr} = \frac{2T}{\tau_{scr}\pi d_m^2} \qquad \text{(Q2.3)}$$

TABLE Q2.4 TYPICAL DRIVESHAFT SPECIFICATIONS

	Compact Cars	Medium-sized Cars	Light Trucks	Medium-Sized Trucks	Heavy-Duty Trucks
Length (m)	1.14	1.40	1.91	1.91	2.54
Diameter (mm)	70	75	95	100	114
Static torque (kN·m)	1.70	2.83	3.39	8.14	10.85
Critical speed (r min^{-1})	12000	7200	6000	3000	5400
Twist rate (° kNm^{-1} m^{-1})	7.5	5.0	3.0	2.0	0.8
Maximum service temperature (°C)	120	120	130	110	90
Minimum service temperature (°C)	−40	−40	−40	−40	−40

TABLE Q2.5 RELEVANT PROPERTIES OF CANDIDATE MATERIALS FOR THE DRIVESHAFT OF A MEDIUM-SIZE CAR

Material	E_{xx} (GPa)	E_{yy} (GPa)	ν_{xy}	ν_{yx}[a]	G_{xy}	ρ (kg m^{-3})	E_{xx}/ρ (10^6 m)	G_{xy}/ρ (10^6 m)
High-strength steel	200	200	0.30	0.300	76	7833	2.60	0.99
7075-T6 aluminum alloy	70	70	0.34	0.340	27	2796	2.55	0.98
Ultra-high-modulus graphite fiber reinforced epoxy resin[b]	303	7	0.25	0.006	6	1688	18.30	0.36
Chopped-glass sheet molding compound[b]	24	24	0.30	0.300	10	1933	1.27	0.53
Kevlar 49 fiber reinforced epoxy resin[b]	86	5	0.33	0.019	2	1384	6.33	0.15

[a] $\nu_{yx} = \nu_{xy}E_{yy}/E_{xx}$

[b] Volume fraction of fiber = 0.65.

where T is the applied torsional moment, or

$$t_{cr} = \frac{2T}{K(\text{TFM})\pi d_m^2} \tag{Q2.4}$$

where TFM (thickness figure of merit) is given by

$$\text{TFM} = \frac{E_{xx}^{3/8}E_{yy}^{5/8}}{(1 - \nu_{xy}\nu_{yx})^{5/8}} \tag{Q2.5}$$

Note that t_{cr} is inversely proportional to TFM.

The values of the relevant parameters for the candidate materials are given in Table Q2.5. On the basis of both specific tensile stiffness and specific shear stiffness, the best materials are the ultra-high-modulus graphite fiber reinforced epoxy resin (UHMGr/Ep) and high-strength steel.

As far as ease of production is concerned, the UHMGr/Ep and the Kevlar 49 fiber reinforced epoxy resin materials have a decided advantage over both the high-strength steel and the aluminum alloy (see Table Q2.6).

TABLE Q2.6 CRITICAL LENGTHS AND THICKNESS FIGURES OF MERIT OF DRIVESHAFTS MADE FROM VARIOUS MATERIALS

Material	Critical Length,[a] L_{cr} (m)	Thickness Figure of Merit (TFM)[b]
High-strength steel	0.748	212
Al alloy	0.745	76
UHMGr/Ep	1.219	29
Kevlar 49/Ep	0.935	15

[a] Obtained from equation (Q2.1) with $C = 33.33$, $d_m = 75$ mm, and $N_{cr} = 7200$ rev min^{-1}. Both d_m and N_{cr} are usual values for medium-sized cars (see Table Q2.4).

[b] Given in equation (Q2.5).

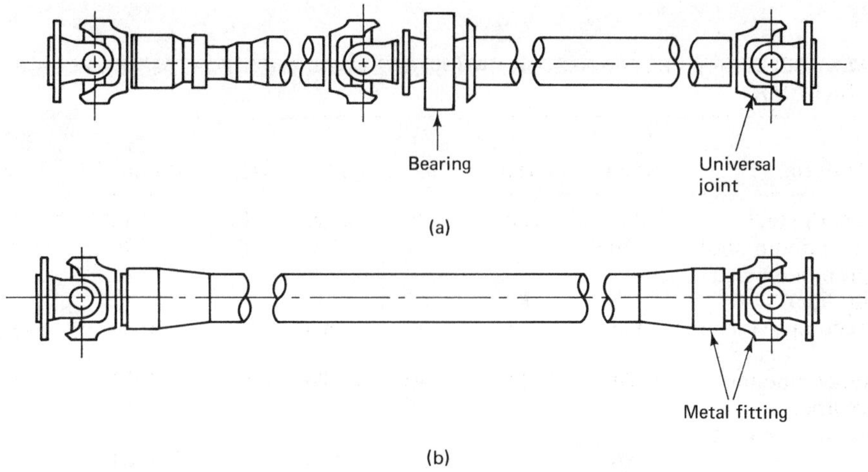

Figure Q2.3 Alternative designs of the driveshaft in a medium-sized car: (a) two-piece metallic driveshaft; (b) one-piece composite-material driveshaft.

This is because of the fact that, since most medium-sized cars have driveshafts between 1.1 and 1.4 m in length (see Table Q2.4), use of the UHMGr/Ep material would ensure a one-piece design, while use of either metallic material would certainly require a two-piece design (Fig. Q2.3), leading to increased cost. Use of the metallic materials, by contrast, would ensure much thinner sections (see Table Q2.6). Also, a driveshaft made of a composite material would need a heat shield near the exhaust system catalysts and a rubber coating that would serve as protection against stones and other debris.

Driveshafts made of filamentary polymer composite materials are installed in vehicles in current service, such as the 1985 Ford Econoline van (Fig. Q2.4) and the 1987 Ford Aerostar van. Details regarding these and other examples are

Figure Q2.4 Fiber reinforced polymer driveshaft of the 1985 Ford Econoline van. (Courtesy of Ford Motor Co.)

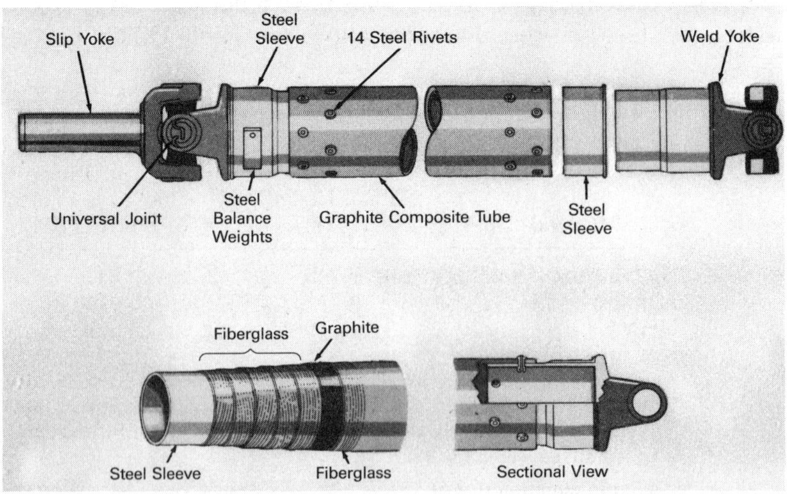

proprietary, but industry insiders say some of these materials are a quick-curing vinyl ester reinforced with graphite and E glass fibers. In addition, the modified Mazda RX-7 racing car has a driveshaft made of a resin reinforced with fibers of glass and carbon. The Renault Quadra car sports a 1.6 m-long driveshaft made of epoxy resin reinforced with 75% carbon fibers and 25% glass filaments. The turbocharged four-cylinder racing Toyota Celicas that captured the International Motor Sports Association's GTO-class championship in 1987 contained filament-wound driveshafts formed from a material consisting of graphite and glass fibers impregnated with an Epron 9000 series resin. Finally, some driveshafts are made of metal-matrix composite materials, such as the aluminum-graphite fiber shaft in the 1988 GMC model GMT 400 pickup truck.

REFERENCE

1. C. Ruegg, "Carbon- and aramid-reinforced plastics in the manufacture of automotive propeller shafts," Ciba-Geigy, Basel, Switzerland, February 1982.

QUESTION 3
Offshore Platforms for Petroleum Exploration

Write an essay on the selection of materials for offshore platforms for petroleum exploration, with special reference to tension leg platforms and filamentary composite materials.

Introduction

Fixed platforms are now clearly recognized to be very dependable, cost-effective, and efficient support systems for offshore oil and gas drilling and production operations conducted in above water (Fig. Q3.1). At the end of 1988, there were nine "pioneering" platforms in the United States standing in water depths greater than 240 m. All are steel, pile-supported, jacket-type structures.

Figure Q3.1 The Beryl B offshore platform. (Courtesy of *Petroleum Engineer International*)

TABLE Q3.1 FIXED PLATFORMS IN THE GULF OF MEXICO IN WATER GREATER THAN 200 m IN DEPTH

Platform Name	Operator	Water Depth (m)	Jacket Placed	Jacket Mass (10^3 t)	Cost ($\$ \times 10^6$)
Cognac	Shell	313	1977–78	30.5	275
Cerveza	Union	285	June 1981	18.1	90
C. Ligera	Union	282	June 1982	13.6	60
Tequila	Zapata	201	May 1984	6.3	N.A.
Boxer	Shell	229	June 1986	13.5	100
Snapper	Standard	263	August 1986	18.6	250
Grand Canyon I	Mobil	232	November 1986	13.8	250
Bullwinkle	Shell	411	Mid-1988	45.0	500

N.A., not available.

The first of these deepwater structures, Hondo, stands in 260 m of water off the California coast, in the Santa Barbara Channel. The other eight platforms are in the Gulf of Mexico (Table Q3.1).

Recent Deepwater Developments

The industry has the capabilities to extend fixed-platform construction into deeper water and to more severe conditions as the search for oil and gas continues. Notwithstanding this, it is obvious that at some water depths, other concepts will prove to be more economical.

These new concepts are the guyed tower; the tanker-based floating production, storage, and off-loading system (TBFPS); semisubmersible-based FPS (SSFPS) with flexible risers; and the tension leg platform (TLP).

Late in 1983, Exxon USA announced completion of the world's first commercial guyed-tower platform and commencement of drilling from it. This tower is located in 305 m of water in Mississippi Canyon Bloc 280 in the Gulf of Mexico, 161 km southeast of New Orleans. The recent developments in TBFPS may be gleaned from the data in Table Q3.2.

TABLE Q3.2 PROGRESS IN TANKER-BASED FPS

Design Parameter	Cadloa, Philippines (1980–81)	Tazerka, Tunisia (1980–81)	Gulf of Mexico (1986–88)
Water depth (m)	94	140	457[a]
Number of wells	4	6	8–10
Field design life (y)	10	4–5	10
Design wave height (m)	17	18	23

[a]Conoco's Jolliet FPS was scheduled to be installed in 563 m of water in mid-1989.

Source: S. Karve, "Tanker-based production eyed for Gulf of Mexico use," *Petroleum Engineer 58* (5), p. 44, 1986.

The offshore petroleum industry has wide experience with SSFPS. Examples of such platforms now in use are the Penrod 72 (operated by Penrod Drilling Corp., Dallas, Texas) and the Zapata Arctic (Zapata Offshore Co., Houston, Texas).

The Tension Leg Platform (TLP)

TLPs, designed for operation in water up to 550 m deep, are a form of semi-submersible structure that use tubular steel mooring lines for connection of the foundation templates to the seabed. Under all loading and environmental conditions, the legs are kept in tension by the natural buoyancy of the platform, which exceeds its weight. This excess buoyancy provides the restoring force to keep the structure afloat. Lateral plane movements (such as surge, sway, and yaw) are compliantly restrained, while vertical plane movements (such as heave, pitch, and roll) are stiffly restrained.

Typically, a TLP is made up of a deck, a hull, the mooring system, and the well template. The TLP has two basic advantages over the fixed structure. First, the length-to-cost ratio is favorable: There is a substantial cost difference, for example, between installing mooring tethers in 500 m of water and building steel towers or concrete pillars for the same depth. The cost of a TLP project is relatively insensitive to the depth of the water. Second, because a TLP is a floating structure, most of the assembly can take place in sheltered, near-shore water before towing out to the production site. This reduces assembly costs.

The world's first commercial TLP, the 29-m-draft TLP for the Conoco Hutton field, was installed in 1984. Set in 148 m of water, the TLP has a maximum capacity of 17.5×10^6 l (110,000 barrels) of oil per day from a 32-slot preset template (Figure Q3.2).

Figure Q3.2 A tension leg platform in Conoco's Hutton field. (Courtesy of Conoco, Inc.)

The latest TLP is also owned by Conoco; installed in 563 m of water in the Gulf of Mexico, it has 4 columns and 12 mooring lines, and has a capacity of 5.6×10^6 l (35,000 barrels) of oil per day.

In 1985, some French workers reported the successful design of the PLTB 1000, a concrete TLP, for drilling and production in the Mediterranean Sea and offshore California in water depths of 500 to 1000 m [1]. The PLTB 1000 has a concrete hull made of five columns in the shape of inverted truncated cones for hydrodynamic reasons and a steel base that includes a drilling template for up to 30 wells, which are connected to the platform deck by individual risers made of filamentary composite materials.

The demand for inexpensive lightweight TLPs of high structure reliability raises challenging questions in material selection, corrosion protection, and fracture mechanics.

Material Selection

The materials now being used are high-strength steels. However, advanced, lightweight filamentary composite materials offer attractive possibilities in the fabrication of several critical components of these offshore structures.

In the present application, high-strength steel means normalized carbon-manganese microalloyed steel with a specified minimum yield strength of 345 MPa and an ultimate tensile strength of between 480 and 620 MPa. Examples of this type of steel are the British Standard Institution's BSI 4360-50E, the American Petroleum Institute's API 2H Grade 50, and the American Society of Testing and Materials' ASTM A633 Grades C and D. In addition, precipitation-hardened steels, such as ASTM A710 and A736, with high weldability, high fracture toughness, and yield strengths of 450 to 690 MPa, are potential candidates. Filamentary composite materials are particularly attractive in these applications for a number of reasons. First, they have high values of specific stiffness and specific strength. Risers, for example, require low axial stiffness and can be fabricated from a material consisting of fibers of either low-modulus graphite, glass, or Kevlar 29 embedded in a matrix of epoxy resin. Mooring lines, by contrast, require high axial stiffness and thus may be fabricated using ropes made of a material consisting of fibers of high-modulus graphite or Kevlar 49 embedded in a matrix of epoxy resin. Second, filamentary composite materials have a high resistance to corrosion by seawater. Thus their use leads to reduced operational costs in that costs for corrosion protection for steel structures are eliminated. Third, use of filamentary composite materials would ensure that the heave natural period be no more than about 3 s. This would, for example, reduce the dynamic amplification of the load on mooring lines caused by wave action. Some structures in a TLP that could be made of filamentary composite materials are indicated in Fig. Q3.3.

A number of unresolved issues about filamentary composite materials make offshore structural designers wary of using them. Among these issues are (a) material characterization and properties (evaluation of new matrices, strength and stiffness anisotropy, fatigue properties in seawater), (b) damage tolerance and flaw sensitivity, (c) fabrication variables and their effect on quality and production rate of components, and (d) the lack of a substantial material properties database.

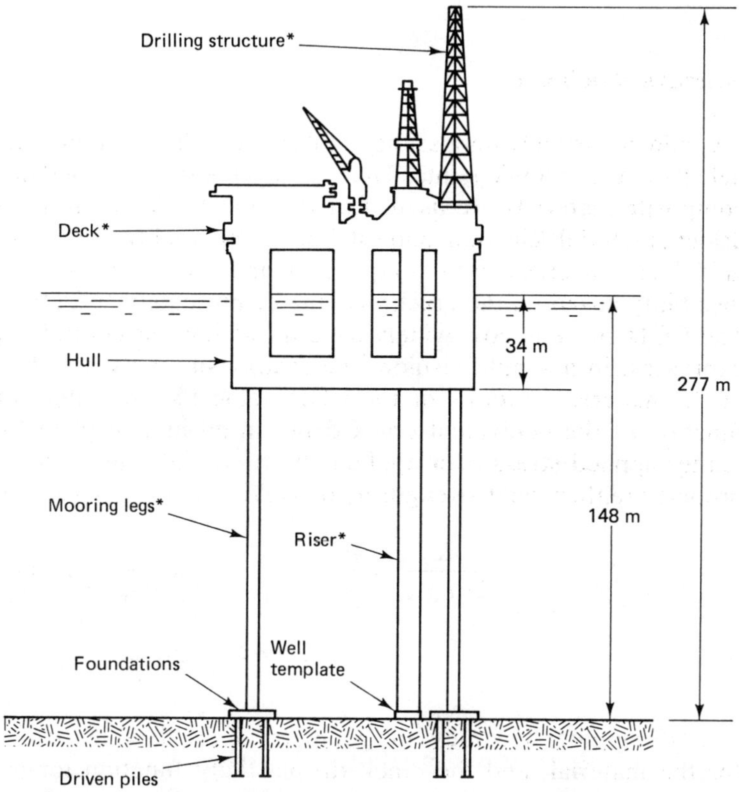

Figure Q3.3 Schematic representation of the tension leg platform in Conoco's Hutton field; asterisk indicates component that may be made of filamentary composite materials. (Courtesy of Hutton Unit Partners*)

*For list of members, see Acknowledgement, page xv.

Weight as a Major Design Consideration

The TLP is a compliant structure that is very sensitive to weight. The Hutton TLP, for example, has a hull weight–to–deck weight ratio of 1.3 and a mooring pretension–to–deck weight ratio of 0.65, with displacement of 62 kt, deck weight of 26 kt, hull weight of 26 kt, and mooring pretension of 13 kt. Table Q3.3 compares the weights and costs of lightweight filamentary composite and steel mooring lines based on Conoco's Hutton. It is clear that the development of an economical deepwater TLP is critically dependent on reducing its weight.

TABLE Q3.3 WEIGHT AND COST OF THREE CANDIDATE MATERIALS FOR MOORING LINES IN A TLP

Material	Weight (in 610 m of water) (kg m^{-1})	Cost (June 1987 prices) (\$ m^{-1})	Potential Savings in Weight over Steel (t)
Steel	383	6,888	—
Graphite fiber reinforced Kevlar	37	19,908	3400
Kevlar 49 rope	46	7,872	3300

Fracture Mechanics

It would be virtually impossible to fabricate a TLP without introducing some weld defects in the metallic joints. There is also the question of damage tolerance when composite materials are used. Fracture mechanics is thus a key element in the design, material selection, and establishment of fabrication limitations. There exist guidelines for crack growth equations and for establishing conditions for crack instability based on the crack opening displacement (COD) design methodology. The COD curve is now widely used in offshore structural mechanics because it expresses, in a simple fashion, the relationship between the fracture toughness of the material in terms of the COD value (Δ), the allowable crack size as a function of the equivalent crack depth in an infinite plate (α_∞), applied loading (either applied stress σ or applied strain e), and a material yield initiation point property (either yield strength σ_y or yield strain e_y). This relationship is

$$\frac{\Delta}{2\pi e_y \alpha_\infty} = \frac{e^2}{e_y} \qquad \text{for } \frac{e}{e_y} \leq 0.50$$

$$= \frac{e}{e_y} - 0.25 \qquad \text{for } \frac{e}{e_y} \geq 0.50$$

For greater accuracy, the geometry of the flaw, the stress-strain relationship for the material, and the crack tip plasticity function for each design must be considered. Other issues to be addressed include the effects of the operational loading spectrum, service medium, and welding or other fabrication residual stresses on crack propagation rates.

Corrosion Protection Measures

Examples of acceptable measures for a TLP include a highly reliable impressed-current cathodic protection system for the hull, a sprayed-metal anodic coating for the submerged mooring and tubular lines, and other coatings such as ethyl zinc silicate primer and high-build urethane elastomer for other components such as the topsides and the seabed anchor system.

As far as the impressed-current system is concerned, the main requirements are proper current distribution and reliability, which in turn depend on (a) excellent measurement or simulation of the current and the voltage distributions over the structure, (b) continuous in situ monitoring of these distributions, and (c) the availability of reliable rectifers with wide current and voltage ranges. The protection system must also be redundant enough to allow for small anode loss without an attendant depreciation in the required current and voltage distributions.

The required current density to achieve a particular steel-seawater potential is a function of the availability of oxygen at the steel-seawater interface. Prior to time-dependent film formation and marine growth colonization, the governing factors are probably the seawater velocity and the oxygen concentration. The current density requirements, prior to polarization and film formation, have been deduced theoretically (Table Q3.4).

TABLE Q3.4 CALCULATED CURRENT DENSITY REQUIREMENTS FOR CATHODIC PROTECTION OF STEEL IN SEAWATER (NORTH SEA WATER, 7°C)

Linear Seawater Velocity ($m\ s^{-1}$)	Current Density ($mA\ m^{-2}$) At Oxygen Concentration Of		
	6 ppm	8 ppm	10 ppm
0.0	68	91	114
0.3	78	105	131
0.4	82	109	136
0.6	89	118	148
1.0	102	136	170
2.0	136	182	227
4.0	205	273	341

Source: V. Ashworth, Seminar. Hot Risers, Managing the Corrosion Problem, Paper 2 (1979).

There is a dearth of reliable field measurements of steel-seawater current density requirements. Some authorities indicate general levels of current density required for some geographic locations; those from one such authority appear in Table Q3.5. These values are best available estimates and should be reviewed prior to application in a given location. There is enough indication, though, that the mean values exceed the minimum current densities required to maintain polarization on a well-protected structure under normal circumstances but are below the minimum values required in protection during winter storm conditions [2].

Table Q3.6 summarizes American, British, and Norwegian recommendations for cathodic protection current density and potential of steel offshore structures in seawater (at ambient temperature).

TABLE Q3.5 RECOMMENDED MINIMUM DESIGN CURRENT DENSITY REQUIREMENTS FOR THE PROTECTION OF A FIXED OFFSHORE STEEL PLATFORM

Location	Minimum Current Density ($mA\ m^{-2}$)		
	Initial	Mean	Final
North Sea (northern)	160	120	100
North Sea (southern)	130	100	90
Persian Gulf	120	90	80
India	120	90	80
Australia	120	90	80
Brazil	120	90	80
Gulf of Mexico	100	80	70
West Africa	120	90	80
Indonesia	100	80	70

Source: Det Norsk Venitas, Oslo, Norway. Technical Note TNA 703. Rev. 1, 1981.

TABLE Q3.6 INTERNATIONAL RECOMMENDATIONS FOR CATHODIC PROTECTION OF OFFSHORE STEEL PLATFORMS

Source of Recommendation	Current Density ($mA\ m^{-2}$)	Potential (V, versus Ag/AgCl)
American: NACE RP-06-75	86–130	−0.80
British: U.K. Department of Energy Project 340 1982	Initial 320–4300 Equilibrium, 20–70	−0.80
Norwegian: NTNF Project 1830.5585 1982	Initial, 2000 Equilibrium, 30–70	−0.80

Sources: National Association of Corrosion Engineers, Standard RP-01-76, "Corrosion control on steel fixed offshore platforms associated with petroleum production," Houston, Texas, 1976; U.K. Department of Energy, Amendment 5 to "Offshore installations: Guidance on design and construction," London, 1982; Norwegian Petroleum Directorate, "Guidelines for the inspection and secondary structures of production and shipment installation and underwater pipeline systems," Oslo, 1982.

As far as the mooring and riser protection system is concerned, a sealed flame-sprayed aluminum (FSA) coating is preferred because (a) it is light, inexpensive, simple to apply, and provides long-term protection, and (b) it increases the corrosion fatigue life of the steel components (Fig. Q3.4).

Several areas on the TLP in addition to the risers are vulnerable to corrosion and hence require adequate protection. One important aspect of any protection system selected is the need for a package suitable for various application and site conditions around the world. The coating types used to protect these components are summarized in Table Q3.7.

Economic Factors

A comparison of the relevant economic factors for five deepwater offshore structural concepts for location in 305 m of water in the North Sea is given in Table Q3.8.

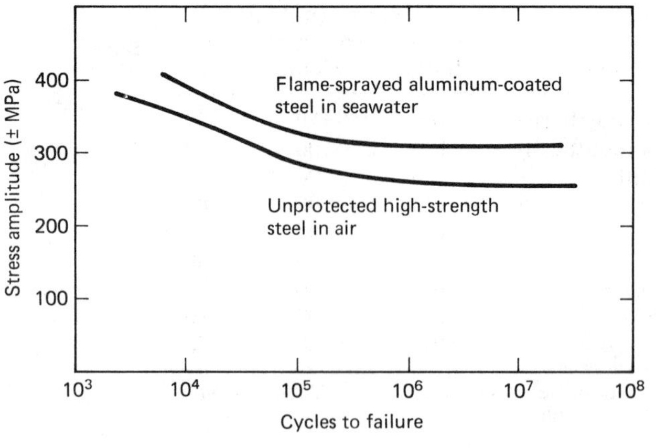

Figure Q3.4 Corrosion fatigue curves for some steels.

TABLE Q3.7 SUMMARY OF COATING SYSTEMS SELECTED FOR VARIOUS COMPONENTS ON CONOCO'S HUTTON TLP

TLP Components	Coating Systems
Topsides	Ethyl zinc sulfate primer; urethane epoxy tie coat; urethane epoxy build coat; urethane acrylic finish; nonslip epoxy/polyurethane deck coating
Hot process equipment	Zinc metal spray; silicone aluminum
Hull splash zones	Glass fiber reinforced epoxy resin
Hull interiors	
Seawater mass-compensation cavities	High-build urethane pitch epoxy; low-temperature-cured epoxy/polyurethane
Pathways and corridors	Colored nonslip urethane epoxies; inorganic zinc silicate primer
Dielectric shields[a]	Solvent-free, high-build urethane elastomer
Anchor templates	Solvent-free, high-build urethane elastomer

[a]Located between the current-emitting anodes and the hull; their function is to provide impressed current to protect the steel hull cathodically.

Source: "Corrosion protection in the offshore oil industries: Some recent developments," *Anti-Corrosion 32* (3), p. 4, 1985.

TABLE Q3.8 FIVE DEEPWATER STRUCTURAL CONCEPTS FOR LOCATION IN THE NORTH SEA IN 305 m OF WATER

Parameter	Fixed Platform	Guyed Tower	TLP[a]	Stiff-Leg Tanker-based FPS	Semisubmersible-based FPS with Flexible Risers
Structural weight (t)	48,000	23,000	63,500	Tanker + 4500	Semisubmersible + 2700
Platform installed cost ($\$ \times 10^6$)	300	450	600	150	150
Annual maintenance cost ($\$ \times 10^6$)	100	150	50	20	20
Platform installed cost per unit structural weight ($\$ \text{ m}^{-1}$)	6250	19,565	9450	—	—
Lead time to positive cash flow (y)	5	5	5	2.5	2

[a]Using only metallic components.

Source: W. B. Bleakley, "What structures to expect for deepwater developments," *Petroleum Engineer International 56,* 1984.

The use of lighter filamentary composite materials in TLPs will reduce the cost per unit weight, and that cost reduces substantially in waters in excess of 450 m.

REFERENCES

1. C. Sparks and J. P. Manesse, "Concrete TLP tie-down procedure," *Petroleum Engineer International 57* (9), pp. 34–40, 1985.
2. R. F. Crundwell, paper presented at UK Corrosion/84, Wembley, England, 1984; B. Richter, *Germanischer Lloyds 32,* pt. 10 and 11, 1980.

QUESTION 4
Use of Polymers in Optical Applications

Enumerate the advantages and disadvantages of polymers as materials for use in various optical applications. Include in the presentation details of the relevant properties of some candidate optical polymeric materials.

ADVANTAGES

- Low density
- Low raw material cost
- Impact resistance—one of the most attractive properties of all polymers; some optical polymers are practically unbreakable
- Excellent configuration flexibility, thereby simplifying assembly; this property is particularly useful in aspherical lens systems (Fig. Q4.1)
- High transmittance of incident light
- Safety involved in their use because when optical polymers break, the fragments tend to be large and obtuse and hence pose less danger
- Ease of application of surface texturing (which is used to disperse light or create a contrasting appearance) without the use of special decorative procedures, such as etching or controlled abrasion

DISADVANTAGES

- Low abrasion resistance; scratch-resistant coatings are, however, available for some optical polymers, such as polymethylmethacrylate (PMMA) and

Figure Q4.1 The curved surfaces of spherical and nonspherical lenses.

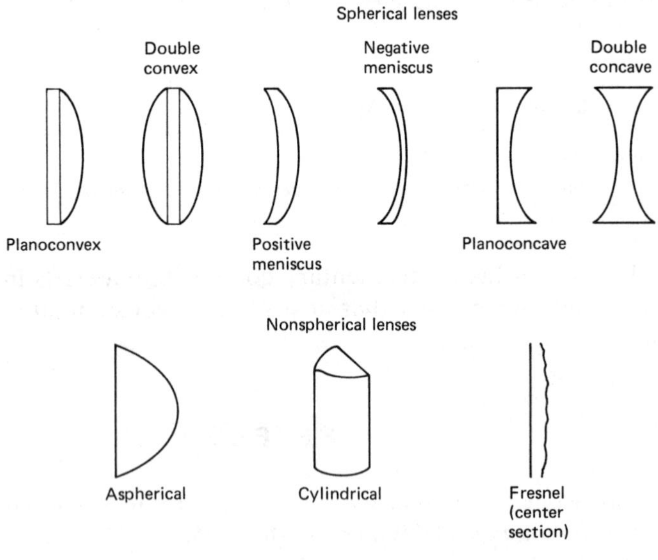

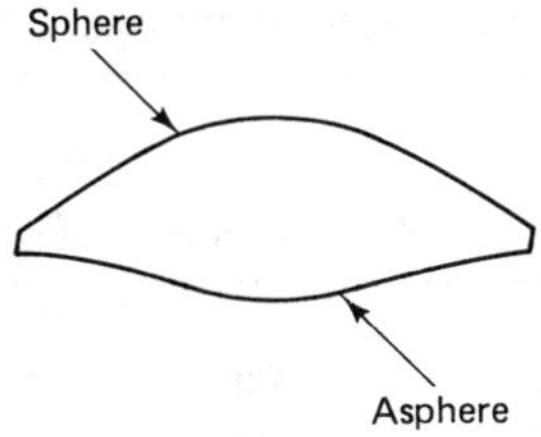

Figure Q4.2 Combination of spherical and aspherical contours in a single lens.

polycarbonate, but the coatings are expensive and necessary in only a few applications, such as ophthalmic lenses

- Poor tolerance of large temperature functuations; this, though, might not be a serious limitation because most optical systems do not normally operate in temperatures beyond the thermal limits of most optical polymers
- Narrow choice of raw materials—this means less design freedom because of a restricted range of refractive indices and dispersions (dispersion is the variation of refractive index with wavelength and is expressed by the Abbe value);* however, creative use of polymer aspherics (Fig. Q4.2) often compensates for this drawback.

CANDIDATE OPTICAL POLYMERS

The principal ones are as follows:

- Polymethylmethacrylate (PMMA) is very hard, abrasion-resistant, mechanically stable, and very widely used.
- Polystyrene is also a widely used optical polymer, but the molding cost far outstrips the resin cost. It can be used for lens designs requiring thin cross sections, thereby reducing resin cost and offsetting the high processing cost. Its low Abbe value allows it to be used in color-corrected designs.
- Polycarbonate is a very tough polymer that gives good performance over a wide temperature range (−140°C to 120°C). Its high impact strength makes it particularly suited for use in streetlight lenses, construction lights, and roadwork warning lights. Because the polymer cannot be easily machined, injection molding is almost invariably used.
- A copolymer consisting of 70 wt % polystyrene and 30 wt % PMMA. Its refractive index is difficult to control because it is a composite material, but many optical designs compensate for this limitation. This copolymer is particularly suitable for thin lens systems.

The relevant properties of these optical polymers are given in Table Q4.1. Other candidate optical polymers and their properties are presented in Table Q4.2.

*Abbe value or V number = $(n_D - 1)/(n_F - n_c)$ where n_D, n_F, and n_c are the indices of refraction of the material at 0.5893 μm, 0.4861 μm, and 0.6563 μm, respectively.

TABLE Q4.1 RELEVANT PROPERTIES OF FOUR PRINCIPAL OPTICAL POLYMERS

Property	PMMA	Polystyrene	Polycarbonate	Methylmethacrylate Polystyrene Copolymer
Refractive index				
At $n_D = 0.589\ \mu m$	1.491	1.590	1.586	1.564
At $n_F = 0.486\ \mu m$	1.497	1.604	1.593	1.574
At $n_c = 0.656\ \mu m$	1.498	1.584	1.576	1.558
Critical angle (°)	42.1	39.0	39.1	39.8
Coefficient of thermal expansion (10^{-6} °C^{-1})	64.8	39.4	68.4	64.8
Maximum continuous use temperature (°C)	79	79	116	79
Density (kg m^{-3})	1190	1060	1200	1090
Hardness, Rockwell *M* (kg mm^{-2})	97	90	70	75
Izod impact strength (J m^{-1} notch)	0.66	0.58	23.93	—
Dielectric strength (MV m^{-1})	19.70	19.70	15.75	17.72
Dielectric constant				
At 60 Hz	3.7	2.60	2.90	3.40
At 1 MHz	2.2	2.46	2.88	2.90
Power loss factor				
At 60 Hz	0.05	0.0002	0.0070	0.006
At 1 MHz	0.03	0.0002	0.0075	0.013
Volume electrical resistivity (Ω·m)	10^{16}	10^{14}	8×10^{14}	10^{13}

TABLE Q4.2 CHARACTERISTICS OF SOME OPTICAL POLYMERS

Optical Polymer	Dielectric Constant at 10 kHz	Tensile Strength (MPa)	Izod Impact Strength (J m^{-1} notch)	Density (kg m^{-3})	Coefficient of Thermal Expansion (10^{-6} °C^{-1})
Styrene acrylonitrile	3.0	76[a]	0.83	1070	67
Allyl glycol polycarbonate	3.8	38[a]	0.50	1350	112
Polysulfone	3.0	70[b]	2.00	1245	56

[a] At break point.

[b] At yield point.

QUESTION 5

Diaphragm for an Artificial Blood Pump

The development of cardiac assistance devices and other instrumentation and components that come in contact with the cardiovascular system requires materials that will perform well in a physiological environment. One particularly difficult application is the diaphragm for the blood pump in an artificial heart. This application is to be made of elastomer that will undergo cyclic deformations and/or flexing while in contact with blood. Even a 3-y implantation of a heart will involve in excess of 100 million cycles (equivalent to a frequency of 1 Hz).

(a) *State the other desirable properties of a material for use in making the pump diaphragm.*

(b) *Compare and contrast the properties of two named elastomers for this application.*

(c) *Give a short list of other candidate elastomers.*

(a) The other desirable properties required of a material for use in making the pump diaphragm are high flexibility, high strength, high ductility, long in vivo durability, high reliability and safety, excellent biocompatibility with blood and tissues, and resistance to degradation by the enzymes papain and urease, which are present during the inflammatory response [1]. Papain is known to be closely related to cathespsin B, a thiol endopeptidase, which is released by the cells of the inflammatory response. Urease is found in leukocytes. The pump diaphragm material should also be resistant to propagation of surface-initiated cracks, which could lead to premature failure [2]. Cracks may also serve as sites for thrombus formation or as initiation points for calcification.

Figure Q5.1 presents a cross section of the widely used artificial heart, the Jarvik 7, showing the position of the blood pump diaphragm relative to the other components.

(b) The two named elastomers to be compared for this application are Hexsyn, a sulfur-vulcanized hydrocarbon rubber, essentially a polyhexene, and Pellethane, an extruded thermoplastic polyether urethane elastomer. A compilation of the relevant properties of these two materials is given in Tables Q5.1 and Q5.2. Here are some comments on each.

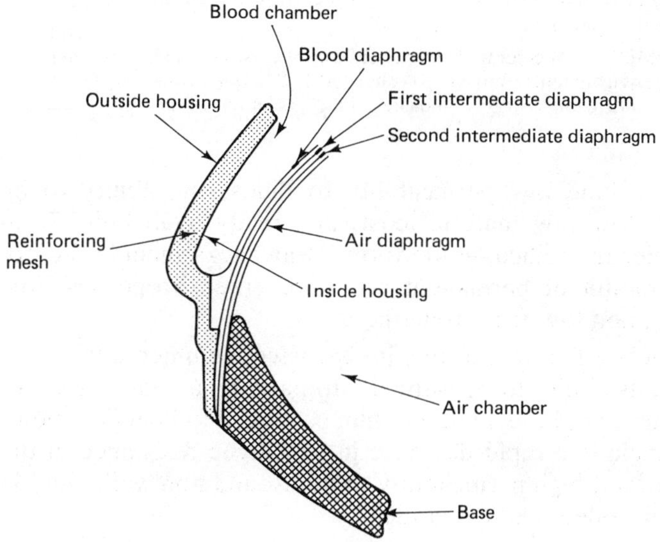

Figure Q5.1 Cross section of the Jarvik 7 artificial heart.

TABLE Q5.1 PROPERTIES OF TWO ELASTOMERS FOR CARDIAC ASSIST DEVICES[a]

Property	Hexsyn	Pellethane
Tensile stregnth (MPa)	12	54
Elongation (%)	284	380
Modulus of elasticity, 100% (MPa)	3.14	5.03
Density (kg m^{-3})	1047	1114
Tear strength, Trouser (kN m^{-1})	2.63	9.28
Tear energy (J)	3.03	16.61
Glass transition temperature (°C)	−52	−57
Single-cycle tension permanent set (%)		
100% strain	5.7	13.3
20% strain	3.6	3.2
Fatigue test permanent set (%)		
100% extension	22	17
150% extension	43	45
Stress relaxation, at 100% strain (kPa)	149	165
Permeability (mg·mm s^{-1} m^{-2})	25	375

[a]All values at average blood temperature, 37°C.

Sources: Goodyear Tire and Rubber, Akron, Oh. (for Hexsyn); Upjohn Co., Torrance, Calif. (for Pellethane).

TABLE Q5.2 FATIGUE RESISTANCE OF CANDIDATE ELASTOMERS[a]

Material	Uncut Specimens	Cut-Initiated Specimens[b]
Hexsyn	190	19
Pellethane	210	7

[a]Expressed as % strain at 50% failure, in blood at 37°C.

[b]Specimens were cut with a razor blade to give a precise 1-mm cut in the gage center of the specimen.

Source: C. R. McMillin, "Accelerated fatigue testing of elastomers in blood and related environments," in M. Szycher (ed.), *Biocompatible Polymers, Metals, and Composites,* Lancaster, Pa.: Technomic Publishing Co., 1983.

- Hexsyn: Attributes include low permeability to fluids, the ability to be compression-molded, low raw material cost, and ready availability from more than one supplier in medical-grade form. Drawbacks include the potential for a large amount of permanent set, poor creep properties, low modulus of elasticity, and low tear strength.
- Pellethane: The attractive features of this inexpensive, commercially available material include its ability to be formed into parts by a wide variety of methods, such as extrusion, blow molding, film casting, and injection molding. Disadvantages include a rapid decrease in its fatigue resistance in the presence of stress raisers, high permeability to fluids, and nonavailability in a specific, closely controlled medical grade.

(c) Other adequate elastomers for this application include Biomer*, which is polyether urethane [3]; Avcothane 51, a block copolymer of 90% polyether urethane and 10% polydimethyl siloxane [2]; and Santoprene, a commercial sheet thermoplastic rubber based on polypropylene.

REFERENCES

1. S. K. Phua et al., "Biodegradation of a polyurethane in vitro," *J. Biomed. Mater. Res. 21,* pp. 231–246, 1987.
2. D. J. Parkins et al., *In Vivo Degradation of a Polyurethane,* St. Paul, Minn.: Cardiac Pacemakers, Inc., 1981.
3. K. Hayashi et al., "Mechanical stability of elastomeric polymers for blood pump applications," *J. Biomed. Mater. Res. 19,* pp. 179–193, 1985.

QUESTION 6

Coatings for Surfaces Exposed to a Low Earth Orbit Environment

The U.S. space shuttle occupies a low earth orbit (LEO) in the performance of its mission. This LEO presents a hostile environment to a number of external components of the vehicle, the most vulnerable being those made of organic-based materials. Some examples are the solar photovoltaic arrays, the payload bay television cameras, and the reflector of the KU-band radar and communications system.

(a) *Write briefly on the low earth orbit environment.*

(b) *What are the main material requirements for a coating to protect such components?*

(c) *List some candidate materials.*

(a) Although altitude, time of day, and solar activity have major effects on the chemical state and energetics of the upper atmosphere, some general features may be discerned. At typical U.S. space shuttle altitudes (200–500 km), although the atmospheric temperature is high (≥400°C, daytime), there is little conductive heat transfer because the atmospheric pressure is low (~0.1 mPa at 200 km). The temperature of the material of the fuselage of a space vehicle in an LEO environment is determined by its radiative energy balance. Typically, materials of a space vehicle are subject to temperatures that range from −50°C to +70°C. Photochemical effects modify the composition of the atmosphere such that above 250 km, atomic oxygen is the principal component (Figs. Q6.1 and Q6.2), with a typical daytime concentration of 7×10^{14} atoms m^{-3}. A space vehicle with an orbital velocity of 28,000 km h^{-1} will also affect its local environment by providing high-collision energies for interactions at the gas-surface boundary that enhance the reactivity of the gaseous species, particularly those involved in bond scission

*Used in the Jarvik 7 artificial heart.

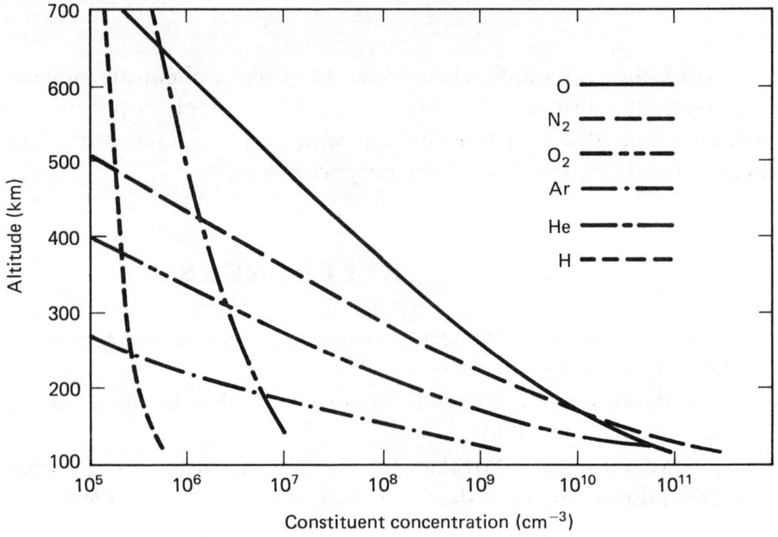

Figure Q6.1 Chemical composition of the atmosphere at low earth orbit (LEO) altitudes under nominal or standard conditions [1] (courtesy of Martin Marietta Corp., Orlando, Fla.).

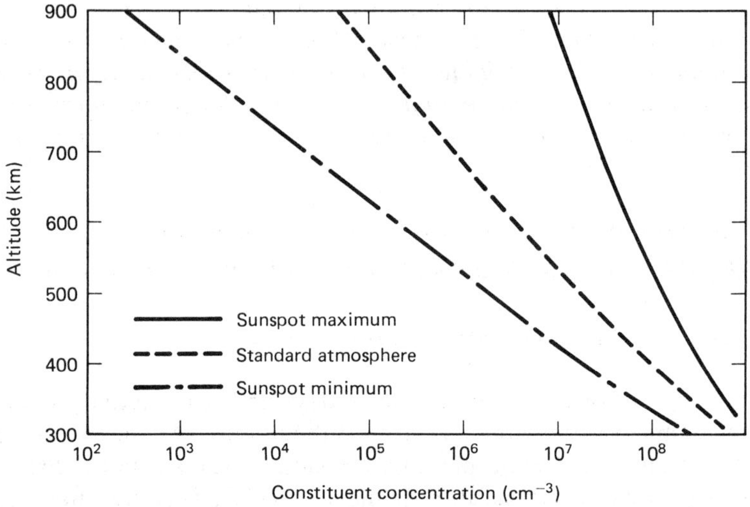

Figure Q6.2 Atomic oxygen species concentration profiles [2] (courtesy of the American Institute of Aeronautics and Astronautics, New York).

reactions. The local environment may also be modified through (1) electrical charging of surfaces, (2) passive charging from the plasma environment, or (3) ram effects, which increase the local pressure ahead of forward-facing surfaces.

Most qualitative information, and essentially all reaction rate data, currently used to quantify atomic oxygen interactions with surfaces were obtained on two material science experiments carried out on space shuttle flights STS-5 and -8. Material loss was found to be dependent on the number of oxygen atoms striking a unit area of surface during a given time period, called the fluence. This fluence is a function of atmospheric density, orbital velocity, surface attitude relative to the velocity vector, and duration of exposure.

(b) There are three main material requirements:

- Resistance to erosion by atomic oxygen species, that is, minimum loss of mass or thickness
- Ease of field application, maintenance, and repair
- Structural compatibility with the body of the component

(c) The prime candidate materials are siloxane copolymers and perfluoronated polymers (such as polytetrafluoroethylene and fluoroethylene propylene). This list is based on data on reaction efficiencies of various materials obtained from STS-5 and -8 experiments (Table Q6.1). Reaction efficiency is a measure of the susceptibility of the material to attack by atomic oxygen species and is expressed as the volume of material lost per incident atomic oxygen atom.

In addition, further research has generated the following candidate materials:

- McGhan Nusil CV 1144 (a silicone coating, offshoot from studies carried out at Hughes Aircraft Co., El Segundo, Calif.)
- Segmented polymide/siloxane copolymers (developed by workers at Virginia Polytechnic Institute and State University, Blacksburg, Va.)
- Metal coating or cladding, such as aluminum-clad graphite/epoxy tubes suitable for the tubular truss structure of the proposed permanent U.S. space station (developed by Lockheed Missiles and Space Co., Inc., Sunnyvale, Calif.)
- Ion-beam sputter-deposited thin films of Al_2O_3, SiO_2/fluoropolymer mixture (developed by NASA-Lewis, Cleveland, Ohio)
- Alloy 1145-H19 full-strain-hardened aluminum foil, developed by Boeing, of Seattle, Wash., and suitable for use on tubes fabricated with graphite/epoxy prepreg tape

TABLE Q6.1 REACTION EFFICIENCIES OF VARIOUS ORGANIC-BASED MATERIALS UPON EXPOSURE TO ATOMIC OXYGEN SPECIES

Material	Reaction Efficiency (10^{-30} m^3 atom^{-1})
Tedlar polyvinylidene fluoride	3.2
Kapton polymide	3.0
Polyethylene	3.7
Polysulfone	2.4
1034C graphite epoxy resin composite	2.1
Teflon polytetrafluoroethylene	<0.05
Teflon fluoroethylene propylene	<0.05
DC6-1104 silicone copolymer	2.00[a]

[a]Units of g m^{-2} for STS-8 experiment results.

Source: L. Leger, J. Visentine, and B. Santos-Mason. "Selected materials issues associated with Space Station," *SAMPE Quarterly 18* (2), p. 50, 1987.

It is anticipated that future U.S. space shuttle flights will provide an opportunity for flight-testing these and other coating materials (as part of the NASA-sponsored Long-Duration Exposure Facility) in order to provide much needed information and new insights into the performance of coatings in LEO environment.

REFERENCES

1. W. W. Vaughan, "Natural environment design criteria for the space station definition and preliminary design," 1st rev., NASA-TM-86460, September 1984.
2. L. Leger et al., "STS Flight 5 LEO effects experiment: Background description and thin film results," paper presented at the AIAA Shuttle Environment and Operations Meeting, Washington, D.C., p. 3, 1983.

QUESTION 7

Resistance of Metals to Erosive Wear

In solid-particle erosive wear, material is removed by an impinging stream of particles. Studies to develop an understanding of this phenomenon have been largely motivated by concern about reduced lifetimes and failures of mechanical components used in erosive environments, such as in pipelines carrying sand slurries, in petroleum refining, and in aircraft gas turbine compressor blades.

Broadly speaking, two approaches have been taken to solve the problem of the lack of predictive relations for the resistance of materials to solid-particle erosive wear.

Outline these approaches, and illustrate them with published results.

The first of these approaches is based on thermal property considerations. Since it is plausible that the high-velocity impact of particles on a metal surface may lead to considerable local heating, correlations have been sought between erosion resistance of a target material and several parameters derived from its thermal properties. Ascarelli [1] suggested the use of a "thermal pressure" parameter, which is equal to $\alpha K \, \Delta T$, where α is the coefficient of thermal expansion, K is the bulk modulus of the target material, and ΔT is the difference between its melting temperature and the ambient temperature. Ascarelli suggested that the solid-particle erosive wear rate of a metal is inversely proportional to its thermal pressure parameter. Although this relationship is reasonably valid for some published erosion rate data [2], it breaks down when tested against other data sets [3]. Malkin [4] used the specific melting energy, $\rho \, \Delta H/AW$, where ρ is the density of the target material, ΔH is the enthalpy change of the metal in going from the solid state at ambient temperature to the liquid state at the melting point, and AW is the atomic weight of the metal, as the relevant predictive factor. Malkin argues that by analogy with grinding, this direct correlation (between specific melting energy and erosion rate) is attributable to the extreme strains and strain rates that characterize the erosive wear process, thereby leading to near-adiabatic plastic

TABLE Q7.1 VOLUME EROSIVE WEAR RATES,[a] MATERIAL PROPERTIES, AND EROSIVE WEAR PREDICTORS FOR NINE ANNEALED METALS

Metal	Volume Erosion ($mm^3\ kg^{-1}$)	σ_y (MPa)	UTS (MPa)	e_f	ρ ($kg\ m^{-3}$)	C_p ($J\ kg^{-1}\ K^{-1}$)	ΔT[b] (K)	$\rho Cp\ \Delta T$ ($MJ\ m^{-3}$)	ΔW ($MJ\ m^{-3}$)
Al alloy 1100	180	35	90	0.45	2,700	917	617	1528	28
Ag	110	55	125	0.48	10,500	235	936	2310	43
Cu	80	69	220	0.45	8,900	385	1058	3625	65
Fe	96	126	283	0.47	7,890	456	1502	5404	96
Mg	440	103	193	0.15	1,741	246	626	268	22
Mo	60	345	549	0.55	10,200	255	2582	6716	246
Pb	540	9	17	0.52	11,340	126	302	432	7
Sn	700	9	15	0.45	7,300	226	207	341	5
Ta	50	271	343	0.45	16,600	151	2971	7447	138

[a] Erosive particles are quartz sand at 45° impact angle at a velocity of 76 $m\ s^{-1}$. Source of data is Finnie et al [2].

[b] Ambient temperature = 298 K.

deformation of the metal. Dimensional analysis [5] yielded a correlation between erosion rate and a parameter of the form

$$\rho^{1/3} k T_m \, \Delta H_m/(AW)^{1/3}$$

where k is the thermal conductivity, T_m is the melting temperature, and ΔH_m is the enthalpy of melting of the target metal.

Hutchings [6] has argued that spherical particles impacting obliquely onto a metal surface cause plastic depressions or craters, leading to the formation of a fragile lip of the metal at the end of the crater. The lip forms the main source of metal that may be eroded by the impacting particles, and its formation is facilitated by localization of deformation, at the metal surface, which is thermal in origin. From these arguments, Hutchins deduced that the volume of the lip, and hence the solid-particle erosive wear rate, is dependent on the product $\rho Cp \, \Delta T$, where Cp is the specific heat of the target material and ΔT is the difference between its melting point and the ambient temperature.

Notwithstanding temperature rises due to friction heating and adiabatic effects, it has been argued that solid-particle erosive wear is essentially a mechanical process governed by local deformation and removal of material from the surface. This is the main thrust of the second approach. Thus Finnie and colleagues [2] showed that this erosive wear rate varies approximately inversely with indentation hardness for many, though not all, pure target metals. However, these workers also found practically identical levels of solid-particle erosive wear resistance for a number of steels in various heat-treated conditions with hardnesses varying by a factor of about 8. Pure iron exhibited only slighty less erosion resistance than

Figure Q7.1 Variation of erosion volume with two erosion prediction parameters.

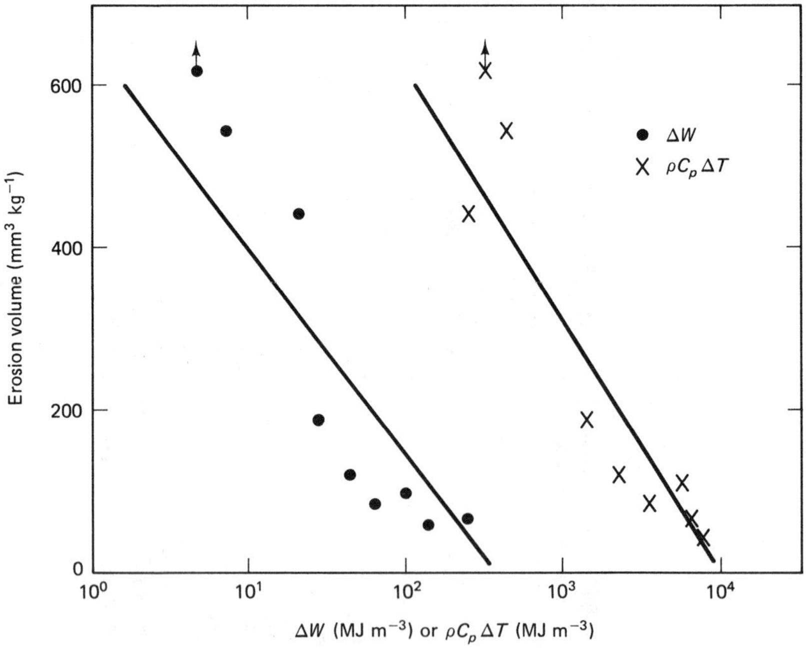

these steels did, and changing the hardness, by cold-working or annealing, was also found to have only a very marginal effect on the erosion resistance. Recently, Rickerby [7] proposed that the relevant predictor of the solid-particle erosive wear rate is the mechanical energy density required to cause failure in a tensile test, which is equal to the work done per unit volume ΔW (that is the area under the stress strain curve) and is estimated to be $0.5(\sigma_y + \text{UTS})e_f$, where σ_y is the yield strength of the material, UTS is its ultimate tensile strength, and e_f is its strain at failure.

A test of the predictive capabilities of a thermal parameter $\rho Cp\ \Delta T$ and a mechanical parameter ΔW is carried out with respect to the volume erosive wear data (Table Q7.1) for a number of annealed metals. The results are given in Fig. Q7.1. The scatter in the results is almost the same irrespective of the predictor, indicating that either predictor could be used in estimating erosive wear of pure metals. A proper resolution of this uncertainty awaits the generation of a bank of detailed solid-particle erosive wear rate results for a wide assortment of metals and alloys.

REFERENCES

1. P. Ascarelli, "Relation between the erosion of solid particles and the physical properties of metals," Tech. Rep. No. 71-47, U.S. Army Materials and Mechanics Research Center, 1971.
2. I. Finnie, J. Wolak, and Y. Kabil, "Erosion of metals by solid particles," *J. Mater. 2,* pp. 682–700, 1967.
3. Y. A. Tadolder, "Influence of abrasive grain geometry on the solid particle erosion of metals," *Tr. Tallin. Politek. Inst., Ser. A 237,* pp. 15–22, 1966.
4. S. Malkin, "Correlation between solid particle erosion of metals and their melting points," *Wear 68,* pp. 391–396, 1981.
5. W. H. Jennings, W. J. Head, and C. R. Manning, "A mechanistic model for the prediction of ductile erosion," *Wear 40,* pp. 93–112, 1976.
6. I. M. Hutchings, "Prediction of the resistance of metals to erosion by solid particle," *Wear 35,* pp. 371–374, 1975.
7. D. G. Rickerby, "Correlation of erosion with mechanical properties in metals," *Wear 84,* pp. 393–395, 1983.

QUESTION 8
Covers and Absorbers of a Flat-Plate Solar Collector

A flat-plate solar heat collector (Fig. Q8.1) consists essentially of a transparent cover, an absorber, tubes or ducts for the heat transfer fluid, and an insulated case to reduce the thermal losses to the environment.

Discuss the main requirements for materials to be used as covers and absorbers in such collectors, using water as the heat transfer fluid, with special reference to polymers and glasses.

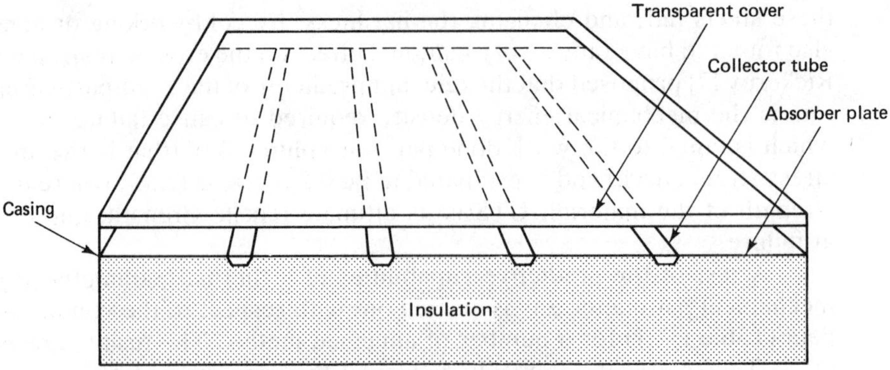

Figure Q8.1 Various parts of a flat-plate solar collector.

Transparent Cover

The cover is selected for the highest transmittance of air-mass-2* solar radiation spectrum and the lowest transmittance of the infrared reradiated from the absorber. The cover should minimize heat loss by conduction and convection, and because of this, it may consist of separate sheets and may include a honeycomb structure. In addition, the cover protects the absorber from mechanical and weathering damage arising from wind, snow, temperature variations, dirt, and so on. Thus the cover material must have adequate mechanical strength and durability.

Glasses and polymers are the two competing material types to be used for the cover. Polymers have a much greater variety in composition and structure than glasses and consequently have a wider range of optical, mechanical, and endurance properties.

In comparison with glasses, polymers have a lower absorptance for thermal radiation, although they have some infrared absorption bands related to certain functional groups. Except for polymethylmethacrylate (PMMA) and polycarbonate (PC), infrared transmittance of polymers is some five times greater than that of glasses (Fig. Q8.2). The simpler polymers, such as polyethylene, have long-wave transmittance almost equal to that of solar radiation. Such long-wave transmittance can be reduced by incorporation of a suitable filler, such as glass fibers, but these also reduce the solar transmittance by about 10% (through interface scattering effects; Fig. Q8.2). The transmission of infrared can be reduced through the use of heat mirror coatings, such as tin oxide or indium oxide, but these mirrors also reduce solar transmittance substantially, by as much as 20% in the case of tin oxide, for example.

As far as mechanical strength is concerned, most glasses can easily support normal weather-imposed loads, in addition to their own weight, over long spans and wide temperature ranges. Polymers have adequate strength at 25°C but lose much of that strength at temperatures in excess of the glass transition point. The

*Air mass is the ratio of the optical thickness of the atmosphere through which beam radiation passes to the optical thickness if the sun were at the zenith. Air-mass-2 is the above ratio for a zenith angle of 60°.

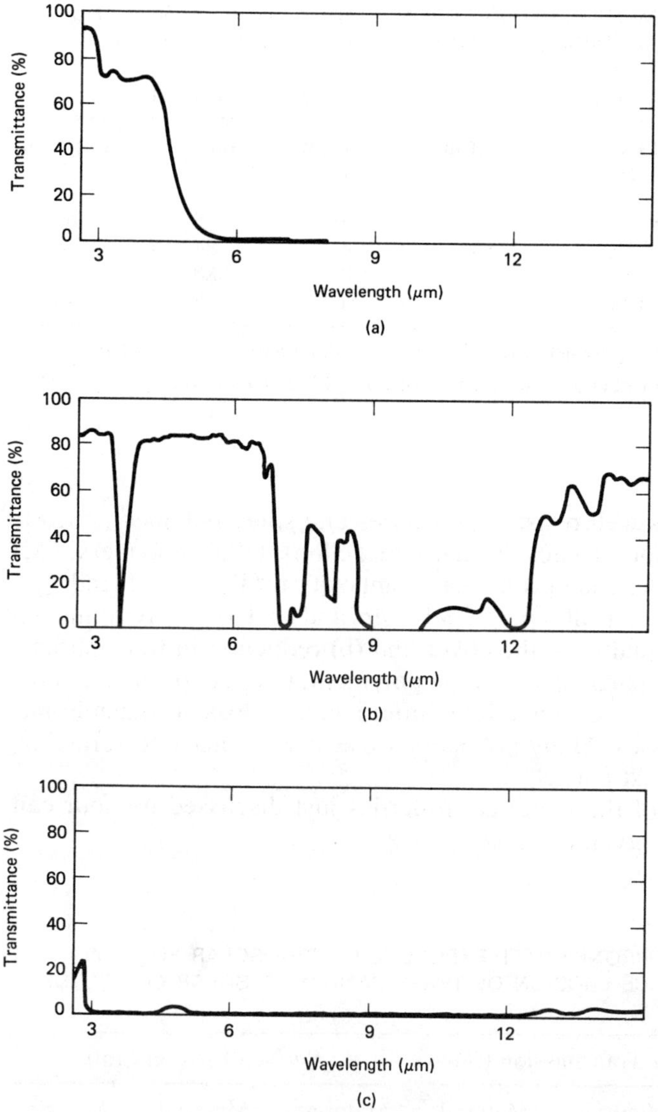

Figure Q8.2 Infrared transmittance curves: (a) glass (0.71 mm); (b) Tedlar (0.05 mm); (c) fiberglass (0.96 mm).

mechanical properties of polymers are also critically dependent on a number of factors, such as processing, presence of additives, and strain rate.

High impact strength is desirable in the cover material in order that the cover can withstand occasional hailstorms, stones, and other impacting objects. Glasses are very prone to brittle fracture, although the use of tempered glass is likely to increase cover life considerably. Thermoplastics have good impact strength, but the thermosetting plastics have as poor an impact strength as glasses.

With respect to durability or weatherability, the prime factors causing breakdown are the ultraviolet portion and heating effects of solar radiation. The outdoor deterioration of polymers is mainly the result of photooxidation activated by solar ultraviolet. One group of polymers, though, is completely transparent and hence stable to ultraviolet. Examples are PMMA and polyvinylchloride (PVC).

TABLE Q8.1 EFFECT OF ENVIRONMENTAL EXPOSURE ON TOTAL COLOR CHANGE AND LIGHT TRANSMISSION OF THREE CANDIDATE SOLAR COLLECTOR COVER MATERIALS

Time In Fluorescent Weatherometer (h)	Total Color Change (%)			Light Transmission (%)		
	Material 1[a]	Material 2[b]	Material 3[c]	Material 1	Material 2	Material 3
500	2.0	3.5	7.5	90	85	60
1000	3.5	5.0	15.0	90	84	—
1500	4.5	7.5	10.0	88	80	—
2000	4.6	19.0	27.5	86	75	—

[a]SunLite (premium glass fiber reinforced polymer, FRP); [b]SunLite regular FRP; [c]Ordinary FRP.

Source: J. J. White, "Weatherability of fiberglass solar collector covers," *Polymer News 2* (5), p. 242, 1977.

Natural weathering is well resisted by glasses and some polymers, such as PMMA and the fluorocarbon plastics (Tedlar and Teflon). Other polymers, such as FRP, weather easily, depending on their composition (Tables 8.1 and 8.2). Industrial pollutants may attack all types of polymers and lead to (a) environmental alterations to the surface quality of the cover and (b) reduction in transmittance as a result of deposition of particulates. Most polymers, because of their electrostatic properties, easily attract dust and thus suffer a greater loss of transmittance due to this effect than glasses. Many polymers are also very prone to scratching and abrasion from these dust particles.

The values of some of the material properties just discussed for four candidate cover materials are given in Table Q8.3.

TABLE Q8.2 EFFECT OF ENVIRONMENTAL EXPOSURE ON THE SOLAR ENERGY TRANSMISSION AND SURFACE EROSION OF THREE CANDIDATE SOLAR COLLECTOR COVER MATERIALS

Years In Southern Florida	Solar Energy Transmission (%)			Surface Erosion (μm)		
	Material 1[a]	Material 2[b]	Material 3[c]	Material 1W[d]	Material 1	Material 3
0.25	85	85	72	—	—	—
0.50	83	83	62	—	—	—
0.75	80	80	55	—	—	—
1.00	75	75	45	0.50	0.75	1.60
2.00	—	—	—	0.50	0.75	2.75
3.00	—	—	—	0.60	0.90	—
4.00	—	—	—	0.70	1.00	—

[a,b,c]See Table Q8.1; [d]Material 1 plus a proprietary high-temperature coating.

Source: J. J. White, "Weatherability of fiberglass solar collector covers," *Polymer News 3* (5), p. 243, 1977.

TABLE Q8.3 SOME PROPERTIES OF FOUR CANDIDATE MATERIALS FOR THE COVER OF A FLAT-PLATE SOLAR THERMAL COLLECTOR

Material	Refractive Index	Normal Incident Long Wave Transmittance ($\lambda = 2.5 - 4.0\ \mu m$) (%)	Normal Incident Light Short Wave Transmittance ($\lambda = 0.4 - 2.5\ \mu m$) (%)	Density (kg m^{-3})	Tensile Strength (MPa)	Impact Strength (J)	Maximum Operating Temperature (°C)	Chemical Resistance
Glass, 3 mm thick	1.52	3	84	2489	40–90	0	400	Resistant to alkalis, weak acids, and oils
Glass fiber reinforced polyester (FRP), 0.63 mm thick (50% glass fiber)	1.54	6	78	1400	75–79	14	93	Resistant to most acids and to alkalis
Polymethylmethacrylate (PMMA), 3 mm thick	1.50	2	91	1189	55–75	11	82–93	Highly resistant to most chemicals
Polycarbonate (PC), 3 mm thick	1.59	6	85	1200	55–65	22	121–132	Resistant to weak acids and to aromatic solvents

Absorber

The absorber should be selected on the basis of collection efficiency and the following material properties:

- Thermal conductivity
- Density
- Coefficient of linear expansion

All common structural plumbing materials have been used for absorbers in solar thermal systems of various configurations related to the properties of the materials employed. Metals (copper, aluminum, and steel), which have relatively high thermal conductivities, are always used in a fin-and-tube configuration since heat collected across the fin can readily flow to the tube and into the working fluid. For polymers, glasses, or crystalline ceramics, the thermal conductivity is quite low, so the fin-and-tube shape would be inappropriate. An acceptable geometry when using any of these materials for absorbers is one in which the irradiated area consists only of conduit walls. Then the heat must flow only through these walls. In other cases, the tube is transparent so that the light passes through it to be absorbed by a black liquid.

Some of the various absorber plate geometries are shown schematically in Fig. Q8.3.

Because metals normally have high reflectance and low absorptivity and emissivity, it is necessary to coat a metal absorber in order for it to collect solar radiation effectively. When using polymers or glasses as absorber materials, suitable light-absorbing particles must be incorporated within them or else they must also be coated. A comparison of the properties for some candidate absorber materials is given in Table Q8.4.

There are two main types of absorber coatings. The first type is a nonselective coating of high absorptivity for short-wavelength radiation. Examples are camphor black and flat black paints, such as Nextel black paint (Fig. Q8.4). The painted surfaces behave almost like black bodies, with absorptivity and emissivity near unity. The use of the coatings result in high losses due to emittance of infrared radiation (the magnitude of such losses depends on the temperature of the absorber). The coating is relatively thick, and the light is absorbed within it and converted to heat, which is then transferred to the metal; however, it must not be too thick, or its low thermal conductivity will diminish the heat transferred. To be effective, the coating must adhere tightly to the metal so that heat is conducted into the metal. Coatings may be applied to a glass or a polymer, but with different degrees of integrity of adherence.

The second type is a selective coating, which has a high absorptivity for short-wavelength radiation and low emissivity in the infrared region (Fig. Q8.4). The emittance is usually specified for temperatures close to ambient; however, emissivity rises with temperature because of the shift in the thermal spectrum to lower wavelengths. The inability to radiate infrared is related to the fact that the

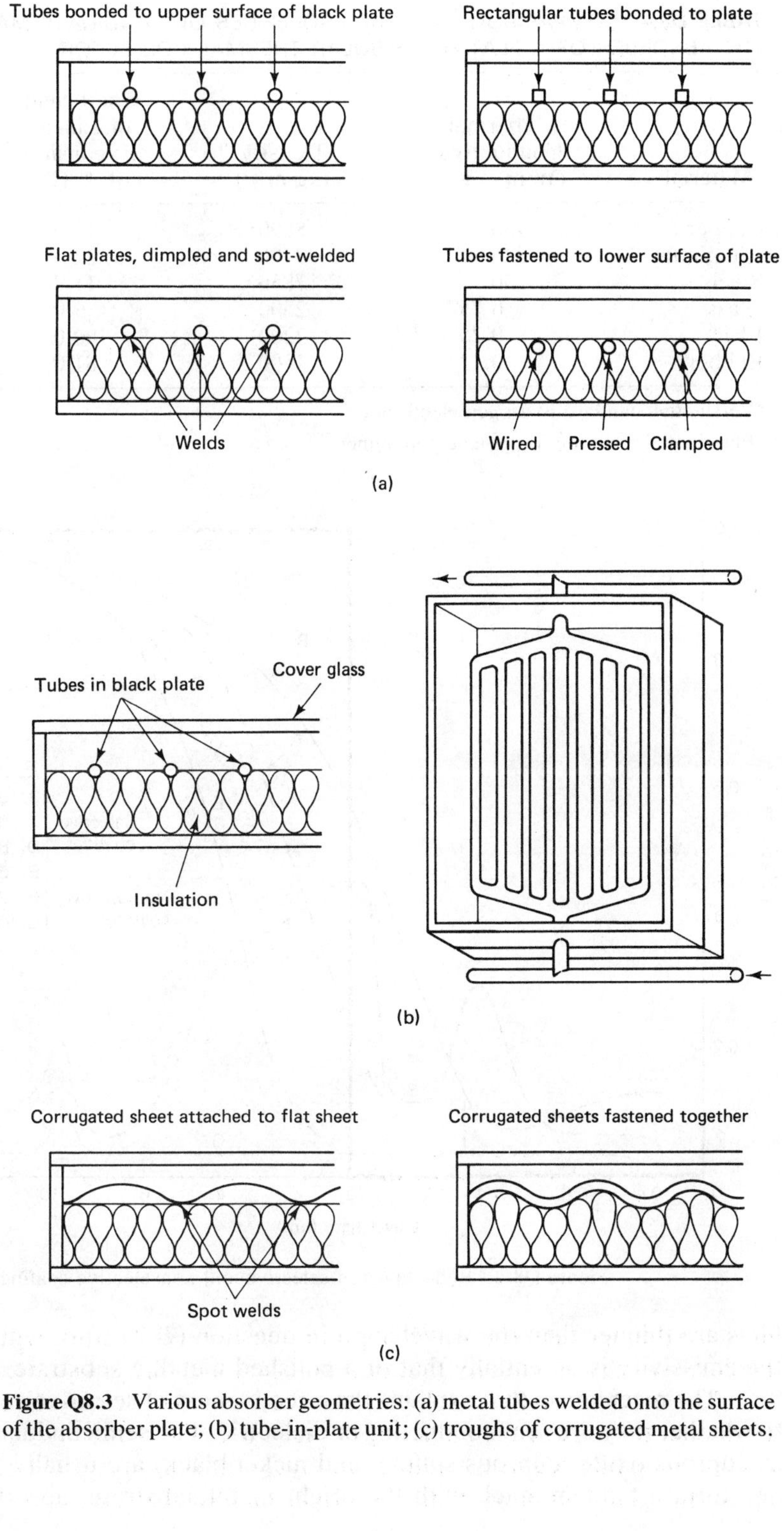

Figure Q8.3 Various absorber geometries: (a) metal tubes welded onto the surface of the absorber plate; (b) tube-in-plate unit; (c) troughs of corrugated metal sheets.

TABLE Q8.4 COMPARISON OF SOME PROPERTIES OF CANDIDATE MATERIALS FOR THE ABSORBER OF A FLAT-PLATE SOLAR THERMAL COLLECTOR

Material	Thermal Conductivity ($W\ m^{-1}\ K^{-1}$)	Density ($kg\ m^{-3}$)	Coefficient of Linear Expansion ($10^{-6}\ °C^{-1}$)	Electrode Potential in Water[a] (V)
Copper	400	8690	17.0	−0.34
Aluminum	220	2700	22.5	+2.30
Steel	50	7850	11.7	0
Glass	0.75	2500	9.0	—
FEP[b]	0.20	2200	100.0	—
Rubber	0.12	1500	81.0	0

[a]Versus the standard hydrogen electrode.

[b]Fluorinated ethylene–propylene copolymer.

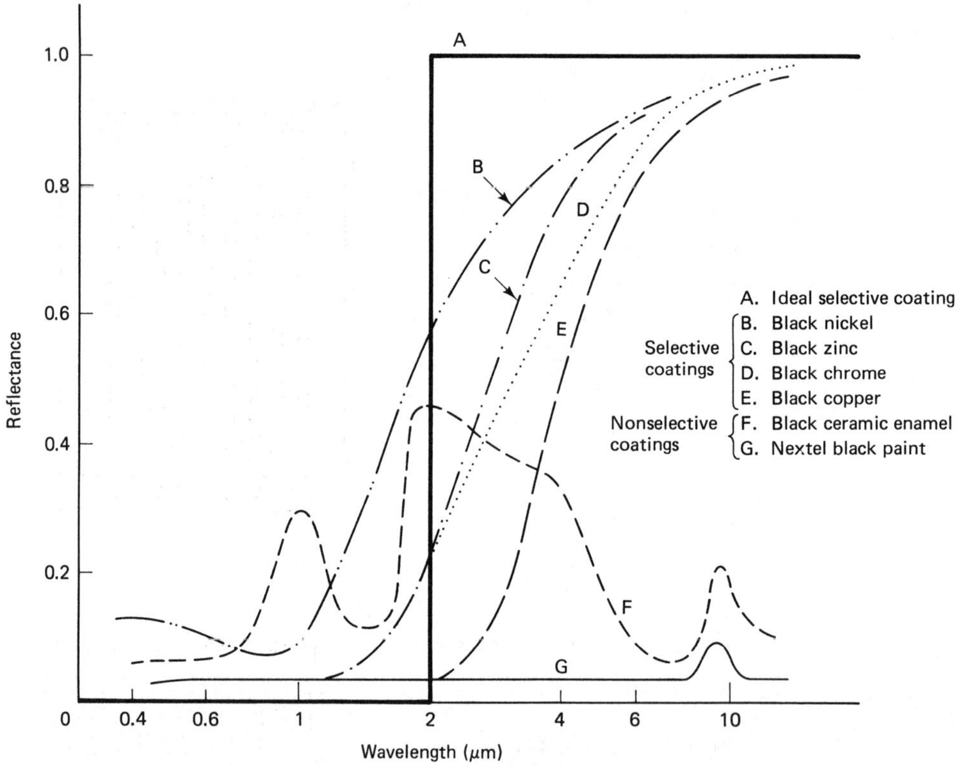

Figure Q8.4 Reflectance of selective and nonselective coatings.

films are thinner than the wavelength in question (7–25 μm), with the result that the emissivity is essentially that of a polished metallic substrate.

The mechanism that explains the effectiveness of selective coatings is related to whether they are semiconducting or dielectric. The semiconductive layers (such as cuprous oxide, cuprous sulfide, and nickel black) are usually 2 to 5 μm thick and form a tandem stack with the bright metal substrate, absorbing the visible

TABLE Q8.5 PROPERTIES OF ABSORBER COATINGS ON A COPPER SUBSTRATE

Coating	Type	Absorptivity	Emissivity
Black chrome	Electroplated	0.96	0.11
Black copper	Conversion coating	0.89	0.17
Nextel	Paint	0.98	0.89
Enersorb	Urethane paint	0.97	0.90

light and transferring the heat by lattice vibration and possibly by electronic conduction. The dielectric layers (such as cupric oxide and chromium black) are usually less than 1 μm thick and form a multilayer interference stack (with an extremely thin intermediate metallic layer), causing selected wavelengths to undergo multiple reflections and hence to be almost totally absorbed by the metal substrate.

The selective coatings are usually produced by one of three techniques:

(a) direct application to reflecting metallic substrates (an example is black chrome coating on copper solar collectors), (b) controlled oxidation or electrolytic oxidation to give thin conversion layers (an example is a selective copper black coating produced by the Ebanol process), and (c) vapor deposition, in an inert gas, of a metallic coating of extreme microirregularity (an example is a coating consisting of tungsten granules vapor-deposited in argon gas).

Some properties of typical absorber coatings are given in Table Q8.5.

The final aspect to consider is the corrosion of the absorber by the heat transfer fluid. The problem is usually localized corrosion (pitting, crevice corrosion, establishment of galvanic cells). Of the three most popular absorber materials (aluminum, steel, and copper), aluminum is the most sensitive to this kind of attack, and copper is the least sensitive.

Water used in the system must be distilled or deionized, and, preferably, cathodic corrosion inhibitors, such as chromate salts, should be added. The cooling fluid must be monitored periodically to ensure that it remains neutral, and the inhibitors must be replenished. The fluid should have high heat capacity and low vapor pressure, should not be flammable or degradable at maximum stagnant temperatures, and should have low toxicity.

QUESTION 9
Ceramic Monolith Honeycomb Supports

Recently enacted legislative requirements to control emissions motivated the development of ceramic monolithic honeycomb supports of cordierite. Some applications of these honeycombs include ozone abatement catalysis systems in high-flying aircraft, catalytic converters in automobiles, and combustors in residential wood stoves.

(a) *Justify the use of cordierite in such honeycombs in a catalytic converter in an automobile.*

(b) *What special properties are needed in monolithic honeycomb ceramics used* *as molten metal filters for superalloys? Compare the performance of several candidate ceramics.*

(a) The catalytic converter (Fig. Q9.1) is designed to oxidize unburnt hydrocarbons and carbon monoxide to carbon dioxide and water and to reduce oxides of nitrogen to elemental nitrogen. Mainly because of severe spatial constraints and variable engine operating conditions* a monolith reactor consisting of an integral bundle of tubes of a suitable material, the walls of which are coated with a catalyst (usually a platinum-group element), is favored. The honeycomb of channels in a monolith reactor has regular cross sections, but their shape may be circular, hexagonal, square, or triangular, depending on the design and mode of construction (Fig. Q9.2).

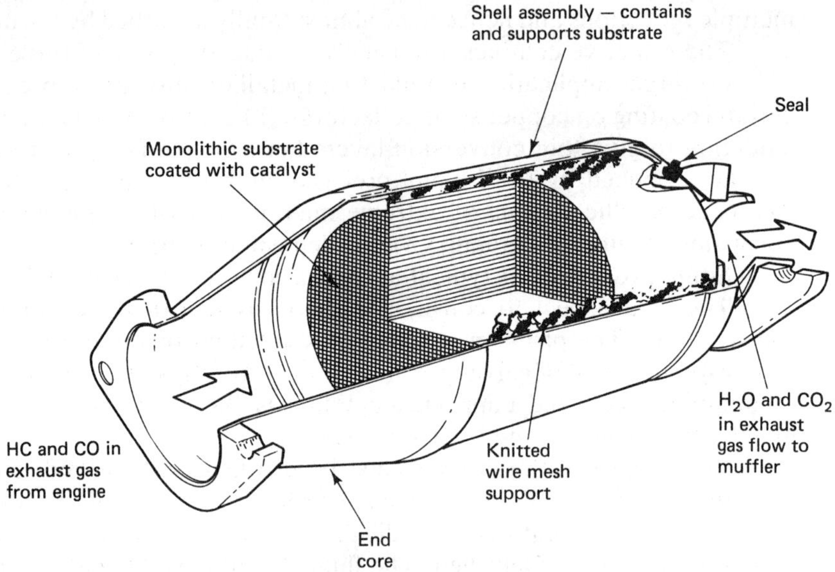

Figure Q9.1 Details of the round type of monolithic catalytic converter for use in the exhaust system of an automobile.

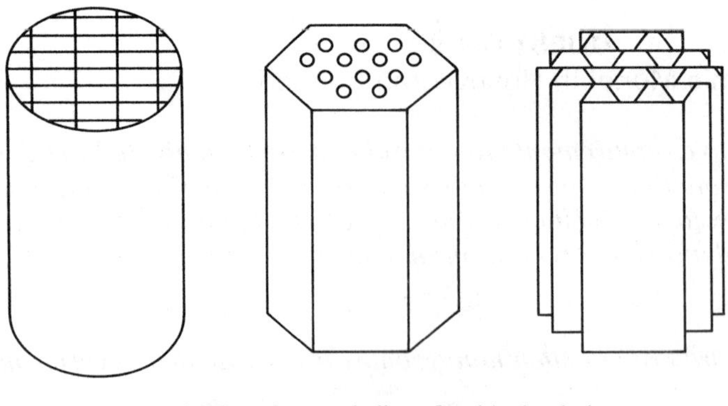

Figure Q9.2 Shapes of the honeycomb of channels of a monolith reactor.

*Otherwise, a shallow fixed-bed tubular reactor containing a mixture of noble metals and promoted iron oxide would be acceptable.

TABLE Q9.1 PROPERTIES OF CORDIERITES FOR MONOLITHIC HONEYCOMB SUPPORTS FOR CATALYTIC CONVERTERS

Property	Cordierite		
	9475	9482	9483
Crushing strength, σ_F (MPa)	21	14	14
Poisson's ratio, ν	0.25	0.25	0.25
Modulus of elasticity, E (GPa)	28	16	16
Coefficient of linear expansion (10^{-6} K^{-1})	1.0	0.85	0.80
Thermal shock resistance,[a] R_2 (K)	563	772	820

[a] $R_2 = \sigma_F(1 - \nu)/E\alpha$; see Chapter 7.

Sources: Anon, "CELCOR® Honeycomb catalyst supports," p. 3, Corning Glass Works, Corning, New York; I. M. Lachman, Corning Glass Works, Corning, New York, Private communication, 1987.

A primary requirement in an automobile catalytic converter is that the material of which the honeycomb support is made must be able to survive the thermal stresses arising from rapid heat-ups and steep thermal gradients. That is, the ceramic material must have very high thermal shock resistance. This explains the use of cordierite—a magnesium aluminosilicate—which combines a low modulus of elasticity, low thermal expansion (due to the orientation of the cordierite phase developed after firing), high fracture strength, and high melting temperature (~1459°C) with excellent compatibility with the catalyst system. The values of the relevant material properties and hence the thermal shock resistance for three varieties of cordierite are given in Table Q9.1.

(b) In this application, the materials need a combination of high thermal shock resistance and good refractoriness. This may be achieved by using a ceramic material based on the concept of developing microcracks in combinations of highly refractory phases [1]. This leads to a high thermal shock–resistant ceramic of excellent refractoriness coupled with acceptable levels of strength. Two such materials are ceramic 9481 (which is mullite plus an aluminum titanate phase with one crystallographic axial expansion that is highly negative and thus yields the microcracked structure) and ceramic 9449 (which contains zirconia with a large and unusual phase transformation at about 1200°C, thereby leading to a microcracked structure [1].) The relevant properties of these materials are given in Table Q9.2.

TABLE Q9.2 TWO CERAMICS FOR MONOLITHIC HONEYCOMBS FOR MOLTEN METAL FILTERS

Property	Ceramic	
	9481	9449
Melting temperature (°C)	1700	1900
Crushing strength (MPa)	30	90
Coefficient of linear expansion (10^{-6} K^{-1})	2.6	7.9

Source: Corning Glass Works, Corning, N.Y.

REFERENCE

1. I. M. Lachman and R. N. McNally, *Ceram. Eng. Sci. Proc.* 2 (5/6), p. 337, 1981.

QUESTION 10
Impeller of a Hover Lawn Mower

Outline the main considerations in the choice of a material for the impeller of a hover lawn mower.

The main function of the impeller (Fig. Q10.1) is to supply an air cushion for the hover action of the mower. These are the material requirements:

- Low density
- Low noise propagation (the current U.S. noise limit, set by the Occupational Health and Safety Act of 1971, is 100 dB)
- Excellent resistance to missile impact
- High specific stiffness in order to limit blade deflection in service
- Excellent dimensional stability to give good balance in the impeller, little undue mold shrinkage, and minimum creep strain when impeller is rotating
- Low raw material cost
- Excellent resistance to corrosion or cracking due to gasoline

The candidate materials considered are (a) a semicrystalline polymer, acetal polymer; (b) polycarbonate (PC); (c) a special grade of acrylonitrile-butadiene-styrene (SABS) having a high butadiene content; and (d) a special grade of polypropylene with rubber reinforcement to improve impact strength and mineral filler to increase stiffness (SPP). The relevant properties of these materials are given in Table Q10.1.

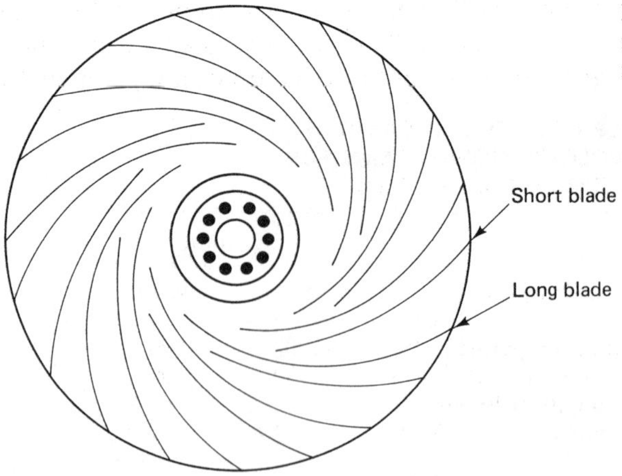

Figure Q10.1 Impeller of a hover lawn mower: plan view showing the arrangement of the blades.

TABLE Q10.1 CANDIDATE MATERIALS FOR THE IMPELLER OF A HOVER MOWER

Property	Acetal	PC	SABS	SPP
Density, ρ (kg m^{-3})	1410	1200	1040	900
Modulus of elasticity, E (MPa)	3.4	2.2	2.5	1.3
Izod impact strength (J m^{-1})	69	160	320	107
Creep modulus (GPa, in 1 year)	0.92	1.50	1.20	0.36
Coefficient of linear expansion ($10^{-6}\cdot C^{-1}$)	100	55	90	110
Unit cost[a] (\$ kg^{-1})	3.44	3.74	2.20	1.25
Specific stiffness, E/ρ (m)	246	187	245	147

[a]June 1988 prices.

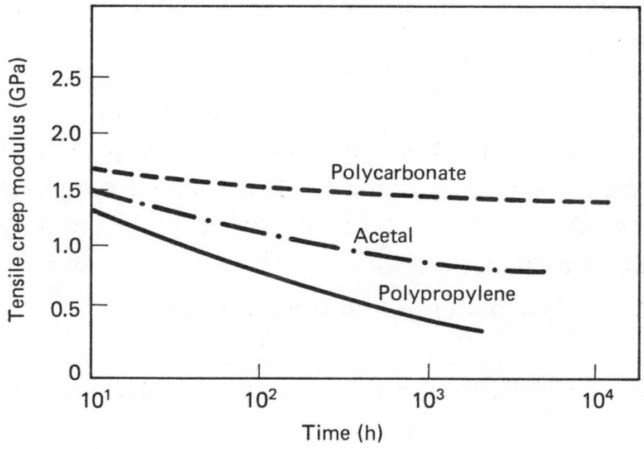

Figure Q10.2 Creep curves for three candidate materials for the impeller of a hover lawn mower.

Acetal polymer has one major disadvantage: difficulty of molding large items from it with acceptable dimensional accuracy. Polycarbonate has poor resistance to environmental stress cracking, while the special reinforced polypropylene has poor creep properties (Fig. Q10.2). The special grade of ABS does not have any of these disadvantages. All the candidate materials have good or acceptable strength, impact resistance, specific stiffness, and corrosion resistance properties. In consideration of these remarks, the best material is the high butadiene ABS.

QUESTION 11
Gas-cooled Nuclear Reactor

Among the many advanced energy systems under development in the major industrialized countries, gas-cooled nuclear reactors offer the potential for a substantial contribution to the satisfaction of those countries' near-term and long-term energy demand. However, these reactors pose enormous problems and major challenges in the use of materials.

Highlight some of these problems in detail, citing suitable materials for structural components in these reactors.

Two subsets of these reactors are under development in the United States, Japan, France, and other advanced nations: (a) High-temperature gas-cooled reactor (HTGR) systems use a graphite-moderated core fueled with uranium and thorium oxide or thorium carbide (Fig. Q11.1). There are at least four variants of this system. (b) Gas-cooled fast breeder reactor (GCFR) systems use a metal-clad plutonium-uranium oxide fuel (Fig. Q11.2). Probably the only major difference between these two types of reactors is the fact that the core in the HTGR is mostly carbon, with uranium-and thorium-fueled regions, while the core in the GCFR is a compact arrangement of fuel assemblies, and the fuel material is highly enriched.

Two features are common to both systems: (a) The maximum temperatures to which structural materials are exposed are very high (Table Q11.1), and (b) high-pressure helium coolant is used to transfer heat from the reactor core to the heat exchangers.

Principal considerations that influence the choice of materials for these structural components are these:

- Tensile creep, low-cycle fatigue, and creep-fatigue interaction
- Interaction of materials with available media or conditions such as helium, steam or water, temperature, irradiation (in fast reactor systems), and the effect of such interaction on the mechanical properties of the materials
- Thermal aging and embrittlement
- Ease of fabrication and raw material and fabrication costs

Figure Q11.1 Schematic diagram of a high-temperature gas-cooled reactor power plant.

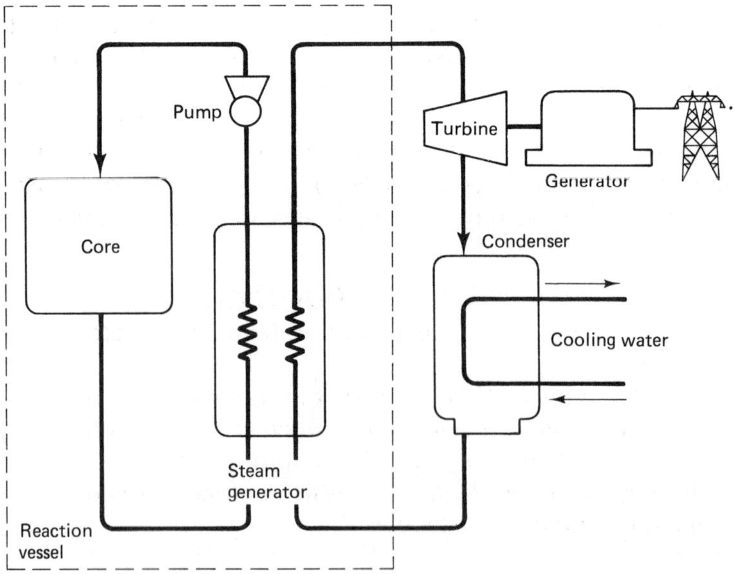

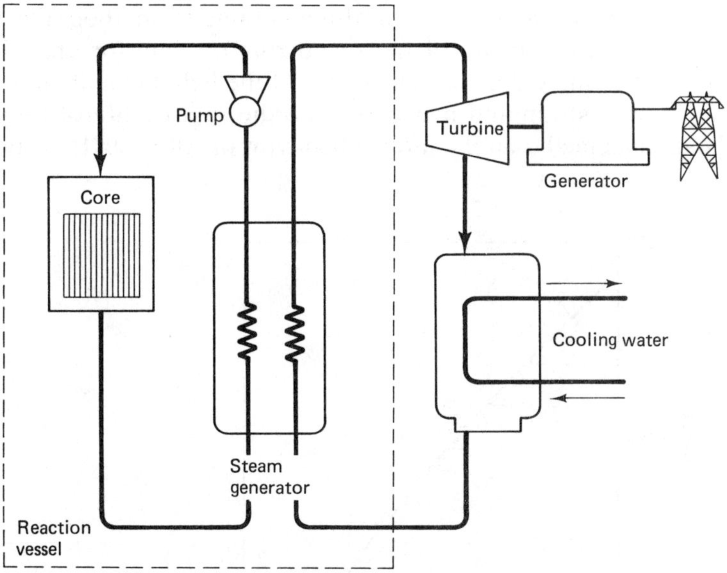

Figure Q11.2 Schematic diagram of a gas-cooled fast reactor system.

TABLE Q11.1 MAXIMUM NOMINAL TEMPERATURES TO WHICH STRUCTURAL COMPONENTS ARE EXPOSED IN GAS-COOLED NUCLEAR REACTORS

	Maximum Nominal Temperature (°C)	
Reactor Type	Hottest Thermal Barrier	Hottest Regions of Heat Exchangers
HTGR-SC (steam cycle)	750	750
HTGR-PH (process heat)	900	850
HTGR-GT (gas turbine)	850	600
HTGR-SM (steam)	600	500
GCFR	650	550

The major alloys of interest for gas-cooled reactor structural applications include (a) high-temperature steels, such as 2.5% Cr–1% Mo, 9%–12% Cr, Type 304 stainless, and Type 316 stainless; (b) nickel-based and iron-based superalloys, such as Alloy 800H, Hastelloy H, and Inconel 617; and (c) structural ceramics, such as silicon nitride and silicon carbide.

Let us examine the performance of some metallic alloys on the basis of the material requirements just outlined.

High-Temperature Behavior

By way of example, the long-term creep rupture behavior of Hastelloy X in air may be extrapolated (while exercising due care) from its short-term behavior, by means of the well-tried Larson-Miller parametric method (Fig. Q11.3). Table Q11.2 presents a comparison of the creep rupture data for annealed 2.25 Cr–1% Mo steel in helium and in air. The effect of the helium medium, it seems, is to reduce the rupture strain and hence shorten the rupture life of the alloy. The effect of these same media on the fatigue behavior of Alloy 800H is shown in Table Q11.3.

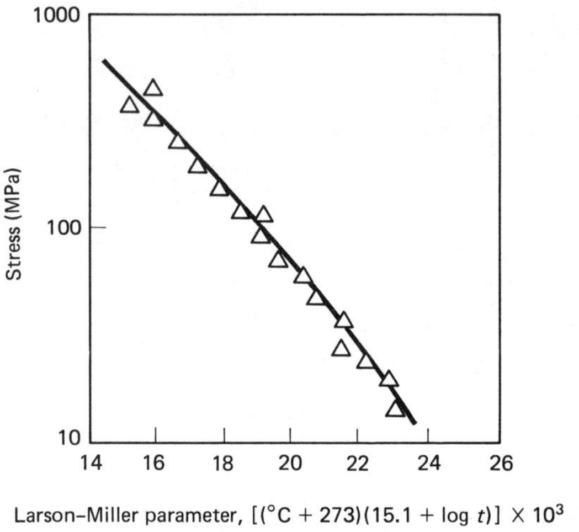

Figure Q11.3 Typical Larson-Miller master curve for Hastelloy X alloy stress rupture in air (t is rupture time in hours).

TABLE Q11.2 CREEP RUPTURE DATA OF ANNEALED 2.25% CR-1% MO ALLOY IN AIR AND IN HTGR HELIUM (4.8 mm DIAMETER SPECIMENS; 595°C; 72.4 MPa)

	Strain (%)	
Time (h)	In air[a]	In HTGR Helium[b]
0	0.0	0.0
1000	1.5	3.5
2000	2.0	5.5
3000	2.5	8.0
4000	3.5	—
6000	5.0	—
8000	19.0	—

[a]Ruptured at 8067 h

[b]Ruptured at 3700 h

Source: G. Y. Lai, and R. J. Wolwowicz, "Creep-rupture behavior of 2¼ Cr-1 Mo steels, Alloy 800 H, and Hastelloy X in a simulated HTGR helium environment," GA-A15572, General Atomic Co., USA, 1979.

TABLE Q11.3 CYCLICAL STRESS DATA OF ALLOY 800H IN VARIOUS MEDIA AT 650°C

Number of Cycles to Cause A Drop Of 5% From The Maximum Tensile Stress	Total Axial Strain Range (%)	
	In Air	In Vacuum + HTGR Helium
1000	1.2	1.5
4000	0.7	0.9
10,000	0.6	0.7

Source: D. I. Roberts, S. N. Rossenwasser, and J. F. Watson, "Material selection for gas-cooled and fusion reactor applications," *J. Mater. Energy Sys. 3,* pp. 54–65, 1981.

Interaction with Cooling Fluids

There have been some very detailed studies of these interaction mechanisms as they relate to HTGR helium. It has been shown that some or all of the impurities that may exist in gas-cooled reactor systems (Table Q11.4) can exert oxidizing or reducing and carburizing or decarburizing effects on the major alloying elements in the high-temperature iron- and nickel-based superalloys. Carburization is of great importance in gas-cooled reactors that have a large hot graphitic core (Table Q11.5), whereas in the practically graphite-free GCFRs, with their high moisture level, the potential problem is decarburization. Oxidation may be a serious problem in some of the heat exchangers in gas-cooled reactor systems, particularly those that employ water or steam as a secondary coolant fluid (Table Q11.6).

Thermal Aging and Embrittlement

As a result of the microstructural changes that many of the alloys of interest undergo at the operational temperatures of these reactors, thermal aging and

TABLE Q11.4 IMPURITY LEVELS EXPECTED IN THE PRIMARY COOLANT OF GAS-COOLED NUCLEAR REACTORS

	Impurity Level (ppm)		
Impurity	HTGR-SC	HTGR-GT and HTGR-PH	GCFR
H_2O	10	<1	4000
H_2	200	500	1000
CO	10	50	<1
CH_4	20	50	<1
CO_2	<1	<1	—
N_2	15	15	—
O_2	<<<1	<<<1	—

Source: Same as for Table Q11.3.

TABLE Q11.5 PREDICTED CARBURIZATION BEHAVIOR OF STEELS IN HTGR HELIUM

Temperature (°C)	Predicted Carburization Depth In 140,000h (mm)			
	Alloy 800H	Inconel 617	Hastelloy X	IN738
550	0	0	0	0
650	40	40	40	10
750	260	200	200	20
850	800	700	500	60

Source: D. I. Roberts, S. N. Rossenwasser, and J. F. Watson, "Materials selection for gas-cooled and fusion reactor applications," *J. Mater. Energy Sys. 3,* pp. 54–65, 1981.

TABLE Q11.6 OXIDATION RATE OF CANDIDATE HEAT EXCHANGER MATERIALS IN SUPERHEATED STEAM (STEAM TEMPERATURE = 540°C; STEAM PRESSURE = 10.5 MPa)

Time (h)	Weight Gain (mg cm^{-2})		
	2.25% Cr-1% Mo Steel	9%–12% Cr-Mo Steels	Inconel 800
0	0.0	0.0	0.0
2500	5.0	4.8	3.8
7500	8.0	7.0	5.0
12500	9.0	8.0	6.0

Source: J. R. D. Stefano, Internal Report, Oak Ridge National Laboratory, Oak Ridge, Tenn., 1978.

TABLE Q11.7 EFFECTS OF THERMAL AGING AT 650°C ON THE CHARPY IMPACT TOUGHNESS OF SOME HIGH-TEMPERATURE MATERIALS

Time (h)	Charpy Impact Toughness At Room Temperature (J)		
	Alloy 800H	Inconel 617	Hastelloy X
2000	72	80	25
4000	60	50	18
8000	60	58	15
12000	60	60	15
16000	60	60	15

Source: D. I. Roberts, S. N. Rossenwasser, and J. F. Watson, "Material selection for gas-cooled and fusion reactor applications," *J. Mater. Energy Sys. 3,* pp. 54–65, 1981.

TABLE Q11.8 TEMPERATURE-TIME ENVELOPE FOR THE ONSET OF RECRYSTALLIZATION IN 20% PRESTRAINED ALLOY 800H

Exposure Time (h)	Temperature (°C)
10	850–1000
1000	730–850
100,000	650–750

Source: G. Y. Lai, and A. B. Smith, "Recrystallization behavior of Alloy 800 H," in *Characterization of Materials for Service at Elevated Temperatures,* MPC-7, New York: American Society of Mechanical Engineers, 1978.

embrittlement can occur, sometimes to the extent of undermining the structural integrity of the component. The effect of thermal aging on the toughness of three candidate alloys is shown in Table Q11.7. It is also essential to avoid the onset of recrystallization in service (Table Q11.8).

Ease of Fabrication and Costs

Because many of the structural components of these reactors are large and complex in shape, it is necessary that the materials employed have low raw material costs and be relatively easy to fabricate, in order to keep fabrication costs to a minimum.

QUESTION 12
Aircraft Engine Compressor Blades

Present generalized guidelines for the selection of a material to be used to fabricate an axial-flow compressor for an aircraft turbine engine. Comment on the suitability of metal-matrix composite materials for this application.

The development of the gas turbine engine for aircraft propulsion, in terms of the improvement of engine thrust over time, is shown in Fig. Q12.1. Schematic representations of the components of the two main forms of axial-flow compressors are given in Fig. Q12.2, and the details of one aircraft engine that incorporates an example of one such compressor are shown in Fig. Q12.3. This engine has been placed in a number of aircraft in current use, such as the Boeing 707 and 727 and the McDonnell Douglas DC8. The requirements for aircraft engine compressor blade needs can best be discussed with reference to current or envisaged aircraft operational zones (Fig. Q12.4). For flight speeds in excess of Mach 3 and altitudes above 24 km, the ramjet or scramjet and/or rockets will be the main propulsive engines (as in envisioned in the transatmospheric vehicle; see Chapter C1, Part 3). These engines will probably not require mechanical compressors but will use some form of booster to get them up to the required speed. At the other

Figure Q12.1 Improvement in the gas turbine engine for use in commercial aircraft, 1950–1986.

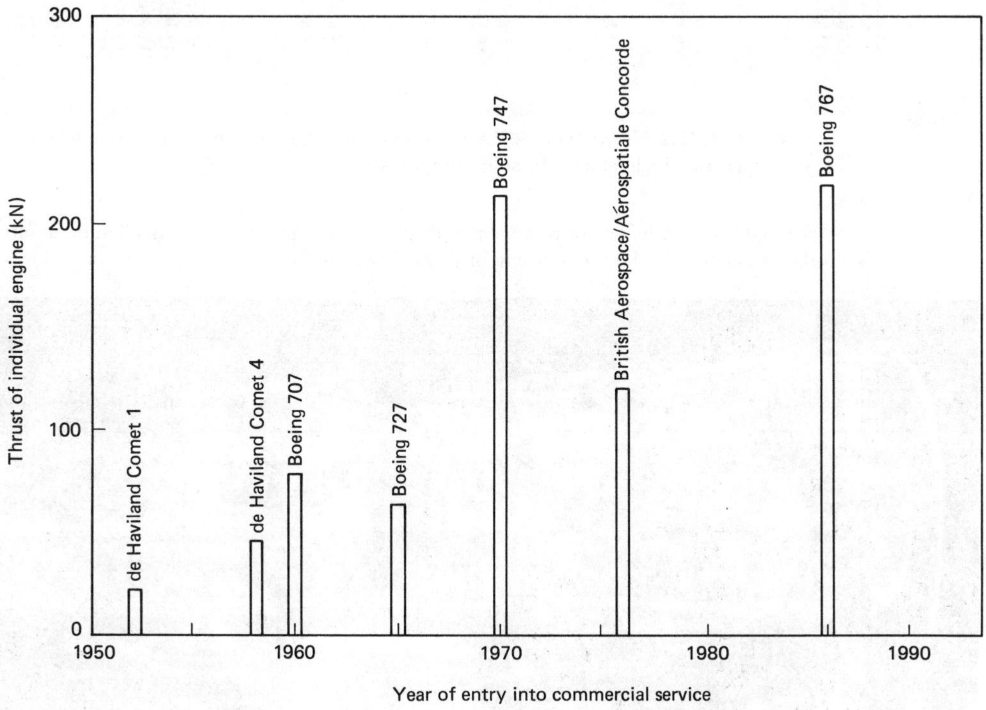

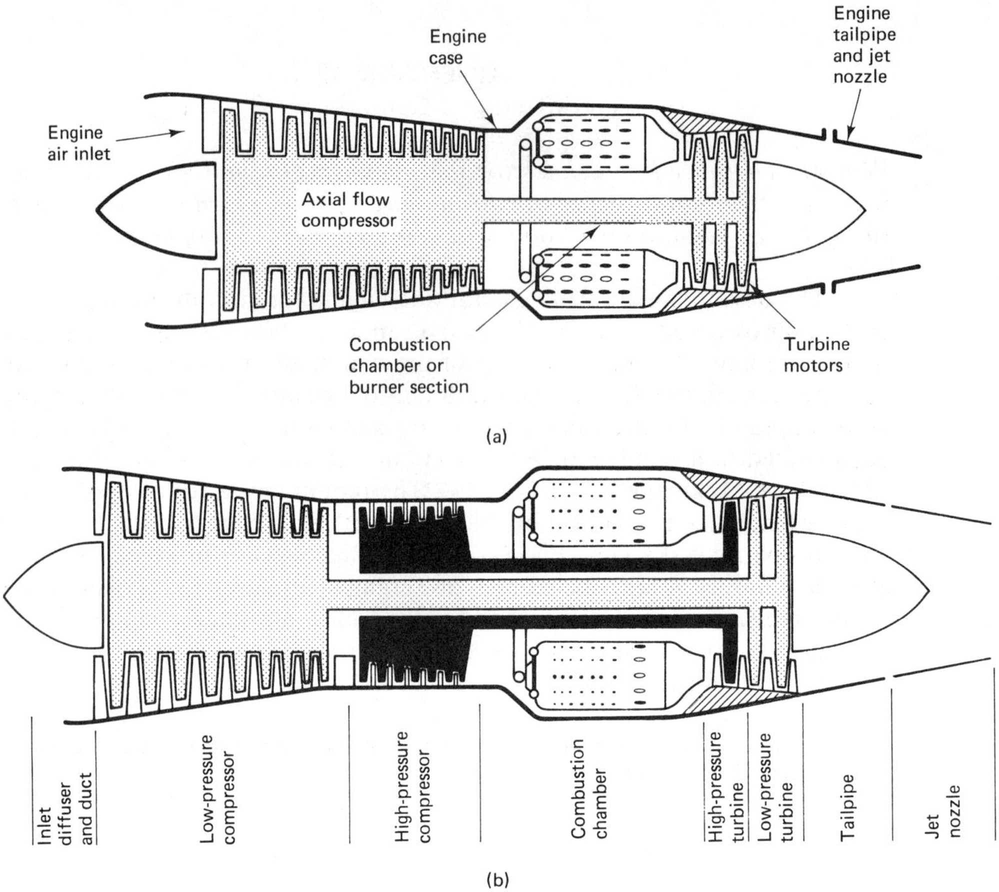

Figure Q12.2 Two types of axial-flow compressor in an aircraft gas turbine engine: (a) single-stage; (b) split-compressor.

Figure Q12.3 Arrangement of a split-type axial-flow compressor as installed in an aircraft turbojet engine, the J-57. (Courtesy of Pratt & Whitney)

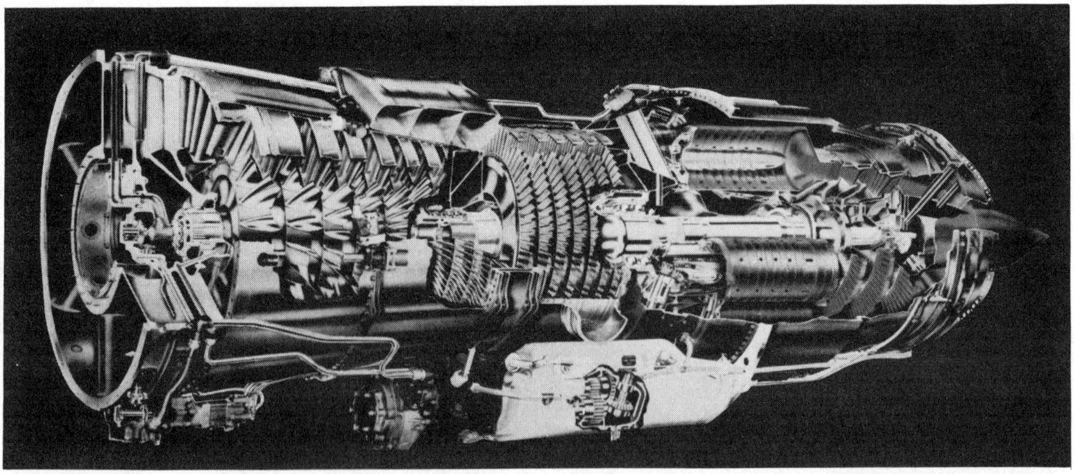

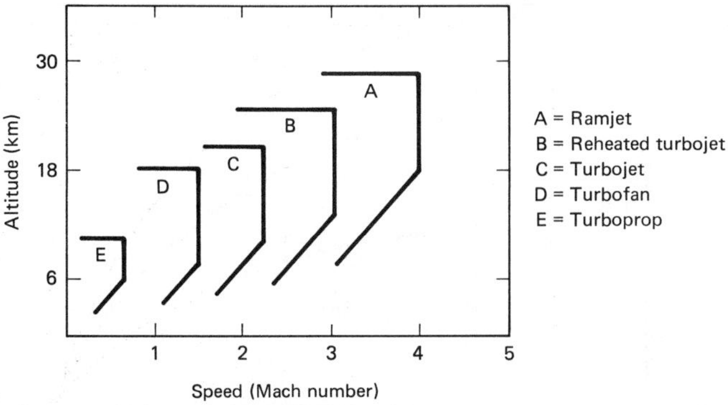

Figure Q12.4 Propulsion system zones of operation.

end of the speed-altitude spectrum, for speeds below Mach 0.7 and altitudes below 10 km, the turboprop will remain the main propulsive engine, and direct lift (vertical takeoff and landing and short takeoff and landing applications) will be used in the altitude range from sea level to about 2.4 km.

The steady-state stresses in a compressor blade are due to gas and centrifugal bending moments, although in a rotating blade, the centrifugal bending moment can be made to counteract the gas bending moments by proper orientation of the blade's axis. The major steady-state stress in the blade airfoil (Fig. Q12.5) is the centrifugal tensile stress, which has a maximum value at the root of the blade and is given by

$$\sigma_{\max} = 6\left[1 - \left(\frac{D_h}{D_t}\right)^2\right]\phi U_t^2 \rho$$

where D_h is the hub diameter, D_t is the tip diameter, ϕ is the blade taper stress reduction factor, U_t is the blade tip speed, and ρ is the density of the blade material. It can be shown that ϕ is a function of the diameter ratio, D_h/D_t. Thus

$$\frac{\sigma_{\max}}{\rho} = f\left(\frac{D_h}{D_t}\right) U_t^2$$

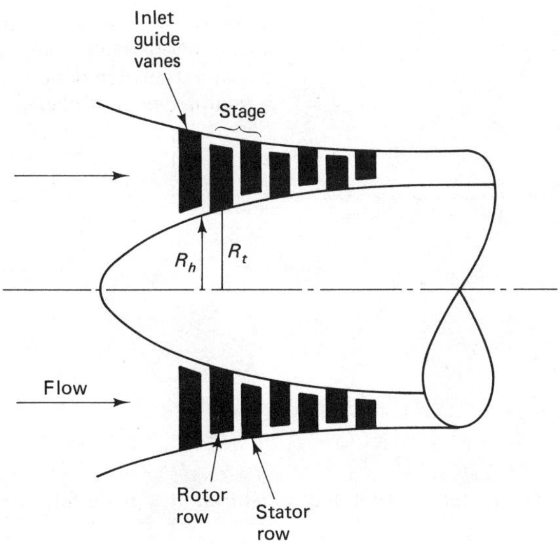

Figure Q12.5 Cross section of an axial-flow compressor (R_t, tip radius; R_h, hub radius).

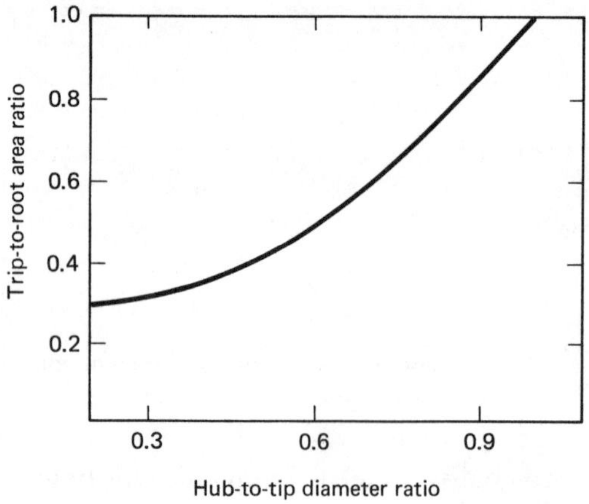

Figure Q12.6 Empirical fit between tip area–to–root area ratio and hub diameter–to–tip diameter ratio for an airfoil.

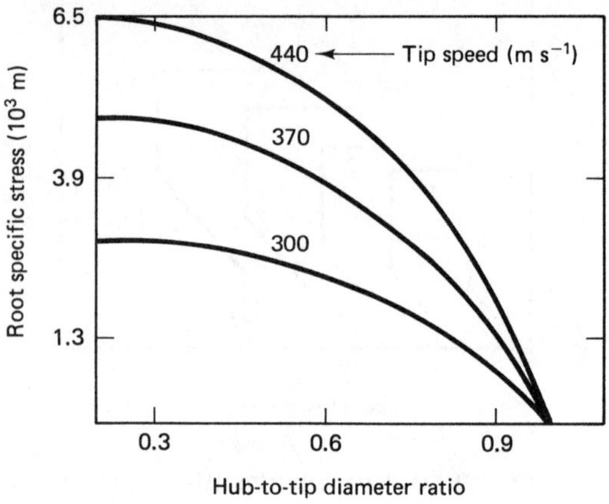

Figure Q12.7 Dependence of root specific stress on the hub diameter–to–tip ratio for various values of tip speed.

where the function $f(D_h/D_t)$ may be developed from the blade profile area distributions and the ratio of the blade tip area to that of the blade root. Employing a linear area distribution and an empirical fit between the tip area–to–root area ratio and the hub diameter–to–tip diameter ratio (Fig. Q12.6), the variation in root centrifugal stress per blade mass (i.e., root specific stress) with the hub diameter–to–tip diameter ratio is obtained. Since most aircraft engines operate at nearly constant blade rotational speeds, this root specific stress variation may be easily obtained for various rotor blade tip speeds (i.e., various types of aircraft engines, in effect; see Fig. Q12.7). Since this root specific stress must be resisted by the material, the one requirement in a blade material is high tensile strength combined with low density, that is, high material specific tensile strength, and its retention at elevated temperatures.* The variation of specific strength with temperature for a number of candidate metal-matrix composite materials is given in Fig. Q12.8.

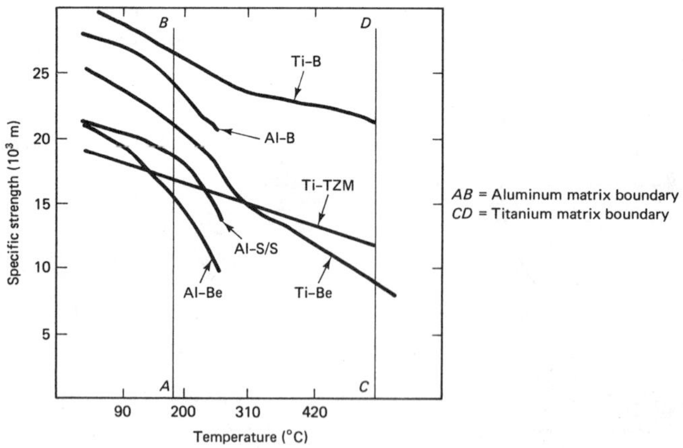

Figure Q12.8 Specific strength of aluminum and titanium composite materials at various temperatures (at 25 vol % reinforcement of fiber).

*The ratio of material specific strength to root specific stress, at a given temperature, is, in fact, a factor of safety.

A second requirement is a combination of high modulus of elasticity E and low density, that is, a high specific modulus. The high value of E means increased stiffness and hence lower elastic deflections for a given load. A high material specific modulus also affects the magnitude of the resonant frequency of the blade by affording blade resonances with the lower engine-exciting orders to be more easily avoided within the engine operating range. In addition, the values of E and ρ affect the level of the vibratory stresses on the blade. For a given excitation level, defined as the product $\Delta_t f$ (where Δ_t is the amplitude of the tip of the blade and f is the frequency of the blade), the maximum stress in a vibrating blade is given by

$$\sigma_{\max} = \frac{2\pi y}{k} f \Delta_t \sqrt{E\rho}$$

where y is the distance from the neutral axis of the cross section of the blade to the outermost fibers of the section and k is the radius of gyration of the root section. Thus the vibratory stresses decline with decreasing value of the material parameter $\sqrt{E\rho}$. The values of E/ρ and $\sqrt{E\rho}$ for some metal matrix composite materials are presented in Tables Q12.1 and Q12.2.

A third requirement in the blade material is a high value of the melting point. This translates to high resistance to creep failure.* It is always desired, in aeroen-

TABLE Q12.1 SOME ROOM-TEMPERATURE PROPERTIES OF CANDIDATE METAL-MATRIX COMPOSITE MATERIALS FOR AIRCRAFT COMPRESSOR BLADES

Material	ρ (kg m^{-3})	E[a] (GPa)
Aluminum alloy reinforced with 25 vol % stainless steel wire [ASS]	3751	131
Aluminum alloy reinforced with 25 vol % boron filaments [ABF]	2574	130
Aluminum alloy reinforced with 30 vol % silicon carbide filaments [ASC]	2823	107
Titanium alloy reinforced with 30 vol % TZM[b] wires [TTZM]	6062	165
Titanium alloy reinforced with 10 vol % silicon carbide whiskers [TSC]	4373	120
Ti-Al alloy reinforced with 25 vol % boron filaments [TABF]	3585	228
Ti-Al alloy reinforced with 25 vol % beryllium wire [TABW]	3210	—

[a]Transverse to fibers.

[b]TZM is produced by vacuum-arc melting of carbon deoxidized molybdenum with titanium and zirconium in a water-cooled mold. These additives produce a dispersion of fine carbon particles that improve strength and high-temperature stability.

*The temperature beyond which creep failure becomes a possibility (called the absolute homologous temperature) is, for most materials, 0.35 to 0.50 times the absolute melting point.

TABLE Q12.2 SOME COMPUTED ROOM-TEMPERATURE PROPERTIES OF CANDIDATE METAL-MATRIX COMPOSITE MATERIALS FOR AIRCRAFT COMPRESSOR BLADES

Material[a]	E/ρ (10^6 m)	$(E\rho)^{0.5}$ (10^6 N·m$^{-2.5}$)
ASS	3.56	69.43
ABF	5.15	57.29
ASC	3.86	54.43
TTZM	2.77	99.06
TSC	2.80	71.75
TABF	6.48	89.35
TABW	—	—

[a]See Table Q12.1.

gine design, to increase the temperature to which the compressor blades are subjected (i.e., the turbine inlet temperature)* since this enhances engine performance in a number of ways, such as increased thermal efficiency, for a given pressure ratio (Fig. Q12.9), decreased specific fuel consumption (Fig. Q12.10), and increased specific engine power output (Fig. Q12.11). The specific creep rupture strengths of cast nickel alloy and one metal matrix composite based on this alloy are given in Table Q12.3.

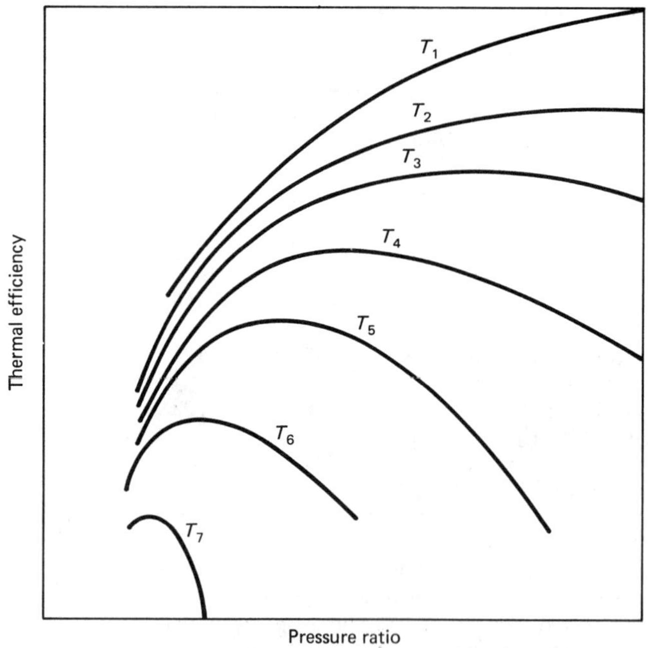

Figure Q12.9 Dependence of thermal efficiency on pressure ratio for various values of turbine inlet temperature ($T_1 > T_2 > T_3 > T_4 > T_5 > T_6 > T_7$).

*This temperature has increased from about 700°C in water-cooled engines made in the 1940s to about 1400°C in turbojet engines made in 1988. Some of this increase is attributable to improvements in materials, but blade cooling (introduced in 1960) has also played a very significant role.

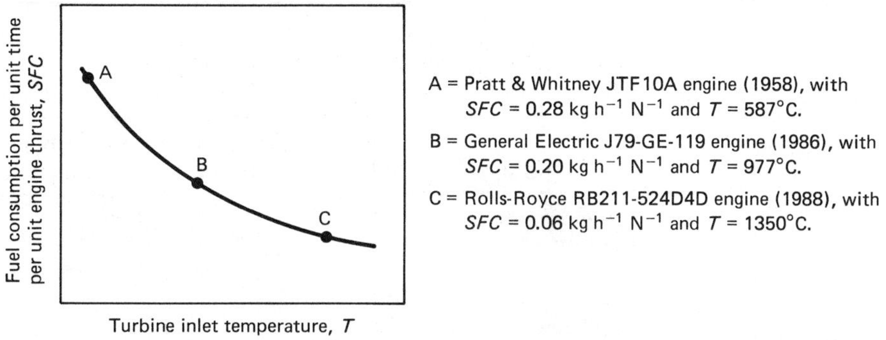

Figure Q12.10 Relationship between engine fuel consumption and turbine inlet temperature.

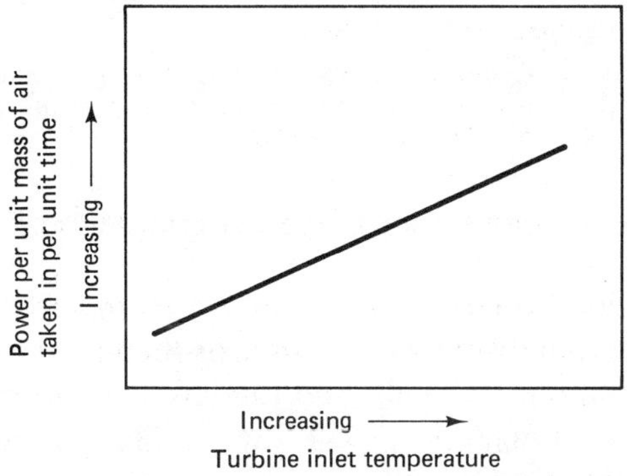

Figure Q12.11 Relationship between turbofan engine power and turbine inlet temperature.

TABLE Q12.3 SPECIFIC CREEP RUPTURE STRENGTH OF NICKEL ALLOYS AT 1093°C

	Specific Creep Rupture Strength (10^3 m)	
Creep Rupture Time (h)	Unreinforced Cast Nickel	Nickel Alloy Reinforced with 70 vol % Tungsten Wire and 2 vol % Thoria Wire
100	0.93	2.12
1000	0.49	1.59

Source: D. W. Petrasek, and R. A. Signorelli, "Tungsten alloy fiber reinforced nickel-base alloy composites for high-temperature turbojet engine applications" in *Composite Materials: Testing and Design,* ASTM STP 460, Philadelphia: American Society for Testing and Materials, p. 412, 1969.

A fourth requirement in the blade material is high resistance to thermal fatigue arising from fluctuating thermal stresses generated in start/stop cycles (Table Q12.4). A blade material must therefore have a low value of the coefficient of thermal expansion α. This is so because the strain suffered (due to stress generated) in a component subjected to cyclical thermal loading is given by

$$\epsilon = \alpha \Delta T - \frac{\sigma_y}{E}$$

where σ_y is the yield strength of the material.

TABLE Q12.4 REPRESENTATION OF A TYPICAL AIRCRAFT FLIGHT CYCLE

Flight Mode	Duration (min)	Engine Rating or Turbine Entry Temperature[a]
Ground mode	0	1.00
Take off	1	2.00
Climb	24	2.75
Cruise	60	2.30
Flight idle	15	1.23
Approach	6	1.85
Reverse thrust	1	2.60
Ground idle	0	1.00

[a]With respect to ground mode conditions.

Source: J. S. Ponsford, and G. K. Waddington, paper presented at the AGARD Conference on Engine Cycle Durability by Analysis and Testing, Lisse: Netherlands, June 1984.

Other requirements in a compressor blade material are as follows:

- High resistance to vibrational fatigue stresses arising from any or all of various types of aerodynamic excitation, such as high-amplitude fixed wake, self-excited flutter, rotating stall cell, and random excitation
- High resistance to impact loading caused, for example, by rocks drawn into the air inlet during taxiing of the plane
- Good notch sensitivity, that is, high fracture toughness
- High resistance to mechanical fatigue stresses
- Resistance to erosion
- Resistance to high-temperature oxidation by hydrogen sulfide, sulfur dioxide, and various combustion products; in effect, the material should have a high value of the enthalpy of formation of sulfide layers at the operational blade temperature
- Good corrosion resistance

QUESTION 13

Shield of an Underground Power or Utility Cable

Discuss the considerations governing the material for the shield of underground power, utility, or telephone cables.

The shield material may be described as follows:

- Economic and abundant
- Highly conductive
- Structurally flexible
- Easy to fabricate

- Resistant to mechanical damage
- Resistant to gopher penetration
- Very resistant to corrosion by the soil

By far the most important of these requirements is soil corrosion resistance. Many metals, such as copper and steel, undergo uniform corrosion in underground applications. The corrosion rate of copper is low in most soils, so copper is generally an excellent shield material, and service life is long and predictable. Under severe conditions, such as in highly acidic soils or soils with high concentrations of sulfide, however, corrosion of copper can occur more rapidly. Other materials, such as aluminum, are subject to localized corrosion in underground applications. The pitting and/or crevice corrosion rates of these materials can be high, and service life is unpredictable.

Important soil properties relevant to underground corrosion are soil resistivity, composition, pH, and aeration. Soil resistivity is widely regarded to be an indication of soil corrosivity, the latter usually increasing with decreasing resistivity (Table Q13.1).

The factors that govern resistivity of a soil include (a) its concentration of soluble salts, (b) its moisture content, and (c) the presence of aggressive ions, such as chloride and sulfide.

The role of pH corrosion is shown schematically in Fig. Q13.1. For most

TABLE Q13.1 RELATIONSHIP BETWEEN THE RESISTIVITY AND CORROSIVITY OF SOILS

Resistivity (Ω·m)	Corrosivity
Below 5	Very corrosive
5–10	Corrosive
10–50	Moderately corrosive
20–100	Mildly corrosive
Above 100	Barely corrosive

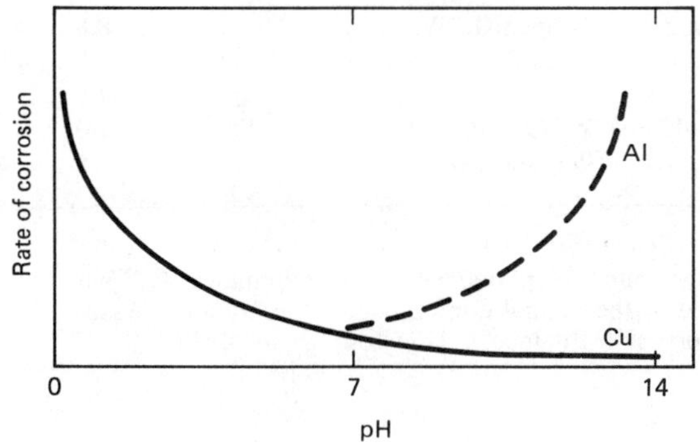

Figure Q13.1 Relationship between the pH and the corrosivity of soils.

metals, the corrosion rate increases with decreasing pH of the soil. In strongly acidic soils (pH < 5), corrosion rates for materials that are generally resistant (such as copper) are significantly higher. This is because of an increase in the rate of hydrogen in reduction in these soils. Thus, except for amphoteric metals such as aluminum, which are subject to higher corrosive rates in alkaline soils, more alkaline soils are less corrosive.

Aeration has a pronounced effect on soil corrosion because oxygen reduction is the principal cathodic reaction in soils. For most materials, corrosion is under cathodic control, and the rate is a direct function of the degree of aeration. The mechanisms of oxygen diffusion (regarded as the rate-determining process in a soil [1]) consists of three aspects: (a) the gaseous convection of oxygen in pores of the soil, (b) the gaseous diffusion in the pores of the soil, and (c) the diffusion of oxygen in a continuous liquid film (or corrosion product). Any one of these transport steps may be the limiting factor for corrosion in the soil.

Differential aeration cells (due to stones, aggregates, lumps of soil, crevices formed by shielding jackets, etc.) are readily set up in a soil. Metals such as aluminum, which are normally passive in oxygen-containing solutions, are susceptible to corrosion due to differential aeration. Metals such as copper, which do not passivate in oxygen-containing solutions, are less prone to this type of attack.

The candidate shield materials are (a) AISI 430 stainless steel metallurgically bonded with copper on both sides (Cu/stainless steel/Cu), (b) pure copper, (c) a copper-based alloy, and (d) aluminum.

The corrosion resistance of these shield materials was studied in an extensive cooperative program between the National Bureau of Standards and the Rural Electrification Administration (the so called NBS-REA study) in a number of soils (Table Q13.2). The results, summarized in part in Table Q13.3, may be used as a guide in selecting shield materials for a particular site.

TABLE Q13.2 SOME CHARACTERISTICS OF THE SOILS IN FIVE OF THE SITES IN THE NBS-REA STUDY

Site	Soil Type	Location	Resistivity ($\Omega{\cdot}$m)	pH
A	Loam	Toppenish, Wash.	4	8.8
B	Loam	Loch Raven, Md.	52	5.8
C	Clay	Cape May, N.J.	3	4.0
D	Sand	Wildwood, N.J.	300	7.3
E	Marsh	Patuxent, Md.	3	7.1

Source: W. F. Gerhold and J. P. McCann, "Corrosion evaluation of underground telephone cable shielding materials," paper presented at the annual conference of the National Association of Corrosion Engineers, Houston, Texas, 1976.

TABLE Q13.3 PERFORMANCE OF THREE CABLE SHIELD MATERIALS AFTER SIX YEARS' EXPOSURE TO VARIOUS SOILS IN THE NBS-REA STUDY

Shield Material	Geographic Site[a]	Location along cable[b]				
		I	II	III	IV	V
Al (75 μm thick)	A	9[c]	9	9	6	9
	B	9	9	9	9	9
	C	0	1	5	3	3
	D	9	9	9	9	9
	E	0	1	5	3	5
Cu (130 μm thick)	A	9	9	10	10	10
	B	9	9	9	9	9
	C	0	4	9	5	9
	D	—	—	—	—	—
	E	8	8	9	9	5
Cu/430 stainless steel/Cu (each 50 μm thick)	A	9/10/9[d]	9/10/9	9/10/9	9/10/9	9/10/9
	B	9/10/9	9/10/9	9/10/9	9/10/9	9/10/9
	C	9/10/9	9/10/9	9/10/9	9/10/9	9/10/9
	D	9/10/9	9/10/9	9/10/9	9/10/9	9/10/9
	E	4/9/9	3/9/9	9/10/9	9/10/9	9/10/9

[a]See Table Q13.2.

[b]See Fig. Q13.2.

[c]For explanation of performance rating, see Table Q13.4.

[d]Denotes the performance of the outer copper cladding, the steel core, and the inner copper cladding, respectively.

Source: W. F. Gerhold and J. P. McCann, "Corrosion evaluation of underground telephone cable shielding materials," paper presented at the annual conference of the National Association of Corrosion Engineers, Houston, Texas, 1976.

Figure Q13.2 Diagrammatic representation of a cable specimen for soil burial tests (I, exposed window; II, exposed ring; III, underjacket; IV, 13 mm from window; V, 13 mm from ring) [2].

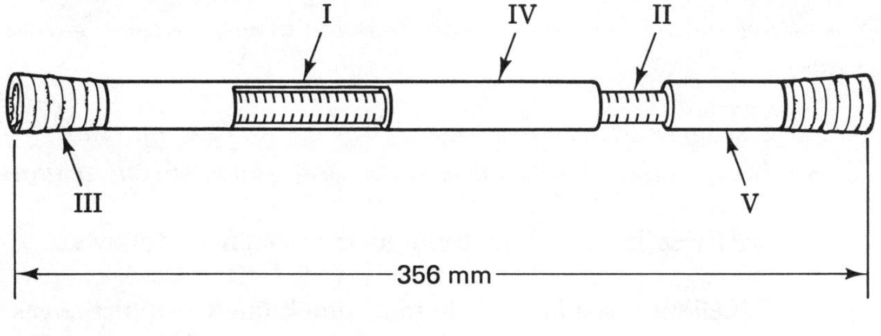

TABLE Q13.4 PERFORMANCE RATING CODE FOR THE EVALUATION OF THE UNDERGROUND CORROSION OF SHIELDS IN CABLE SPECIMENS

Rating Code	Performance	Type of Corrosion
10	Excellent	Unaffected; no evidence of corrosion.
9	Excellent	Some rust on surface.
8	Very good	Uniform metal attack, rust, and/or slight localized attack.
6	Good	Pitting causes 2 to 5 perforations in shield.
5	Fair	Many pits, leading to perforation of shield; <5% of shield consumed by corrosion; extensive delamination of metallurgically-bonded metals.
4	Poor	Severe corrosion, ranging from pitting to perforation of shield; 5% to 10% of shield consumed by corrosion; severe corrosion of anodic part of metallurgically-bonded metals.
3	Poor	Severe corrosion, ranging from pitting to perforation of shield; 10% to 25% of shield consumed by corrosion.
1	Very poor	Severe corrosion, shield is close to electrical discontinuity due to perforation in it and consumption of metal by corrosion.
0	Very poor	Severe corrosion, shield is electrically discontinuous due to consumption of metal by corrosion.

Source: W. F. Gerhold and J. P. McCann, "Corrosion evaluation of underground telephone cable shielding materials," paper presented at the annual conference of the National Association of Corrosion Engineers, Houston, Texas, 1976.

REFERENCES

1. N. D. Tomashov, *Theory of Corrosion and Protection of Metals,* New York: Macmillan, 1956, p. 414.
2. W. F. Gerhold and J. P. McCann, "Corrosion evaluation of underground telephone cable shielding materials," paper presented at the annual conference of the National Association of Corrosion Engineers, Houston, Texas, 1976.

QUESTION 14
Ovenproof Casserole

An "ovenproof" casserole is to be made from a suitable glass or glass-ceramic material.

(a) *List three important properties of the chosen material.*
(b) *Using relevant data on material properties, specify a suitable material.*

(a) Three important material properties are as follows:

- Excellent resistance to thermal shock due to step changes in external temperature; this is reflected in a high value of the thermal shock resistance parameter $R_2 = \sigma_F(1 - \nu)/E\alpha$ (see Chapter 7)

- Excellent resistance to spalling; this is reflected in a high value of the spalling resistance index SRI $= \sigma_F k / E C_p \rho \alpha$
- High thermal conductivity

(b) The relevant data for four candidate materials are given in Table Q14.1. On the basis of the best combination of high values of R_2, SRI, and k, the best material is 96% silica glass. However, because of its high cost,* the most realistic choice is either Pyroceram or Pyrex. Indeed, in late 1987, Corning Glass Works, of Corning, N.Y., introduced under the trade name Visions, a new range of cookware made of Pyroceram.

*An oven-to-table dish (150 mm in diameter, 75 mm in depth, and 3 mm in thickness) made of this material would cost $150 (June 1988 prices).

TABLE Q14.1 PROPERTIES OF FOUR CANDIDATE "OVENPROOF" CASSEROLE MATERIALS

Material	Coefficient of Linear Expansion, α (10^{-6} °C^{-1})	Thermal Conductivity, k (W m^{-1} °C^{-1})	Poisson's Ratio, ν	Modulus of Elasticity, E (GPa)
96% silica glass (Vycor)	0.5	1.50	0.17	69.0
Borosilicate glass (Pyrex)	3.0	1.30	0.21	67.6
Pyroceram 9608 (glass-ceramic)	1.3	0.87	0.25	86.0
Soda lime glass	9.0	0.70	0.22	69.0

Material	Fracture Strength, σ_F (MPa)	Density, ρ (kg m^{-3})	Specific Heat, C_p (J kg^{-1} °C^{-1})	R_2 (°C)	SRI (m^2 °C·s^{-1})
96% silica glass (Vycor)	42	2200	840	1010	9.9×10^{-4}
Borosilicate glass (Pyrex)	65	2240	753	253	2.5×10^{-4}
Pyroceram 9608 (glass-ceramic)	180	2500	979	1208	5.7×10^{-4}
Soda lime glass	70	2440	990	88	33.0×10^{-6}

QUESTION 15
Moderator of a Heat Transfer Nuclear Reactor

Give a brief description of the material selection considerations for the moderator of a heat transfer nuclear reactor for aerospace propulsion.

TABLE Q15.1 APPROXIMATE PERFORMANCE LIMITS OF THE MAJOR SPACE PROPULSION SYSTEMS

System	Thrust per Weight	Specific Impulse (s)
Chemical rocket	2.000	4×10^2
Consumable nuclear rocket	6.000	1×10^5
Heat transfer nuclear rocket	0.300	1×10^2
Thermonuclear engine	0.040	1×10^3
Low pressure nuclear rocket	0.002	1×10^3

The approximate performance limits of a number of competitive nuclear systems for aerospace propulsion, some of them only conceptual at this time, are summarized in Table Q15.1. The most developed of these systems is the solid-core nuclear heat exchanger or heat transfer nuclear rocket (Fig. Q15.1). The main parts of this system are (a) a container to hold hydrogen, preferably in the liquid state; (b) a heat exchanger reactor consisting of fissionable fuel in a matrix reflector and pressure shell around it; and (c) the pumping plant (pump, turbine, and associated valving).

Certain requirements in a moderator material are a high melting point and good mechanical properties (and their retention at elevated temperatures). These

Figure Q15.1 Nuclear rocket engine cycles [1] (reprinted with permission from McGraw-Hill).

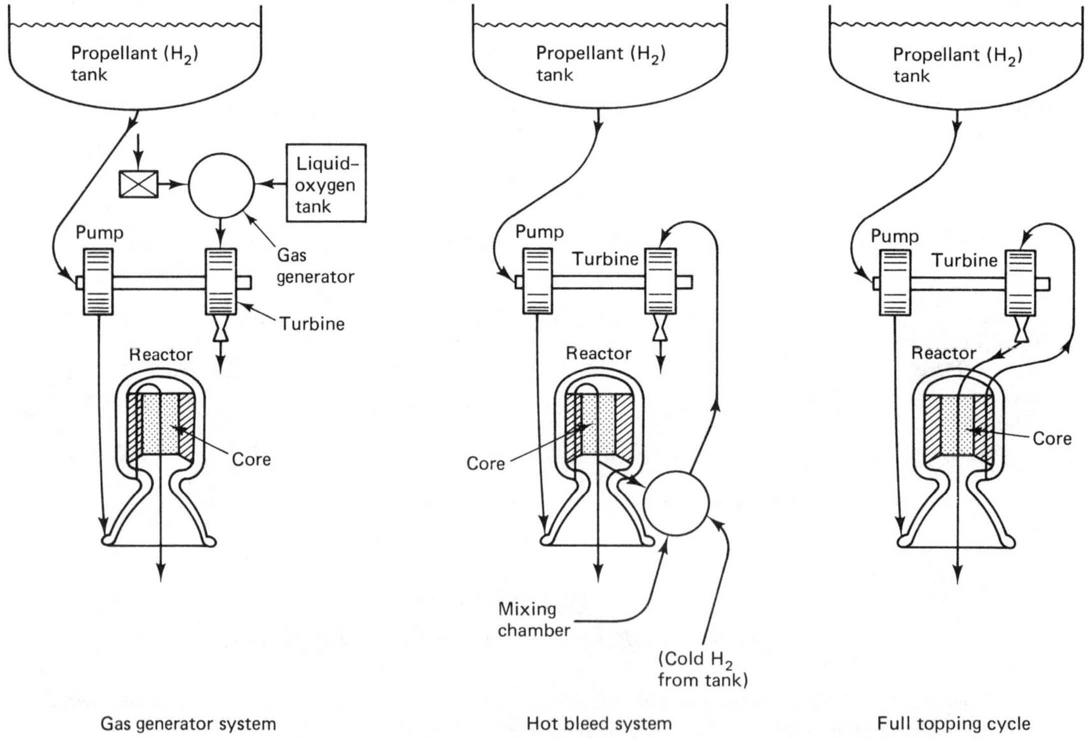

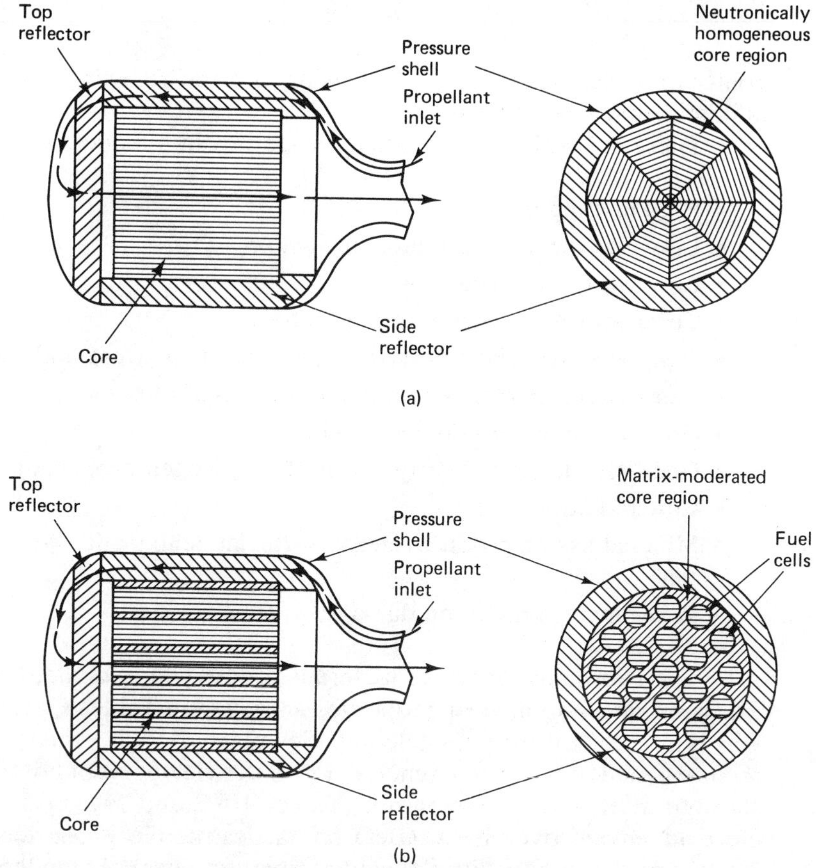

Figure Q15.2 Scheme of solid-core rocket reactors: (a) homogeneous type; (b) heterogeneous type (cylindrical geometry with axial flow in both cases).

requirements, however, are related to the type of reactor (Fig. Q15.2). In the homogeneous core-moderated reactor, the neutron thermalization is achieved primarily by the core, hence by the fuel element material. In this case, the moderator material must be capable of operating at high temperatures. In matrix-geometry cores, moderation is by the material between the bundles of fuel elements. In externally reflected cores, moderation is by thermalizing material surrounding or enclosing the reactor core. In the latter two cases, the moderator material need not have the stability at elevated temperatures required of a moderator in the first-mentioned type of reactor.

Other requirements are related to the nuclear properties, specifically (a) low atomic mass and a small neutron absorption cross section, both of which are needed if neutron thermalization is to be achieved without excessive loss by absorption of neutrons; (b) strong slowing-down power, which is indicative of a high fractional neutron energy loss per unit path length occurring in scattering collisions in the epithermal-energy range;* and (c) a high moderating ratio, which

*This is inversely proportional to the total random-walk distance a neutron must travel from birth at fission energies to thermal energies.

points to a high probability of a fission neutron reaching thermal energy in an infinite medium of the material.

These are additional desirable properties in a moderator material:

- High melting point
- Low neutron absorption cross section
- Low vapor pressure
- Compatibility of fuel with the matrix
- High thermal conductivity and good thermal stress and shock resistance
- Easy fabrication to very close dimensional tolerances
- Good resistance to creep failure
- Good resistance to corrosion by the hydrogen propellant
- Low radiation damage
- Minimal fission product release to the launching pad, the atmosphere, or the sea
- Resistance to radiation damage

The two main candidate materials are beryllium oxide, BeO, and graphite. Their physical and nuclear properties are given in Tables Q15.2 and Q15.3, from which is seen that from a nucleonic viewpoint, BeO is preferable. Its use would require a moderate volume reactor. From the viewpoint of both melting point and strength at elevated temperature (Table Q15.2 and Fig. Q15.3), graphite has a clear advantage over BeO. (BeO has a destructive phase temperature around 2050°C, further limiting its elevated-temperature use). At moderate temperatures,

TABLE Q15.2 PHYSICAL PROPERTIES OF TWO MODERATOR MATERIALS

Property	BeO	Graphite[a]
Melting point (°C)	2533	3616
Density (kg m^{-3})	2700	1.65–1.85
Modulus of elasticity (GPa)		
At 5°C	276	2.7 (N), 8.1 (P)
At 1120°C	207	3.5 (N), 9.0 (P)
Coefficient of linear expansion (10^{-6} °C^{-1})		
At 4°C	1.67	1.67 (N), 0.6 (P)
At 1060°C	3.34	2.22 (N), 1.11 (P)
Thermal conductivity (W m^{-1} K^{-1})		
At 560°C	36.34	—
At 840°C	—	52–86 (N), 31–62 (P)
Poisson's ratio	0.35	0.2–0.3
Atomic mass (g mole^{-1})	25.0	12.0
Thermal stress parameter,[b] T, at 840°C (m W^{-1})	4.68×10^{-4}	1.04×10^{-5}

[a] N, normal to grain orientation; P, parallel to grain orientation.

[b] $T = \alpha E/[\sigma_a(1 - \nu)k]$, where α is the coefficient of linear expansion, E is the modulus of elasticity, σ_a is the short-term tensile strength, ν is the Poisson's ratio, and k is the thermal conductivity. The lower the value of T, the higher the resistance of the material to thermal stress.

TABLE Q15.3 NUCLEAR PROPERTIES OF TWO MODERATOR MATERIALS

Property	BeO	Graphite
Average long energy change per scattering collision, *a*	0.172	0.158
Macroscopic epithermal-scattering cross section, *m* (cm^{-1})	0.640	0.400
Macroscopic thermal-absorption cross section, *M* (cm^{-1})	6×10^{-4}	3.7×10^{-4}
Slowing-down power ($= a \times m$) (cm^{-1})	0.11	0.06
Moderating ratio ($= am/M$)	183	170

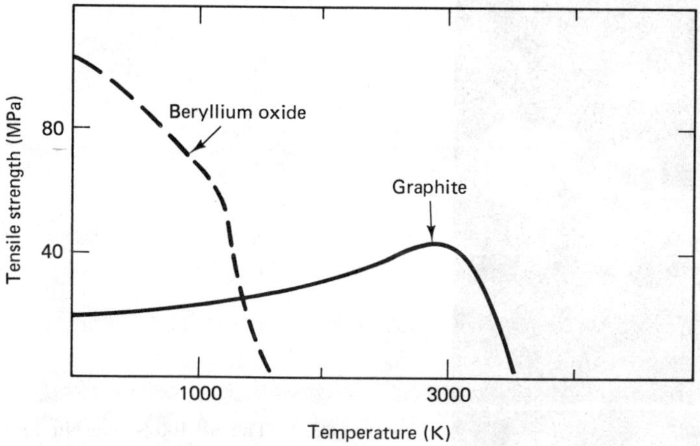

Figure Q15.3 Effect of temperature on the tensile strength of beryllium oxide and graphite.

BeO is brittle and is a relatively poor conductor of heat. This latter characteristic is a definite limitation when considering the removal of heat generated within the moderator by neutron slowdown and gamma-ray absorption. When volumetric heat generation rate (a thermal stress phenomenon) is considered, the data in Table Q15.2 show that graphite is preferred.

An extremely important consideration that could affect the selection of graphite is its susceptibility to corrosion by hydrogen at high velocity (Table Q15.4). This attack becomes appreciable above about 1500°C, at which point redeposition within the gas passages becomes probable, which in turn would increase pressure drop and so cause hot spots and engender serious loss of performance. Thus if graphite is used, appropriate thick diffusion-resistant coatings or barriers, such as zirconium carbide, are mandatory. However, the thickness of these coatings must be such that there is little possibility of (a) problems of differential thermal expansion between the coating and the matrix, (b) introducing a large temperature gradient in the coating and hence creating a thermal stress problem or (c) any large increase in the mass of the reactor.

In connection with the problem of radiation damage, although the flux is high ($>10^{16}$ neutrons cm^{-2} s^{-1}), the lifetime is short, so integrated flux is small. Furthermore, temperatures are so high that radiation damage tends to be annealed out to a large extent.

TABLE Q15.4 EFFECT OF TEMPERATURE ON THE CORROSION OF COMMERCIAL-GRADE GRAPHITE IN HYDROGEN

Temperature (°C)	Mass Loss (mg cm^{-2} min^{-1})
1400	0.80
1800	4.00
2200	16.00
2600	160.00

Source: D. P. Macmillan, "High-temperature materials for rocket reactors," *Nucleonics 19,* p. 88, 1961.

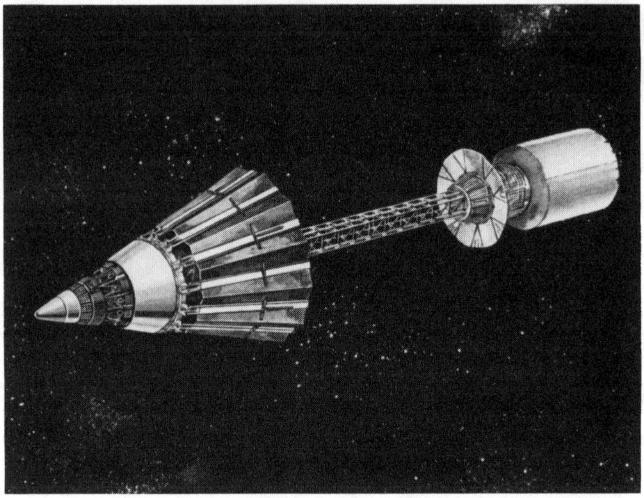

Figure Q15.4 The SP-100 Space Nuclear Reactor Power Supply. (Courtesy of General Electric)

A number of companies working in the field of nuclear propulsion systems for aircraft and space-borne vehicles scored notable successes in the late 1980s. General Electric's SP 100 space nuclear reactor power supply system (Fig. Q15.4) is a fast-spectrum liquid-lithium-cooled reactor design that may be used to power various Strategic Defense Initiative (SDI) systems and is considered an appropriate power source for a space-based wide-area surveillance radar system. The Dynamic Isotope Power System* (Rocketdyne Division of Rockwell International) is another strong candidate to power SDI satellites. Other systems are Star C and Star M (General Atomics, San Diego, Calif.).

REFERENCE

1. R. W. Bussard and R. D. De Laner, *Fundamentals of Nuclear Flight.* New York: McGraw-Hill, 1965, p. 427.

*In late 1987, the U.S. Department of Energy awarded contracts to the Rocketdyne Division of Rockwell International to build two ground demonstration units of this system.

QUESTION 16
Valves in an Automobile Engine

Most of the valves and the valve spring retainers in many automobile engines are currently made of alloy steels or nickel-based superalloys. Demands for increased fuel economy have led to the consideration of titanium alloys for these applications.*

(a) *State the service conditions under which these components operate.*
(b) *List the important requirements in materials to be used for intake and exhaust valves.*
(c) *Comment on the proposed use of α/β titanium alloys as valve train materials.*

(a) The service conditions for these components (Fig. Q16.1) are as follows:

- High temperature and cyclic stresses in both the intake and exhaust valves can lead to thermal fatigue either in the valve fillet region (in the intake

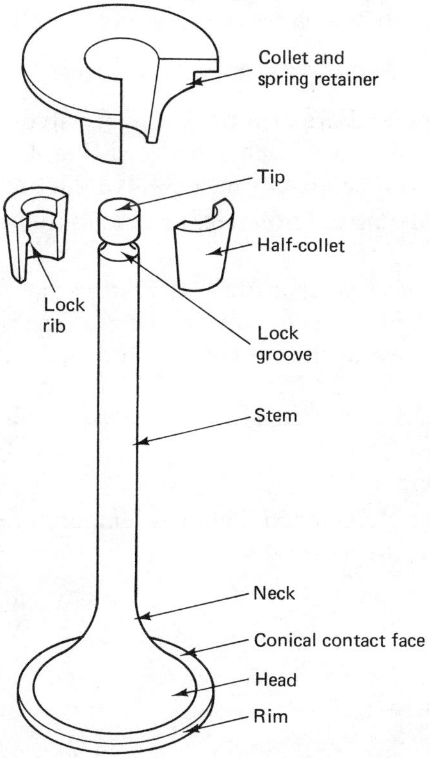

Figure Q16.1 Poppet valve spring components.

*About 80% of North American requirements for exhaust valves, which include essentially all light- to medium-duty engine applications, are satisfied by steel alloy 21-2N (Cr, 20.5; Ni, 2.2; Mn, 8.25; C, 0.55; N, 0.30; Si, 0.25; Fe, balance). Low-carbon steel 1541H (C, 0.4; Mn, 1.5; Fe, balance) is ideally suited for intake valves. Appropriate nickel-based superalloys include Nimonic 80A and Pyromet 31.

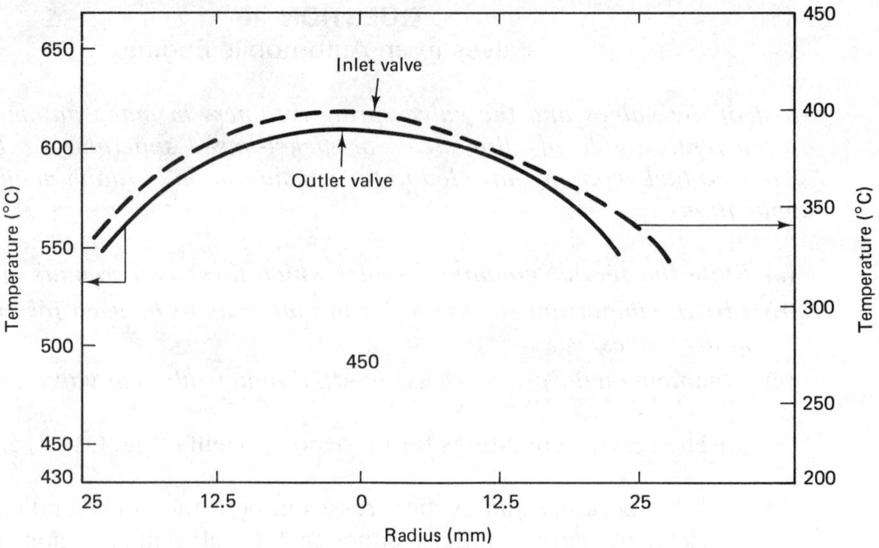

Figure Q16.2 Finite element analysis of the valve head temperature gradient in a firing Cummins NTC-350 engine in cylinder 2 (adapted from [1], pp. 5 and 6).

valve) or in the stem-blend area (in the exhaust valve). The valve temperatures in a specific engine have been calculated (Fig. Q16.2). In both intake and exhaust areas, creep can occur in the valve fillet region.

- Corroding and/or oxidizing environments can lead to problems in the valve seat region.
- Sliding friction can cause wear in the valve stem due to its motion in the valve guide, at the tip of the valve due to the rocker arm, and in the valve spring keeper groove due to the rubber action of the valve spring keeper.

(b) Desirable properties in materials for an intake valve are as follows:

- Resistance to oxidation
- High tensile strength (at room and elevated temperatures)
- High creep rupture strength
- High hot hardness
- Low creep elongation
- Good wear resistance
- Good impact strength

Desirable properties for an exhaust valve are these:

- Fatigue strength (smooth and notched)
- Resistance to corrosion by lead oxide (in leaded fuel applications)

- Resistance to corrosion by sulfur compounds (in unleaded and diesel fuel applications)
- Resistance to oxidation (only at elevated temperatures and in the absence of lead and sulfur compounds)
- High tensile strength (at room and elevated temperatures)
- High creep rupture strength
- High hot hardness
- Low creep elongation
- Good wear resistance
- Good impact strength

(c) The α/β titanium alloys are considered the most appropriate alloys for consideration as potential valve materials. This is so in spite of their high cost (6 to 8 times the price of alloy steels). Use of these alloys results in improved performance because they offer a unique combination of excellent mechanical properties (especially at elevated temperatures; see Table Q16.1), and good corrosion resistance. Lighter-weight valve train components permit the use of lower spring loads, thus reducing friction and increasing fuel economy.

These values may be fabricated from α/β titanium alloys using a two-step hot forging process (carried out at temperatures in the α/β region for intake valves and in β region for exhaust values) or casting methods such as hot isostatic pressing or investment casting, or powder metallurgical procedures.

TABLE Q16.1 THE INFLUENCE OF TEMPERATURE ON THE YIELD STRENGTH AND THE STRESS TO A SPECIFIED PLASTIC STRAIN OF THREE TITANIUM ALLOYS

	Yield strength (MPa)			Stress to produce 0.2% total plastic strain in 100 h (MPa)		
Temperature (°C)	Ti-6246 $(\alpha + \beta)$[a]	Ti-6242S $(\alpha + \beta)$	Ti-6242S (β)	Ti-6246 $(\alpha + \beta)$	Ti-6242S $(\alpha + \beta)$	Ti-6242S (β)
200	980	800	650	—	—	—
400	830	640	500			
500	750	580	480	200	300	400
600	—	450	400	—	50	80

[a] Microstructural condition

Sources: P. A. Blenkinsop, "Developments in high temperature alloys", in G. Luetjering, U. Zwicker, and W. Bunk (eds.) *Titanium Science and Technology,* Oberursel, West Germany: DGM, pp. 2323–2338; D. Eylon, et al., "High temperature titanium alloys—a review", *J. Metals 36* (11), pp. 55–62, 1984.

REFERENCE

1. R. P. Worthen and T. N. Tunnecliffe, "Temperature controlled engine valves," SAE Technical Paper No. 820501. Warrendale, Pa.: Society of Automotive Engineers, 1982, pp. 5 and 6.

QUESTION 17
Flexure Mode Electromechanical Transducer

Flexure-mode electromechanical transducers use the transverse piezoelectric effect. Examples include an omnidirectional speaker (tweeter), optical scanners, vibrational fans, and variable-aperture diaphragms.

(a) *Briefly explain the principle of these devices and, for a specified type, suggest the relevant criteria for the selection of a material for its construction.*

(b) *Compare candidate materials on the basis of these criteria.*

(a) Many of these devices are based on the theory of flexure of beams. They are usually constructed in a bimorph or multimorph fashion. The electrical connections are made so that when an electric field is applied across the bimorph leads, one piezoelectric sheet will extend longitudinally while the other will contract longitudinally, resulting in the bending of the free end.

For a series-mode bimorph (Fig. Q17.1), the polarization vectors of the piezoelectric sheets opposite each other, and the small-angle deflection of the free end Δ_f is given by [1]

$$\Delta_f = 1.5 d_{31} \frac{VL^2}{t^2} \tag{Q17.1}$$

where d_{31} is the piezoelectric strain coefficient of the transducer material; L and t are the length and thickness, respectively, of the bimorph; and V is the applied voltage.

The fundamental resonance frequency f_0 of an ideal bimorph is given by [2]

$$f_0 = 0.162t \frac{\sqrt{E/\rho}}{L^2} \tag{Q17.2}$$

where E is the modulus of elasticity of the bimorph material and ρ is its density.

When devices that require both large motion and high speed are considered, it is useful to define, as a material figure of merit, the deflection bandwidth product (DBWP), which is the product of Δ_f and f_0. Thus, from equations (Q17.1) and (Q17.2),

$$\text{DBWP} = \frac{0.243 V d_{31} \sqrt{E/\rho}}{t}$$

Figure Q17.1 Bimorph configurations: (a) series; (b) parallel.

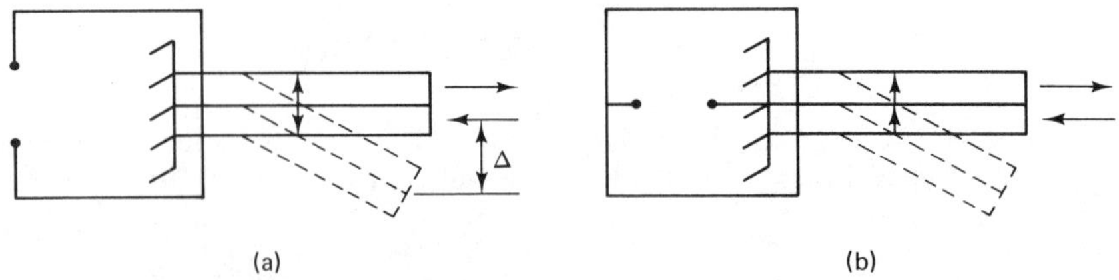

(a) (b)

TABLE Q17.1 VALUES OF FIGURE OF MERIT FOR THE CANDIDATE BIMORPH MATERIALS

Material	E (GPa)	d_{31} (10^{-12} mV^{-1})	ρ (kg m^{-3})	F (MV m^{-1})	DBWP (m$^{-0.5}$)
Polyvinylidene fluoride, PVF_2	3	30	1780	30.0	0.091
Barium titanate, $BaTiO_3$	116	−59	5550	2.2	−0.046
Lead zirconate-titanate, PZT-4	139	−123	7500	1.0	−0.041

Defining the maximum (or allowable) applied electric field as $F = V/t$ yields

$$\text{DBWP} = 0.243 F d_{31} \sqrt{\frac{E}{\rho}}$$

(b) The values of DBWP for three candidate piezoelectric materials when used as cantilever-mounted bimorphs are given in Table Q17.1, from which is seen what the preferred material is PVF_2.

REFERENCES

1. Y. Hatanaka, S. Okamoto, and R. Nishida, *Advanced Electron. Electron Physics 52,* p. 31, 1979.
2. J. K. Lee and M. A. Marcus, *Ferroelectrics 32,* p. 93, 1981.

QUESTION 18
Use of Beryllium in Space Shuttle Orbiter Components

At the initial design phase of the orbiter of the U.S. space shuttle, there was no plan to use beryllium in either structural or system components. When the design of the horizontal flight test orbiter was completed, an analysis indicated that the orbiter was overweight. Subsequently, beryllium was recommended for three areas of the orbiter:

- *The crew module windshield retainers, spacers, and beams*
- *The navigation base in the crew module*
- *The umbilical external tank doors in the aft fuselage*

(a) *Explain why beryllium was selected for these applications, and state some disadvantages associated with the use of this material.*

(b) *List some other innovative or potential uses of beryllium in aerospace applications.*

(a) The reasons that beryllium was chosen for these applications (Fig. Q18.1) are its high values of (1) specific stiffness, and a high retention of this property at elevated temperatures, (2) specific heat, (3) thermal conductivity, and (4) melting point. Another attraction of beryllium for use in these applications is its low coefficient of thermal expansion.

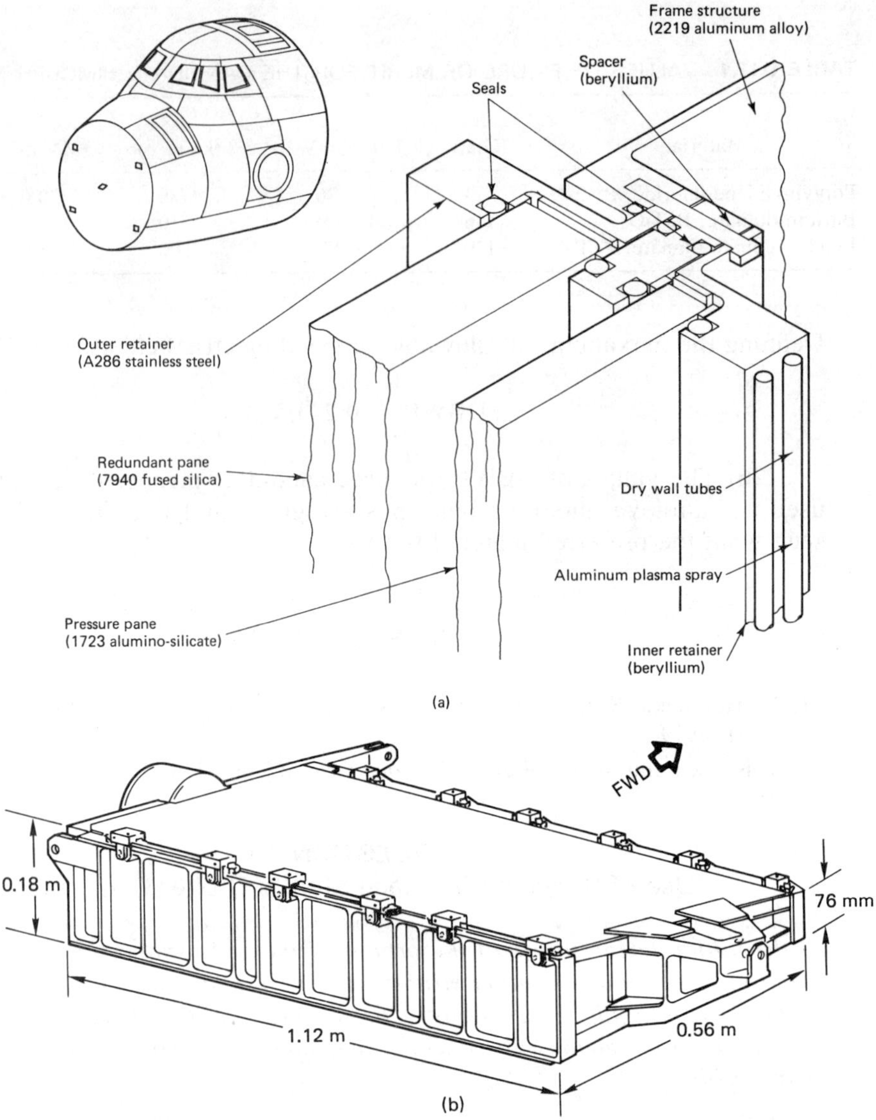

Figure Q18.1 Applications of beryllium in the orbiter of the U.S. space shuttle: (a) crew module windshield retainers, spacers, and beams; (b) navigation base assembly (Courtesy of Rockwell International).

The main disadvantages of beryllium are (1) high unit raw material cost, (2) limited ductility, (3) high notch sensitivity, (4) difficulty of machining without surface damage, and (5) toxicity.

For a comparison of beryllium and some other candidate materials for the stated applications, see Tables Q18.1 and Q18.2.

(b) Other actual and potential innovative uses of beryllium in aerospace include these:

- Parts of the proposed U.S. space station. In this structure, beryllium will find uses in the power system, which is the "solar dynamic" energy concept that uses concentrated solar energy to power a Stirling cycle engine. Beryllium may be used in the power piston and gas spring cylinder in this engine. As envisioned by several contractors, the material could also be used in the solar collector array as the primary structural material.

- Heat shields on Project Mercury and Project Gemini capsules and interstage spacers (joining the reentry vehicle to the guidance and control sections) on Minuteman missiles.
- Torque tubes in the control system of the F-111 tactical fighter.
- Antenna material in data-gathering satellites.

TABLE Q18.1 ROOM-TEMPERATURE VALUES OF VARIOUS PROPERTIES OF CANDIDATE MATERIALS

Material	E (GPa)	α (10^{-6} K^{-1})	ρ (kg m^{-3})	Melting Point (°C)	Specific Heat (kJ kg^{-1} K^{-1})	Thermal Conductivity (W m^{-1} K^{-1})	Cost[a] (\$ kg^{-1})
Beryllium	304	10	1855	1285	1.93	180	400.0
Structural aluminum (7075-T6)	72	23	2796	617	0.90	150	1.5
Titanium (Ti-6Al-4V)	116	9	4540	1647	0.61	6	15.0
Structural steel (SAE 4340)	207	15	7861	1327	0.50	40	1.6

[a]June 1988 prices.

TABLE Q18.2 VALUES OF E/ρ (in 10^6 m) FOR CANDIDATE MATERIALS AT VARIOUS TEMPERATURES

	Temperature (°C)					
Material	25	93	204	316	427	538
Beryllium	16.7	16.5	16.4	16.3	16.2	14.5
Steel	2.7	2.6	2.5	2.4	2.3	2.2
Nickel-cobalt alloy	2.6	2.4	2.1	2.4	2.0	2.1
Titanium	2.6	2.3	2.1	2.0	1.7	1.4
Aluminum	2.4	2.4	2.1	1.4	—	—
Magnesium	2.4	2.3	1.6	1.0	—	—

QUESTION 19
Insulation Material for a Missile Case

Air-to-air tactical missiles must have near-hypersonic capability to overtake and destroy supersonic aircraft. Concomitant with high missile velocities is high aerodynamic heating. One feasible solution of this problem involves the use of high–temperature–resistant materials for the fuselage of the missile, coupled with insulation material to protect internal components.

(a) *State the main criteria on which selection of an insulation material for use in a missile fuselage structure depends.*

(b) *What are some of the candidate materials for this application?*

(a) The criteria are as follows:

- Low density (less than 1000 kg m^{-3})
- Low thermal conductivity (less than 0.144 W m^{-1} K^{-1})
- Adequate strength to resist aerodynamic shear forces, bending loads and vibration forces, and rough handling loads such as those arising from impact and abrasion
- Resistance to various environmental conditions such as humidity, heat, cold, vacuum, moisture at low temperatures, rain, fungus, rot, solvents (hydraulic and cooling fluids, jet fuels, etc.), and animal and insect attack
- Ready availability and ease of access of source
- Good resistance to missile storage temperatures (−30°C to 70°C)
- Resistance to surface boundary temperatures during missile captive flight (about −18°C to 180°C)

(b) These are some candidate materials:

- Granulated cork and binder of phenolic resin, with polyol plasticizers
- Aluminum silicate fiber paper, with inorganic binder
- Asbestos paper, reinforced with phenolic microballoons and resin

QUESTION 20
Use of Plastics in Packaging

Write a concise essay on the use of plastics in packaging. Include an outline of the main requirements for the use of plastics in packaging foods and drinks. Explain the following choices:

- *Polyethylene terephthalate for grocery bags*
- *Oriented polypropylene for easy-open bags and wraps*
- *Low-density polyethylene for bread bags*
- *Nylon for flexible packaging for meats and cheeses*

By any measure, packaging is a large market, with 1985 sales of some \$7.5 billion. The market share of plastics in these applications (as represented by containers, closures, coatings, and films) in that year was 16%, surpassing glass (at 10%) but trailing paper (paperboard and paper), which had a 42% share. The important point, however, is that the plastic packaging industry is very dynamic, with the material making significant strides in sales volume, sales value, and market share (Figs. Q20.1, Q20.2, and Q20.3). Much of the success of plastics in

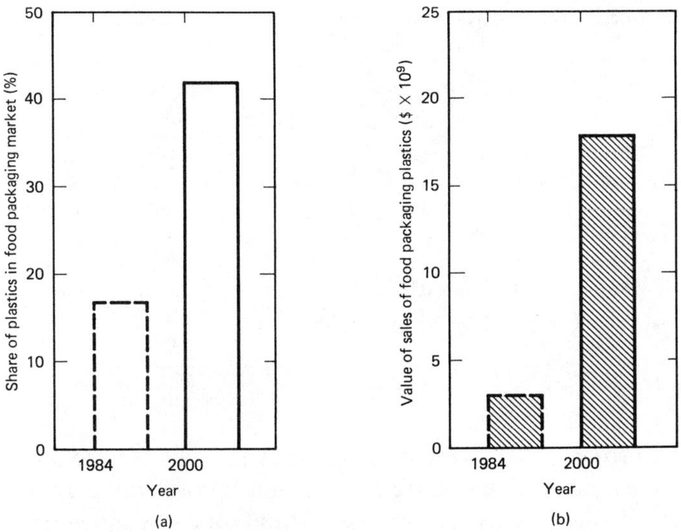

Figure Q20.1 Projected increase in the use of plastics in food packaging, 1984–2000: (a) market share; (b) sales. Source: Society of the Plastics Industry, Washington, D.C.

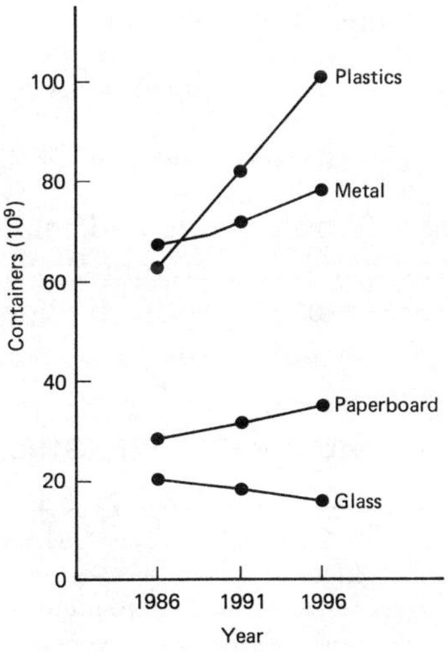

Figure Q20.2 Growth in the use of plastics vis-à-vis other packaging materials for single-serve food containers. (Courtesy of Business Communications Co., Inc., Norwalk, Conn.)

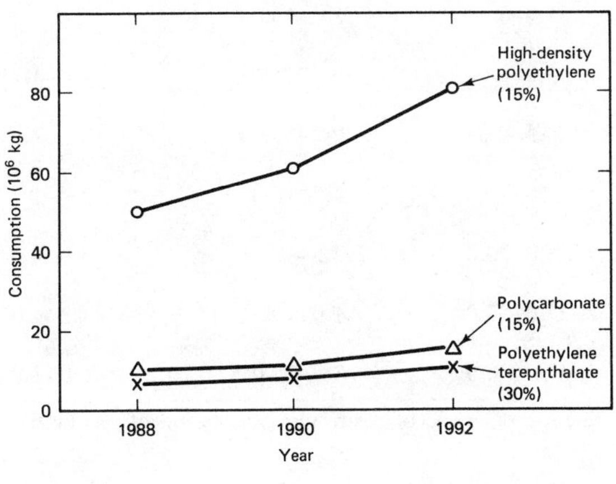

Figure Q20.3 Growth of the use of three plastics for the bottling of drinking water (assumed annual growth rate in parentheses).

TABLE Q20.1 CONSUMPTION OF FLEXIBLE FOOD PACKAGING POLYMERS IN THE UNITED STATES (kg × 10^6)

Market	PVDC emulsion coatings			Extrudable PVDC			Barrier resin EVOH		
	1981	1986	1991	1981	1986	1991	1981	1986	1991
Unprocessed meats	—	—	—	2.22	2.40	2.72	—	0.23	0.41
Dry foods and mixes	0.36	0.27	0.05	—	—	—	—	0.23	0.34
Processed meats	0.86	0.82	0.34	0.18	0.18	0.30	—	0.07	0.41
Cheeses	1.45	1.54	0.59	0.05	0.02	0.38	—	0.01	0.83
Snack foods	3.77	2.68	0.68	—	—	0.73	—	0.27	1.82
Cereals	1.00	1.00	0.14	—	—	—	—	0.02	1.00

Source: duPont de Nemours, Wilmington, Del.

packaging applications derives from the growing use of barrier plastics. These are resins with low permeability to gas and moisture that, when incorporated into packaging, protect the contents from spoilage (in the case of food) or environmental attack and degradation. In the past, such protection was obtained through the use of solvent and emulsion coatings. The trend now, though, is to use extrudable barrier resins, such as polyvinylidene chloride (PVDC) and ethylene vinyl alcohol (EVOH); see Table Q20.1.

Strength Properties

The first property is strength (tensile, tear, and impact). All of these types of strength vary both within and between different types of resins. The strength of films further varies with direction of extrusion. If choosing a resin based only on tensile strength, PET and OPP would be the best selection (Table Q20.2).

TABLE Q20.2 TENSILE, TEAR, AND IMPACT STRENGTHS OF SOME PACKAGING POLYMERS

Polymer	Tensile Strength Ratio (vs. LDPE)	Tear Strength Ratio (vs. OPP)	Impact Strength Ratio (vs. Nylon or HDPE)
Low-density polyethylene (LDPE)	1.0–3.5	25.0–100.0	1.8–3.5
High-density polyethylene (HDPE)	3.0–9.0	10.5–75.0	1.0–1.7
Linear low-density polyethylene (LLDPE)	3.5–9.0	30.0–150.0	2.2–3.2
Polyvinylchloride (PVC)	2.2–18.0	—	3.4–4.8
Nylon	6.0–19.0	4.5–18.0	1.0–1.7
Polyvinylidene chloride (PVDC)	8.0–20.0	2.5–4.5	3.0–4.0
Oriented polypropylene (OPP)	27.0–37.0	1.0–1.8	1.2–4.0
Polyethylene terephthalate (PET)	27.0–40.0	3.5–25.0	6.0–7.8

Spread in ratios reflects the influence of grade of resin and orientation of production (in or normal to machine direction).

As far as tear strength is concerned (this parameter usually varies inversely with tensile strength), the polyethylenes—all relatively poor in tensile strength—demonstrate strong resistance to tearing, with the best materials being LLDPE and LDPE (Table Q20.2). PET is probably the only high-tensile-strength resin also to have a high tear strength. PVC is excluded here because its tear strength varies too widely and unpredictably to allow accurate comparison.

There is a much smaller variation in the impact strength of resins, it having a little bearing on either tensile or tear strength (Table Q20.2). The best materials are PET and PVC.

PET has a high tear strength and a very high tensile strength, hence its use in grocery bags. The use of PET in food and drinks packaging is increasing rapidly, with an expected 500% increase in the period 1986 to 1991.

OPP has a very high tensile strength and a very low tear strength, hence its use in easy-open bags and wraps.

Barrier Properties

The second consideration involves barrier properties (resistance to permeation by oxygen, water vapor, and carbon dioxide). Many foods may be seriously degraded if exposed to oxygen or if they lose water vapor. Also, carbon dioxide barriers are important for carbonated drinks (soft drinks and beer) and similar products. Plastics industry sources all agree that new barrier technology will be the path through which plastics will further penetrate the metal- and glass-container markets.

EVOH and PVDC are by far the most effective barriers to oxygen and carbon dioxide (Fig. Q20.4). Except for PVDC, water vapor barrier properties are essentially the inverse of the oxygen and carbon dioxide barrier properties for most of the common resins (Fig. Q20.4). LDPE and HDPE perform excellently as moisture barriers. Hence, LDPE is well suited for use as bread bags.

Packaging-Content Interactions

Although in the past, the U.S. Food and Drug Administration and other federal agencies, as well as state governments have banned or restricted the use of certain polymers (such as acrylonitrile) for food and drink containers,* in general, most common packaging resins are considered safe for such uses, but the contents of a package can adversely affect the stability and properties of the packaging material (Table Q20.3). Nylon has a very good imperviousness to oils and grease and is thus admirably suited for packaging meats and cheeses.

Other Properties

Other desirable properties belong to the category of aesthetics and consumer acceptance. These properties include (1) printability (since most printing on plas-

*For example, Kentucky has a ban on the sale of plastic soft drink and beer containers with a capacity of 830 ml or less, and in April 1988, Suffolk County, N.Y., proposed a ban on the use of polystyrene and polyvinylchloride for grocery bags and food containers.

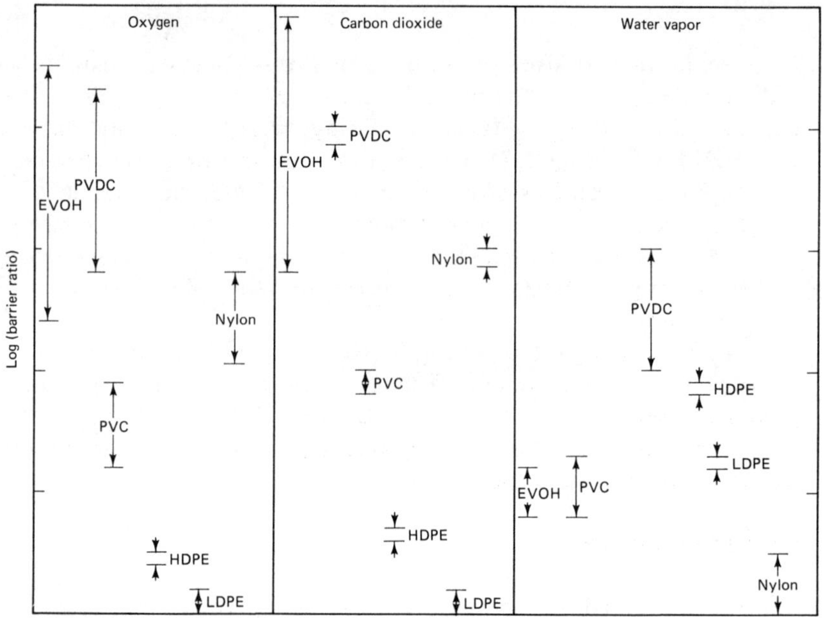

Figure Q20.4 Effectiveness of six polymers as barriers to oxygen, carbon dioxide, and water vapor. (The barrier ratio of a polymer is the ratio of the transmission rate of the diffusing species through that polymer to the rate through the polymer with the highest transmission rate.)

TABLE Q20.3 QUALITATIVE RATING OF THE RESISTANCE OF SELECTED PACKAGING RESINS TO VARIOUS CORROSIVE CONTENTS

	Content			
Resin	Oils and Greases	Acids	Alkalis	Solvents
HDPE	Excellent	Acceptable	Good	Good
LDPE	Acceptable	Acceptable	Good	Good
Nylon	Excellent	Unacceptable	Good	Good
PET	Excellent	Good	Good	Good
Polypropylene	Excellent	Acceptable	Good	Good
Polystyrene	Acceptable	Acceptable	Good	Good
PVC	Good	Good	Good	Acceptable
PVDC	Excellent	Good	Good	Acceptable

tics is multicolored, excellent control of color tones is needed to reproduce all the subtle chromatic variations), (2) scuff resistance, and (3) clarity (as is found in PET, PVC, PVDC, and crystalline forms of polypropylene).

An appreciation for the extent of the penetration of plastics as packaging materials in a wide range of markets may be obtained from a few recent examples:

- Hot-fillable PET containers for tomato-based products (spaghetti sauces)
- PVC bottles for beer

- Polypropylene bottles for spices and seasonings
- A line of shelf-stable entrees in dual ovenproof trays, reportedly made of coextruded amorphous PET with no additional barrier
- PET bottles for cooking oil
- Carbonated soft drinks in see-through PET cans

QUESTION 21
Use of High-Density Polyethylene in Containers

Discuss the choice of high-density polyethylene (HDPE) for containers to hold certain aggressive fluids, such as bottles for herbicides and pesticides, drums for lubricating oils, and tins for charcoal lighter fluids and paint thinners.

Blow-molded HDPE containers seem at first to be a poor choice for containing aggressive fluids, for two main reasons. First, hydrocarbons and other small-molecule, active ingredients in chemicals, fuel additives, and penetrating oils migrate quickly through large-molecule HDPE-blown container walls. The problem of permeation is greatest when nonpolar solvent products are the contents. This is because polyethylene is itself a nonpolar petroleum distillate product. By contrast, aggressive contents that are polar in nature (alcohols, ketones, and such fluid products as windshield wiper, brake, and antifreeze solutions) require no supportive barrier when they are contained in HDPE. They do, however, require to be stored in containers that are made of materials that have good environmental stress crack resistance (ESCR), especially when long shelf life is anticipated and when long-term dimensional stability is a key requirement.

Second, U.S. government regulations prohibit the shipping of flammable liquids in polyethylene (PE) containers that exceed 4% permeation weight loss at 18°C. This practically precludes the use of HDPE in many household and industrial applications.

HDPE, does, however have its advantages as far as its use in containers for aggressive fluids is concerned:

- Excellent moisture barrier properties
- Low density, leading to high specific stiffness and specific strength
- Ease of manufacturing of tough, dent-free, reusable blown containers, hence its use in 208-l (55-gal) drums to hold bulk lubricating oils
- Excellent resistance to corrosion, hence its use in janitorial chemical storage containers
- Ease of design changes to make products more attractive or serviceable, hence its use as automobile fuel tanks*
- Low cost

*This application was introduced in late 1986 by Kantex, Canada.

In addition to these properties, a wide spectrum of multilayer barrier technology and surface treatment options now exists, leading to chemically resistant treated HDPE. In 1986 alone, three new multilayer barrier procedures were launched. DuPont de Nemours, of Wilmington, Del., introduced its solar RB laminar barrier process, which involves dispersion of nylon "platelets" in a discontinuous band through the HDPE matrix to create a circuitous-path barrier that slows penetration of hydrocarbons. Monsanto Chemical Co., applied its own technology in the form of a three-layer HDPE-/adhesive-tie/nylon coextruded multilayer bottle. Continental Can, introduced the Conolene II bottle, in which the barrier layer (consisting of ethylene vinyl alcohol or nylon) is sandwiched between tie, regrind, exterior, and interior HDPE layers in a six-ply structure.

Methods to impart barrier properties to HDPE through appropriate surface treatment of containers are becoming more sophisticated and successful. Two approaches are in common use. The first is fluorination (Fig. Q21.1), in which fluorine gas blown during molding modifies the inner unpigmented layer of a two-ply coextruded HDPE container or fluorine is used in a batch process to create a surface-reaction barrier on the inner and outer surfaces of the HDPE container. The second approach is sulfonation (sulfur trioxide gas introduced in the downstream operation creates a surface-reaction barrier in the monolayer HDPE container).

The net effect of the combination of the inherently attractive properties of HDPE outlined earlier and the efficacy of the various surface treatment and/or multilayer barrier technologies is to produce a treated HDPE material that is acceptable for the containment for various aggressive fluids (Tables Q21.1 and Q21.2). Treated HDPE thus has enormous potential for mass-volume replacement of glass, metal, and higher-priced, higher-performance polymers in these applications.

Figure Q21.1 A schematic diagram of the in-line fluorination system (courtesy of Air Products and Chemicals, Inc., Allentown, Penn.).

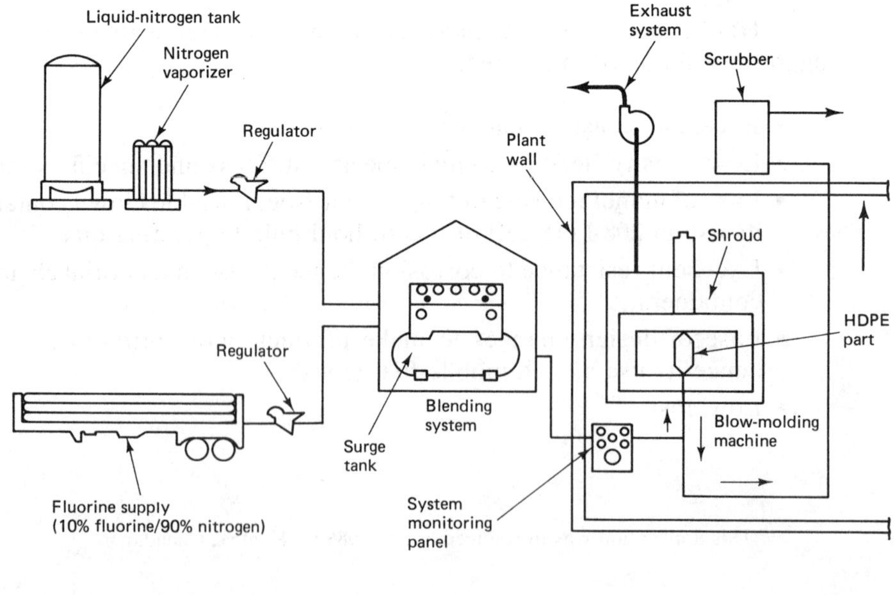

TABLE Q21.1 PERMEATION DATA FOR COMMERCIAL LIQUIDS PACKAGED IN HDPE

Liquid	Loss in weight (%)		Improvement Due to Barrier (%)
	Untreated Container	Fluorinated Container[a]	
Killmaster pesticide	26.90	0.09	99.7
Gulflite charcoal lighter fluid	4.56	0.13	97.1
Turpentine	3.92	0.00	100.0
STP gas treatment fluid	9.10	0.00	100.0
PineSol household cleaner and disinfectant	6.57	0.40	93.9

[a] Airpak containers (Air Products and Chemicals, Inc., Allentown, Pa.).

Source: D. M. Buck, et al., "Treated HDPE containers resist solvent permeation," *Plastics Eng. 42* (4), p. 35, 1986.

TABLE Q21.2 WEIGHT LOSS OF TREATED AND UNTREATED HDPE CONTAINERS HOLDING GASOLINE AT 50°C

Time (d)	Weight loss (%)	
	Control (Untreated) Container	Fluorinated Container
0	0	0.0
7	20.0	1.5
14	36.0	2.5
21	45.0	3.0
28	54.0	3.5

Source: D. M. Buck et al., "Treated HDPE containers resist solvent permeation," *Plastics Eng. 42* (4), p. 34, 1986.

QUESTION 22

Prediction of Resistance to Cavitation Erosion

The phenomenon of cavitation erosive wear or cavitation material damage has assumed increasing significance, probably as a result of the trends in engine design toward higher rotational speeds and, in some cases, higher rates of change of cylinder pressure.

(a) *Write a brief note on "cavitation erosion."*

(b) *Discuss the use of target material properties as a predictor of the resistance of the material to this type of attack.*

(a) In regions of high velocity and low pressure in a constricted fluid system, bubbles of fluid are produced where the fluid is locally in a state of tension. These bubbles are eventually convected into a region of higher pressure, where they reach a maximum size and then implode. The potential energy associated with a bubble at its maximum size is progressively converted into kinetic energy of the surrounding fluid during the collapse. This energy is concentrated into a progressively smaller volume. For an ideal fluid, at the moment of complete collapse, the velocity and associated pressure of these bubbles are infinite. For a real fluid, though, the velocity is reduced by the effects of fluid incompressibility, and the maximum pressure achievable is limited by the adiabatic compression of dissolved gas.

It has been calculated that for water, the bubble collapse pressures are comparable to the ultimate tensile strength of many engineering materials. Thus in ductile materials, pitting occurs*, while in brittle materials, cracking results. The material loss occurs from these localized sites and eventually over a larger area. This phenomenon is known as cavitation erosion. The four stages of cavitation erosion are shown schematically in Fig. Q22.1.

The resistance of a material to this phenomenon is related to a number of factors, namely (1) type of cavitation (hydrodynamic, vibratory, etc.), (2) intensity of cavitation (fluid velocity, cavitation number),† (3) corroding nature of the fluid (although cavitation erosion is essentially a mechanical process, a corrosive medium has a synergistic effect on erosion rate), and (4) target material properties.

Cavitation erosion wear is common in pumps and turbines, pipe systems, hydrofoils, and propellers. It has been reported in the high-pressure turbopumps in the main engines of the U.S. space shuttle. The function of these pumps is to feed liquid oxygen into the main combustion chamber. The conditions in this application are very conducive to cavitation erosion: simple stage, mixed flow; at the inlet to the pump, the pressure is only 2.6 MPa, but after passing through

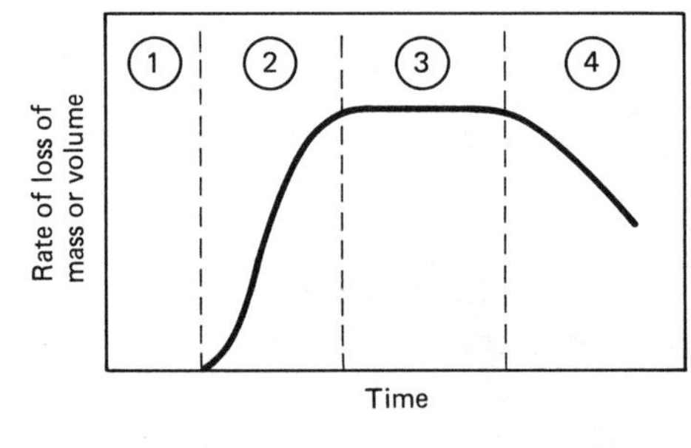

Figure Q22.1 The four stages of erosion.

*Because of the pitting, it has been argued that cavitation is not due to the bubble pressure only, since the stress required to produce a pit is at least the hardness of the material. For most metals, hardness is about three times ultimate tensile strength. This observation has led to an alternative mechanism for cavitation erosion. The pits, it is postulated, are due to fatigue loading by the bubble pressures at a typical frequency of 1 Hz.

†Cavitation number CN $= 2(p - p_v)/\rho U^2$, where p is the static pressure, p_v is the saturated vapor pressure at the bulk fluid temperature, ρ is the fluid density, and U is the fluid velocity. Generally, the severity of cavitation erosive wear may be reduced by increasing CN by reducing velocity, for example.

the turbine blades (rotating at 26,000 r min^{-1}), the discharge pressure is 28 MPa; at 100% power, the mass flow rate of the liquid oxygen is 485 kg s^{-1}.

(b) A large number of studies have attempted to relate the cavitation erosive wear rate of a material, under specified conditions, to its mechanical properties.

Thiruvengadam and Waring [1] used the concept of strain energy to fracture (defined as the area under the tensile stress strain curve). Such an idea is based on the supposition that all materials removed during the erosion had failed, so that the external forces must have done work at least equal to the strain energy to fracture of the eroded volume. Hobbs [2] proposed that the response of a material to cavitation erosive wear is by way of elastic deformation, plastic deformation, and/or fracture. He thus considered the proof resilience [(proof strength or yield strength $\sigma_y)^2/(2 \times$ modulus of elasticity E)], ultimate resilience [(ultimate tensile strength UTS)$^2/2E$], and the work to fracture {[proof or yield strength + (UTS − proof or yield strength)/3] × elongation el} as the relevant material parameter.

Other researchers [3, 4] have suggested that cavitation erosive wear resistance (reciprocal of the cavitation erosive wear rate) is a function of ultimate tensile strength of the target material only.

James and Blount [5] proposed using the dynamic indentation parameter, DIP = UTS$^2/EHel$, where H is the hardness of the material. Mousson, Laird and Hobbs, and Lichtman and Weingram [6] have all suggested the hardness of the target material as the most suitable correlating parameter. Woods and colleagues [7] presented the relevant material properties as $(\sigma_f)^{-2} + (\%RA)^{-3}$, where $\%RA$ is the percentage reduction in cross-sectional area of the specimen at fracture and σ_f is its fracture strength. The role that grain size of the target material plays in its cavitation erosive wear has been discussed at some length by Preece [8]. Arguing that fracture toughness determines the resistance of a material to rapid crack growth, leading to rupture and eventual removal of the material from the surface undergoing erosion, Akhtar and Stowell [9] have suggested that the fracture toughness of a target material is a reasonable predictor of its cavitation erosive wear behavior.

It seems unlikely that cavitation erosive wear would be predicted by a single target material property. It would be instructive, therefore, to test the predictability of parameters that contain a combination of target material properties. This test is carried out for some target material–liquid systems. The first one is the cavitation erosive wear of 0.2 wt % carbon steels, grades G and K (see Table Q22.1 for compositions), in distilled water maintained at 22°C, as per ASTM G32-77. The results and the relevant material properties are given in Table Q22.2. The

TABLE Q22.1 COMPOSITION OF GRADES G AND K 0.2 wt % CARBON STEELS (in wt %)

Grade	C	Mn	Si	Cr	Ni	Cn	V	S	P	Fe
G	0.19	1.30	0.70	0.20	0.15	0.19	0.15	0.018	0.016	Balance
K	0.23	0.87	0.38	0.13	0.28	0.23	—	0.021	0.012	Balance

Source: A. Akhtar and V. A. Stowell, "Cavitation erosion of hydraulic turbine steels containing 0.2 pct carbon and 1 pct Mn," *J. Mater. Energy Sources 4* (1), pp. 58–63, 1982.

TABLE Q22.2 CAVITATION EROSIVE WEAR RATES AND MATERIAL PROPERTIES OF 0.2 wt % C STEELS

Steel	Cavitation Erosive Wear Rate (mg h^{-1})	UTS (MPa)	E (GPa)	el	Brinell Hardness, H (kg mm^{-2})	Area under Tensile Stress Strain Curve (arb units)	Yield Strength (MPa)
G (stress-relieved)	31	634	200	0.18	92	13.1	420
G (normalized)	37	555	200	0.18	84	16.2	345
K (stress-relieved)	51	505	200	0.29	78	20.6	279
K (normalized)	37	543	200	0.29	20	21.8	309

Source for tensile stress-strain curve data: A. Akhtar and V. A. Stowell, "Cavitation erosion of hydraulic turbine steels containing 0.2 pct carbon and 1 pct Mn," *J. Mater. Energy Sources 4* (1), pp. 58–63, 1982.

TABLE Q22.3 CAVITATION EROSIVE WEAR RATES AND MATERIAL PROPERTIES OF 0.2 wt % C STEELS

Steel	Area under Tensile Stress Strain Curve, P_1 (arb units)	Proof Resilience, P_2 (kPa)	Ultimate Resilience, P_3 (MPa)	Work to Fracture, P_4 (MPa)	DIP, P_5 ($\times 10^3$)
G (stress-relieved)	13.1	441	1.00	101	12.37
G (normalized)	16.2	298	0.77	87	10.38
K (stress-relieved)	20.6	195	0.64	125	5.75
K (normalized)	21.8	238	0.74	135	25.91

Source for tensile stress-strain curve data: A. Akhtar and V. A. Stowell, "Cavitation erosion of hydraulic turbine steels containing 0.2 pct carbon and 1 pct Mn," *J. Mater. Energy Sources 4* (1), pp. 58–63, 1982.

values of the predictor parameters to be tested against erosive wear rate are given in Table Q22.3 and are plotted in Fig. Q22.2.

The second study is that of the cavitation erosive wear behavior of three high-strength alloys in mercury at 260°C. The results are presented in Table Q22.4.

The indications, from the results in Fig. Q22.2, are that parameters that include ultimate tensile strength, yield strength, modulus of elasticity, and tensile elongation (P_2, P_3, and P_5) are reasonably good predictors of the cavitation erosive wear resistance of materials. A qualified judgment, based on statistical treatment of the results in Fig. Q22.2, is that the "best predictor" seems to be the ultimate resilience, implying that plastic deformation and ductile rupture play the dominant role in the cavitation erosive wear of these materials. A plot of the cavitation erosive wear data in terms of P_3 for the high-strength materials-mercury system (Table Q22.4) is therefore obtained (Fig. Q22.3).

The tendency to universalize these results must be resisted. It is well known that good correlations have been limited to either a single class of materials or similar materials [3, 10]. The search for universal correlations continues.

TABLE Q22.4 CAVITATION EROSION BEHAVIOR, IN MERCURY AT 260°C, AND MATERIAL PROPERTIES OF THREE CANDIDATE MATERIALS

Material	Cavitation Erosive Wear Rate (mg h^{-1})	UTS (MPa)	E (GPa)	UTS^2/2E = P_3 (MPa)
304 stainless steel	20.83	638	179	1.14
Mo-0.5Ti	43.16	580	296	0.57
C6-1Zr	125.78	173	100	0.15

Source for wear rate data: A. Akhtar and V. A. Stowell, "Cavitation erosion of hydraulic turbine steels containing 0.2 pct carbon and 1 pct Mn," *J. Mater. Energy Sources 4* (1), pp. 58–63, 1982.

Figure Q22.2 Variation of mass loss due to cavitation erosion with various predictive parameters for carbon steels G and K in distilled water at 22°C.

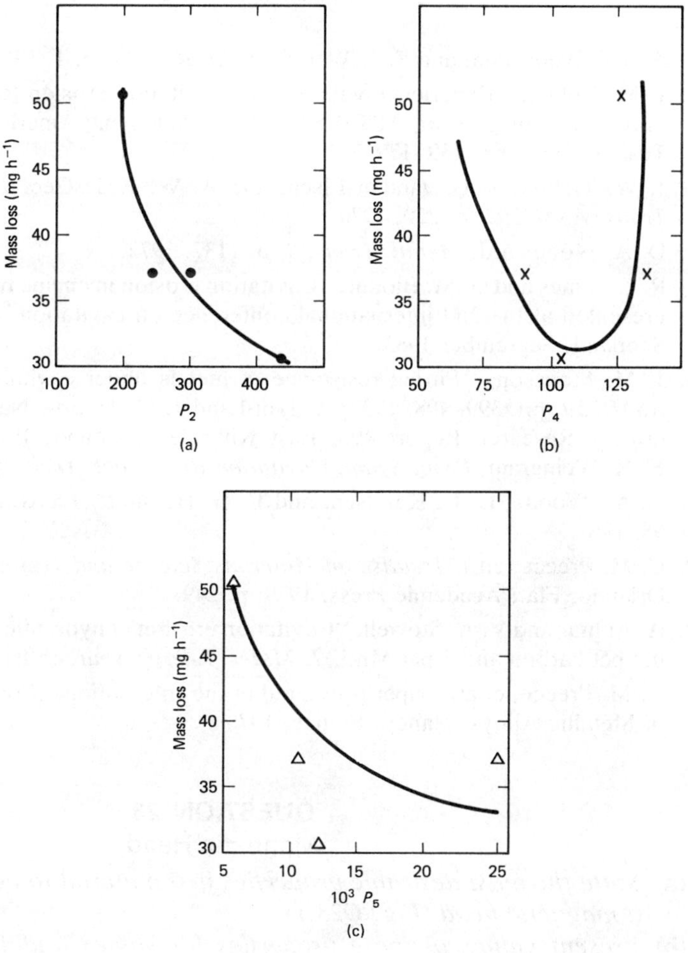

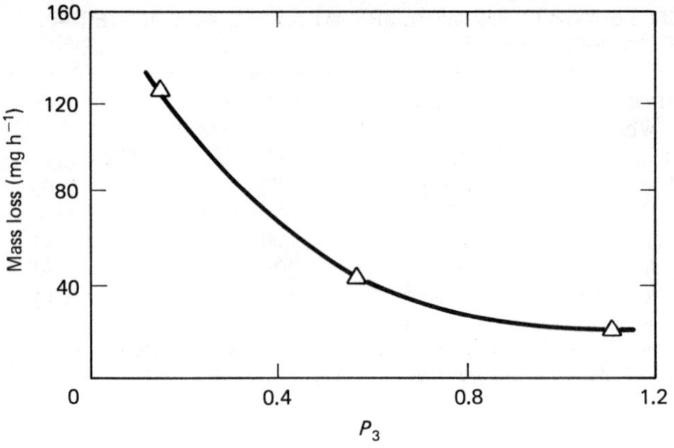

Figure Q22.3 Variation of mass loss due to cavitation erosion with predictive parameter P_3 for three materials in mercury at 260°C.

REFERENCES

1. A. Thiruvengadam and S. J. Waring, *Ship. Res. 10,* p. 1, 1967.
2. J. M. Hobbs, "Experience with a 20-kc cavitation erosion test," in *Erosion by Cavitation or Impingement,* ASTM STP 408. Philadelphia: American Society for Materials Testing, 1967, pp. 159–179.
3. J. W. Tichler, J. B. vanden Elsen, and A. W. J. de Gee, *J. Lub. Tech., Series F, Trans. ASME 92,* p. 220, 1970.
4. D. A. Woodward, *Metall. Trans. 3,* p. 1137, 1972.
5. R. C. James and G. M. Blount, "Cavitation erosion in engine bearing materials," paper presented at the 2nd international conference on cavitation, Heriot-Watt University, Scotland, September 1983.
6. J. M. Mousson, "Pitting resistance of metals under cavitation conditions," *Trans. ASME 59,* pp. 399–408, 1937; A. Laird and J. M. Hobbs, National Engineering Laboratory Research Report 496, East Kilbride, Scotland, 1971; J. Z. Lichtman and E. R. Weingram, *Proc. Symp. Cavitation Res. Facil. Tech.,* p. 185, 1964.
7. G. M. Woods, L. L. Knudsen, and F. G. Hammitt, *Trans. ASME, Series D, 89,* p. 98, 1967.
8. C. M. Preece (ed.), *Treatise on Materials Science and Technology: Vol. 16, Erosion.* Orlando, Fla.: Academic Press, 1979, p. 249.
9. A. Akhtar and V. A. Stowell, "Cavitation erosion of hydraulic turbine steels containing 0.2 pct carbon and 1 pct Mn," *J. Mater. Energy Sources 4*(1), pp. 58–63, 1982.
10. C. M. Preece, et al., paper presented at the International Conference on the Strength of Metallic Alloys, Nancy, France, 1976.

QUESTION 23
Magnetic Head

(a) *State the most desirable properties in a material to be used to manufacture a magnetic head (Fig. Q23.1).*

(b) *Present values of these properties for some candidate materials for this application.*

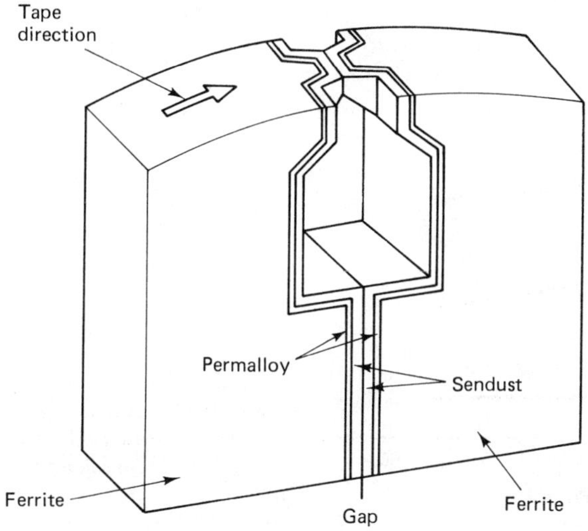

Figure Q23.1 Sample metal-in-gap magnetic head.

(a) The most desirable material properties are these:

- A high saturation magnetization; this allows the material to do the most work. (To ensure this, the material must have a high initial relative magnetic permeability.)
- The magnetic induction–applied magnetic field hysteresis loop (Fig. Q23.2) must be small enough to keep energy losses, when the component is cycled through an alternating field, to a minimum.
- A low remanence (Fig. Q23.2).
- High hardness, thereby ensuring adequate surface wear resistance.
- A moderate value of electrical resistivity, σ. (If σ is too low, current may leak from the field as eddy currents, which may cause heating of the device.)
- Magnetostriction (magnetically induced reversible elastic strain; see Fig. Q23.3) must be practically zero.

(b) The values of some of these properties for a number of candidate materials are given in Table Q23.1.

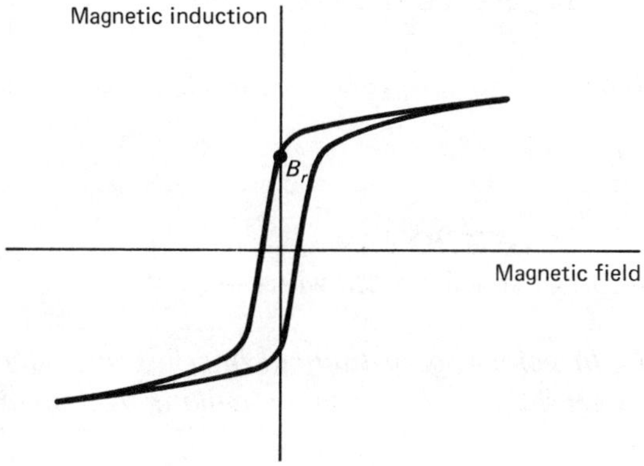

Figure Q23.2 Hysteresis loop for a soft magnetic head material (B = remanence).

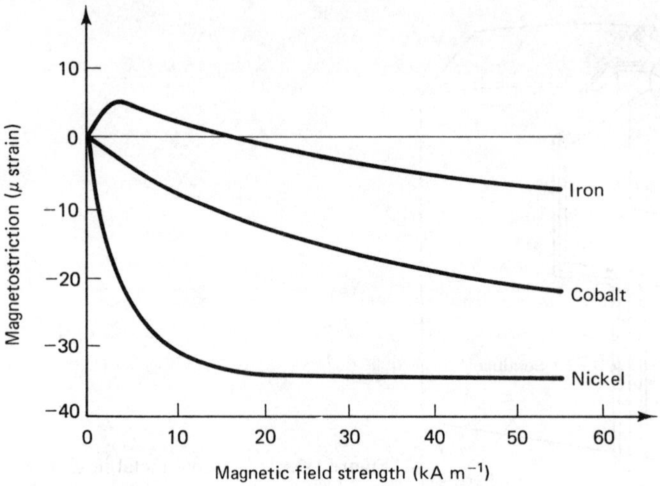

Figure Q23.3 Magnetostrictive behavior of iron, cobalt, and nickel [1] (reprinted with permission from McGraw-Hill.)

TABLE Q23.1 PROPERTIES OF SIX CANDIDATE MATERIALS FOR A MAGNETIC HEAD

Material	Composition (wt %)	Initial Relative Permeability (μ_i)	Saturation Induction (T)	Vickers Hardness (kg mm^{-2})	Electrical Resistivity ($\mu\Omega$·cm)
Basic Permalloy	77 Ni, Cu, Mo, Fe	20,000	0.80	120	60
Age-hardenable Permalloy	77 Ni, Cu, Mo, Fe 5 Nb, 5 Ti	40,000	0.50	240	90
Sendust	85 Fe, 9.6 Si, 5.4 Al	25,000	1.10	450	88
—	84 Fe, 16 Al	8,000	0.90	280	145
—	66 Co, 4 Fe, 30 (Mo, Si, B)	20,000	0.55	900	130
—	69 Co, 6 Mn, 25 (Si, B)	10,000	0.80	900	130

REFERENCE

1. W. F. Smith, *Principles of Materials Science and Engineering*. New York: McGraw-Hill, 1986, p. 614.

QUESTION 24
Radome of a Tactical Missile

The increase in carry and flight velocities of tactical missiles over the years has increased the demand on materials to be used in making radomes for these weapons.

Present an account of the main requirements for suitable materials for this application, and give the values of some of these properties for specified candidate materials.

The main property requirements for a material for a radome (Fig. Q24.1) are as follows:

- Low dielectric constant
- Low loss tangent
- Satisfactory achievement of the electrical transmissions specifications, which can be established by comparing, through superposition, the electrical property data (dielectric constant and loss tangent) with the flat plate transmission analysis, as given in Fig. Q24.2, assuming a maximum specified transmission loss, such as 0.5 dB at 35 Hz
- High relative stability of the first two listed properties under a variety of environmental conditions that would be encountered in radome operation: temperature ranges from −15°C to 70°C and relative humidity that can be as high as 100% for sustained periods
- High temperature resistance, as reflected in its heat deflection temperature; a minimum required value is at least 230°C to meet the peak recovery temperatures experienced during the carry mode (Fig. Q24.3)
- Ability to impart mechanical integrity to the structure, expressed as the ability to withstand bending moments and pressure loads that would be imposed on it during flight, especially during the operational mode when maneuver loads are present

When subjected to a bending moment M, the radome wall thickness t is given by $t = FS(M/\pi r^2 \sigma_a)$, where FS is the factor of safety, r is the radius of the

Figure Q24.1 The radome of a high-performance aircraft. (Courtesy of Brunswick Corp.)

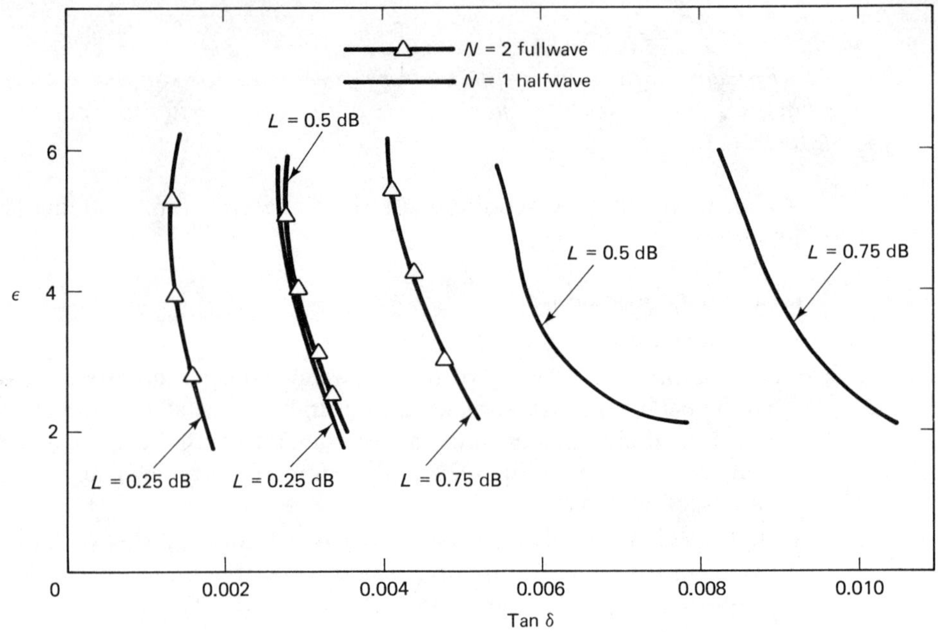

Figure Q24.2 Transmission loss bounds as a function of dielectric constant and loss tangent (typical curves derived from flat-plate analysis at normal incidence; L = transmission loss; ϵ = dielectric constant; tan δ = loss tangent; N = wall thickness order—the higher the order, the thicker the wall)

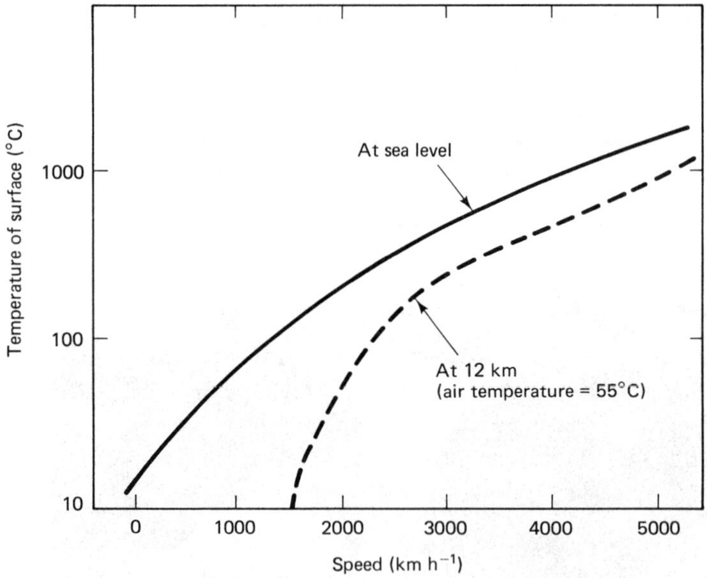

Figure Q24.3 Aerodynamic heating of a rapidly traveling object, such as a missile, in air [1].

radome base, and σ_a is the allowable strength of the radome material. The variation of t with σ_a is given in Table Q24.1.

In addition to the bending loads, there are air loads (resulting from the pressure distribution) that could lead to buckling failure. In this case, the section thickness t is obtained from the relationship $t = (4\sigma_{cr}r^3/E)^{0.333}$, where σ_{cr} is the

TABLE Q24.1 RADOME DESIGN THICKNESS AS A FUNCTION OF RADOME MATERIAL ALLOWABLE STRENGTH (for M = 140 Nm, r = 75 mm, and FS = 1.5)

σ_a (MPa)	t (mm)
3.50	3.39
7.00	1.70
10.50	1.13
14.50	0.82

critical Euler buckling air pressure and E is the modulus of elasticity of the radome material. The variation of t with σ_{cr} and E is given in Table Q24.2.

The results in Tables Q24.1 and Q24.2 show that the controlling properties are the allowable strength and the modulus of elasticity of the material.

Two additional desirable properties are these:

- Acceptable resistance to erosion arising from rain, dust, or other particles acting severally or jointly
- Low raw material and fabrication costs

(b) The materials usually employed are fiber-filled, fiber-reinforced, or non-filled thermoplasts or thermosets that can be fabricated by injection molding, compression molding, or resin impregnation of the appropriate fabric with subsequent lamination.

These are some popular candidate materials:

- Polyphenylene sulfide resin reinforced with glass fibers (for example, Ryton R-4)
- Fluorocarbon resin reinforced with glass fibers (for example, UF-1006)
- Polyester resin reinforced with mica (for example, VITUF 5172)
- Polyamide-imide resin reinforced with glass fibers (for example, Torlon)

The values of some of the relevant parameters for these materials are given in Tables Q24.3, Q24.4 and Q24.5.

TABLE Q24.2 RADOME DESIGN THICKNESS REQUIREMENTS DUE TO EXTERNAL PRESSURE VARIATION (for r = 75 mm)

σ_{cr} (MPa)	E (GPa)	t (mm)
1.00	7.0	6.24
0.50	7.0	4.95
0.33	7.0	4.31
1.00	0.7	13.43
0.50	0.7	10.66
0.33	0.7	9.29

TABLE Q24.3 PERFORMANCE OF VARIOUS MILLIMETER-WAVE RADOME CANDIDATE MATERIALS

Material	Dielectric Constant at Room Temperature and Frequency of 35 GHz		Loss Tangent at Room Temperature and Frequency of 35 GHz		Deflection Temperature at 1.8 MPa (°C)	Erosion Resistance Measured by Crater Volume Loss (10^{-6} cm^3) When Particle Velocity Is between 610 m s^{-1} and 690 m s^{-1}
	As Received	After Exposure[a]	As Received	After Exposure[a]		
Polyphenylene sulfide resin reinforced with 40 vol % glass fibers	3.8	3.9	0.002	0.010	260	259
Fluorocarbon resin reinforced with 30 vol % glass fibers	3.6	3.6	0.022	0.025	238	409
Polyester resin reinforced with mica	3.5	3.8	0.007	0.036	193	—
Polyamide-imide resin reinforced with 30 vol % glass fibers	4.0	4.2	0.023	0.037	274	25

[a]Conditions: relative humidity of 95% while raising temperature from 24°C to 66°C in 2 h, holding at this temperature at a steady rate back to room temperature over 16 h while keeping relative humidity above 85%. This cycle is repeated 10 times over a total exposure period of 240 h.

Source: R. A. Mayor and A. Ossin, "Cost-effective high-performance plastics for millimeter wave radome applications," 24th National Symposium and Exhibition of SAMPE, San Francisco, Calif., May 1979, p. 1578.

TABLE Q24.4 MECHANICAL PROPERTIES OF MILLIMETER-WAVE RADOME CANDIDATE MATERIALS

	Polyamide-imide–10% glass fiber		Polyphenylene sulfide–40% glass fiber	
	As Received	After 120 d[a]	As Received	After 120 d[a]
Tensile strength (MPa)				
−54°C	241	221	131	138
24°C	197	185	123	121
149°C	134	96	52	52
232°C	66	63	11	22
Flexural strength (MPa)				
−54°C	265	257	204	186
24°C	194	214	161	163
149°C	148	130	90	90
232°C	70	77	16	24

[a]Exposure to central Florida weather (August to December).

Source: R. A. Mayor, E. A. Welsh, and A. Ossin, "Material selection for cost-effective millimeter wave radomes," *SAMPE Q.*, p. 3, April 1980.

TABLE Q24.5 RAIN EROSION MASS LOSS DATA FOR RYTON AND TORLON MATERIALS (SPEED OF RAINDROPS = 800 km h^{-1})

	Mass loss (g)	
Time (min)	Ryton	Torlon
10	0.15	0.002
20	0.70	0.006
30	1.00	0.008

Source: R. A. Mayor, E. A. Welsh, and A. Ossin, "Material selection for cost-effective millimeter wave radomes," *SAMPE Quarterly*, p. 6, April 1980.

REFERENCE

1. H. E. Bennett, "The need for better optical materials," *Adv. Opt. Mater. SPIE* 505, pp. 2–3, 1984.

QUESTION 25
Cab of a High-Speed Passenger Train

To cope with the huge increase in the number of people choosing to travel by train and to cut traveling times, the nationalized railroad authority of a Western

European country has been designing and testing a new class of electric locomotives. Three materials were considered for the cab of such a train: aluminum, steel, and glass fiber reinforced-epoxy resin (GFRP). The final choice was GFRP.*

(a) *Give reasons for this choice.*

(b) *Outline the advantages and disadvantages of this material for this application.*

(a) Aluminum was rejected because of the cost of hand-beating panels, galvanic corrosion problems when in contact with steel, and poor impact resistance. Steel was rejected because of its high density, problems with welding it, and the high investment costs associated with low-production quantities (typically, only about 150 cab units are required).

(b) The advantages of GFRP are directly related to the design specifications for these cabs:

- The shape has to be aerodynamically efficient (streamlined) to reduce the effects of shock waves in front when passing other trains and when entering a tunnel or a station.
- The cab has to be as light as possible, in order to keep power consumption to a minimum and have low-weight trains, which means lower railroad track maintenance costs.
- The method of manufacturing has to be suitable to a small production run.
- The structure has to protect the driver from missiles thrown deliberately by vandals or accidentally induced by the passage of the train.

The advantages of using GFRP for this application are as follows:

- Low weight
- Economical process for small production runs
- Large area, irregularly curved parts required in small production numbers easily obtained by the efficient hand lap-up operation
- Noncorrosive and chemically inert
- Good impact resistance (see Table Q25.1†)
- Parts require minimum finishing operation, because much trimming can be done when the molding is in the semicure stage, thereby reducing post-molding operations costs
- Local thickness variations can be accommodated without any major molding problems

*For example, the Class 91 of British Rail, the nationalized railroad authority of the United Kingdom (Figure Q25.1).

†Although the data in this table are not for GFRP but for glass-based rovings and mat, the impact behavior of these two sets of materials is similar.

Figure Q25.1 The InterCity Class 91 electric locomotive. (Courtesy of British Rail)

TABLE Q25.1 PENETRATION SPEEDS THROUGH VARIOUS MATERIALS FOR POINTED AND FLAT-ENDED 65 g, 16 mm DIAMETER MISSILES

	Penetration speed (m s^{-1}) in				
Missile Mass Per Area (kg m^{-2})	Glass, Woven Rovings	Glass, Chopped-strand Mat	Glass, Cross-plied Rovings	Mild Steel	Aluminum
3 (pointed missile)	70	22	42	—	37
6 (pointed missile)	98	50	76	—	50
9 (pointed missile)	—	82	118	64	60
3 (flat-ended)	34	22	29	—	—
6 (flat-ended)	66	40	50	38	74
9 (flat-ended)	90	70	78	58	100

Source: J. Batchelor and P. J. Garrington, "Impact considerations and performance for fiber-reinforced polymers on British Rail trains," Paper presented at the BPF Reinforced Plastics Congress, Brighton, England, November 1980.

The disadvantages of using GFRP for this application are as follows:

- Variable quality of parts because the lap-up process is manual; success is dependent on operator skill
- Long cycle times are needed, usually one large molding per day
- High labor costs per unit of production

The complete cab consists of a shell, floor, and bulkhead, together with various fittings such as panels, doors, windows, and an instrument console. The shell and bulkhead are made of GFRP skins with a core of polyurethane (PU) foam (Fig. Q25.2). The floor consists of GFRP-skinned rigid polyvinylchloride foam, and the other fittings are GFRP laminates.

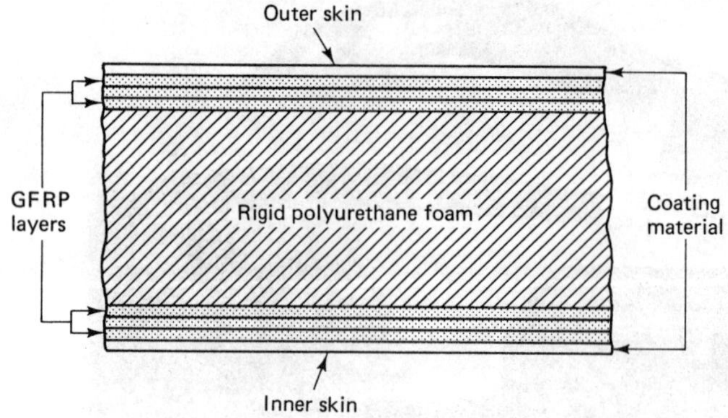

Figure Q25.2 Sandwich construction of the shell and bulkhead of the cab of a high-speed passenger train.

QUESTION 26
Hybrid Voltage Regulator

Discuss the design requirements of a hybrid voltage regulator. Outline the important material selection factors, and suggest the best materials and method of mounting for the silicon power transistor chip.

The design, development, and manufacturing of high-performance power hybrid products (voltage regulators, overvoltage protection modules, etc.) that are both economical and reliable is a major challenge in electric power engineering. Materials for such devices must be selected on the basis of certain electrical, mechanical, thermal, and chemical characteristics, and such properties must be compatible with each other.

The hybrid voltage regulator is shown in Fig. Q26.1. It consists of two subassemblies: a power section and a control section. The former consists of a power transistor and a power resistor and is used to regulate the load. This section must be capable of dissipating high power (from 85 to 500 W, depending on the model). The control section contains all the required semiconductors, resistors, and capacitors and is used to control the voltage as well as the provide protection against both excess temperature and current.

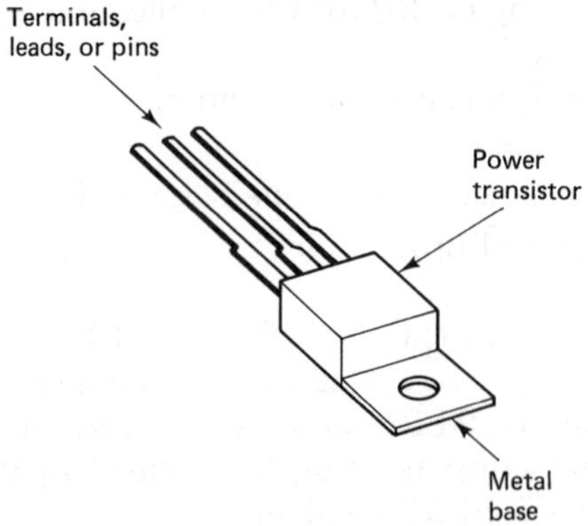

Figure Q26.1 Power transistor, metal base, and terminals or pins of a hybrid voltage regulator.

The major design considerations of the hybrid assembly are as follows:

- Heat transfer in the power section
- Heat capacity of the power section
- Thermal expansion of the power section material
- High service temperature (80°C to 150°C)
- Electrical isolation for both control and power plates
- Raw material and processing costs

Power Section Thermal Design

Two parameters are important in this design: heat transfer and heat capacity.

1. Heat transfer may be expressed in the closed form by the approximate expression [1]

$$\Theta = \frac{1}{2k(a - b)} \ln \left[\frac{a\,(2t + b)}{b\,(2t + a)} \right] \tag{26.1}$$

where Θ is the thermal resistance, k is the thermal conductivity of the material, a and b are the dimensions of the rectangular heat source at the surface of the material whose thermal resistance is being calculated, and t is the thickness of the material.

For a square heat source

$$\Theta = \frac{1}{ka(2t + a)} \tag{26.2}$$

Equations (Q26.1) and (Q26.2) are based on the assumption of heat flow through a spreading 45°. For layers of differnt materials (as in Fig. Q26.2), the thermal resistance of each material can be calculated separately and then summed to give the total thermal resistance.

Figure Q26.2 Structural details of a hybrid voltage regulator.

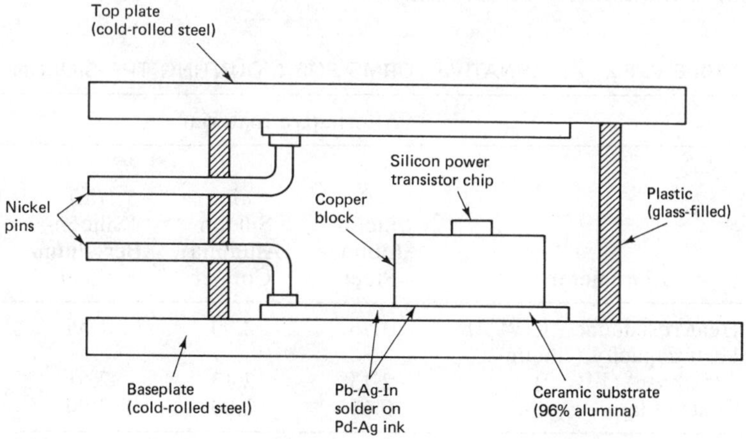

TABLE Q26.1 VALUES OF PROPERTIES OF VARIOUS MATERIALS

Material	k ($W\ m^{-1}\ K^{-1}$)	ρ ($kg\ m^{-3}$)	Cp ($J\ g^{-1}\ K^{-1}$)	Cost[a]
Silicon	0.84	2300	0.70	N.R.[b]
Copper	3.20	8900	0.39	$1.40 kg^{-1}
96% Al_2O_3	0.30	4000	1.13	$77.50 m^{-2}
Cold-rolled steel	0.55	7800	0.45	$0.50 kg^{-1}
BeO	2.20	2900	1.25	$121.15 kg^{-1}

[a]June 1988 prices.

[b]N.R., not required for subsequent calculations.

2. The heat capacity H of the material is given by

$$H = \rho C p V$$

where V is the volume of the material.

Let us compare the thermal performance and cost of the voltage regulator design shown in Fig. Q26.2 using three alternatives based on different materials. (The values of the relevant properties are given in Table Q26.1.) These alternatives are as follows:

- Mounting the silicon chip onto the alumina on steel as the baseplate
- Mounting the silicon chip onto the alumina on copper as the baseplate
- Mounting the silicon chip directly onto beryllium on a steel baseplate
- Mounting the silicon chip onto copper onto alumina on a steel baseplate

It is evident from Table Q26.2 that alternative (iv) is the preferred option, combining the lowest thermal resistance, a high heat capacity and a moderate cost. Note that alternatives (iii) and (iv) have the same thermal resistance and heat capacity but the cost of alternative (iii) is much higher than that of alternative (iv) mainly because of the high cost of beryllia. Another disadvantage of alternative (iii) is the toxicity of beryllia.

TABLE Q26.2 ALTERNATIVE FORMS FOR MOUNTING THE SILICON

	Alternative material set			
Parameter	(i) Silicon-Alumina-Steel	(ii) Silicon-Alumina-Copper	(iii) Silicon-Beryllium-Steel	(iv) Silicon-Copper-Alumina-Steel
Heat resistance (°C W^{-1})	3.76	2.80	1.55	1.48
Heat capacity (silicon + its base) (J °C^{-1})	2.13	2.13	2.51	2.47
Cost[a] ($ kg^{-1})	0.65	2.00	2.00	1.00

[a]Excluding the cost of the silicon; normalized with respect to silicon-copper-alumina-steel.

REFERENCE

1. L. Balents et al., "Design considerations for power hybrid circuits," Proc. 1969 ISHTM Microelectronics Symposium, pp. 323–341, 1969.

QUESTION 27
Transducers in Medical Ultrasonic Devices

Briefly discuss the use of transducers in medical ultrasonic devices, specifically from the viewpoint of the piezoelectric material in the active parts of such transducers, the acoustical energy vibrations, and the important characteristics of suitable materials for use in these transducers.

Ultrasonic diagnostic equipment is now widely used in medicine, due to the merits of noninvasion and good real-time displays of body tissues. Among the many areas in medicine where ultrasonics are now routinely used are the following:

- Obstetrics (to detect the fetal heart in the tenth week of pregnancy, localize the placenta, and detect multiple fetuses)
- Cardiology (to observe all parts of the heart in motion)
- Internal medicine (to visualize internal organs such as the thyroid, gallbladder, etc.)
- Neuromedicine (to find cerebral infractions, display brain midline, etc.)

To obtain sharp, well-defined ultrasonograms, high-sensitivity transducers with good impulse responses—that is, high-efficiency, low-ripple, wideband transducers—are desired.

A schematic representation of the setup for the use of ultrasonics to observe the human brain (called transducal echoencephalography) is shown in Fig. Q27.1. Some ultrasonograms of the human head are shown in Fig. Q27.2.

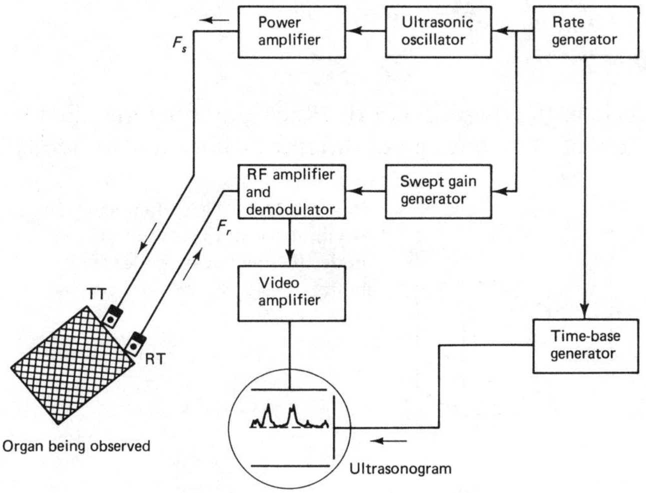

Figure Q27.1 Typical instrumentation for ultrasonic measurement (TT = transmitting transducer; RT = receiving transducer; F_s = radiated ultrasonic wave; F_r = reflected wave).

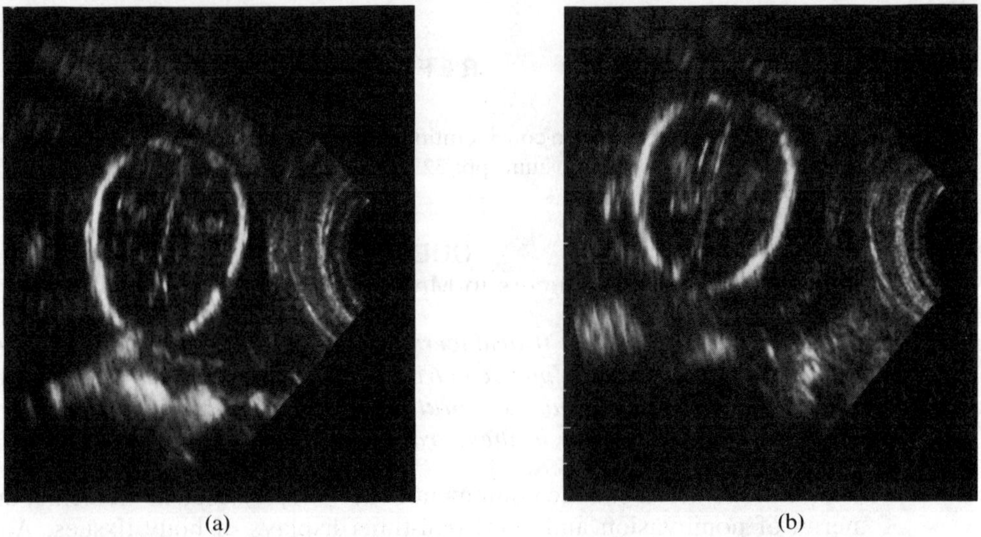

Figure Q27.2 Ultrasonograms of the head of an 18-week-old human fetus: (a) biparietal level; (b) choroid plexus level. (Courtesy of Douglas L. Brown, M.D., Department of Radiology, University of Tennessee, Memphis)

Typically, the transducers for these ultrasonic devices contain an active part (a piezoelectric material), which may consist of many small elements, and a passive part (the so-called matching layers). The piezoelectric material converts the electrical energy into acoustical energy (and vice versa), which appears as vibrations. Under normal conditions, element vibration in one direction only is required because spurious modes in other directions may cause artifacts and energy loss. The mode of vibration is directly related to the geometry of the piezoelectric material element.

There are three important modes of vibration (Fig. Q27.3):

- Plate mode: the width w and the length l are much greater than the thickness t
- Thickness mode: $L \gg t$ and w, and $w < 0.7t$
- Bar mode: $t \gg L$ and w

In each case, the poling axis is in the direction of thickness, and the metallization is orthogonal to this axis. Each of the modes gives different values for the acoustic

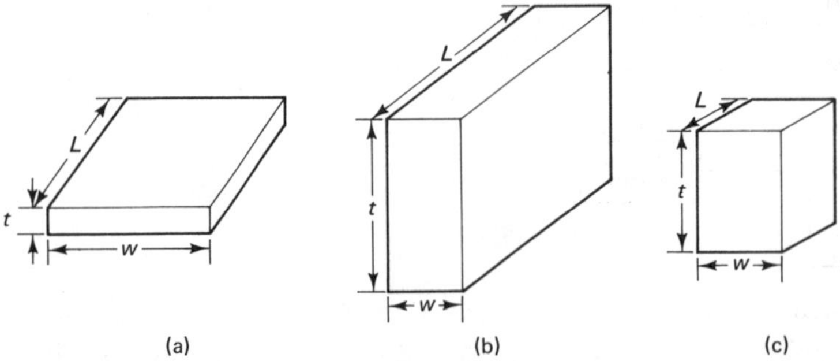

Figure Q27.3 Three important modes of vibration of a transducer: (a) plate mode; (b) thickness mode; (c) bar mode.

velocity (and hence antiresonant frequency) and coupling factor. For example, the acoustic velocity and coupling factor in the thickness and bar modes are, respectively, 15% less and 50% more than in the plate mode.

The performance of a transducer depends to a large extent on the matching layers, the backing impedance, and the properties and geometry of the piezoelectric active elements. Some relevant properties of two candidate piezoelectric materials (one ceramic and the other polymeric), as far as their use in ultrasonic transducers is concerned, are compared and commented on in Table Q27.1.

TABLE Q27.1 VALUES OF SOME RELEVANT MATERIAL PARAMETERS FOR PZT and PVDF[a]

Parameter	PZT5A	PVDF	Comments
Acoustic impedance, ρc (kg m^{-2} s^{-1})	30×10^6	4.3×10^6	This means that mismatch with soft tissues in the human body [$\rho c = 1.5 \times 10^6$] is much lower with PVDF.
Quarterwave matching layer impedance (10^6 kg m^{-2} s^{-1})	5	2	This low acoustic impedance is found in commercially available plastics, whereas for matching with PZT, the matching layer has to be composed of different materials.
Electromechanical coupling factor, k_{33}	0.5–0.7	0.2	The low value for PVDF indicates that a large amount of electrical energy is needed to supply the necessary amount of mechanical energy, and vice versa.
Other coupling modes, k_{spurious}	High	Low	PZT needs special handling, such as subsplicing of rectangular elements, to cancel spurious modes.
Piezoelectric stress constant, e_{33} (C m^{-2})	15	0.16	The main effect of this parameter is found in the transformation factor 2α and hence in the value of the electrical impedance Z_e (Fig. Q27.4); i.e., $Z_e(\text{PVDF})/Z_e(\text{PZT}) = 2 \times 10^3$.
Dielectric constant	600	6	This constant determines the transducer capacitance, C_o. Comparing transducers with identical area and resonant frequency gives $C_o(\text{PVDF})/C_o(\text{PZT}) = 0.02$. Although this seems to be an advantage for PVDF, it has to be considered in combination with Z_e, leading to the result that in spite of its low capacitance, tuning of a PVDF transducer is even more necessary than with PZT transducers.
Mechanical loss tangent, $\tan \delta_m$	0.004	0.10	PZT has low internal damping and therefore requires external backing and matching layers to obtain a short impulse response and a wide bandwidth.
Dielectric loss tangent, $\tan \delta_e$	0.002	0.25	The loss tangent is very large in PVDF, resulting in an appreciable amount of electrical power being dissipated in the transducer itself.

[a]PZT, lead zirconate titanate; PVDF, polyvinyldene fluoride.

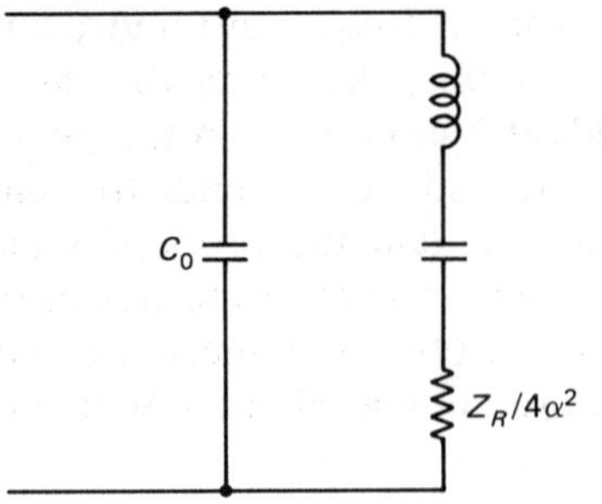

Figure Q27.4 Simplified equivalent circuit of an air-backed transducer near the resonance frequency (C_0 is the capacitance of the transducer, Z_R is the acoustical load impedance, and 2α is the transformation factor and is equal to $e_{33}A/t$, where A is the area of the transducer and t is its thickness).

TABLE Q27.2 SOME PROPERTIES OF PVDF AND P(VDF-TrFE)

Property	PVDF	P(VDF-TrFE)
k_{33}	0.20	0.30
e_{33}	0.16	0.23
$\tan \delta_m$	0.10	0.05
$\tan \delta_e$	0.25	0.15

A new and potentially attractive candidate material has recently been introduced, the copolymer P(VDF-TrFE). The coupling factor k_{33} and the piezoelectric stress constant e_{33} of this material are 50% higher than those of PVDF, while the losses are reduced by a factor of 2 (Tabel Q27.2). As a result, the transmission loss of transducers using P(VDF-TrFE) is reduced considerably.

QUESTION 28
Heat Exchanger in an OTEC Plant

Write a brief essay on the material requirements for the heat exchangers in a closed-cycle ocean thermal energy conversion (OTEC) system. Make special reference to the problem of biofouling.

A large quantity of thermal energy is stored in the tropical and semitropical seas of the world as a result of solar radiation. The key to harvesting these enormous energy resources lies in the evolution of feasible power generation systems, ones that will operate efficiently and economically in the small temperature difference that exists between the warm surface water and the cold water a few meters below it. These systems are referred to ocean thermal energy conversion (OTEC) systems. Typical surface temperatures are in the range of 25°C to 30°C, and typical cold water temperatures are in the range of 3°C to 8°C (Fig. Q28.1).

Oceanographic criteria to be satisfied for sites considered for the location of OTEC plants are shown in Table Q28.1. Among the most attractive and potentially viable locations for OTEC plants are Keahole Point, Hawaii (in the Pacific Ocean), St. Croix and the Virgin Islands (in the Caribbean), sites 300 km west of

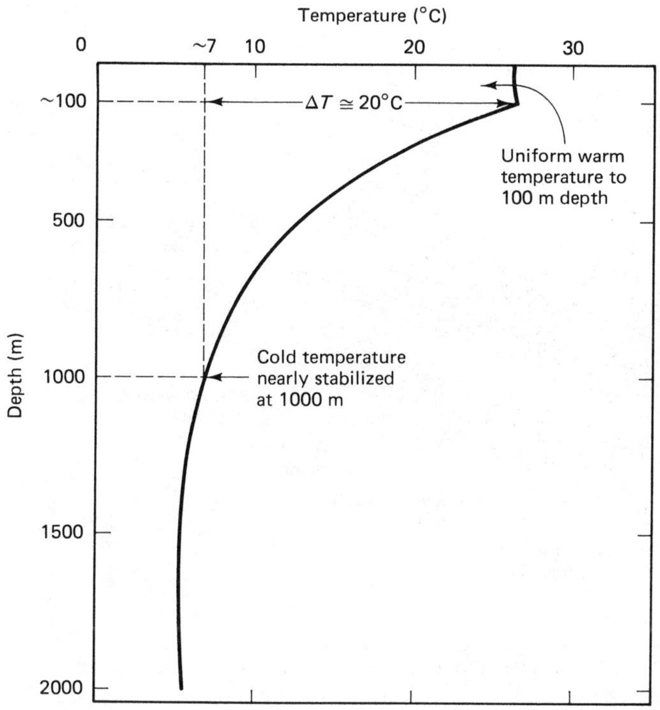

Figure Q28.1 Typical temperature profile in the tropical seas.

TABLE Q28.1 OCEANOGRAPHIC CRITERIA FOR SATISFACTORY OTEC LOCATION SITES

- Surface currents with speeds of about 932 m h^{-1} or less and no substantial subsurface currents; deep currents with speeds of 373 m h^{-1} or less
- Deep bathymetry, which ensures that there is no possibility of encountering an under-ocean peak extending up to about 1 km below the surface
- Available water temperature differences between the surface and the deep cold water exceeding 20°C; surface temperature of 25°C or greater
- Normal winds of 13–30 km h^{-1}
- Normal sea states ≤ 5 (wave height ≤ 4 m)

Tampa and Panama City, Fla. (in the Gulf of Mexico), and various sites in the Japan Sea, such as Kumejima (on an island in the Okinawa Prefecture).

The technical feasibility of the OTEC concept has been demonstrated in only a few plants. The "Mini-OTEC" program was operational from August to November 1979. The plant was mounted on a barge located about 2 km off Keahole Point, Hawaii (Fig. Q28.2) and was designed to produce 50 kW of gross electrical power at an average seawater temperature of 21°C [1]. The world's largest demonstration plant to date (100 kW gross electrical power) was operational from October to December 1981 in the Republic of Nauru, located in the central Pacific Ocean.

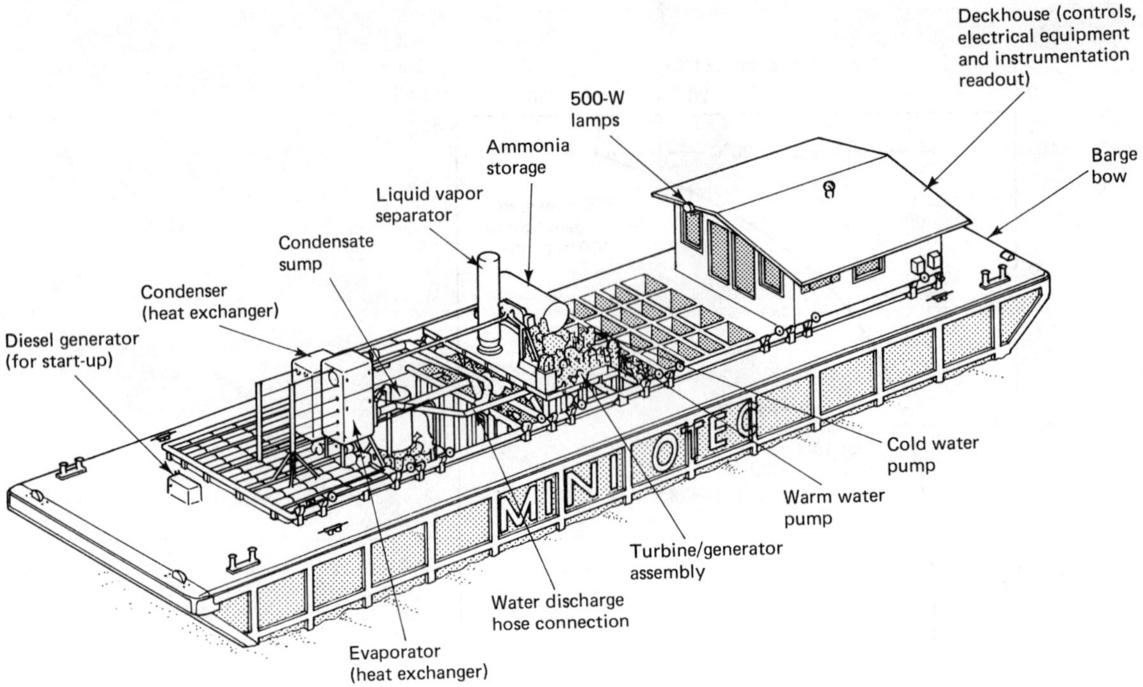

Figure Q28.2 Aerial view of Mini-OTEC plant (courtesy of Lockheed Missiles and Space Company, Sunnyvale, Calif.).

Classification of the techniques used for OTEC plants can be based on one of two criteria:

- Choice of the cycle: open cycle [Fig. Q28.3(a)], where the motive fluid is steam produced from vaporization of the water from the hot source, or closed cycle [Fig. Q28.3(b)], which uses an intermediate fluid that the hot water vaporizes and is condensed after use by the cold source
- Choice of the site: land-based or open-sea

The heat exchangers are the most important components of any OTEC plant and account for 25% to 50% of total plant costs. The mean temperature differences between the seawater and the working fluid are quite small (about 4°C to 5°C), and thus large areas of heat transfer area are required (typically on the order of 5–8 $m^2\ kW_e^{-1}$ (net). For example, in the 100-kW_e OTEC demonstration plant in Nauru, the total evaporator and condenser heat transfer area is 809 m^2 [1]. The overall heat transfer coefficient U is given by the expression

$$\frac{1}{UA} = \frac{1}{h_r A_r} + \frac{x}{k A_m} + R_f + \frac{1}{h_w A_w}$$

where

h_r = heat transfer coefficient of the working fluid
x = tube wall thickness
A = reference area

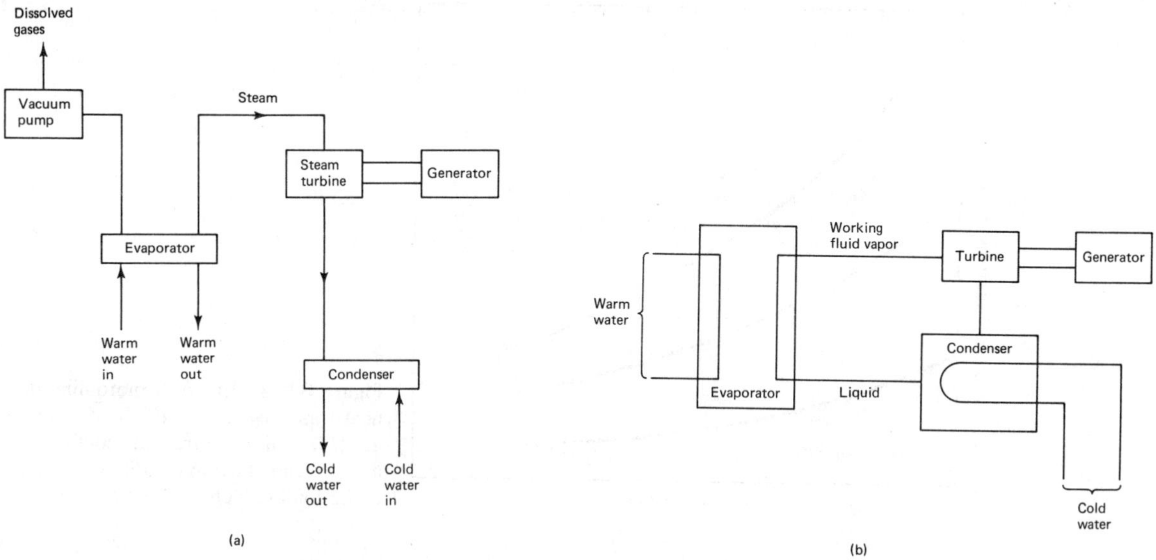

Figure Q28.3 Schematic representations of OTEC systems: (a) open cycle; (b) closed cycle.

A_r = area on the refrigerant side
A_w = area on the water side
A_m = mean area of the tube normal to heat flow
k = thermal inductivity of the tube material
R_f = resistance due to biofouling or scale formation on the seawater side

The primary requirement for the heat exchanger material is thus low thermal resistance (that is, high thermal conductivity). Other desirable properties are these:

- High resistance to corrosion by seawater (the relevant corrosion types are general, pitting, crevice, and galvanic)
- Excellent resistance to both stress corrosion cracking and corrosion fatigue
- High resistance to attack by various concentrations of ammonia in the seawater
- Low density
- Ease of fabrication
- Good resistance to erosion by the fast-moving seawater
- Low raw material cost

In addition to these requirements, the material used must have excellent resistance to biofouling. The efficiency of the heat exchangers is dependent on the maintenance of a surface with a high heat transfer coefficient. Materials exposed to seawater typically develop a layer of attached organisms, a phenomenon generally known as biofouling. A biofouling film can contain certain dead cells and cellular debris, organic secretions, and inorganic precipitates, in addition to

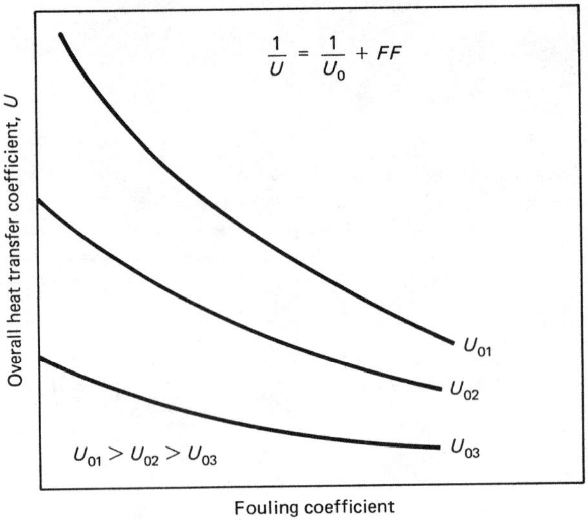

Figure Q28.4 Effect of biofouling on heat transfer in an OTEC heat exchanger (U_0 is the heat transfer coefficient under clean start-up conditions, and *FF* is the fouling factor or coefficient).

living cells. Metal surfaces in contact with seawater may be further altered by the formation of a corrosion product layer of scale. The resulting biofouling and/or corrosion-derived deposits have poorer thermal conductivities than the original metal, introducing resistance to heat transfer between the seawater and the working fluid. This phenomenon of biofouling is illustrated in Fig. Q28.4. One method of controlling biofouling is through the use of chlorine. The chlorine is generated by in situ electrochemical production of sodium hypochlorite and is kept below the Environmental Protection Agency (EPA) level of 200 ppb. The effect of chlorination on biofouling may be gleaned from the data in Table Q28.2.*

TABLE Q28.2 EFFECT OF CHLORINATION ON BIOFOULING OF A TITANIUM TUBE IN SEAWATER

Time (d)	Reciprocal of heat transfer coefficient ($m^2\ kW^{-1}$) × 10^4	
	Free Fouling Conditions	Chlorination (0.05 $ppml^{-1}h^{-1}d^{-1}$)
50	0.13	
60	0.43	
70	0.88	0.88
80	—	0.50
90	—	0.20
100	0.00	0.00
110	0.30	—
120	0.43	—

Source: J. B. Darby, Jr., "Ocean thermal energy conversion—materials issues," *J. Mat. for Energy Systems* 6 (2), p. 132, 1984.

*Current guidelines set by the U.S. Environmental Agency on allowable discharge concentration of chlorine residuals are so stiff as to restrict use of bulk chlorination. One alternative biofouling control technique now receiving attention is targeted chlorination, in which a high dose chlorine solution (0.5 to 2 ppm) is applied to small segments of the condenser inlet tubesheet for short periods (5–15 min.).

TABLE Q28.3 GENERAL PERFORMANCE, IN SEAWATER, OF CANDIDATE MATERIALS FOR AN OTEC HEAT EXCHANGER

	Corrosion resistance				Erosion Resistance	Biofouling Resistance	Stress Corrosion Cracking (SCC) Resistance[a]	Corrosion Fatigue Resistance
	General	Pitting	Galvanic	Crevice				
Titanium alloys (grades 1 and 2)	E	E	E	G–E	E	P	E	E
Specialty stainless steels, such as AL-6X, Type 325, Monit, and Nitronic 50	E	E–F	VG	G–F	G	P	E	F
Aluminum alloys, such as 5052, 3003, and 7072 clad	VG	F–P	P	F–P	F–P	P	E	P
Copper-nickel alloys	E	E	G	E	G	G	E	E

E, excellent; VG, very good; G, good; F, fair; P, poor.

[a]SCC in ammonia is a serious problem.

Sources: Lockheed Missiles & Space Company, Inc., Ocean Systems, Sunnyvale, Calif.; TRW Defense and Space Systems Group, Rendondo Beach, Calif.; OTEC Program, Ocean Energy Technology Division, U.S. Department of Energy; Materials Science and Technology Division, the Argonne National Laboratory, Argonne, Ill.; and the Department of Mechanical Engineering, University of Hawaii.

Another method of biofouling control is sponge ball cleaning, as is used in fossil fuel–based and nuclear power plants.

The best candidate heat exchanger materials currently available are the titanium and aluminum alloys, special seawater-service stainless steels, and copper-nickel alloys. The experimental performance of these materials is summarized in Table Q28.3.

REFERENCE

1. Y. Mochida, S. Kawano, T. Takahata, and M. J. Miyoshi, of Solar Energy Engineering, Trans. ASME *106,* 187-192 (1984).

QUESTION 29
Plume Impingement

(a) *Write a brief note on the phenomenon of plume impingement with regard to military rocket-launching systems and communications satellites.*

(b) *Comment on suitable candidate materials for one specified design.*

(a) There is an increasing trend toward the use of highly aluminum-filled solid propellant motors for shipboard and ground rocket-launching systems (as are found in the U.S. Navy's Vertical Launching System and the U.S. Army's Patriot antiaircraft missile; see Fig. Q29.1). A principal design consideration in

Figure Q29.1 The Patriot antiaircraft missile, at launch. (Photo Courtesy of the Raytheon Company.)

these systems is rocket exhaust plume impingement, which leads to particle erosion. The main ingredient of the exhaust plume is α – alumina particles. It has been estimated, for example, that the Army's Patriot generates plumes with a nominal particulate content of about 23% [1]. Consequences of exhaust plume impingement in such systems include (1) additional forces and torques exerted on the rocket, (2) surface heating, (3) contamination, and (4) electrical charging.

Space vehicles, such as communications satellites, which require jets for orbit adjustment and/or attitude control have satellite surfaces, antennas, solar panels, and other components positioned within the range of the exhaust plume of the jets. Plume impingement on any of these surfaces will have two effects:

- It will cause disturbance twisting moments, which must be absorbed and/or controlled by the satellite attitude control system.
- It will introduce an inefficiency due to drag (during any orbit-adjust maneuver), thus increasing the fuel load of the vehicle for a specified design life.

The Satcom communications satellites, the first of which was placed in geosynchronous orbit in December 1975, are structured such that the solar array panels are affected by significant impingement due to the inclination control jets. Necessary corrections to accommodate this was then incorporated into the final design of the attitude control subsystem and for the allocation of fuel.

Rocket plume impingement problems have been reported in the separation of the launcher stage of the European Space Agency's Ariane rocket, which powers communications satellites into geosynchronous orbit for various clients.

(b) The choice of materials for the coat of the solid propellant motor of a military rocket system is considered.

The candidate materials have to be evaluated on the basis of relevant performance-to-weight and performance-to-cost indices. The critical characteristic that describes the performance of a material under plume impingement of particle erosive wear conditions is its erosive wear resistance (EWR). A first-order approximate expression for EWR is given by [2]

$$\mathrm{EWR} = \frac{C\dot{m}U^2}{\rho \cdot \mathrm{MRR}} \qquad \mathrm{J\ g^{-1}}$$

where $\dot{m}$ is the mass flux of impacting particles, U is the impact velocity of the particles, ρ is the density of the material, MRR is the material recession rate (or erosion rate), and C is a constant. Thus the more resistant the material is, the lower is MRR and the higher is EWR. The parameter EWR, therefore, relates the material mass loss to the kinetic energy of the impacting particles and in that sense is analogous to the effective heat of ablation, a parameter used in ablative analysis where the primary recession mechanism is convective and/or radiative heating. EWR is thus a plausible performance-to-weight index.

The performance-to-cost index MCI = EWR per unit cost of the material p.

Suitable candidate materials for this application include these polymer composites:

- Silica–phenolic resin; end-grain laminate
- Rubber-modified glass fibers–phenolic resin; end-grain laminate
- High-density asbestos fibers–phenolic resin
- Compression-molded E glass fibers–phenolic resin
- Cast asbestos, fiber-filled–phenolic resin.

Material costs may be obtained from current price lists, catalogs, or trade journals. All the materials, however, require additional processing to obtain final part form. Thus the total cost per mass is complicated by the varying labor times associated with the different fabrication processes applicable to different materials. For example, compression-molded E glass fiber–phenolic resin is available as a preimpregnated sheet compound that can be cut, laid up in flat sheets, and compression-molded at temperature to final part. By contrast, the fabrication of parts from silica–phenolic resin is complicated by the fact that hand layups are required to produce the required 90% (end-grain) shingle-angle configuration.

Detailed recession rate data must be obtained on either full-scale systems or specially designed solid rocket motor simulators.* Using those results and total material costs (raw material and fabrication costs), the screening of the candidate materials could be finalized.

TABLE Q29.1 PERFORMANCE INDICES OF FIVE CANDIDATE MATERIALS IN ROCKET EXHAUST PLUME IMPINGEMENT CONDITIONS[a]

Material	Mean EWR ($J\ g^{-1}$)	Mean MCI[b]
1. Silica–phenolic resin	940	0.33
2. Rubber-reinforced glass fibers–phenolic resin	825	0.50
3. High-density asbestos fibers–phenolic resin[c]	730	0.70
4. E glass fibers–phenolic resin	720	1.00
5. Cast asbestos, fiber filled–phenolic resin[c]	700	0.55

[a] In short-run burn simulation rocket motor tests.

[b] Relative to material 4.

[c] These materials are not considered acceptable candidates in view of the current uncertainty surrounding the future use of asbestos.

Sources: Data from various reports, including T. V. Radovich and J. M. Potts, "Development and testing of heatshield materials for rocket launching systems," Orlando, Fla.: Martin Marietta Aerospace, 1979; "Antiballistic missile (ABM) aeromechanical and materials design," vol. 2, Orlando, Fla.: Martin Marietta Corp., 1972; and T. V. Radovich, "Performance of materials in solid propellant rocket motor exhaust environments," paper presented at the 14th Intersociety Conference on Environmental Systems, San Diego, Calif., July 1984.

*An example of such a test program is the Conflow SRMS (solid rocket motor simulator) material screening facility at Martin Marietta Aerospace, Orlando Division, Orlando, Fla.

As an indication of the relative performance of these candidate materials, the best materials on a performance-to-weight basis are the first two (see Table Q29.1). However, they are very expensive (for example, their raw material costs in 1988 were \$40 kg^{-1} and \$20 kg^{-1}, respectively). The best material, based on considerations of plume erosion resistance, cost, and environmental safety, is material 4.

REFERENCES

1. T. V. Radovich, Martin Marietta AeroNaval Systems, Baltimore, private communication, 1988.
2. A. S. Benjamin and R. J. Sargent, "Hypersonic erosion facilities and rain, dust and ice correlation (*U*)," TRW Corporation, Technical Report AFML-TR-73-233, pt. 3, May 1974.

QUESTION 30
Pyroelectric Transducers

A new photocopying process, a system that uses passive infrared intrusion detectors, and new charge-separation filters are all examples of applications that employ a polymeric or ceramic pyroelectric transducer.

Comment on this choice of material.

Pyroelectric detection depends on the absorption of incident infrared radiation by the detector material, raising its temperature and thus producing a pyroelectric signal, the magnitude of which is proportional to the temperature change ΔT between the material and its environment. ΔT is given by

$$\Delta T = \frac{V\epsilon}{\lambda(T)t}$$

where V is the dc pyrolectric voltage, ϵ is the dielectric constant of the detector material, $\lambda(T)$ is its pyroelectric coefficient at temperature T, and t is the thickness of the detector.

Thus desirable properties for the detector material (for V to be as large as possible for a given t) are a large pyroelectric coefficient and a low dielectric constant. In addition, low dielectric loss tangent, low specific heat, low density, high volume resistivity, and a broad usable temperature range are attractive properties.

Thermal losses are important considerations in the design of pyroelectric detectors. If the detector is set up in vacuum conditions, conductive and convective losses through air are negligibly small, while radiative losses can be specified in terms of a thermal diffusion length, L_d (defined as the distance along a slab at which the amplitude of the temperature change has decreased to about 37% of that at the end of thermal excitation), such that

$$L_d = \sqrt{\frac{2k}{C_p\rho\omega}}$$

where k, C_p, and ρ are the thermal conductivity, specific heat, and density, respectively, of the detector material, and ω is the angular frequency of operation. For good pyroelectric response at a given frequency, $t \ll L_d$, thus ensuring that the total thickness of the detector is uniformly heated. In addition to thermal losses, thermal noise and dielectric loss must also be compensated for.

The effect of material properties on the performance of a pyroelectric detector is related to the requisite use of the detector. For high voltage responsiveness—defined as $\eta\lambda(T)/\omega C_p A\epsilon$, where η is the efficiency with which the incident radiation is absorbed and A is the cross-sectional area of the detector—the material figure of merit (the parameter used to compare the candidate materials) is $\lambda(T)/C_p\epsilon$. For high current responsiveness—defined as $\eta\lambda(T)/C_p t$—the relevant material figure of merit is $\lambda(T)/C_p$.

The final figure of merit may be derived from the normalized detectivity D, which is given by [1]

$$D = \frac{\lambda(T)\eta}{2C_p\rho}\left[\frac{1}{t\omega\epsilon(\tan\delta)k'T}\right]^{0.5}$$

where k' is the Boltzmann constant, $\tan\delta$ is the dielectric loss factor, and T is the absolute temperature. Thus a third figure of merit is

$$\frac{\lambda(T)}{C_p\rho}\left(\frac{1}{\epsilon\tan\delta}\right)^{0.5}$$

The values of these figures of merit for three candidate materials are given in Table Q30.1. Although on the basis of the values of two of the three figures of merit PVF_2 performs badly, it is being used in a growing number of pyroelectric detectors [2] because it can be readily prepared in any shape or size and scores high marks on considerations of chemical stability, mechanical strength, power-handling ability, and ease of preparation of ultrathin films.

TABLE Q30.1 PROPERTIES AND FIGURES OF MERIT FOR THREE PYROELECTRIC DETECTOR MATERIALS

Material	$\lambda(T)$ (10^{-10} cm^{-2} K^{-1})	ϵ	C_p (J cm^{-3} K^{-1})
Lithium tantalate ($LiTaO_3$)	190	46	3.2
SBN ($Sr_{0.5}Ba_{0.5}Nb_2O_6$)	600	400	2.3
Polyvinylidene fluoride (PVF_2)	−40	11	2.4

Material	Figures of merit: $\lambda(T)/C_p\epsilon$ (10^{-10} cm J^{-1})	$\lambda(T)/C_p$ (10^{-10} cm J^{-1})	$\lambda(T)/\rho C_p \left(\frac{1}{\epsilon\tan\delta}\right)^{0.5}$ (cm^4 kg^{-1} J^{-1})
Lithium tantalate ($LiTaO_3$)	1.29	59.4	0.050
SBN ($Sr_{0.5}Ba_{0.5}Nb_2O_6$)	0.65	260.9	0.030
Polyvinylidene fluoride (PVF_2)	−1.52	−16.7	−0.009

REFERENCES

1. W. R. Peters, *Proc. Carnahan Crime Countermeas. 78,* p. 141, 1977.
2. U. Korn, Z. Rav-Noy, and S. Shtrikman, "Pyroelectric PVF_2 infrared detector arrays," *Applied Optics 20* (11), pp. 1980–1982, 1981; J. E. Jacobs and S. A. Remily, "Optimized PVF_2 targets for infrared Vidicon camera tubes," *Infrared Physics 19* (1), pp. 1–7, 1979.

QUESTION 31
Thermal Barrier Coatings

Many approaches have been taken to increase the resistance of aircraft gas turbine blades to hot working fluids. One approach involves the use of advanced materials, such as single-crystal alloys, metal fiber reinforced superalloys, and oxide dispersion strengthened superalloys. Another approach is to employ a suitable cooling scheme such as convection, impingement, film or transpiration cooling. A third alternative is to protect the blades with a thermal barrier coating.

(a) *What are the main functions of a thermal barrier coating in this application?*
(b) *Give some examples of thermal barrier coatings in current use.*
(c) *State some factors that affect the amount of thermal protection afforded by such a coating.*
(d) *Briefly describe the most common method currently employed to deposit these coatings.*
(e) *Briefly discuss the major problems associated with the use of a thermal barrier coating in this application.*

(a) The main functions of a thermal barrier coating (TBC) are as follows:

- To increase the life of the blade
- To permit the use of structural materials for the fabrication of the blade that are much cheaper than those currently in use
- To afford the opportunity for the use of simpler and more reliable blade designs
- To allow higher operating temperatures (for increased fuel efficiency of the turbine)
- To allow the use of moderate thermal gradients (a TBC just 130 μm thick could lower the metal temperature of a film-cooled blade by about 200°C; see Fig. Q31.1)

(b) The most common TBC consists of a coat (comprised of ceramic materials, such as zirconia, with additions of cubic phase stabilizers, such as calcia, magnesia, yttria, or ytterbia), held onto the basis metallic blade substrate with an oxidation- and hot-corrosion-resistant metallic bond coating. This bond coating, typically, has the composition *M*-Cr-Al-*X*, where *M* can be nickel, cobalt, or iron, and *X* is a reactive metal, such as yttrium or ytterbium.

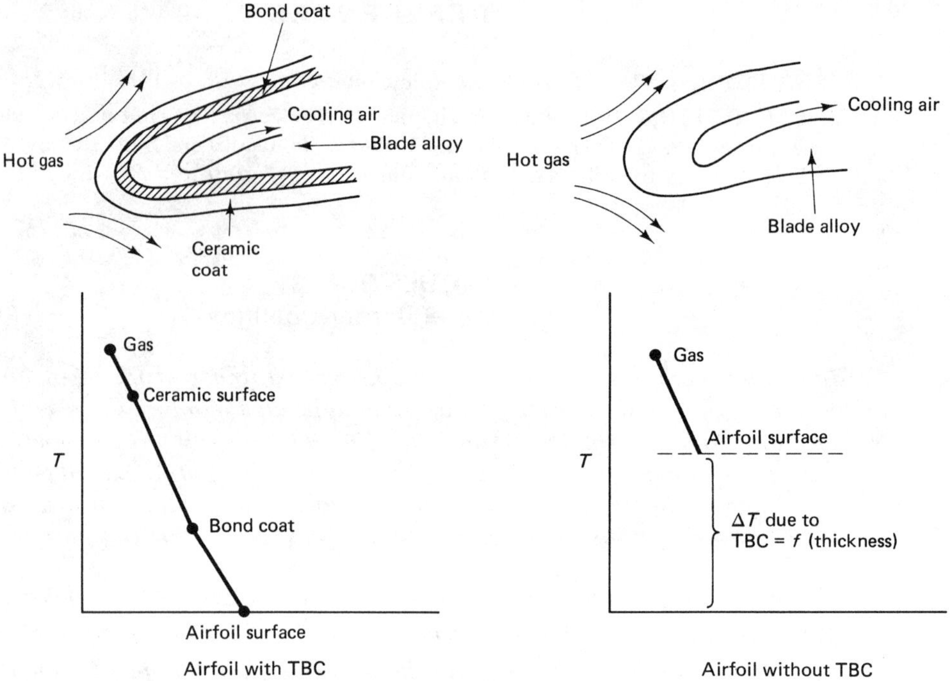

Figure Q31.1 Concept of a thermal barrier coating (TBC) for cooled turbine airfoils, with the corresponding temperature distribution.

Examples of TBC–bond coating combinations now in use are ZrO_2-$12Y_2O_3$/Ni-16Cr-6Al-O-6Y, ZrO-5.4 CaO/Ni-16Cr-6Al-0.6Y, ZrO_2-Yb_2O_3/Ni-18.3Cr-6.4Al-0.22Yb, and ZrO_2-3.4MgO/Ni-16Cr-6Al-0.6Y.

(c) Factors that affect the amount of thermal protection afforded by a TBC include these:

- The heat flux of the impinging hot gas (higher heat fluxes require thinner coatings to achieve a given temperature drop)
- Its thickness
- Its thermal conductivity
- Its composition (recent results [1] indicate that the life of the TBC is critically dependent on the amount of the stabilizer addition; see Fig. Q31.2), as well as its structure and form

(d) The most common TBC deposition procedure currently in use is the plasma spray process. The ceramic powder is fed into a high-temperature, high-velocity arc plasma where the powder particles melt and are propelled to the surface to be coated. The "molten" material then splatters the surface, and a continuous coating builds up. The plasma-sprayed coating quality depends on (1) particle molten state and velocity, on impact, (2) particle chemical reaction with environment during flight, (3) heat transfer control to substrate while spraying, and (4) relative movement of plasma torch to substrate.

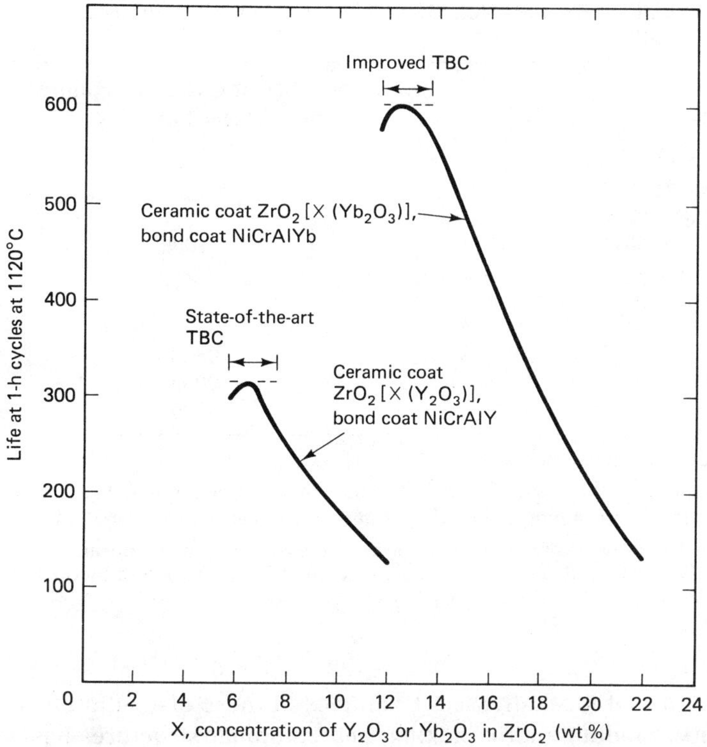

Figure Q31.2 Dependence of the life of a thermal barrier coating (TBC) on the amount of stabilizer added [1] [reprinted with permission from *Aerospace America*].

A newer process, arc spray, is likely to be widely used in the near future. This method has a lower operating cost, higher deposition rates, and better coating quality than the plasma spray coating method.

(e) Three major problem areas have been identified: volume changes accompanying phase transitions, thermal expansion mismatch, and oxidation.

ZrO_2 undergoes two solid-state phase changes:

$$\text{monoclinic} \underset{\sim 950°\text{C}}{\overset{\sim 1170°\text{C}}{\rightleftarrows}} \text{tetragonal} \overset{2370°\text{C}}{\rightleftarrows} \text{cubic} \overset{2680°\text{C}}{\rightleftarrows} \text{liquid}$$

The tetragonal to monoclinic transition is a disruptive martensitic process accompanied by a volume increase of between 3% and 9% that can cause the coating to crack during thermal cycling. Stabilization of zirconium is one measure to use in order to avoid these phase transitions or to limit them to temperatures low enough to freeze in a metastable condition. A study of the Zr_2O_3-Y_2O_3 phase diagram [2] shows that about 17% of zirconia is needed to stabilize the cubic structure fully throughout the temperature range of technical interest of these coatings.

Thermal-stress-induced spalling, due to large thermal expansion mismatch between the coating and the substrate, is another problem. One way to improve spalling resistance relies on the fact that failure is caused by flaws that cause propagation of cracks in the plane of the TBC. The stresses that give rise to such propagation might be reduced by controlling the coating structure so that appropriate discontinuities cause a decrease in the modulus of elasticity of the coating

TABLE Q31.1 THERMAL CYCLIC PERFORMANCE OF THERMAL BARRIER COATINGS WITH VARIOUS MODIFICATIONS

Coating Type	Number of Cycles to Failure in Thermal Stress Test[a]
I. Unmodified PS ZrO_2–20% Y_2O_3	200
Microcracked PS ZrO_2–21% MgO	5200
Segmented PS ZrO_2–20% Y_2O_3	5500
Segmented EB ZrO_2–20% Y_2O_3	>1000
II. Substrate temperature during plasma spraying of ZrO_2–20% Y_2O_3	
−34°C	3500
20°C	9500
316°C	9000
649°C	200

PS, plasma-sprayed; EB, electron beam vapor–deposited.

[a]Thermal cycles in a petroleum fuel–fired burner with gas velocity equal to Mach 0.3. Thermal cycle consisted of heating to 1010°C for 4 min, followed by 2 min of cooling to below 316°C.

Sources: D. L. Ruckle, "Plasma-sprayed ceramic thermal barrier coatings for vane platform," p. 407, and P. E. Hodge, R. A. Miller and M. A. Gedwell, "Evaluation of hot corrosion of thermal coatings," p. 447, in *Proc. Int. Conf. Metall. Coatings*. Lausanne, Switzerland, and New York: Elsevier Sequoia S.A., 1980.

and hence lead to a reduction of local stresses. Examples of these discontinuities include increased porosity, random microcracking, and columnar structures having columns perpendicular to the coating substrate (so-called coating segmentation). Another way to alleviate the problem is through appropriate control of substrate temperature during coating. This procedure provides a better balance between compressive and tensile strains imposed on the TBC during cyclic operation of the turbine. The effects of TBC structure and substrate temperature on the thermal cyclic performance of a TBC may be appreciated from the data given in Table Q31.1.

The third problem involves the oxidation of the bond coat, which contributes to the failure of a TBC. Spraying in air prevents the oxidation of the components and allows the formation of aluminum- and yttrium-oxide stringers, which act as

TABLE Q31.2 OXIDATION PERFORMANCE OF APS (ARC PLASMA-SPRAYED) AND VPS (VACUUM PLASMA-SPRAYED) BOND COATS (CYCLE TEMPERATURE = 1070°C)

Coat Type	Samples	Number of Cycles to Failure
APS	1	100
APS	2	130
VPS	3	700
VPS	4	1250

Source: E. J. Kubel, "Thermal spray technology: from art to science," *Advanced Materials and Processes 132* (6), p. 79, 1987.

diffusion barriers. The net result is a low high-temperature oxidation resistance of the coating and, subsequently, low TBC lifetimes. In addition, spraying in a vacuum produces splatters that are very tightly knit (since there are no trapped air and oxide particles), thereby allowing interdiffusion between the nickel and the aluminum layers in the coating. This results in improved high-temperature oxidation resistance of the TBC. Some recent results on this question of oxidation resistance of the bond coat are presented in Table Q31.2.

REFERENCES

1. C. A. Stearns, "Thermal-barrier coatings," *Aerospace America 27,* pp. 30–31, May 1987.
2. H. G. Scott, "Phase relationships in the zirconia-yttria system," *J. Mater. Sci. 10,* p. 1527, 1975.

QUESTION 32
Wear-resistant Coatings for Cutting Tools

The use of coatings for cutting tool materials has been commonplace since about 1970.

(a) *What are the main factors that govern the choice of a coating?*

(b) *Name three refractory compounds that have emerged as successful coating materials, and write down the governing chemical reactions leading to their formation from the relevant precursors. Also give the values of two important properties of these compounds.*

(a) The main factors are these:

- High temperature stability
- Chemical stability
- High crater wear resistance
- High hardness-edge retention
- High abrasion resistance
- Low coefficient of friction

(b) Three such refractory materials are titanium carbide (TiC), titanium nitride (TiN), and aluminum oxide (Al_2O_3). These are the relevant chemical reactions:

$$TiCl_4(g) + CH_4(g) - 1050°C, H_2 \longrightarrow TiC(s) + 4HCl(g)$$
$$TiCl_4(g) + \tfrac{1}{2}N_2(g) + 2H_2(g) - 1050°C \longrightarrow TiN(s) + 4HCl(g)$$
$$Al_2Cl_6(g) + 3H_2(g) + 3CO_2(g) - 1050°C \longrightarrow Al_2O_3(g) + 6HCl(g) + 3CO(g)$$

The standard free energy of formation of these coatings and their hardness values are given in Table Q32.1.

TABLE Q32.1 ENERGY OF FORMATION AND HARDNESS OF THREE REFRACTORY COATING MATERIALS

Coating Material	Standard Free Energy of Formation (kJ mole^{-1}) At 298 K	At 1000 K	Vickers Hardness (kg mm^{-2})
Al_2O_3	−1587	−1365	2000
TiN	−311	−244	2000
TiC	−181	−172	2900

QUESTION 33

Use of Cast Alloys in High-Temperature Pyrolysis Furnaces

The production of ethylene from raw material feedstocks, such as ethane, naphtha, and gas, is carried out in a pyrolysis furnace. Selecting materials for such a furnace requires careful consideration of the high operating temperatures. Material strength requirements, especially creep rupture strength and extension, must be examined, as well as ductility and resistance to the effects of chemical atmospheres.

Discuss this statement with reference to the use of cast alloys in the construction of a high-temperature pyrolysis furnace.

A pyrolysis furnace (Fig. Q33.1) basically consists of an insulated box containing coils in the form of a number of parallel passes of tubular members that enter and leave through the furnace wall, creating a flow path through the furnace. The outside of the coils is heated by burners located on the side walls or at the bottom of the furnace. Petroleum feedstock is supplied to the inside coils, is mixed with steam, and is decomposed in the radiant zone to give ethylene. The incoming

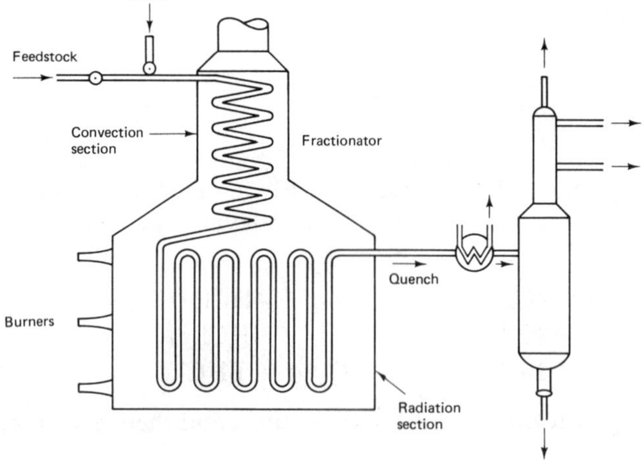

Figure Q33.1 Main components of a typical pyrolysis furnace.

TABLE Q34.1 CAPACITIES OF MSFD PLANTS IN SAUDI ARABIA

Name of Project	Production Capacity (10^3 m^3 d^{-1})	Start-up Date
Jubail I	136	1982
Jeddah IV	220	1982
Khobar II	150	1984
Jubail III	945	1984
Asir I	95	1987

The main parameters that control cost and output are (a) the temperature drop allowed in each stage, (b) the overall flash range, that is, the difference between brine inlet temperature to the first stage and discharge temperature at the last stage, and (c) the stage heat transfer coefficients.

Over 85% of the world's installed desalting capacity is provided by MSFD plants, with some of the largest of these plants being located in the arid and semiarid regions of the Gulf of Arabia (Table Q34.1).

In May 1987, a new desalination process, developed by a small U.S. company (Reliable Water Company, Inc., Billerica, Mass.) was announced. The process is based on a reverse osmosis filtration technique and an energy recovery system (Fig. Q34.2). All the processes are controlled by a sophisticated artificial intelligence program that runs on a standard personal computer. Company officials reported that the new system could produce water suitable for drinking or industrial uses at 33% to 67% of the cost of systems that use conventional desali-

Figure Q34.2 Schematic representation of a new desalination process. (Courtesy of Reliable Water Co., Inc., Billerica, Mass.)

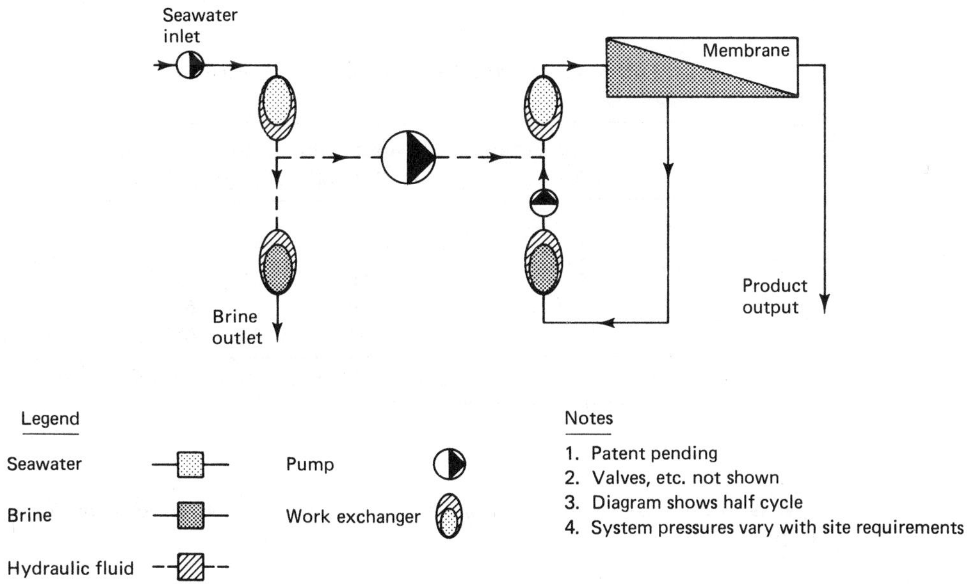

nation methods, such as MSFD. The company plans to market its new process in small commercial units that would produce about 115 m^3 of water per day, enough to supply a small hotel's water needs.

The largest and perhaps most important item in an MSFD plant is the heat exchanger. It has been shown that to produce about 4 l of potable water per day, about 9000 mm^2 of condensing surface is required [1]. This large surface area makes the choice of material for the condenser tubes critical. It has been suggested that the very economic viability of a MSFD plant depends on this choice since tube material cost is between 20% and 33% of total plant cost, depending on the material used and the location of the plant. The three most important considerations in this choice are the mechanical properties, corrosion resistance, and raw material and fabrication costs of the material.

On the basis of the above-stated considerations, the performance of the 2778-H aluminum alloy* may be summarized thus:

- It has good mechanical properties with ultimate tensile strength of 294 MPa, 0.2% proof strength of 280 MPa, and Brinell hardness of 79 MPa.
- It has a very low corrosion rate (on the average 0.2 g m^{-2} d^{-1} under dynamic conditions [2]). Long-term studies have shown the pitting resistance to be excellent (Table Q34.2). Pitting is not a problem at pH $\geq$ 7 and, at pH 6, pitting would not be anticipated until about 125°C.
- Its use leads to substantial savings in the initial investment in a plant. The material and installation costs of a tube (outer diameter = 25 mm, wall thickness = 1.5 mm) of alloy 2778-H and 90Cu-10Ni are approximately $1.50 and $3.00, respectively. Thus for a plant with a capacity of 75,000 m^3 d^{-1}, the estimated tube costs using alloy 2778-H and 90Cu-10Ni are $22.5 million and $45 million, respectively. This lowered cost using alloy 2778-H results in about a 10% reduction in final water cost.

TABLE Q34.2 PITTING DATA OF ALUMINUM ALLOY 2778-H

Exposure Time (10^3 h)	Maximum Pit Depth (μm)	Average Pit Depth (μm)
1	—	—
2	<51	43
4	56	46
8.5	70	61
10	65	60

Source: Z. Ahmad, paper presented at the 3rd Middle East Corrosion Congress, Bahrain, May 1983.

*Nominal composition (in wt %): 0.71Si, 0.19Fe, 0.01Cu, 0.02Mn, 0.60Mg, 0.29Cr, 0.02Zn, <0.02Ti, balance Al.

The following are some of the problems associated with the use of aluminum alloy 2778-H.

- It has a tendency to stimulate anhydrite scaling at the high temperature side of brine heater tubes.
- Onset of pitting, by heavy metal ion contamination (particularly mercury and copper) is a serious problem, although studies on surface conditioning by controlled water chemistry show some promise.
- Galvanic corrosion caused by contact with other metals occurs but could be minimized by careful selection of materials.

On the whole, though, 2778-H is to be preferred to the Cu-Ni alloys (which are used fairly extensively in the manufacture of condenser tubes of MSFD plants) because a large number of Cu-Ni tubes fail after only a small fraction of their predicted lives. For example, corrosion and erosion were found to be the causes of failure in 80% of the Cu-Ni heat exchanger tubes in the brine heater in the Jeddah I desalting units in Saudi Arabia. These tubes, which failed in 1979, were reportedly replaced at a cost of $7 million at that time [3].

REFERENCES

1. E. T. Wanderer and M. W. Wei, paper presented at the Metal Congress, Chicago, 1966.
2. G. Venkateswaran and E. Galow, Report GKSS 83/E/48. G.K.S.S. Forschungszentrum, Geesthacht GmbH, West Germany: 1983.
3. Z. Ahmad, paper presented at the 3rd Middle East Corrosion Conference, Bahrain, May 1983.

QUESTION 35
Advanced High-Temperature Fuel Cells

Advanced high-temperature fuel cells based on molten carbonate and solid oxide electrolytes show special promise for the efficient conversion of hydrocarbon fuels to electricity. Such promise, however, cannot be realized without the development of improved cell materials.

(a) *State the functional requirements of the various components of these fuel cells.*

(b) *Give a summary of the candidate materials for use in fabricating these components.*

(a) The selection of suitable cell materials is dependent on the functional requirements of the various components of the fuel cell. These components (Fig. Q35.1) are as follows:

- The anode, at which the fuel reacts with an ionic oxidizing species

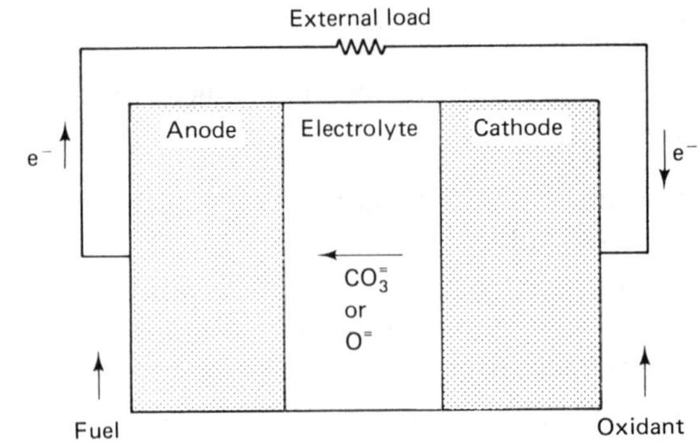

Cell	Anode Reaction	Ionic Transport	Cathode Reaction
MCFC:	$H_2 + CO_3^= \rightleftarrows H_2O + CO_2 + 2e^-$ $CO + CO_3^= \rightleftarrows 2CO_2 + 2e^-$	$\leftarrow CO_3^=$	$CO_2 + \frac{1}{2}O_2 + 2e^- \rightleftarrows CO_3^=$
SOFC:	$H_2 + O^= \rightleftarrows H_2O + 2e^-$ $CO + O^= \rightleftarrows CO_2 + 2e^-$	$\leftarrow O^=$	$\frac{1}{2}O_2 + 2e^- \rightleftarrows O^=$

Figure Q35.1 Schematic diagram of the operation of a fuel cell (MCFC, molten carbonate fuel cell; SOFC, solid oxide fuel cell).

- The cathode, at which oxygen from the oxidant gas is reduced by the electrons to form the ionic species
- The electrolyte, which transports the ionic oxidizing species from the anode to a neighboring cell or to an external load and incorporates the ionic species from the cathode
- The bipolar separator sheet or cell interconnector, which conducts electrons between the anode of one cell and the cathode of the next cell in an electrical series

The general principle of the operation of a fuel cell is shown schematically in Fig. Q35.1. Fuel, such as hydrogen or carbon dioxide, is fed to the anode, where it is catalytically oxidized, while an oxidant (air) is fed to the cathode, where it is catalytically reduced. Electrons liberated by the anode oxidation reaction are conducted through an outer circuit to the cathode for the reduction reaction. The electron flow through the outer circuit produces dc electricity. The fuel cell generator system, shown in Fig. Q35.2, consists of three main compo-

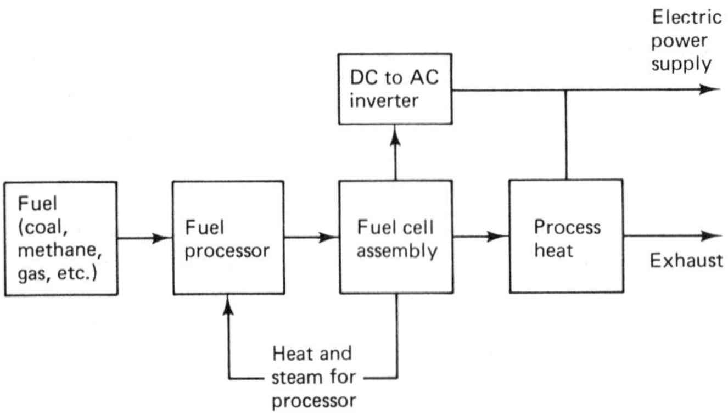

Figure Q35.2 Schematic diagram of the operation of a fuel cell power generation system.

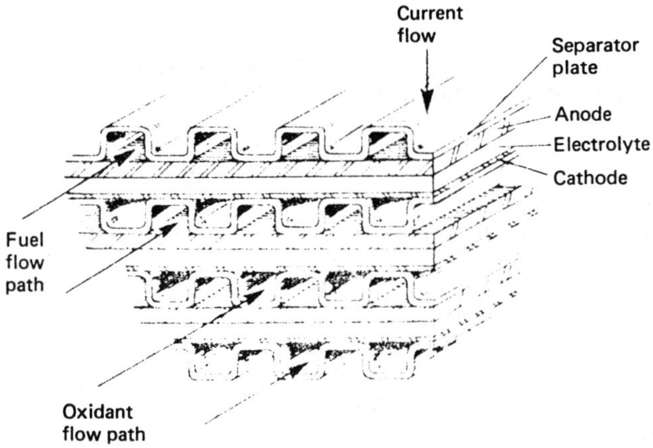

Figure Q35.3 Schematic section of a stack of carbonate cells.

nents. The fuel processor converts a hydrocarbon fuel to a hydrogen-rich gas that is fed to the fuel cell stacks of the power section. This section combines oxygen from the air with the hydrogen from the fuel processor to generate dc electricity. The power conditioning section (inverter) changes the dc electricity to ac electricity.

The distribution of the four main components of a fuel cell (anode, cathode, separator, and electrolyte) in a molten carbonate fuel cell is shown in Fig. Q35.3. Two types of a solid oxide fuel cell are shown in Figs. Q35.4 and Q35.5.

(b) The functional requirements of the various materials for these four components are summarized in Table Q35.1. The candidate materials currently being used or being investigated for these components are given in Table Q35.2.

Figure Q35.4 Cross section of a tubular solid oxide fuel cell. (Courtesy of Westinghouse Research and Development Center, Pittsburgh, Pa.)

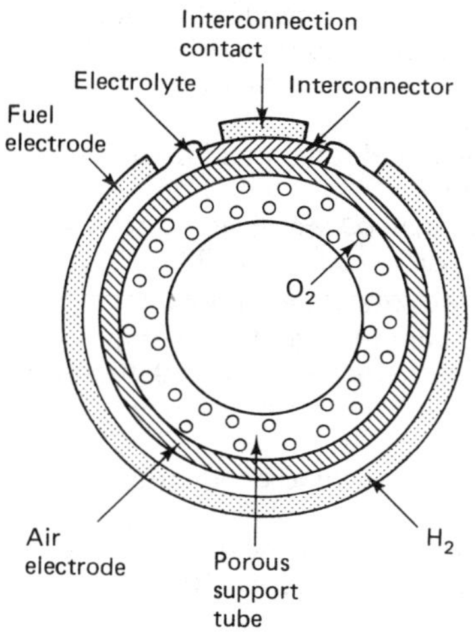

Figure Q35.5 A monolithic fuel cell. (Courtesy of Argonne National Laboratory, Argonne, Ill.)

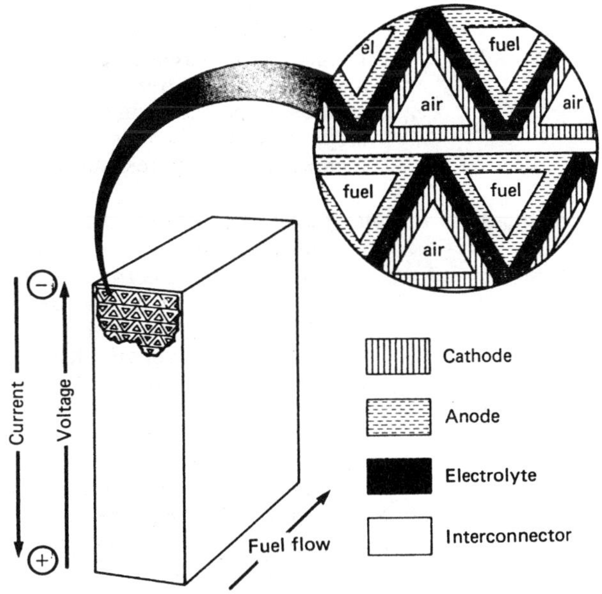

TABLE Q35.1 FUNCTIONAL REQUIREMENTS OF THE MATERIALS FOR VARIOUS COMPONENTS OF ADVANCED HIGH-TEMPERATURE[a] FUEL CELLS

Component	Material Requirements
Anode	Good electronic conductivity Stability in reducing atmospheres Ability to provide electrocatalytic surface Good porosity, to allow passage of gaseous reactants and products
Cathode	Good electronic conductivity Stability in oxidizing atmospheres Ability to provide electrocatalytic surfaces Good porosity, to allow passage of gaseous reactants and products
Electrolyte	Good ionic conductivity Imperviousness to gaseous reactants Stability in oxidizing and reducing atmospheres
Interconnector	Good electronic conductivity at cell operating temperature[a] Imperviousness to gaseous reactants Stability in oxidizing and reducing atmospheres

[a]Nominal operating temperatures of MCFC and SOFC are 650°C and 1000°C, respectively.

TABLE Q35.2 CANDIDATE MATERIALS FOR THE VARIOUS COMPONENTS OF ADVANCED HIGH-TEMPERATURE FUEL CELLS

Component	Molten Carbonate Fuel Cell (MCFC)	Solid Oxide Fuel Cell (SOFC)
Anode	Ni stabilized with $LiAlO_2$ to prevent excessive sintering of the structure, with associated losses in porosity and surface area, at the MCFC operating temperatures of 873 K to 973 K. Cobalt has also been used. Efforts are in progress to use Cu as a low-cost alternative material. Ceramics, to provide sintering response and to increase the compressive creep resistance.	Metallic nickel or cobalt supported by a porous Y_2O_3-stabilized ZrO_2 matrix. The ZrO_3-Y_2O_3 phase supports the Ni or Co catalyst phase, inhibits coarsening of the metallic catalyst particles, and provides an anode thermal expansion coefficient close enough to that of the other cell materials. Electronic and ionic conducting ceramics. Ceramic matrix composites such as Al_2O_3-ZrO_2 and Al_2O_3-BN.
Cathode	NiO doped with ~3 at % Li, to provide the required electronic conductivity and electrocatalytic activity. Problems, such as cell shorting during prolonged operation of the fuel cell, have recently been reported. The search is on for alternative cathode materials or for other cell designs that will extend the lifetimes of NiO cathodes. Electronic conducting ceramics, such as $LiFeO_2$ or $LiMnO_2$ doped with suitable cations.	The perovskite strontia-doped lanthanum manganite of composition $La_{1-x}Sr_x MnO_{3+x}$, where $0.10 \leq x \leq 0.15$. An example is Sr-doped $LaMnO_3$. Electronic and ionic conducting ceramics. Ceramic matrix composites, such as Al_2O_3-ZrO_2 and Al_2O_3-BN.

Component	Molten Carbonate Fuel Cell (MCFC)	Solid Oxide Fuel Cell (SOFC)
Electrolyte	Usually a mixture of alkali carbonates supported in a porous, particulate matrix of $LiAlO_2$ by capillary action. Generally, this mixture is $LiCO_3$-K_2CO_3, Li_2CO_3-Na_2CO_3, or ternary Li_2CO_3-Na_2CO_3-K_2CO_3 compositions. $LiCO_3$-K_2CO_3/$LiAlO_2$ composites.	ZrO_2 doped with yttria or calcia, to stabilize the cubic crystal structure of the former and to create vacancy defects for enhancement of ionic transport. A typical composition is ZrO_2 with about 17 wt % Y_2O_3.
Interconnect	Type AISI 316 austenitic stainless steel. Because stainless steels experience more severe hot corrosion attack in the presence of molten carbonate films and Cu and Ni are stable against this electrochemical oxidation, these latter materials are now being used. Ni or Cu clad stainless steel composites.	Lanthanum chromite ($LaCrO_3$) acceptor doped with Mg or Sr.

QUESTION 36
Devices for Measuring Static Magnetic Fields

Two basic approaches exist for the measurement of static magnetic fields with conventional single-mode silica and doped silica optical fibers: the Faraday rotation approach and the magnetostrictive jacket or stretcher approach.

Write a brief essay on use of each of these approaches for the stated application.

The Faraday Effect

The Faraday effect is a well-known optical phenomenon. If an external magnetic field H is applied to a suitable medium such that there is a field component parallel to the direction of the linearly polarized light in the medium, it is found that the direction of polarization of the emergent light is rotated through an angle θ such that

$$\theta = V' \int_L H \, dl$$

where L is the path traversed by the light in the medium and V' is a property of the medium called the Verdet constant.

An external magnetic field H is applied longitudinally to a fiber to produce a rotation in the direction of linear polarization. Smith [1] demonstrated that this technique can be used to measure high electric currents (up to 1 kA) without the danger of short-circuiting cables. However, the Verdet constant of most doped silica fiber materials is sufficiently small ($\sim 1.5 \times 10^{-2}$ min A^{-1}) that only large currents and magnetic fields can be detected. Yariv and Winsor [2] suggested incorporation of paramagnetic ions into optical fiber materials to raise the Verdet constant to much higher levels. Rare-earth ions can also be incorporated into the silica, but constraints on solubility of these ions in the glass and the optical absorption introduced by these ions limit the enhancement that can be achieved in practice.

Magnetostriction

This straightforward method is capable of much greater sensitivity in magnetic field measurement. The fundamental principle of sensor operation is associated with the measurement of the longitudinal strain produced in an optical fiber bonded to or jacketed by a magnetostrictive material or using a magnetostrictive toroid (Fig. Q36.1).

Magnetostriction is the change in dimension of a ferromagnetic material when it is placed along the axis of an applied magnetic field. In practice, a magnetostrictive jacket on single-mode optical fibers is incorporated into a fiber interferometer (Fig. Q36.2). Since about 1980, the main focus of research has been the identification of materials with good magnetostrictive properties* and the incorporation of such materials into fiber sensors by appropriate bonding or coating methods.

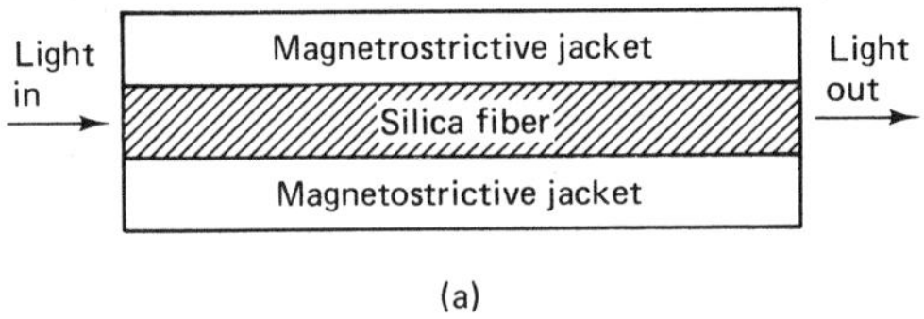

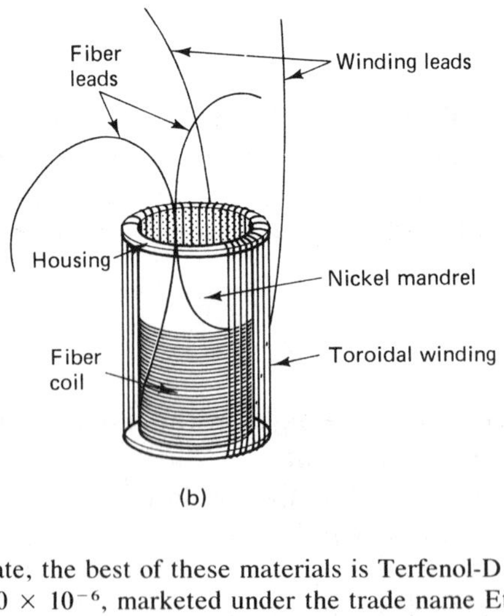

Figure Q36.1 (a) Basic geometry of a silica fiber bonded to a magnetostrictive jacket; (b) a magnetostrictive nickel jacket.

*To date, the best of these materials is Terfenol-D ($Tb_{0.3}Dy_{0.7}Fe_{1.93}$), with a magnetostriction of 1500 to 200 $\times 10^{-6}$, marketed under the trade name Etrema (compare Terfenol with materials in Table Q36.2).

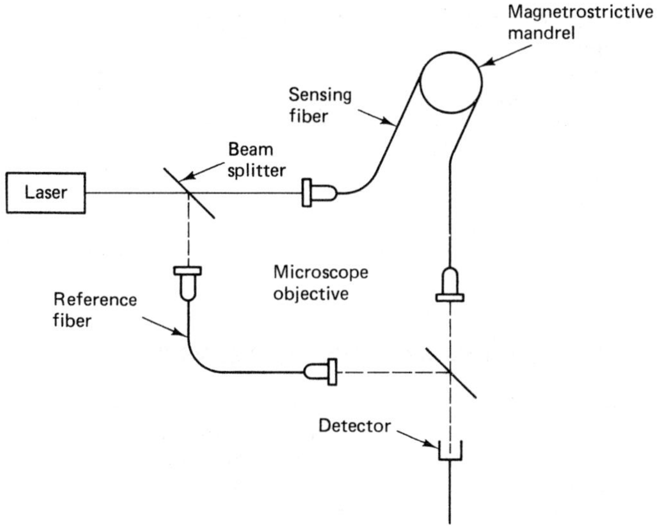

Figure Q36.2 Interferometer employed for magnetic measurements.

Magnetostrictive materials of interest are either selected crystalline metals or metallic glasses, such as alloys 2605SC and 2605CO (Allied Chemical Corp., Morristown, N.J.). The magnetostrictive coefficient λ_s is given as a fractional dimensional change Δl divided by the original dimension l. The change in optical path length generated by this dimensional change is detected by an all-fiber interferometer capable of detecting phase shifts as small as 10^{-6} rad. Data on a typical interferometer response of a sensor are presented in Table Q36.1.

Some typical magnetostrictive constants for some crystalline materials are given in Table Q36.2.

For iron-nickel alloys, longitudinal magnetostriction can either be positive (expanding dimension) or negative (contracting dimension), with the 80 wt % Ni–20 wt % Fe alloy yielding a practically zero dimension change, the 60 wt % Ni–40 wt % Fe alloy exhibiting a very high positive magnetostrictive effect, and pure nickel showing large contraction. Based on considerations of fabrication, corrosion resistance, and cost, pure nickel is a convenient material for use in fiber magnetostriction devices. However, its properties are greatly affected by the presence of impurities, processing, and final heat treatment. Iron-cobalt alloys

TABLE Q36.1 THE DEPENDENCE OF THE OUTPUT OF A DUAL-ARM INTERFEROMETER CONTAINING A METALLIC GLASS SENSOR ON THE APPLIED MAGNETIC FIELD (BIAS FIELD = 0; T = 296 K)

A.C. Magnetic Field (T)	Interferometer Output (mV)
10^{-8}	0.15
10^{-7}	1.00
10^{-6}	10.00
10^{-5}	100.00

Source: T. G. Giallorenzi et al., "Optical sensor technology," *IEEEJ. Quant. Electronics QE-18* (4), p. 637, 1982.

TABLE Q36.2 MAGNETOSTRICTIVE CONSTANTS FOR SOME CRYSTALLINE MATERIALS

Material	$\lambda_{100}(\times 10^6)$	$\lambda_{111}(\times 10^6)$	$\lambda_s{}^a(\times 10^6)$
Nickel	−46	−24	−33
60 wt % Ni–40 wt % Fe	27	22	24
80 wt % Ni–20 wt % Fe	9	0	4
$Co_{0.8}Fe_{2.2}O_4$	−590	120	−164

[a] $\lambda_s = (2\lambda_{100} + 3\lambda_{111})/5$

TABLE Q36.3 SENSITIVITIES OF MAGNETIC FIELD SENSORS AT 1 kHz

Type Of Sensor	Minimum Detectable Magnetic Field ($T\ m^{-1}$)
Fiber optic	
Evaporated Ni film (1.5 μm thick)	1.0×10^{-8}
Bulk Ni jacket	8.0×10^{-11}
2605CO metallic glass strip	2.3×10^{-12}
SQUID under laboratory conditions	10^{-5} to 10^{-16}

also offer large values of magnetostriction, with the 60–70 wt % Co alloy providing the maximum response for most applied magnetic fields.

The development of a compact, high-sensitivity, room-temperature magnetic field sensor would represent a significant advance. The most sensitive devices currently available for the measurement of small magnetic fields are superconducting quantum interference devices (SQUIDs), which have extremely high sensitivity under laboratory conditions (Table Q36.3) but often are substantially less sensitive under actual field conditions. These devices are limited in their utility by the fact that they require cryogenic instrumentation for their operation.*

Applications of magnetic field sensors include magnetic antennas for detection of electromagnetic signals and a magnetooptic compass to provide magnetic heading.

REFERENCES

1. A. M. Smith, "Polarization and magnetooptic properties of single mode optical fiber," *Applied Optics 17,* pp. 52–56, 1978.
2. A. Yariv and H. Winsor, "Proposal for the detection of magnetostrictive perturbation of optical fibers," *Optics Letters 5* (3), pp. 87–89, 1980.

*At least until the new high-temperature superconducting materials become usable.

QUESTION 37
Potting Material for an Electrical or Electronic Component

Discuss the thermomechanical criteria for the selection of a potting material for an electrical or electronic component that is installed outdoors.

Potting refers to a polymeric material that is cast around an electronic or electrical component, such as solar cells. It is needed primarily to protect the component from environmental hazards, such as solar radiation (ultraviolet, infrared), pollution, dirt, salt spray, rain, snow, hail, lightning, wind, temperature cycling, animal and insect attack, humidity cycling, and vandalism. It is also used to provide support from vibration and to serve other functions, such as being a dielectric or a thermal conduction medium.

The main design requirement is for the pottant to allow the component to withstand thermal cycling. Thermal cycling can lead to cracking of the pottant or failure of the component due to the high stresses generated in it. Such stresses are mainly due to a mismatch in the coefficients of thermal expansion of the pottant and the component, and their magnitude are highly dependent on the geometry of the potted assembly.

There are two ways of applying the pottant: If it surrounds the component, as in Fig. Q37.1(a), it is called external or substrate-bonded potting; if the component surrounds the pottant, as in Fig. Q37.1(b), it is called internal or superstrate-bonded potting.

Figure Q37.1 Potting geometries: (a) external or substrate-bonded potting; (b) internal or superstrate-bonded potting.

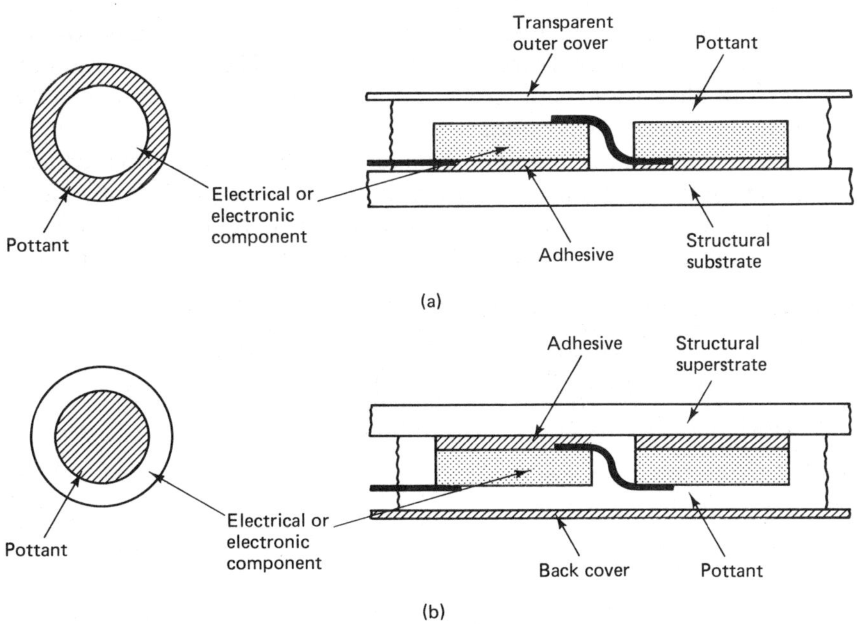

In external potting, the pottant is not restrained in all three dimensions, and the stresses developed, due to mismatch in the coefficient of linear expansion α, are small and are given by

$$\Delta\sigma = CE_p\,\Delta T(\alpha_p - \alpha_c)$$

where C is a constant, E_p is the modulus of elasticity of the pottant, ΔT is the temperature change from the stress-free temperature state, and p and c refer to the pottant and the component, respectively. Normally $\alpha_c \ll \alpha_p$, whence

$$\Delta\sigma = CE_p\alpha_p\,\Delta T$$

Thus for a given temperature change, the induced stresses are directly proportional to the material property product $E_p\alpha_p$. Note, however, that E_p and α_p are themselves functions of temperature.

In internal potting, changing the temperature by ΔT from the stress-free level leads to equal and isotropic dilational strains (given by $\alpha_p\,\Delta T$) that, if not relieved or accommodated by the rigid components, result in an internal pressure (a hydrostatic stress) in the pottant equal to $3K_p(\alpha_p - \alpha_c)\,\Delta T$, where K_p is the bulk modulus of the pottant. Recalling the relationship between K and E and noting that, in general, $\alpha_c \ll \alpha_p$ leads to the value of the generated stress being given by

$$\Delta\sigma' = \alpha_p\,\Delta T\frac{E_p}{1 - \nu_p}$$

where ν_p is the Poisson's ratio of the pottant.

If the temperature is decreased below the stress-free temperature level, the pottant is under a tensile hydrostatic stress and the component is in a compressive state of stress. When the temperature of the potted assembly is increased above the stress-free level, the reverse occurs.

One other consideration in the choice of materials for the pottant is its resistance to thermal-stress-induced failure. The appropriate material properties are the ultimate tensile strength (UTS), the modulus of elasticity, and the coefficient of linear expansion.

To summarize, candidate materials are evaluated on the basis of two indices of performance: (a) $E_p\alpha_p$* (a low value of this index is desirable) and (b) $UTS_p/E_p\alpha_p$ (a high value of this index is desirable). The values of these indices for various materials are given in Table Q37.1, which shows that the best material is GE RTV

TABLE Q37.1 PROPERTIES OF THREE CANDIDATE POTTANT MATERIALS

Material	Modulus Of Elasticity, E (kPa)	Ultimate Tensile Strength, UTS (kPa)	Coefficient Of Linear Expansion, α (10^{-6} °C^{-1})	$E\alpha$ (Pa °C^{-1})	UTS/$E\alpha$ (°C)
GE RTV[a] 655	690	241	200	138	1746
Uralane LV 5753	2,070	1,379	130	269	5124
Scotchcast 281	448,175	11,377	55	24,650	462

[a]Room-temperature vulcanizable

*Because ν_p has an almost constant value of 0.25 to 0.33 for most pottant materials, $\Delta\sigma'$ is also, approximately, dependent on the material property product $E_p\alpha_p$.

655 silicone resin on the basis of criterion (a) and Uralane LV polyurethane according to criterion (b). Either material would be an excellent "compromise" choice.

QUESTION 38

Connecting Rod in a Four-Stroke Spark-Ignition Internal Combustion Engine

The idea of using filamentary composite materials to manufacture the moving parts of an automobile engine has long been attractive to automotive engine designers.

Discuss the choice of these materials for one such part, the connecting rod in a four-stroke spark-ignition internal combustion engine.

This component (Fig. Q38.1) is subjected to particularly demanding operating conditions (Table Q38.1) and has therefore proved to be the most difficult engine component to design successfully in filamentary composite materials.

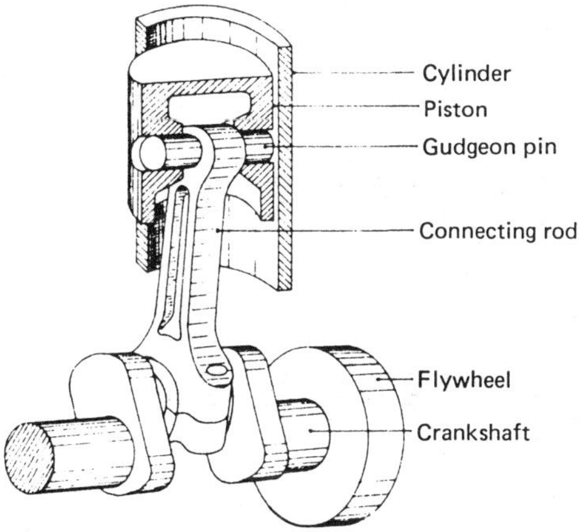

Figure Q38.1 Pictorial view of some parts of an internal combustion engine.

TABLE Q38.1 OPERATIONAL PARAMETERS OF A CONNECTING ROD OF A FOUR-STROKE SPARK-IGNITION INTERNAL COMBUSTION ENGINE (Length = 100 mm)

Operational Parameter	Value
Maximum load	40 kN (compressive), 25 kN (tensile)
Maximum extension	160 μm (compressive), 100 μm (tensile)
Maximum distortion of big- and small-end eyes	10 μm
Minimum durability	3000 h
Minimum number of load cycles	3×10^8
Minimum temperature range	−30°C to 180°C (small-end eye), −30°C to 140°C (big-end eye)

TABLE Q38.2 CANDIDATE MATERIALS FOR A CONNECTING ROD OF A FOUR-STROKE SPARK-IGNITION INTERNAL COMBUSTION ENGINE

Material	Tensile Modulus, E (GPa)	Maximum Compressive Strength, σ_c (MPa)	Density, ρ (kg m^{-3})	Maximum Service Temperature (°C)	Unit Cost (\$ t^{-1})	Specific Stiffness, E/ρ (10^6 m)	Specific Compressive Strength, σ_c/ρ (10^3 m)
Steel	210	1000	7890	800	880	2.71	12.92
Aluminum alloy	71	280	2700	350	1,200	2.68	10.57
Silicon carbide ceramic	310	490	3200	1000	5,700	9.88	15.61
S glass fiber reinforced polymer laminate[a]	56	540	1997	220	5,000	2.86	27.56
Carbon fiber reinforced polymer laminate[a]	138	600	1661	220	130,000	8.47	36.82

[a]Fiber volume fraction = 0.60.

These service requirements, together with the need to withstand hot engine fluids (engine oil, engine oil additives, etc.), place severe restrictions on the types of materials that can be used for connecting rods. Some of these materials are compared in Table Q38.2, which shows that only steel, the ceramic, and the continuous carbon fiber reinforced polymer (CFRP) are suitable. Kevlar 49 fiber reinforced polymer is not considered because of its poor compressive strength and creep properties.

A typical CFRP connecting rod is shown in Fig. Q38.2. The rod components are divided into a compression part and a tension part mainly to allow the carbon fibers to be placed in the direction of the greatest load. Compression molding is the most suitable manufacturing process for the big- and small-end eyes, pultrusion is suitable for the part to bear compressive loading, and filament winding is best for the parts to be in tension. All fibers are to run axially. The force acting on the piston, due to combustion chamber pressure, is transmitted through the gudgeon pin, small end, and compression part to the crankshaft via the big-end assembly. The tensile forces, due to inertia loading, are transmitted to the tensile part of the rod through the loops of the wound fiber at the small end of the rod.

The problem of obtaining the correct fiber orientation in the eyes has yet to be solved. In one of the first CFRP connecting rods (designed and built by engineers at the Institute of Plastics Research in Aachen, West Germany, and tested for Volkswagen as part of the Composites for Highly Stressed Components in the Internal Combustion Engine research project, which is promoted by the West German Ministry for Research and Technology), the big- and small-end eyes are made of steel.

Figure Q38.2 Composite-material connecting rod for an automobile.

QUESTION 39
Atmospheric-Pressure Fluidized-Bed Combustor

(a) *Write a short account of the two major material issues in an atmospheric-pressure fluidized-bed combustor.*

(b) *State other material-related issues of concern in such a combustor.*

Since 1970, there have been concerted efforts on the part of industry and government in the United States to develop fluidized-bed combustion as an alternative to conventional coal combustion. The impetus for this interest came from two sources: (1) the passage of the Clean Air Act of 1970 and the resulting

TABLE Q39.1 SUMMARY OF THE DESIGN SPECIFICATIONS FOR THE MAJOR U.S. AFBC FACILITIES (EARLY 1988)

Site (Start-up Date)	Bed Dimensions	Combustion Conditions[a]	Heat Transfer/ Working Fluid
TVA Shawnee Steam Plant near Paducah, Ky. (mid-1982)	Main bed, 3.8 m × 5.5 m (4 compartments); start-up bed, 3.6 m × 1.9 m	Temp. = 840°C SV = 1.2–3.6 ms^{-1} Fuel: Bituminous coal	Closed cycle/steam Rate = 51 t h^{-1}
Babcock & Wilcox EPRI, B&W Alliance Research Center (April 1978)	1.8 m × 1.8 m total (four compartments)	Temp. = 790°C–900°C SV = 1.2–4.3 m s^{-1} Fuel: Bituminous coal	Closed cycle/steam Rate = 0.9–4.5 t h^{-1}
Great Lakes Naval Station, near Chicago (December 1981)	2.4 m × 5.3 m total (six compartments)	Temp. = 800°–815°C SV = 2.1 m s^{-1} Fuel: Bituminous coal	Closed cycle/steam Rate = 22.7 t h^{-1}
Georgetown University, Washington, D.C. (July 1979)	1.7 m × 5.9 m total (two compartments)	Temp. = 840°–865°C SV = 1.8–2.4 m s^{-1} Fuel: Bituminous coal	Closed cycle/steam Rate = 45.4 t h^{-1}
East Stroudsburg State College, Pa. (September 1984)	6.7 m × 1.6 m total (six compartments)	Temp. = 870°C SV = 1.8–2.4 m s^{-1} Fuel: Anthracite culm	Closed cycle/steam Rate = 18.2 t h^{-1}
Wilkes-Barre, Pa. (February 1984)	5.5 m × 4.3 m total (one compartment)	Temp. = 820°C SV = 1.2–1.8 m s^{-1} Fuel: Anthracite culm	Closed cycle/steam Rate = 27 t h^{-1}

[a]Temp., temperature; SV, supercritical velocity.

restrictions on SO_2 emissions, making the in situ retention of sulfur in the bed a very attractive feature, and (2) the oil crises of 1973–74 and 1980–81.

Of the three concepts now being studied—atmospheric fluidized-bed combustion (AFBC), entrained flow or circulating fluidized-bed combustion (CFBC), and pressurized fluidized-bed combustion (PFBC)—AFBC is the most developed and is favored by many boiler manufacturers. Table Q39.1 contains a summary of the major coal-fired atmospheric fluidized-bed combustion projects in the United States.

An atmospheric-pressure fluidized-bed combustor (APFBC) (Fig. Q39.1) has a bed that is usually fairly shallow (typical expanded bed height is 1 m), because of the efficiency penalties associated with the pressure drop across the bed.

(a) The two major materials issues are as follows:

1. *Corrosion of the in-bed heat exchanger.* The corrosion of the heat exchanger in the gas pass is dependent on the form and quantity of the alkali salts in the gas stream. The alkali release is less in an APFBC train than in a conventional pulverized-coal power plant (PCPP), the ash is solid, and the gas temperature is considerably lower than in a PCPP boiler. Thus corrosion should be much less of a problem than in a PCPP boiler.

This corrosion in an APFBC is characterized by the following features:

- At metal temperatures in excess of about 600°C, heat exchanger alloys suffer from internal sulfidation, a phenomenon that eventually leads to a destruction of the protective surface layer (oxide). This enhances oxidation.

Figure Q39.1 Schematic representation of an atmospheric-pressure fluidized-bed combustor.

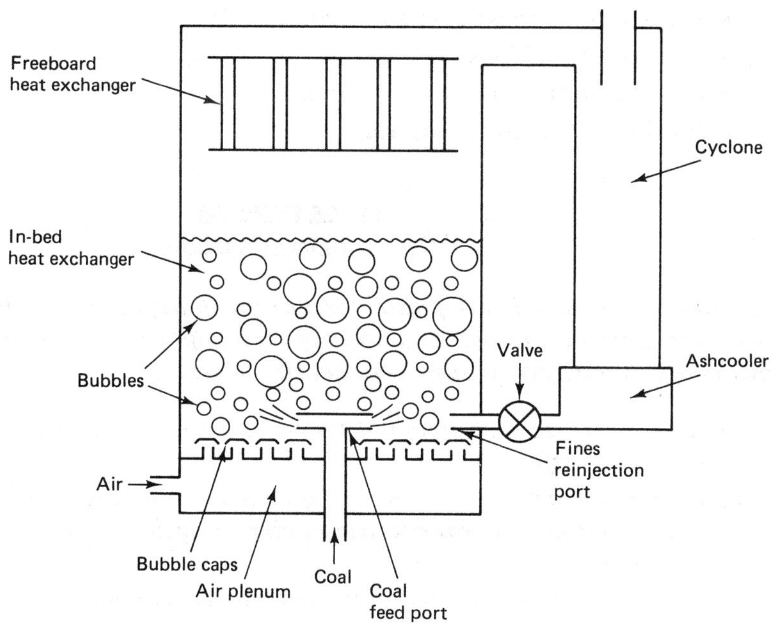

- The extent of the corrosion appears to vary with alloy position in the bed and also from bed to bed, but all beds suffer from sulfidation or oxidation attack.
- Sulfidation and oxidation are generally absent if there is no acceptor and if there is no significant amount of calcium in the coal ash itself.
- Corrosion is worse in sections adjacent to a coal feed port.
- The use of high-chlorine coals results in further acceleration of attack of some alloys and encourages sulfidation of low-alloy steels at lower temperatures.

The best material to resist in-bed corrosion is type 347 or type 304 stainless steel at normal tempratures or type 347 stainless steel at very high temperatures.

2. *Erosion of the in-bed heat exchanger.* Mechanistically, this erosion has been found to be accounted for by any of the following factors:

- In-bed particles having high kinetic energies but low velocities
- Wear by particles "plied" onto the metal surface by a stack of other particles
- Relatively fast-moving particles in the wake of bubbles
- Particles impacting onto the metal surface as a result of bubble collapse
- Fast-moving particles thrown up by the bubble collapsing at the bed surface
- Various flow patterns in the bed, such as "gulf streams"

The best method of containing in-bed erosion is by attaching plates made of type 310 stainless steel to the eroding surfaces using straps.

(b) Other material-related issues are these:

- Corrosion of freeboard heat exchanger
- Erosion of freeboard heat exchanger
- Erosion of cyclones
- Erosion of fines transport equipment
- Erosion of the hot solids valve

QUESTION 40
Thermoelectric Cooler

Thermoelectric coolers are particularly useful in applications where precise temperature control is desired. Examples of such applications are infrared detectors, avionics systems, gyroscopes, and accelerometers.

(a) *Give a concise description of the most common manifestations of the thermoelectric effect.*

(b) *Present a simple derivation of a figure of merit for the evaluation of materials to be used in a thermoelectric cooler, in order to achieve maximum heat pumping.*

(c) *Comment on the suitability of some named materials for this cooler.*

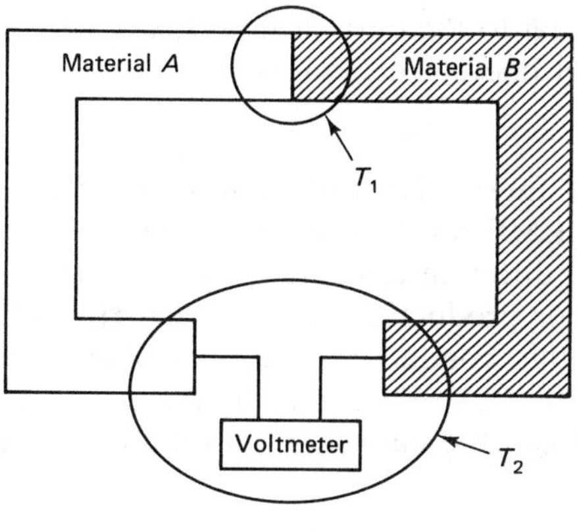

Figure Q40.1 The Seebeck effect (the voltmeter reads the Seebeck voltage).

(a) When two dissimilar materials are in contact, a potential difference is set up between them. If a second junction is made by joining their opposite ends, thereby establishing a closed circuit, the contact potential difference at the second junction is equal and opposite to that at the first junction, provided the two junctions are at the same temperature. This means that the net electromotive force (EMF) in the circuit is zero. If the junctions are at different temperatures, the two contact potential differences are different, and an EMF will exist, producing a current in this circuit. This EMF is called the thermoelectric EMF or voltage (Fig. Q40.1), and the effect is called the Seebeck effect. The thermoelectric voltage V_s is directly proportional to the temperature difference between the two junctions ΔT, with the constant of proportionality α being the Seebeck coefficient; that is,

$$V_s = \pm\alpha\,\Delta T$$

The converse of the Seebeck effect is the Peltier effect, in which current flowing through a junction between dissimilar materials causes either the absorption or production of heat, depending on the temperatures of the individual junctions (heat flows into the junction at the lower temperature); see Fig. Q40.2. The Peltier

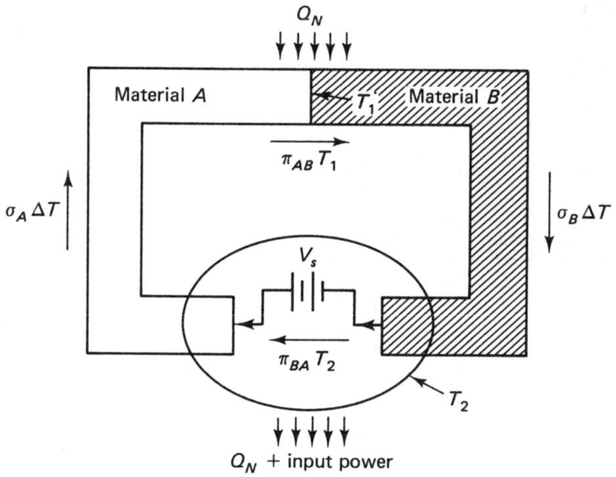

Figure Q40.2 A thermoelectric device showing the Peltier and Thomson voltage.

heat effect Q_p is proportional to the current flow I, with the constant of proportionality π being the Peltier coefficient; that is,

$$Q_p = \pm \pi I$$

Thomson, as a result of a thermodynamic analysis of the Seebeck and Peltier effects, predicted that an EMF must exist between different parts of the same material if they are held at different temperatures. Thus the Thomson effect has to do with the absorption or production of heat occurring when a current flows along a single, homogeneous conductor across or along which a temperature gradient exists (Fig. Q40.2). This effect may be put in the form of an equation that links heat transferred Q_T, the current I, and the temperature gradient, ΔT, thus

$$Q_T = \sigma I \, \Delta T$$

where σ is called the Thomson coefficient.

(b) The thermoelectric cooler (Fig. Q40.3) consists of materials A and B, forming two junctions, the cold one at temperature T_c and the hot one at temperature T_h. The Seebeck coefficient, electrical resistivity, and thermal conductivity of each material are designated α, ρ, and k, respectively. The net heat absorbed at the cold junction is equal to the Peltier heat less one-half of the usual Joule heat less the heat absorbed due to leakage from the hot to the cold junctions; that is,

$$Q_N = \alpha I T_c - 0.5 I^2 R - K \, \Delta T \qquad \text{(Q40.1)*}$$

where R is the total resistance of the couple† (i.e., the pair of junctions) and is given by

$$R = \frac{\rho_A L_A}{A_A} + \frac{\rho_B L_B}{A_B} \qquad \text{(Q40.2)}$$

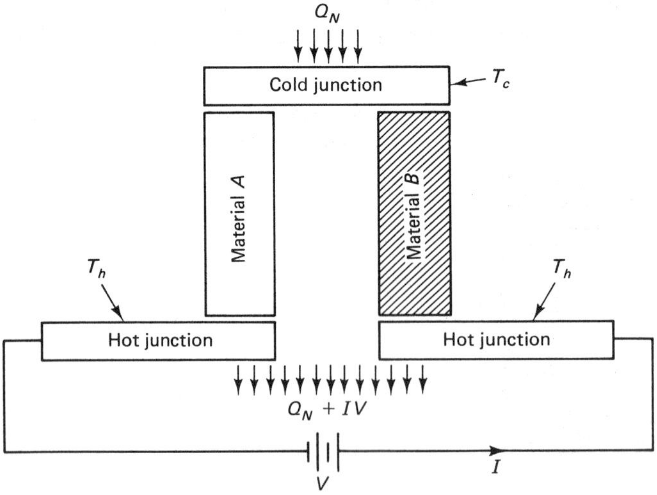

Figure Q40.3 A thermoelectric cooler.

*It can be shown that $\rho = \alpha T$.

†Junction resistance is zero.

K is the overall heat conductance of the couple and is given by

$$K = \frac{k_A A_A}{L_A} + \frac{k_B A_B}{L_B} \qquad \text{(Q40.3)}$$

The Seebeck coefficient of the couple α is given by

$$\alpha = |\alpha_A| + |\alpha_B|$$

The current that produces the maximum heat pumping capacity I_m is obtained by differentiating equation (Q40.1) to yield $I_m = \alpha T_c/R$. When I_m is inserted in equation (Q40.1), the maximum net heat absorbed at the cold junction Q_{NM} is given by $Q_{NM} = (\alpha^2 T_c^2/2R) - K/\Delta T$. For optimum performance of the cooler, Q_{NM} should be zero; thus the maximum temperature gradient obtainable ΔT_m has a magnitude of $\alpha^2 T_c^2/2\text{KR}$. This means that to maximize ΔT_m, the couple characteristic KR must be as low as possible. In other words,

$$\frac{d(\text{KR})}{d(A_A/A_B)} = 0 \qquad \text{(Q40.4)}$$

From equations (Q40.2) and (Q40.3) and putting $L_A = L_B$ (a usual design reality),

$$\text{KR} = k_A\rho_A + \frac{A_B}{A_A} k_B\rho_A + \frac{A_A}{A_B} k_A\rho_B + k_B\rho_B \qquad \text{(Q40.5)}$$

From equations (Q40.4) and (Q40.5),

$$\frac{A_A}{A_B} = \left(\frac{\rho_A k_B}{\rho_B k_A}\right)^{0.5} \qquad \text{(Q40.6)}$$

Putting equation (Q40.6) in equation (Q40.5) yields the optimum value of KR, KR_o; that is,

$$\text{KR}_o = \beta = k_A\rho_A + 2\sqrt{k_A\rho_A k_B\rho_B} + k_B\rho_B \qquad \text{(Q40.7)}$$

Thus the maximum temperature gradient obtainable ΔT_m becomes

$$\Delta T_m = \frac{\alpha^2 T_c^2}{2\beta}$$

This means that the relevant figure of merit (FOM) of the cooler materials is given by

$$\text{FOM} = \frac{\alpha^2}{\beta}$$

(c) The relevant properties of the materials in two candidate couple systems and their FOM values are given in Table Q40.1. It is clear that semiconducting materials are far superior to metals for thermoelectric applications.

TABLE Q40.1 MATERIALS AND FIGURES OF MERIT FOR A THERMOELECTRIC COOLER DESIGNED TO MAXIMIZE HEAT PUMPING

	Couple 1		Couple 2	
Property	Material *A* (*n*-type bismuth telluride)	Material *B* (*p*-type bismuth telluride)	Material *A* (tungsten)	Material *B* (copper)
Seebeck coefficient (V °C^{-1})	2.3×10^{-4}	-2.1×10^{-4}	5×10^{-6}	3.98×10^{-6}
Electrical resistivity (Ω·m)	10^{-5}	10^{-5}	0.55×10^{-8}	1.7×10^{-8}
Thermal conductivity (W m^{-1} °C^{-1})	1.70	1.45	167	390
FOM	308×10^{-5}		0.65×10^{-5}	

QUESTION 41
Optical Components of a Thermal Imaging System

Discuss the main optical requirements for materials to be used in optical components of a thermal imaging system.

Thermal imaging systems rely on detection of radiation from warm bodies that have emission spectra peaked in the infrared (IR). Since the system operates in the earth's atmosphere, selective transmission through the air essentially limits the application of IR optics to three bands: 0.8–1.3 μm, 3–5 μm, and 8–13 μm. Thermal imaging is now used, for example, in advanced weapons and certain medical diagnostic and industrial processing systems. The key material requirements are the optical characteristics (refractive index, transmission, dispersion), the physical characteristics (solubility, melting point, thermal expansion), and the ease of fabrication.

Optical Requirements

1. *Transmission and reflectance.* The transmission is measured on plane parallel-cut and polished discs with an IR spectrophotometer. Taking into account the multiple reflections within the disc, the transmission value T is given by

$$T = \frac{2n}{n^2 + 1}$$

provided that the material is free from absorption. Here n is the refractive index of the material. At each material-to-air boundary, a certain amount of the incident beam is reflected. At normal incidence, this reflected component or reflectance R can be obtained from the equation

$$R = \left(\frac{n - 1}{n + 1}\right)^2$$

Because of the high refractive index of most IR materials, antireflection coatings have special importance. These coatings also improve the transmission through the system.*

2. *Dispersion.* Dispersion is defined as the change in refractive index with wavelength, that is, $\Delta n/\Delta\lambda$. It is an important optical property because it determines whether an IR material can be used for lens design as a "flint glass type" or as a "crown glass type." Using IR materials with different dispersion values, lenses can also be achromatized in the IR.

3. *Homogeneity and birefringence.* No matter what fabrication process is used (such as the Czochralski crystal growth process for single-crystal germanium or silicon, chemical vapor deposition of zinc sulfide and zinc selenide, or sintering of magnesium fluoride), the materials must have excellent homogeneity. This minimizes both refractive index gradients and changes in refractive index.

The materials used must also be free from birefringence. The measurement of birefringence is based on a determination of the constants of elliptically polarized light and can be measured with the aid of a suitable compensator in the orthoscopic beam of a polarizing microscope. The birefringence in optical materials is given in terms of the phase angle shift or lag between the ordinary and extraordinary emergent waves and is expressed in units of nanometers per unit material thickness. For example, the maximal birefringence in the marginal area of a 20-mm-thick monocrystalline germanium disc 136 mm in diameter measured in the spectral band 8–12 μm varies from 800 to 1100 nm per centimeter of material thickness.

Table Q41.1 contains the values of some of these optical characteristics for four candidate materials.

TABLE Q41.1 SOME OPTICAL PROPERTIES OF FOUR CANDIDATE MATERIALS FOR THE OPTICS OF A THERMAL IMAGING SYSTEM

	n		Dispersion (μm^{-1})	
Material	$\lambda = 4\ \mu m$	$\lambda = 10\ \mu m$	$\lambda = 4\ \mu m$	$\lambda = 10\ \mu m$
Germanium	—	4.00	0.0141	0.0008
Silicon	3.42	—	0.0054	—
Zinc selenide	—	2.40	0.0037	0.0057
Calcium fluoride	1.40	—	—	—

	T (%)		R (%)	
Material	$\lambda = 4\ \mu m$	$\lambda = 10\ \mu m$	$\lambda = 4\ \mu m$	$\lambda = 10\ \mu m$
Germanium	—	47	—	36
Silicon	54	—	30	—
Zinc selenide	—	71	—	17
Calcium fluoride	95	—	3	—

*Both T and R are functions of wavelength λ, since the refractive index is a function of λ.

QUESTION 42
Use of Silicon Nitride

An advertisement for GTE technical ceramics and metals in the March 1986 issue of the journal Materials Engineering *proclaimed: "The age of metals is over; think* Si_3N_4*."*

(a) *Give some reasons for the optimism, implicit in this advertisement, for the expanded use of silicon nitride,* Si_3N_4*.*
(b) *Mention some applications for which silicon nitride is particularly suitable.*

(a) Here are some reasons to be optimistic about Si_3N_4:

- Low density
- Good mechanical strength (greater than 700 MPa in bending at temperatures up to about 560°C)
- Excellent resistance to attack by most molten nonferrous metals or common chemicals
- Excellent thermal shock resistance

(b) Applications for Si_3N_4 include turbine blades, pistons, nozzles, unlubricated bearings, housings and gears, missile radomes, turbine shrouds, and electronic substrates.

QUESTION 43
Floor of a Short-Takeoff-and-Landing (STOL) Aircraft

Discuss the criteria for the choice of material for the floor of a short-takeoff-and-landing (STOL) aircraft, such as the de Haviland DHC-7. Suggest some candidate materials.

The primary objective is to provide a serviceable floor at minimum weight. The basic design requirements are as follows:

- Bending moment, with a minimum value of about 1200 Nm per meter width
- Shear load, with a minimum value of about 9000 N per meter width
- Bearing load, about 140 kg on a 19-mm-diameter ball, maximum indentation of about 1.5 mm
- Walkway and entrance panels required to withstand an ultimate load of about 4000 kg m^{-2}; underseat panels subjected to an ultimate load of about 1800 kg m^{-2}
- Point load to withstand about 70 kg on 10 MPa

The material used must have the following properties:

- High strength, to withstand all the loads just described

TABLE Q43.1 STRENGTH AND MODULUS VALUES OF SOME HYBRID FILAMENTARY POLYMER COMPOSITE MATERIALS

Material	Configuration[a] (°)	Tensile Strength (MPa)	Compressive Strength (MPa)	Modulus of elasticity[b] Tension (GPa)	Modulus of elasticity[b] Compression (GPa)
Kevlar/graphite/Kevlar	$0_2/\pm45_2/90$	496	176	48/28	39/21
Boron/graphite/graphite	$0_3/\pm45/90$	856	655	152/58	24/15
Graphite/graphite	$0_4/\pm45_3$	634	614	74/25	72/21

[a] All configurations are symmetrical about the last ply in the sequence. Subscripts denote the number of plies.
[b] First number is longitudinal, and second is transverse (line of action of load with respect to fibers).

- High resistance to corrosive media (all normal aircraft fluids, water, etc.)
- High wear resistance
- High stiffness, to limit deflection
- Excellent fire resistance, to meet the relevant U.S. government specifications [1]

Candidate materials include these:

- A sandwich consisting of a core (nylon paper honeycomb impregnated with phenolic resin) and prepregged Kevlar 49–epoxy facings
- Hybrid constructions of Kevlar 49–graphite facings
- A proprietary material, Fiberlam (Ciba-Geigy, Basel, Switzerland), which consists of unidirectional cross-plied glass-epoxy resin skins and a honeycomb core
- Any other suitable hybrid filamentary polymer composite material (relevant data for some examples appears in Table Q43.1)

REFERENCE

1. "Smoke emission from compartment interior materials," Code of Federal Regulations, FAR 25-853, Federal Register 40 FR 6506, p. 2439, 1975.

QUESTION 44
Laser-Damage-resistant Optical Coatings

There is a growing demand for damage-resistant optical coatings for high-energy laser systems. The process of testing and eventual selection of the most appropriate coating materials is both expensive and time-consuming. It would therefore be of great practical significance if the laser damage could be correlated to some "easy to measure" physical properties of materials.

(a) *What properties of materials may be correlated with their laser damage resistance?*

TABLE Q44.1 OPTICAL CHARACTERISTICS OF COATING MATERIALS FOR UV LASERS

Coating Material	Absorption Band Edge (nm)	Refractive Index at 248 nm	Extinction Coefficient at 248 nm	Laser Damage Threshold [1] ($J\ cm^{-2}$)
ThF_4	$\leqslant 200$	1.59	<0.001	25.00
MgO	200	1.83	0.002	4.00
ZrO_2	230	2.25	0.006	<1.00

(b) *Comment on the suitability of the following coating materials: thorium fluoride, ThF_4; magnesium oxide, MgO; and zirconium dioxide, ZrO_2.*

(a) Research [1] has shown that for single-layer coatings, laser damage resistance is correlated with the following properties of the coating material:

- Coating absorption
- Position of the UV absorption edge
- Refractive index

(b) For the given materials, the values of these relevant properties and the laser damage threshold* are given in Table Q44.1.

The best material is thorium fluoride, ThF_4, which has low values of refractive index, extinction coefficient, and absorption band edge, as well as a high laser damage threshold.

REFERENCE

1. F. Rainer, et al., "Damage thresholds of thin film materials and high reflectors at 248 nm," in *Laser Induced Damage in Optical Materials,* NBC Special Publication 669. Washington, D.C.: National Bureau of Standards, 1982.

QUESTION 45
Gun Barrel Liner

State the major desirable properties of a material to be used as a gun barrel liner. Comment on the suitability of ceramic gun barrel liners as alternatives to existing hard-metal liners.

The major desirable material properties are these:

- High melting point (>1400°C)
- High modulus of rupture (>350 MPa)

*This is the average of the highest fluence that did not cause damage and the lowest fluence that caused damage.

TABLE Q45.1 PROPERTIES OF FOUR CANDIDATE MATERIALS FOR GUN BARREL LINERS

Property	Sintered Silicon Nitride	Transformation-toughened Zirconia	Silicon Carbide	Gun Steel
Melting point (°C)	2800	2700	2800	1450
Modulus of rupture (MPa)	698	620	397	1172
Modulus of elasticity (GPa)	296	207	379	207
Hardness (kg mm^{-2})	2700	1100	2800	400
Density (kg m^{-3})	3200	5600	3200	8000
Coefficient of thermal expansion (10^{-6} °C^{-1})	3.5	10.5	4.0	15.0

- High modulus of elasticity (>190 GPa)
- High hardness
- Low density
- High wear resistance
- High coefficient of thermal expansion

In Table Q45.1, a number of candidate ceramic materials are compared to gun steel on the basis of this list of material properties. It appears that both sintered silicon nitride and transformation-toughened zirconia are promising. But before either ceramic material or any other is adopted for use as small-caliber gun barrel liners, prolonged testing of rifled bores is mandatory.

QUESTION 46
Use of Ceramics in Fusion Devices and Radio Frequency Heating Systems

In magnetically confined fusion devices, ceramics are required for insulators for lightly shielded magnetic coils and for windows for radio frequency (RF) heating systems.

For each application, briefly outline the main considerations relevant to material selection and suggest suitable materials to be used.

Insulation for Magnetic Coils

Fusion reactor design requires large nonsuperconducting magnets to act as choke coils for mirror machines, diverters, and plasma-shaping coils for tokamaks, etc.* Choke and diverter coils have to be proximate to the first wall and will thus be

*For details, see Chapter 28 in Part I.

TABLE Q46.1 RADIATION-INDUCED CONDUCTIVITY AS A FUNCTION OF THE RATE OF ABSORPTION OF IONIZING ENERGY FOR ALUMINA-BASED MATERIALS AT 300 K

	Conductivity (Ω^{-1} m^{-1})			
Ionizing Dose Rate (Gy s^{-1})	Al_2O_3	Al_2O_3 + 0.004 wt % Cr_2O_3	Al_2O_3 + 0.03 wt % Cr_2O_3	Al_2O_3 (Neutron-irradiated)
10	1×10^{-8}	3×10^{-9}	7×10^{-10}	1.5×10^{-11}
100	4×10^{-8}	3×10^{-8}	7×10^{-9}	1.5×10^{-10}
1000	6×10^{-7}	4×10^{-7}	5×10^{-8}	1.5×10^{-9}

Sources: R. W. Klaffky, "Special Purpose Materials" Annual Progress Report for 1979, No. DOE/ER 0048/1, Washington: U.S. Department of Energy, 1980, p. 29; R. W. Klaffky, et al., "Radiation-induced conductivity of Al_2O_3: experiment and theory," *Physical Review B21,* pp. 3610–3634, 1980.

exposed to lifetime neutron irradiation doses on the order of 10^{27} neutrons m^{-2} and ionizing fluxes of 10^4 Gy s^{-1}.

The main considerations are thus high mechanical strength and the requirement that electrical resistivity must not fall below a calculable (design-dependent) level without causing significant shunting of current to the copper conductor. Although radiation-induced conductivity increases dramatically with increasing rate of absorption of ionizing energy (Table Q46.1), the absolute value of the electrical conductivity is reduced either by the presence of chromium oxide as an impurity or as a result of structural irradiation damage. This behavior may be explained in terms of phenomena like creation of electron hole pairs, trapping and detrapping, scattering, and carrier recombination.

The best materials for this application are magnesium aluminate spinel ($MgAl_2O_4$), magnesium oxide (MgO), and aluminum oxide (Al_2O_3).

Windows for RF Heating Systems

Plasma heating is an integral part of most fusion RF reactor concepts. RF energy can be transmitted to the plasma, for example, by coaxial cables (at lower frequencies, e.g., 10^8 Hz) and various types of waveguides (at intermediate and high frequencies, e.g., 10^9 to 10^{10} Hz and 10^{11} to 10^{12} Hz, respectively). Waveguides will require windows to separate the environment of the plasma chamber from that of the RF source, which may include a pressurized dielectric gas such as SF_6. These windows will absorb energy from the beam and will be actively cooled to near-room temperature by a fluid passed between two laminar plates.

The main requirements in the window material are thus structural reliability (because thermal stresses will be high), low loss tangent (to minimize the absorption of microwave energy), high mechanical strength, and high thermal conductivity. All these properties must be retained as much as possible in the atmosphere of neutron irradiation.

The best materials for this application are beryllium oxide (BeO), $MgAl_2O_4$, and silicon nitride (Si_3N_4).

QUESTION 47

Lithium Containment Vessels in Thermonuclear Fusion Reactors

Materials that are resistant to liquid lithium corrosion are of interest for liquid lithium containment for applications such as controlled thermonuclear fusion reactors, advanced batteries, and advanced heat engines.

Comment on the suitability of three categories of candidate materials for such applications.

Pure Iron

Iron has a good resistance to liquid lithium corrosion up to about 1000°C but is prone to intergranular attack if carbon or nitrogen is present either in the iron or in the lithium. Iron carbide may be reduced to lithium carbide, which may lead to internal stress and hence cracking.

Stainless Steels

Stainless steels are more attractive than iron because of their higher strength, but their resistance to corrosion by liquid lithium is not so good. These steels are also prone to preferential dissolution of nitrogen and chromium, leading to a ferritic surface layer. The presence of carbon or nitrogen is also known to increase the rate of mass transfer of nickel or chromium and to give rise to severe grain-boundary attack. Carbon in lithium is known to result in carburization of chromium steels, while carbon in the steel itself results in chromium carbide formation leading to a chromium-depleted zone and making the steel susceptible to the leaching of nickel. Ferritic stainless steels have better resistance to liquid lithium corrosion than austenitic stainless steels but are still prone to the leaching of chromium. Low-carbon, low-chromium steels are superior to steels with higher contents of carbon and chromium.

Refractory Alloys

Refractory alloys have good resistance to lithium even at very high temperatures. Molybdenum and its alloys, for example, give excellent service at temperatures up to 1650°C, while zirconium is suitable up to about 800°C. Both niobium and tantalum give good service performance but are subject to attack by lithium when oxygen is present.

QUESTION 48

Military Armor

The use of ceramic military armor was accelerated by the need for lightweight high-performance armor to protect U.S. ground and air forces during the Vietnam War.

Briefly comment on the suitability of hot-pressed boron carbide, sintered

TABLE Q48.1 PROPERTIES OF CANDIDATE CERAMIC MATERIALS FOR LOW-DENSITY MILITARY ARMOR

Material	E (GPa)	H^a (kg mm^{-2})	ρ (kg m^{-3})	M (10^{15} Pa·m)
Hot-pressed boron carbide	400	3000	2500	480
Sintered alumina	340	1600	3800	143
Hot-pressed titanium diboride	570	3300	4500	418

[a]Knoop hardness.

alumina, or hot-pressed titanium diboride as a candidate material for this application.

No satisfactory theoretical relationship has yet been developed between ballistic performance and the mechanical properties of ceramic armor, but Stiglich [1] has presented an acceptable material figure of merit M given by

$$M = \frac{EH}{\rho}$$

where E, H, and ρ are the modulus of elasticity, hardness, and density of the ceramic armor material, respectively.

The three given candidate materials are compared on the basis of this figure of merit in Table Q48.1. Hot-pressed boron carbide is the best material, but its high raw material cost restricts its use to specialist applications, such as armoring for seats, control systems, and crew members in helicopters. Alumina has the disadvantage of high density but is relatively inexpensive and is widely used as ceramic armor for land-based personnel and vehicles. Hot-pressed titanium diboride has been developed to provide protection for vehicles against very heavy threats, such as armor-piercing antitank ammunition. Its relatively high density is compensated for by its very high modulus of elasticity and hardness.

REFERENCE

1. J. J. Stiglich, U.S. Army Report AD 666-766, 1968.

QUESTION 49
Encapsulant in a Photovoltaic Module

The objective of the encapsulant cover for a photovoltaic module is to protect the optically and electrically active elements from the degrading effects of the terrestial environment. The encapsulation system also serves other functions: structural integrity, electrical resistance to high voltage, and dissipation of thermal energy.

(a) *State the requirements of a cover material to meet the stated objectives for a photovoltaic module to be used as part of the heating and cooling system of a domestic residence.*
(b) *Briefly discuss the suitability of glass for this application.*

(a) The requirements for the material are as follows:

- High optical transmittance
- Good weathering characteristics
- High reliability
- Low cost, high annual production rate, and long lifetime (at least 20 y)
- Low linear expansion and thermal expansion
- Easy workability, as revealed in viscosity/temperature behavior
- High volume and surface resistivity
- High dielectric strength

The electrical properties (the final two) are important because building codes require safety measures in order to prevent electrical shocks to maintenance personnel.

(b) Glass satisfies all the criteria and is therefore the ideal material for this application. However, since glass is available in over 10,000 different types of forms, it is necessary to examine the characteristics of certain named glasses with respect to their suitability.

Experience with encapsulation systems in which glass constitutes one part of the system has shown the major features to be weatherability and design. Two broad categories of glass, soda-lime and borosilicate, have emerged as the leading candidate materials. Examples of the former type include AFG's Sunadex and Solarex and Fourco's Clearite, and examples of the latter type include Corning's 7070 or 7740 (Pyrex) and Schott's Tempax.

The type of glass needed for photovoltaic module applications has low distortion and low solar absorptance. Since iron is a known element that reduces optical transmittance, it is important that the chosen glass have a low ferrous ion content (~0.01% to 0.06%).

The values of all the relevant properties for these two types of glass, each 3 mm thick, are given in Table Q49.1. The effect of temperature on four properties for these glasses is shown in Figs. Q49.1 to Q49.5, while the effect of relative humidity on the surface resistivity of these glasses is given in Fig. Q49.6.

TABLE Q49.1 COMPARISON OF PROPERTIES OF TWO TYPES OF SOLAR CELL COVER GLASS

Material Property	AFG Sunadex Soda-Lime Glass	Corning 7740 Borosilicate Glass
Chemical composition (wt %)		
SiO_2	—	80.5
Na_2O	—	3.8
K_2O	—	0.4
Li_2O	—	—
CaO	—	—
MgO	—	—
Al_2O_3	—	2.2
Fe_2O_3	0.02	—
B_2O_3	—	12.9
ZnO	—	—
TiO_2	—	—
As_2O_3	—	—
Transmittance		
Visible, at 500 nm (%)	91	90
IR, at 1000 nm (%)	90	88
Refractive index	1.519	1.474
Density	2.5	2.23
Thermal conductivity at 273 K ($W\ m^{-1}\ k^{-1}$)	0.91	1.09
Expansion coefficient ($10^{-7}\ °C^{-1}$)	88	32.5
Modulus of elasticity (GPa)	—	63
Poisson's ratio	—	0.20
Annealing point (°C)	550	560
Softening point (°C)	780	820
Weatherability	May weather under some conditions	No weatherability problem anticipated

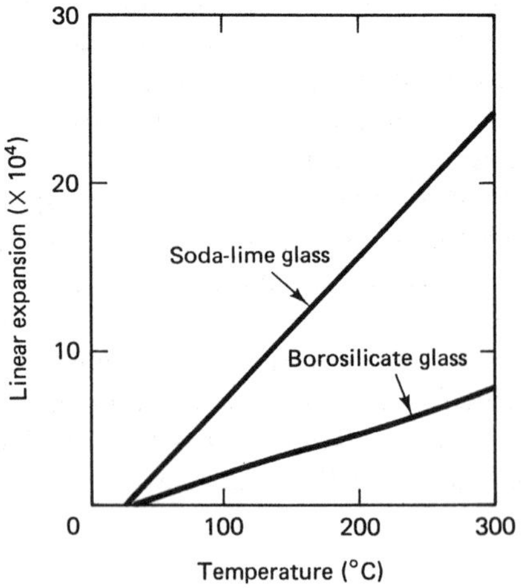

Figure Q49.1 Linear expansion of glasses [1].

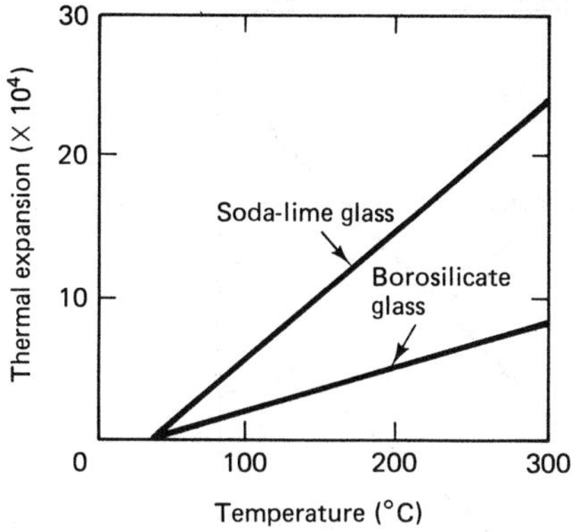

Figure Q49.2 Expansion temperature curves for typical glasses [2].

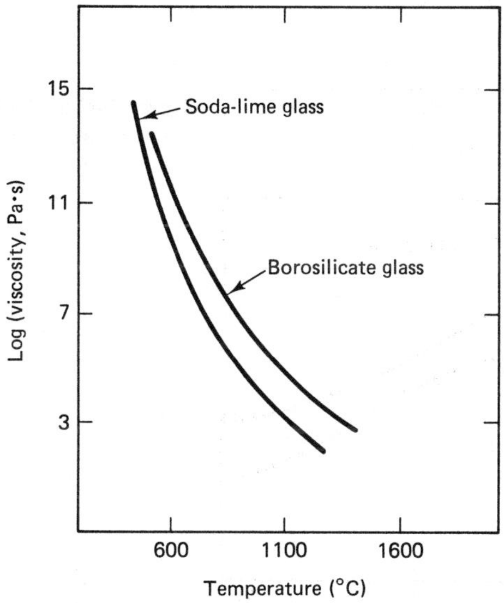

Figure Q49.3 Variation of viscosity with temperature for two glasses [2].

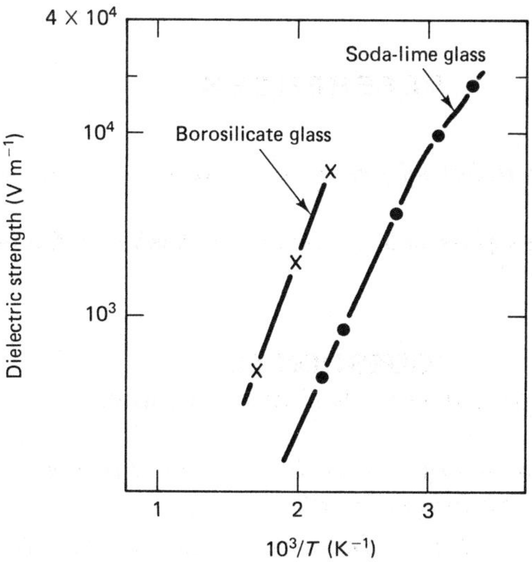

Figure Q49.4 Dielectric strength versus temperature for two glasses [2].

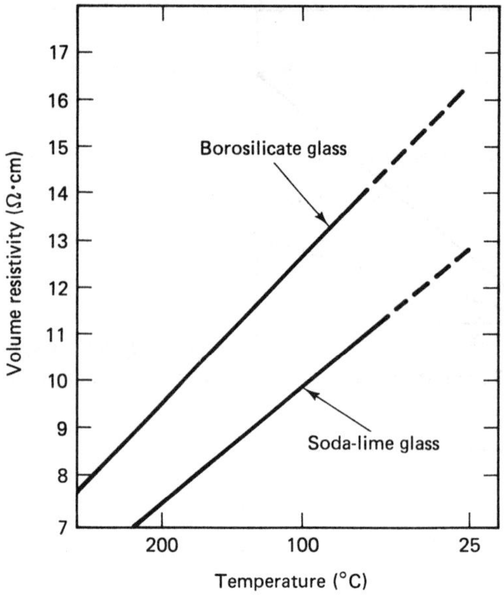

Figure Q49.5 Volume resistivity for two Corning glasses [2].

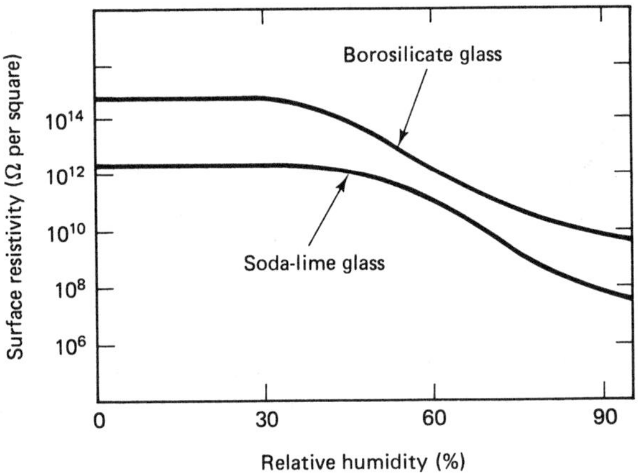

Figure Q49.6 Surface resistivity for two Corning glasses [2].

REFERENCES

1. C. J. Phillips, *Glass: Its Industrial Applications*. New York: Van Nostrand Reinhold, 1960.
2. "Properties of Corning's glass and glass-ceramics families," Corning Glass Works, Corning, New York, 1973.

QUESTION 50
Gear Case in a Military Helicopter

(a) *List the key requirements of a material to be used to construct a gear case for the engine of a turboprop aircraft.*

(b) *For four candidate materials, tabulate the values of the relevant properties.*

TABLE Q50.1 PROPERTIES OF FOUR CANDIDATE MATERIALS FOR THE GEAR CASE OF A TURBOPROP AIRCRAFT ENGINE

Material	ρ (kg m^{-3})	E (GPa)	UTS (MPa)	SS (MPa)
Magnesium	1827	45	234	117
Short boron fiber reinforced epoxy resin[a]	2132	69	258	155
Continuous boron fiber reinforced epoxy resin[a]	2132	190	1241	69
Aluminum	2700	71	105	75

Material	α (10^{-6} °C^{-1})	E/ρ (10^6 m)	UTS/ρ (10^3 m)	SS/ρ (10^3 m)
Magnesium	25.20	2.51	13.06	6.53
Short boron fiber reinforced epoxy resin[a]	3.20	3.30	12.33	7.41
Continuous boron fiber reinforced epoxy resin[a]	4.14	9.08	59.34	3.30
Aluminum	24.00	2.68	3.96	2.83

[a] Volume fraction of fibers = 0.33.

(a) The key requirements of a gear case material are as follows:

- High ultimate tensile strength, UTS
- Low density, ρ
- High tensile modulus, E
- High shear strength, SS
- High specific stiffness (= E/ρ)
- High specific tensile strength (= UTS/ρ)
- High specific shear strength (= SS/ρ)
- Good corrosion resistance
- Suitability for easy machining and molding

(b) The values of the relevant properties of four candidate materials for this application are given in Table Q50.1.

QUESTION 51
Antenna Support Structure of a Communications Satellite

Intelstat V, a latter-generation communications satellite, was placed into geosynchronous orbit in December 1980 to provide global communications for the International Telecommunications Satellite Organization. The three main components of the three-axis stabilized spacecraft are the antenna, communications, and support subsystems modules.

One of the elements of the antenna module is the antenna support structure, which provides the interfaces and support for all spacecraft parts located above the communications module. It consists of the tower truss, the reflector support, and the antenna feed support platform.

(a) *Outline the main service conditions that guide the selection of materials for the antenna support structure.*

(b) *On Intelstat V, a filamentary composite material was selected for this structure. Comment on this choice, and suggest some suitable candidates.*

(a) The main service conditions for the antenna support structure are as follows:

- Large temperature extremes ($-160°C$ to $100°C$)
- Large number of thermal cycles
- High doses of radiation

Thus the materials used for this structure must possess these primary properties:

- Low thermal expansion, on the order of $0.0 \pm 0.5 \times 10^{-6}\ °C^{-1}$
- High specific strength and modulus
- Little variation with temperature in the values the properties just cited; that is, excellent thermal stability
- Excellent resistance to attack by radiation

(b) In general, filamentary polymer composite materials fulfil the requirements very admirably. Some suitable filamentary composite materials are epoxy resin reinforced with fibers of ultrahigh-modulus graphite, high-modulus graphite, or intermediate-strength graphite.

QUESTION 52

Write-once and Erasable Compact Discs

The wave of compact discs (CDs) broke on U.S. shores in 1983. But even though market demand is expected to grow steeply, the opportunity for new players in the CD business, mainly plastic material suppliers, processers, and equipment manufacturers, is ebbing. Industry observers are unanimous in forecasting the next wave of opportunity to be write-once or erasable optical discs.

Comment on this statement, with respect to the present CD market, material requirements for the new erasable discs, and candidate materials.

At the beginning of 1987, North American CD capacity stood at about 55 million discs per year, but through planned capacity extension and production by new manufacturers who entered the market in 1988 and 1989, this capacity is expected to rise dramatically to about 260 million discs per year by 1990 (Table Q52.1).

TABLE Q52.1 PRODUCTION OF COMPACT DISCS IN NORTH AMERICA, 1986–1990

Company	Location	Production capacity (millions of discs per year)	
		Actual 1986	Planned by 1990
Laser Video	Anaheim, Calif.	2	5
	Huntsville, Ala.	—	60
Philips/Dupont Optical	King's Mountain, N.C.	25	60
DADC	Terre Haute, Ind.	25	50
Shape Optimedia	Kennebunk, Maine	—	20
Praxis Technologies Corp.	Ontario, Canada	—	30
Digital Disc Corp., Eastman Kodak, Pioneer, Nestor International Corp., Dennon Digital Industries, Inc., and Digital Equipment Corporation	—	—	30

High-purity polycarbonate (PC) resin, with minimal optical distortion, will continue to be the material of choice for CDs. Consumption of this material in the United States is expected to grow from about 1 million kg in 1986 to about 9 million kg by 1990. Some properties of one popular optical-grade polycarbonate resin are given in Table Q52.2.

Most CDs today are made of PC using injection compression, starve-feed injection, or conventional "straight" injection molding manufacturing methods. All these methods have very rigid processing conditions that include clean-room environments to assure quality control.* The outstanding audio and data storage properties of CDs explain their rapid acceptance. For example, one 121-mm-diameter audio CD can store 75 min of distortion-free music, and the scratchproof discs have an almost infinite life. CDs with "read-only memory" (ROM) formats have substantial data storage capacities. One CD-ROM, with a diameter of 133 mm, can store some 250,000 typewritten pages of text, or as much as the equivalent of 1500 floppy discs.

Thus CD-ROM distributed databases are on the increase. For example, there is now a CD version of the popular *Grolier Encyclopedia.* Both Engineering Information, Inc., and Digital Equipment Corporation offer a series of commercially published engineering and scientific databases, which are updated quarterly. Boeing Aircraft Co. has installed a digital optical storage and retrieval system to

*Innovative manufacturing processes will reduce the need for such stringent environmental conditions. Three examples of such processes are the Monoline cells on "independent profit centers," developed by Rokomay Holland and Technetronics, Inc., West Chester, Pa., a proprietary photo-thermographic stamping operation developed by Com Disc, Inc., Los Angeles, and an injection-molding technology based on compressive filling of the mold and articulated tooling, developed by Galic-Maus Ventures, Columbia Heights, Minn.

TABLE Q52.2 PROPERTIES OF AN OPTICAL-GRADE POLYCARBONATE RESIN[a]

Property	Value
Density (kg m^{-3})	1200
Ultimate tensile strength (MPa)	62
Ultimate tensile strain (%)	90
Izod impact strength[b] (J m^{-1})	587
Rockwell hardness, *M* scale	89
Light transmittance at 550 nm[c] (%)	>90
Vicat softening point, method B (°C)	140
Melt flow rate at 300°C[d] (g 10 min^{-1})	55–60

[a]Merlon CD-2000, manufactured by Mobay Chemical Corp., Pittsburgh, Pa.

[b]Notched 3.18 mm

[c]1.20-mm-thick sample

[d]1.2-kg load

Source: R. Riess and H. Loewer, "Lasers and polycarbonate make music together," *Plastics Engineering 41* (7), pp. 27–29, 1985.

accommodate 2 million engineering drawings. One phase of this system consists of a 12-terminal pilot system serving select groups of engineers and manufacturing personnel building the Boeing 757 aircraft. At General Electric's Steam Turbine Generator Division, a similar system has been operational since 1985. In addition, various organizations, such as NASA and the National Oceanographic and Atmospheric Administration, are experimenting with CD-ROMs for map, chart, and telemetry storage. CD-ROMs will also enable state and local governments and power and telephone companies to create instant charts of underground utility networks and pipelines, for example. The Philips Co. has demonstrated the feasibility of a navigation system that uses satellite transmissions to pinpoint an automobile's exact location on any of a series of maps stored on CDs.

CD technology has one drawback: Formats are prerecorded and cannot be altered. Industry watchers all agree that before they can become competitive with other storage devices, such as floppy discs, optical discs must include (1) a write-once and read-many-times (WORMS) capability (to allow end users to record permanent data themselves) and (2) erasable formats.

WORMS and erasable formats, using polymers as the storage material, are attracting a great deal of research and development attention,* to the extent that some industry insiders predict that erasable discs could be widely used by 1990 and will replace floppy discs by 1999.

*Among the companies working on this technology are GE's Plastics Group in Pittsfield, Mass., Philips du Pont Optical Co. in Wilmington, Del., and Eastman Kodak in Rochester, N.Y. Philips has produced a WORMS disc for the graphic arts and printing industry. Eastman has produced the 6800 WORMS system, which can store more than a terabyte (10^{12} bytes) of information and retrieve it in 12 s or less.

Three erasable technologies are currently being studied:

- *Magnetooptic*. Application of the high-power (20 to 30 mW) laser to the magnetic field of a disc* changes its polarity in the microscopic area, and the low-power (2 to 3 mW) laser beam reads the data by measuring the differences in polarity.
- *Phase change*. The surface polymer changes from crystalline to amorphous on applying a high-power laser beam. The low-power laser reads the data by measuring the difference in reflectivity between the amorphous and crystalline areas.
- *Polymer dyes*. A high-molecular-weight polymer layer changes color when a high-power laser beam impinges on it. Since different colors have different degrees of reflectivity, the low-energy laser compares the reflectivity of the colors formed by the high-power laser, translates them, and thus reads the data.

The erasable optical-disc-grade resin must meet a number of requirements including those needed for CDs, namely low birefringence, outstanding long-term dimensional stability, and incomparable resistance to the premeation of gas and moisture. In addition, the resin must have the following properties:

- Minimal molecular orientation, thereby reducing stresses during processing
- High heat-deflection temperature, at least 150°C, to be able to withstand the high-power laser beams that impinge on it
- Low viscosity or high melt index, to assure fast cycle times (on the order of 12 to 14 s)
- Good shot accuracy and ability to be processed at low temperatures, thereby allowing parts to cool faster (keeping cycle times low) and eliminating oil cooling of molds, which can lead to contamination of the clean-room plant

Apart from optical-grade PC, polymer candidates for these optical erasable discs include acrylic, polyacrylate, polymethylmethacrylate, polyolefins, polyvinyl chloride, polyester, thermoset epoxies, and blends of styrene–maleic anhydride copolymers, polybutadiene-styrene copolymers, and polystyrene.

QUESTION 53
Coatings for Human Teeth

(a) *Give a brief description of enamel, dentin, and cementum, all parts of human teeth, from the viewpoint of materials science. Include a comparison of the values of some relevant material properties.*

(b) *What are the main requirements for materials to be used to coat the human teeth? Compare the performance of natural enamel with some of the competitive candidate materials.*

*Coated with gadolinium terbium iron, terbium iron cobalt, or terbium iron.

(a) Enamel is a smooth, white, semitransparent material forming the outer exposed part of teeth. It varies in thickness, from 0.01 mm to about 2.5 mm over the grinding surfaces of molars. It is extremely hard and brittle and not very strong. The basic "building block" of enamel is a calcified rod or prism that is inclined approximately normal to the dentinoenamel junction. The main constituent inorganic phase (95 wt %) is calcium phosphate, in the form of hydroxyapatite, $Ca_{10}(OH)_2(PO_4)_6$. The organic phase (1 wt %) is made up of protein, carbohydrates, lipids, and other matter. The remaining phase (about 4 wt %) is water, mostly present as a shell surrounding the hydroapatite crystallites.

Dentin, the most important part of the tooth structure, is covered by the enamel in the crown and the cementum in the root. Dentin is a calcified mesh of collagen fibrils surrounding the cellular pathways of the dentin-forming cells. The resultant tubular structure contains about 50% hydroxyapatite (but the crystal size is smaller and the orientation different from that in enamel), 32% collagen, 8% mucopolysaccharides, and 10% water.

Cementum is similar to enamel in that it covers the dentin. It starts where the enamel ends, approximately at the gum level, and covers the root. It is composed of two types of tissue: cellular and acellular.

Some properties of enamel and dentin are given in Table Q53.1,* and the relative positions of these materials and cementum in a human tooth are shown in Fig. Q53.1.

(b) The main material requirements are as follows:

- *High wear resistance*. Enamel may be removed by fracture or chipping of the layer, chemical erosion (due to acidic medication or drinks, dietary oxalate, or high oral hydrogen ion concentration as a result of disease), physical erosion or abrasion (due to idiopathic mechanisms, dentrifices, or toothbrushes; see Fig. Q53.2), and abrasive diets (as those with an especially high whole grain content).

TABLE Q53.1 SOME PROPERTIES OF ENAMEL AND DENTIN

Material Property	Enamel	Dentin
Knoop hardness (kg mm^{-2})	340	65
Density (kg m^{-3})	2200	1900
Modulus of elasticity (GPa)	47	19
Ultimate tensile strength (MPa)	76	41
Ultimate compressive strength (MPa)	414	296
Thermal conductivity (W m^{-1} K^{-1})	0.88	0.59

These are average values; some are quite variable, due to variations in composition and microstructure among different teeth, the inhomogeneous nature of the microstructure, and differences in values reported by different researchers.

*Data on the physical and mechanical properties of cementum is lacking, but it is assumed that cementum has properties like those of dentin and bone, to which it is compositionally similar.

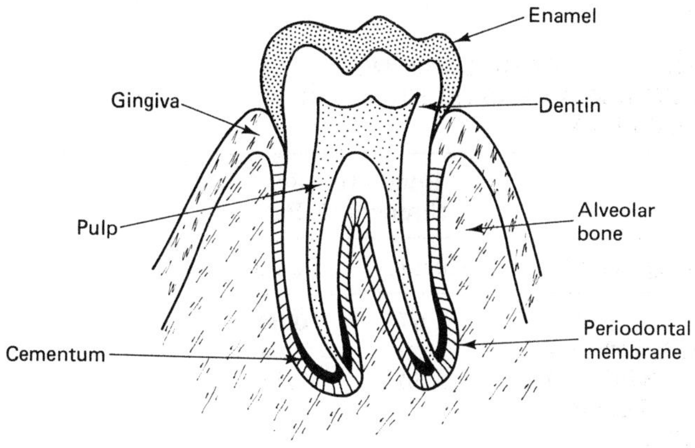

Figure Q53.1 Schematic diagram of a human tooth.

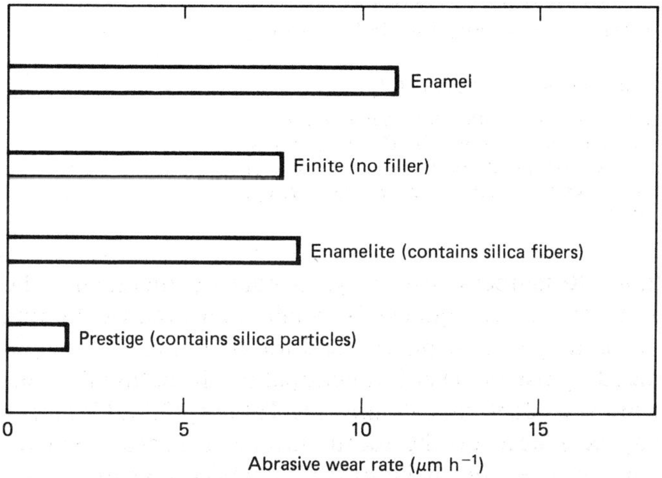

Figure Q53.2 Resistance of various teeth coating materials to abrasive wear caused by brushing with a toothbrush and toothpaste for 86,400 strokes [1].

- *Good insulation of the pulp against temperature extremes.* Two of the problems arising from dysplasia (or disruptions in the enamel surface) are exposure of the comparatively soft dentin and enhanced sensitivity to thermal changes.
- *Aesthetic acceptability.* First, the coating material must be nearly transparent, making it necessary that resins and fillers have about the same refractive indices. Second, the material must be capable of resisting discoloration arising from both extrinsic sources (chromogenic microorganisms in food, drink, tobacco, etc.) and intrinsic sources (blood pigment staining, high-fluoride water, and drugs such as tetracycline administered during tooth development).
- *Excellent bonding.* Adhesive bonding can take place by one of two mechanisms. One is mechanical and is achieved by geometric effects (perosites, roughness, undercuts, etc.) and rheological effects (shrinkage, etc.). The other is chemical and is achieved by primary bonding (ionic, covalent, metallic) and secondary bonding (van der Waals, London, hydrogen). It is now known that most dental materials bond to tooth enamel by means of mechanical forces, the main vehicle being porosities for geometric hooking and

TABLE Q53.2 EFFECT OF COUPLING AGENTS ON BOND (TENSILE) STRENGTH OF ACRYLIC RESIN

Coupling Agent	Bond (Tensile) Strength (MPa) Enamel	Dentin
None	1.67	0.00
NPG-GMA[a] (10%)	5.00	2.64
EDTA[b] + NPG-GMA	5.30	7.55
EDTA + NPG-GMA (after 20 d immersion)	5.20	5.40

[a]N-(2-hydroxy-3-methacryloxypropyl)-N-phenylelycine.

[b]Ethylene diamine tetraacetate, pH 7.

Source: R. L. Bowen, "Adhesive bonding of various materials to hard tooth tissues. IV: Bonding to dentin, enamel and fluorapatite improved by the use of a surface-active comonomer," *J. Dent. Res. 44,* pp. 906–911, 1965.

rheological shrink-fitting. Roughness also plays a part by increasing the potential surface area of contact. Roughness is usually achieved by etching the tooth surface in, for example, phosphoric acid for 1–2 min.

Bonding is enhanced by use of (1) polyfunctional or difunctional "coupling agents" that chelate with dentinal calcium (see Table Q53.2); (2) mordants (metal cations may be placed on the tooth surface to replace or supplement the calcium ions as sites for chelation); (3) calcifying agents, such as a solution of tricalcium phosphate, sodium fluoride, and sodium chloride; and (4) organic polyphosphonates, such as vinyl phosphonic acid and vinylbenzyl phosphonic acid.

- *Biological safety*. The material is to be used in the mouth, in contact with the circulatory system through the dentinal tubules and pulp chamber. Therefore, neither the coating material itself nor any breakdown product can be very toxic either as poisons or allergens.

REFERENCE

1. B. J. Rogers, "Coating materials for teeth," Research Report 77–140, Lee Pharmaceuticals, South El Monte, CA, p. 5, 1977.

QUESTION 54
Precision Optics Components

The race to build "monster" optical telescopes is getting really exciting. A number of groups are currently working toward the construction of "optical giants," with two to three times the light-gathering power of the largest optical telescope in the Western world, the 5-m Hale device situated atop Mount Palomar, Calif.

(a) *List all the relevant requirements in a material suitable for use in constructing the main optics component (the mirror) in these telescopes.*

(b) *Limiting the consideration to nonmetallic materials, select, from a candidate list, the most promising ones.*

(c) *Present the main features of the various approaches currently in use for the creation of precision optical surfaces 8 m or more in diameter.*

(a) It is true that the trend toward larger optical telescopes is one of the most exciting fields in applied physics today. Some relevant details of current developments are given in Table Q54.1.

Thermomechanical loading and distortion considerations require that the mirror materials possess the following characteristics:

- Almost zero coefficient of thermal expansion α, in order to minimize dimensional change due to thermal growth. From structural and systems considerations, the thermal expansivities of the mirror as well as the support structure (including the mirror mounting fixtures) must be nearly equal.

TABLE Q54.1 SOME DEVELOPMENTS IN THE CONSTRUCTION OF LARGE OPTICAL TELESCOPES (as of January 1988)

Location	Mirror Diameter (m)	Comments
Mauna Kea, Hawaii	10	Keck telescope; being built for the California Institute of Technology and the University of California. First light expected in 1995.
Las Campañas, Chile	8	Agreement signed in October 1987 by the Carnegie Institute of Washington, the University of Arizona, and Johns Hopkins University. First light expected in 1990.
Mount Hopkins, Arizona	6.5	The Multiple Mirror Telescope is to be upgraded from 4.5 m to 6.5 m by substituting one mirror for the four now in use.
Mount Graham, Arizona	8	Two spun-cast mirrors mounted side by side form a giant binocular. Joint development by University of Arizona, Ohio State University, University of Chicago, and Italy's Arcetri Astrophysical Observatory.
Cerro Paranal Silla, Chile	16	The Very Large Telescope (VLT) to be built by a consortium of eight Western European nations[a] under the aegis of the European Southern Observatory. Final approval of the project was obtained in December 1987. Estimated cost is $235 million over a 12-y period. The VLT will comprise four 8-m telescopes, working in unison to produce the same results as a single 16-m instrument (see Fig. Q54.1).

[a]Belgium, Denmark, France, Italy, the Netherlands, Sweden, Switzerland, and West Germany.

Figure Q54.1 Model of the four 8-m telescopes of the Very Large Telescope for the European Southern Observatory. (courtesy of European Southern University, La Silla, Chile).

Incompatibility in thermal expansion coefficients can lead to structural misalignment and may consequently necessitate tighter controls for surface-figure retention.

- Microstructural homogeneity and isotropy of α, in order to minimize spatial gradients in α (i.e., $\delta\alpha/dx \rightarrow 0$, where x is the spatial dimension).
- Relative insensitivity of α to temperature change in the region of interest (i.e., $\delta\alpha/dT \rightarrow 0$).
- High thermal diffusivity σ, in order to dissipate the absorbed incident energy.
- Low scattering of light, or else high compatibility with antireflection coatings.
- High specific elastic stiffness (E/ρ), to enable fabrication of a mechanically stable mirror.
- Rapid and easy preparation for final polishing and coating.
- High microyield strength $\sigma_y{}'$, so that all strains in service are totally elastic.
- Using the simple theory of flexure of a plate under uniform loading (arising from self-weight), the magnitude of the maximum sag or deflection Δ for a circular mirror, simply supported continuously around its periphery (Fig. Q54.2), is given by [1]

$$\Delta = \frac{3W\left(\frac{1}{\nu} - 1\right)\left(\frac{5}{\nu} + 1\right)R^2\nu^2}{16\pi E t^3} \qquad \text{(Q54.1)}$$

where W is the total self-weight of the mirror, ν is the Poisson's ratio of the mirror material, R is the outside radius of the mirror, and t is its wall thickness. Taking the average value of the Poisson's ratio of most nonmetallic

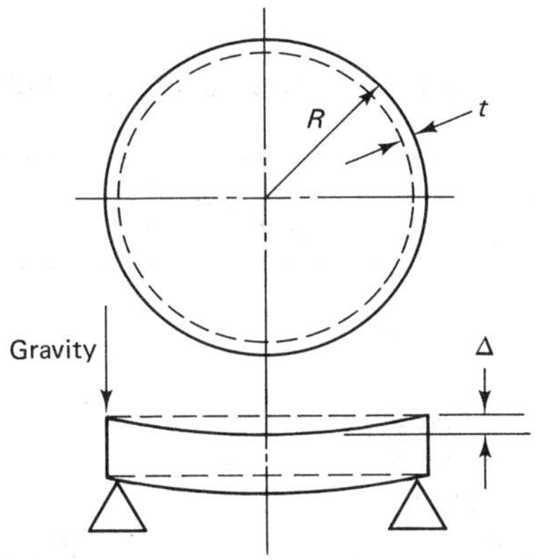

Figure Q54.2 Deflection of a circular solid mirror supported continuously around its edges.

mirror materials to be 0.23,* we have an approximate expression for Δ given by

$$\Delta = \frac{CWR^2}{Et^3}$$

where C is a constant of magnitude 0.24. It is thus clear from equation (Q54.1) that to minimize the central deflection, E must be as high as possible; to minimize self-weight, for a given mirror stiffness (W/Δ), it can be shown that the material property index $(\rho^3/E)^{0.5}$ must be as low as possible.

(b) The values of the relevant properties of some of the more promising materials for these mirrors and comments on them are presented in Tables Q54.2 and Q54.3.

(c) Three approaches toward developing large optical surfaces are currently in use. The *thin-meniscus* approach uses a mirror that has a thickness of only about 2–5 cm. The optical surface is kept at a very accurate level (~0.5 μm) by a set of computer-controlled supports located on the reverse of the mirror that exerts alternating tensile and compressive force at the membrane of material. The magnitude of this applied force is approximately equal and opposite to the distortional gravity forces. The thin meniscus technology has been demonstrated in the 3.5-m New Technology Telescope for the European Southern Observatory.

The second approach is the *segmented mirror* concept. This involves the formation of the precision optical surface as an intricate array of hexagonal pieces that are aligned to a precision of about 100 nm through use of computer-controlled supports. This technology is used in the Keck telescope.

The third approach is the *honeycomb* design in which the precision optical surface is cast by spinning the mold, resulting in a paraboloid shape. Following the cooling of the material and the freezing of the curve, the surface is given a final polishing. The major advantage of this approach is that it completely dispenses with the need for grinding out the basic curve from a slab of the optical material. This approach is to be used in the United States' National New Technology Telescope.

*The value of ν for fused silica is 0.167; for ULE 7971 it is 0.176; for Zerodur 543561 it is 0.24; for CG4 it is 0.121; and for borosilicate glass it is 0.21.

TABLE Q54.2 SOME PROPERTIES OF SELECTED CANDIDATE MATERIALS FOR PRECISION OPTICS COMPONENTS

Material	α (°C^{-1})	σ^a at 298 K (10^{-8} m^2 s^{-1})	Comments
Zerodur quartz glass ceramic	12×10^{-8}	85	—
Borosilicate glass			
Tempax	320×10^{-8} at 300 K	56	—
E6	300×10^{-8} at 300 K	—	—
Quartz TO8E	50×10^{-8} at 300 K	80	—
ULE glass[b]	3×10^{-8} in the 273–473 K range	78	Microstructural and chemical inhomogeneities following fabrication; anisotropy in α
Cordierite glass ceramic[c]	100×10^{-8d}	164	Anisotropy in α
Germania-cordierite glass ceramic[e]	-7×10^{-8} in the 273–373 K range -2.5×10^{-8} in the 310–353 K range	164	Partial substitution of Si^{4+} ions by Ge^{4+} ions in the hexagonal ring structure of cordierite leads to a substantial reduction in α and elimination of its anisotropy

[a]Thermal diffusivity $= k/(\rho c_p)$, where k is the conductivity, ρ is the density, and c_p is the specific heat.

[b]Ultralow expansion glass: 92.5% SiO_2–7.5% TiO_2; registered trademark of Corning Glass Works, Corning, N.Y.

[c]$2MgO \cdot 2Al_2O_3 \cdot 5SiO_2$.

[d]Volume coefficient of thermal expansion $= (\alpha$ along a-axis $+ \alpha$ along c-axis$)/3$.

[e]$2MgO \cdot 2Al_2O_3 \cdot 4SiO_2 \cdot GeO_2$.

TABLE Q54.3 PROPERTIES OF SOME CANDIDATE MIRROR MATERIALS

Material	E (GPa)	ρ (kg m^{-3})	E/ρ (10^6 m)	$(\rho^3/E)^{0.5}$ (N m$^{-3.5}$)
Borosilicate glass (such as Pyrex)	68	2240	3.08	12.5
Fused silica glass (such as Corning 7940)	74	2200	3.43	11.7
ULE glass (such as Corning 7971)	69	2210	3.18	12.2
Germania-cordierite glass ceramic (such as CG4)	61	2695	2.30	17.4
Zerodur glass ceramic	91	2530	3.67	13.0

REFERENCE

1. E. J. Hearn, *Mechanics of Materials, vol. 2,* Oxford: Pergamon Press Ltd., p. 632, 1985.

QUESTION 55
First Wall or Blanket in a Fusion Reactor

Many nuclear engineers and scientists now agree that first wall and blanket lifetime would determine the success of efforts at commercialization of magnetic confinement fusion reactors. That is to say, the structural materials employed must exhibit long, stable life under severe service conditions, and hence low-activation materials should be used wherever possible. Recent results from fission reactor irradiation studies in the United States and Western Europe suggest that the iron-chromium martensitic stainless steels, such as alloy HT-9, are prime candidates for first walls and blankets.

Comment on this choice, and discuss its advantages and disadvantages compared to two other candidate materials.

HT-9 will be compared with annealed type 316 stainless steel and Inconel 718 alloy on the basis of the important material requirements (see Table Q55.1 for alloy compositions).

The first consideration is the effect of high temperature on the strength of the alloy. From the results given in Fig. Q55.1, it is evident that Inconel 718 has superior performance. HT-9 performs better than 316 stainless steel up to about 550°C, after which its strength decreases rapidly.

Thermal stresses are an important source of cyclic fatigue and creep fatigue damage in these reactors, so the thermal shock resistance of the material is important. It is seen from the results in Table Q55.2 that the thermal shock resistance of HT-9 is about 60% that of Inconel 718 but more than double that of 316 stainless steel.

TABLE Q55.1 NOMINAL COMPOSITIONS OF THREE CANDIDATE FUSION REACTOR MATERIALS (in wt %)

	P	Cr	Ni	Mo	W	V	Si
HT-9	—	11.5	0.5	1.0	0.5	0.3	0.3
Type 316 stainless steel	0.05	17.0	12.0	2.5	—	—	1.0
Inconel 718 alloy[a]	0.015	18.5	53.0	3.0	—	—	—

	Ti	Mn	C	Al	S	Fe
HT-9	—	0.5	0.2	—	—	Balance
Type 316 stainless steel	—	2.0	0.08	—	0.03	Balance
Inconel 718 alloy[a]	1.4	0.35	0.08	0.5	0.015	18

[a] Also contains small amounts of boron, zirconium, columbium, cobalt, tantalum, and copper.

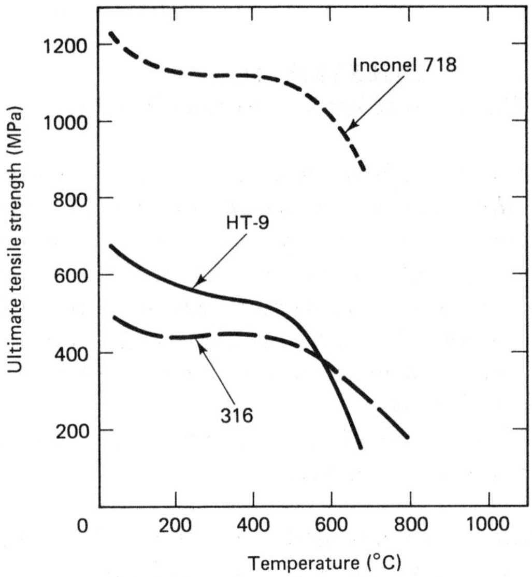

Figure Q55.1 Effect of temperature on the ultimate tensile strength of three candidate first wall and blanket materials.

TABLE Q55.2 THERMAL SHOCK RESISTANCE OF CANDIDATE FUSION REACTOR ALLOYS (at 500°C)

Candidate Alloy	σ_y (MPa)	ν	k ($\mathrm{W\ m^{-1}\ K^{-1}}$)	E (GPa)	α ($10^{-6}\ \mathrm{K^{-1}}$)	R_2 ($\mathrm{kW\ m^{-1}}$)
Annealed type 316 stainless steel	205	0.26	12.8	193	17.8	0.56
HT-9	275	0.26	22.8	200	17.8	1.31
Inconel 718	931	0.34	12.0	200	16.0	2.31

[a]Relevant thermal shock resistance parameter; see Chapter 7 in Part I.

The third important consideration is the resistance of the material to neutron radiation damage. Preliminary results indicate that at greater than 10^{23} neutrons $\mathrm{cm^{-2}}$ (>100 keV),

- Void swelling in HT-9 is about ten times that in annealed type 316 (35%, 580°C at 150 dpa compared to no more than 3.7% at 500°C at 150 dpa [1]).
- The in-reactor creep resistance of HT-9 is better than that of annealed type 316 to temperatures in excess of 600°C [2].
- The tensile ductility retention of HT-9 is superior to that of cold-worked type 316 (particularly at temperatures greater than the irradiation temperature).

Finally, it has been shown [3] that the high-temperature service performance of HT-9 is excellent, as evidenced in the high retention of its creep rupture (Table Q55.3), impact (Table Q55.4), and tensile properties (Table Q55.5). There are still, however, a number of unresolved issues with regard to its use in fusion reactors. These issues include its ductile-to-brittle transition behavior, the effect of the fusion reactor environment on its structural properties, its interaction with coolants, and its weldability and fabricability.

TABLE Q55.3 EFFECT OF 80,000-h, 600°C SERVICE EXPOSURE ON THE CREEP RUPTURE PROPERTIES OF ALLOY HT-9

	Larson-Miller parameter[a]	
Stress (MPa)	Before Service	After Service
50	21600	21000
100	20200	19400
200	18700	18000

[a] Larson-Miller constant of HT-9 is 20.

Source: B. Walser, "The Metallurgical and Mechanical Properties of X20CrMoWV12 1 After Service Exposure in a Fossil Fired Boiler," 222/1511, Winterthur, Switzerland: Sulzer Brothers, 1977.

TABLE Q55.4 EFFECT OF 80,000-h, 600°C SERVICE EXPOSURE ON THE IMPACT PROPERTIES OF ALLOY HT-9 (HALF WIDTH SPECIMENS)

Test Temperature (°C)	Impact energy (J)	
	Before Service	After Service
−45	17–20	6
22	53–55	16–17

Source: D. I. Roberts, S. N. Rossenwasser, and J. F. Watson, "Material selection for gas-cooled and fusion reactor applications," *J. Mater. Energy Sys. 3,* pp. 54–65, 1981.

TABLE Q55.5 EFFECT OF 80,000-h, 600°C SERVICE EXPOSURE ON THE TENSILE PROPERTIES OF ALLOY HT-9

	Ultimate tensile strength (MPa)		0.2% proof strength (MPa)		Elongation (%)	
Temperature (°C)	Before Service	After Service	Before Service	After Service	Before Service	After Service
100	740	670	520	470	20	19
300	670	600	500	430	20	19
500	500	450	400	380	22	20

Source: D. I. Roberts, S. N. Rossenwasser, and J. F. Watson, "Material selection for gas-cooled and fusion reactor applications," *J. Mater. Energy Sys. 3,* pp. 54–65, 1981.

REFERENCES

1. F. A. Smidt, et al., "Swelling behavior of commercial ferritic alloys, EM-12 and HT-9, as assessed by heavy ion bombardment," in *Irradiation Effects on the Microstructure and Properties of Metals,* ASTM STP No. 611. Philadelphia: American Society for Testing Materials, 1976, p. 227.
2. B. A. Chin, et al. "A comparison of the in-reactor creep behavior of selected commercial alloys at a temperature of 593°C," paper presented at the 9th ASTM International Symposium on Effects of Radiation in Structural Materials, Richland, Wash., 1978.
3. J. E. Chafey and L. D. Thompson, "Post-service evaluation of the properties of a 12Cr-1Mo steel superheater tube," paper presented at the Annual Winter Meeting of the ASME, Chicago, November 1980.

QUESTION 56
Use of Ceramics in Automotive Engine Parts

Competitive juices are flowing as American engineers match their skills against those of foreign experts to develop ceramic vehicular engine parts.

Comment on this statement, with special reference to (a) the advantages of ceramic materials for applications in various engine parts and (b) the more recent research and development efforts in this area in the United States, Japan, and Western Europe.

(a) Basically, the future opportunities for the use of ceramics in vehicular engines are related to (1) the very high temperatures these materials can withstand (Fig. Q56.1); (2) their low densities (leading, in part, to lower fuel consumption by the engines); (3) their low thermal expansion, and (4) their high strength at both room and elevated temperatures (see Table Q56.1).* The market potential

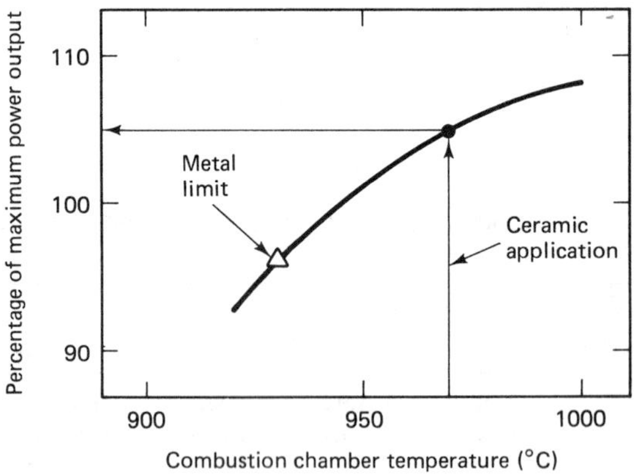

Figure Q56.1 Effect of combustion chamber temperature on the maximum power output of a diesel engine.

*Because ceramics are brittle and contain microcracks, allowance for strength dispersion must be considered by taking a probabilistic approach. Most current design methodologies are based on Weibull's [1] fracture theory, although other weakest-link theories, such as those of Batdorf [2] and Lamon [3], have been attracting some attention.

TABLE Q56.1 COMPARISON OF METALLIC AND CERAMIC MATERIALS FOR ROTORS

Property	Nickel-Based Alloy	Silicon Carbide	Silicon Nitride
Ultimate tensile strength (MPa)			
At 25°C	965	414	760
At 1093°C	365	414	530
Coefficient of thermal expansion (10^{-6} K^{-1})	13	4.3	3.2
Thermal conductivity (W m^{-1} K^{-1})	15	84	17
Density (kg m^{-3})	7900	3200	3200
Melting temperature (°C)	1580	3110	2173
Allowable gas temperature (°C)	1079	1288	1288

for selected ceramic parts in these applications is enormous (Table Q56.2). The specific advantages conferred by the use of ceramics in selected engine components are presented in Table Q56.3.

(b) The following is a summary of the recent research, development, and production efforts in the United States, Japan, and Western Europe. These efforts are grouped under the headings "Gas Turbine Engines," "Advanced Diesel Engines," and "General Developments."

Gas Turbine Engines

In the United States, this work has been driven by the Advanced Gas Turbine (AGT) Program during the period 1971–87. Since then, funding has been provided by the Department of Energy's Advanced Turbine Technologies Applications Program (ATTAP).

Most of the work has been conducted by General Motors' Allison Turbine Division and Garrett Turbine Engine Co. Since 1979, both companies have been competing to design a 74-kW AGT engine that can deliver 18 km l^{-1} of diesel fuel or 15.3 km l^{-1} of gasoline in a 1.36-t vehicle. Both companies scored major successes in 1986. Garrett's single-shaft AGT 101, developed in conjunction with

TABLE Q56.2 MARKET POTENTIAL FOR SELECT CERAMIC PARTS, 1981–2000

Application	Market Potential ($\$ \times 10^9$, 1981 dollars) 1990	2000
Diesel/gas-turbine components for highway vehicles		
Pistons, piston rings, valves	1.2	1.8
Cylinder liners and cylinder head inserts	1.3	2.0

Source: Argonne National Laboratory, Argonne, Ill.

TABLE Q56.3 APPLICATIONS OF CERAMIC MATERIALS IN VEHICLE ENGINE COMPONENTS

Component	Material[a]	Advantage(s)
Disc brake piston	LAS	Elimination of vapor lock
Regenerator bulkhead	LAS	Reduction in engine weight
Heat exchanger (gas turbine), regenerator	MAS	Increase in engine life, reduction in torque needed to drive cams
Piston cap, cylinder liner, hotplate	PSZ	Decrease in fuel consumption
Glow plug (diesel engine), precombustion chamber, mouthpiece	SSN	Increase in rate of ignition

[a]LAS, lithia-alumina-silicate; MAS, magnesia-alumina-silicate; PSZ, partially stabilized zirconia; SSN, sintered silicon nitride.

Ford Motor Co., incorporates ceramics for the combustor, regenerator, turbine, compressor, foil gas bearing, ball bearings, flow transition liners, turbine stator, and diffuser. In 1986, the engine was run for 85 h at 1204°C before carbon from uneven burning in the combustor reportedly damaged the rotor and stator. All other structural ceramic parts reportedly remained in excellent condition. Allison also reported foreign-object damage in its AGT 100, a two-shaft design, containing a ceramic gasifier turbine rotor and scroll assembly (Fig. Q56.2), that was also tested in 1986.

Figure Q56.2 The AGT 100 advanced gas turbine engine. (Courtesy of the Allison Gas Turbine Division, General Motors.)

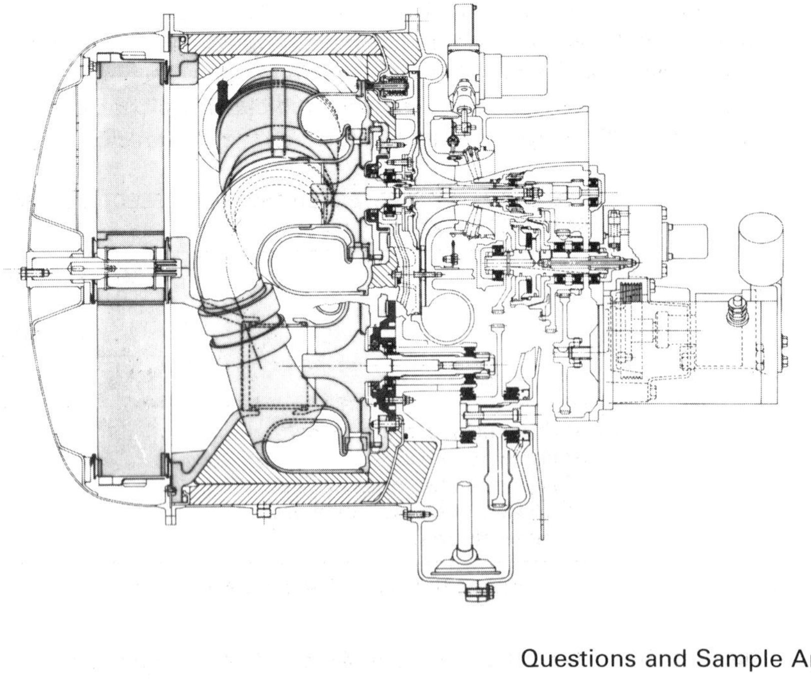

A major boost was given to efforts on ceramic automotive gas turbine research in 1987 with the award of a 5-year contract from the U.S. government to Garrett Auxiliary Power for the development of fabrication methods and production of ceramic component prototypes for this kind of engine. The three subcontractors in this project are Norton/TRW Ceramics, Standard Oil Engineered Materials, and Garrett Ceramic Components. It is anticipated that the completion of this contract will provide enough of a boost to initiate commercial production of ceramic components for this and other automotive engines.

In Western Europe, EUREKA, a consortium of companies from Common Market countries formed in 1986, started the development of a ceramic gas turbine engine under the leadership of Peugeot of France.

In Japan, a prototype sedan with a ceramic coating gas turbine engine was introduced by Nissan at the 1983 Tokyo Motor Show.

Advanced Diesel Engines

In the United States, the principal players are Cummins Engine Co. of Columbus, Ind., and the Ford Motor Co. of Detroit. The former has worked in collaboration with the U.S. Army Tank Automotive Command in Warren, Mich., in investigating the application of ceramic materials in the combustion area and hot gas path of heavy-duty diesel engines, as are found in U.S. Army trucks. The feasibility of the adiabatic turbocompound engine was demonstrated in 1980, which efforts led to a first-generation engine installed in a 4.54-t U.S. Army truck. Subsequently, in 1984, the program completed the prototype development of a Cummins 325 242-kW V-8 commercial truck engine into a 522-kW adiabatic turbocompound engine (Fig. Q56.3).

Figure Q56.3 Schematic representation of an adiabatic diesel engine. (Courtesy of Isuzu Motors.)

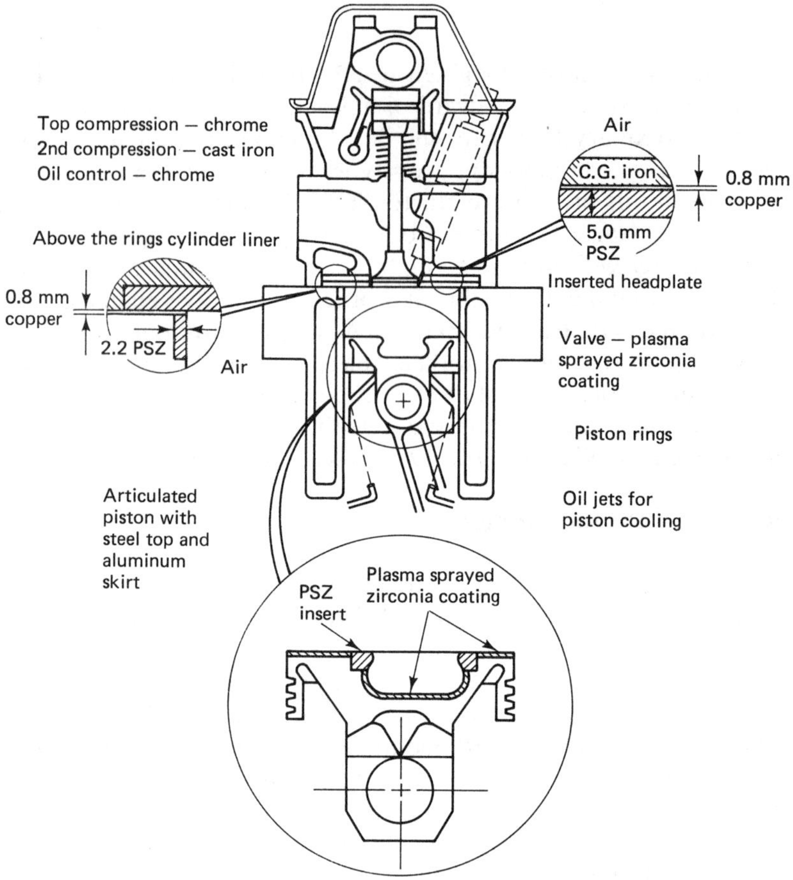

Figure Q56.4 Partially stabilized zirconia (PSZ) piston parts for an uncooled single-cylinder direct-injection diesel engine. (Reprinted with permission; © 1986 Society of Automotive Engineers, Inc.)

In 1987, Ford Motor Co. successfully tested structural ceramic components, including cylinder, piston, and pin, for more than 100 h in a single-cylinder direct-injection diesel engine (Fig. Q56.4) that features a ringless piston with gas squeeze film lubrication. The company reported that heat loss was 30% lower than for a water-cooled engine, fuel consumption is 5% to 9% lower, ignition delay is reduced, and combustion is smoother. Ford has also been among the first automobile companies to adopt ceramics for commercial automative applications. Thus in the 2-l diesel engine of the 1987 Escort, parts of the precombustion chamber were made of ceramics.

In Japan, there have been a number of major developments, some of which are in commercially available vehicles. Kyocera International Corp., for example, produces diesel glow plugs for Isuzu, Mitsubishi, and Toyota, as well as diesel precombustion chambers for Isuzu, Mazda, and Toyota. Nissan is installing turborotors, made by NGK Insulators Ltd., in the 300 ZX Fair Lady models sold in Japan. In addition, Kyocera itself tested uncooled direct-injection, turbocharged, four-cylinder engines (KC 301 and KC 401) for a passenger car. The

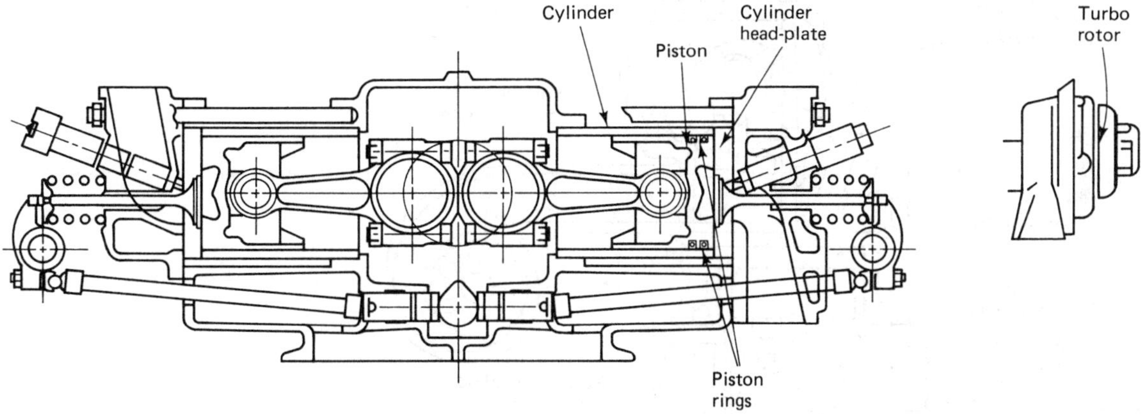

Figure Q56.5 Sectional drawing of a four-cylinder direct-injection diesel engine, the KC 401. (Reprinted with permission; © 1986 Society of Automotive Engineers, Inc.)

engine contains piston, piston rings, cylinder, cylinder head plate, and turborotor (all made of sintered silicon nitride; see Fig. Q56.5). The company reported that all these ceramic components surpassed 100 h of durability testing at full engine load and 3600 r min^{-1}.

Mazda is developing both a diesel and a rotary engine with ceramic parts. The diesel engine uses an all-ceramic swirl combustion chamber, which will make it possible to meet U.S. diesel particulate emission standards without requiring an electronic fuel-injection system. The rotary engine uses a ceramic seal between the rotor and side housing wall, which is expected to increase the thermal efficiency by 10%. Isuzu Motors Ltd., in conjunction with Kyocera, has developed an adiabatic diesel engine with ceramic parts (Fig. Q56.6).

In Western Europe, Porsche of West Germany introduced a ceramic-lined exhaust port system in its 944 model automobile. The system insulates the head from the heat in the exhaust system, protects electronic components, supplies more heat to the turbocharger system to improve its efficiency, and makes possible the use of a smaller radiator (since less heat goes to the cooling fluid).

AE Developments of England produced a reaction-bonded silicon nitride material that was tested in a horizontally opposed two-stroke diesel engine developed at University College, Dublin, Ireland. In addition, AE successfully ran an aluminum piston with zirconia crown inserts in a single-cylinder diesel engine.

General Developments

It is expected that basic research work in ceramics will lead to improved ceramics for applications in vehicular engine parts. Among these efforts are the following:

- The Advanced Materials Development Project, administered through Oak Ridge National Laboratory, Oak Ridge, Tenn., is supporting a series of programs aimed at improving the reliability of monolithic ceramics, developing ceramic-matrix composite materials and ceramic coatings, and improving fabrication and joining technology.

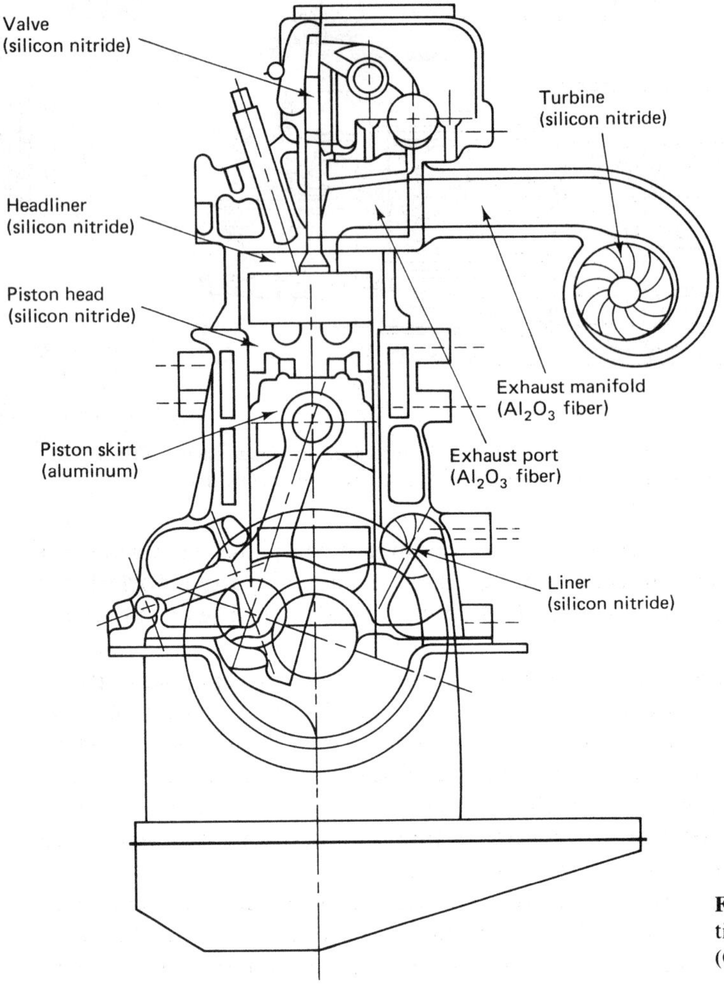

Figure Q56.6 Schematic representation of an adiabatic diesel engine. (Courtesy of Isuzu Motors.)

- Work at Norton/TRW Ceramics* involves virtually all aspects of material research for heat engine ceramics: monolithic and composites, testing methods, and manufacturing processes.
- GTE in the United States has established a special engineering center capable of producing ceramic components for emerging markets, such as metal-matrix composites (silicon carbide whiskers in a silicon carbide matrix) for turborotors.
- A three-year collaborative research program was started in 1987 by the British Ceramic Applications in Reciprocating Engines (CARE) Group of 30 engine component manufacturers and ceramics suppliers. With the support of the British Department of Trade and Industry, the program is pursuing four basic areas of research: substitution, insulation, turbocharging, and materials.

*Ceramic valves made by this company performed trouble free for over 20,000 km of city and highway operation in the engine of a 1988 Oldsmobile Cutlass Ciera automobile.

- The West German Science Ministry, in 1987, started a three-year basic ceramics technology program involving some 20 university- and industry-cosponsored projects. The projects cover silicon carbide and silicon nitride for engine applications and transformation-toughened ceramics.
- A similar government-sponsored Swedish program has projects covering injection molding, grinding and laser treatment of silicon nitride, and plasma spraying.
- The Midwest Technology Development Institute (MTDI), set up by nine midwestern U.S. states in 1985, organized Advanced Ceramics and Composites Partners, a consortium to coordinate corporate-funded ceramics research at midwestern universities and at its own laboratories. At least 11 major U.S. companies are partners in this venture.

REFERENCES

1. W. A. Weibull, "Statistical distribution function of wide applicability," *J. Appl. Mech. 18* (3), pp. 293–297, 1951.
2. S. B. Batdorf and J. G. Crose, "A statistical theory for the fracture of brittle structures subjected to polyaxial stress states," *J. Appl. Mech. 41*, pp. 459–465, 1974.
3. J. Lamon, "Statistical approaches to failure for ceramic reliability assessment," *J. Amer. Ceram. Soc. 71* (2), pp. 106–112, 1988.

QUESTION 57
Fiber-optic Materials

From telephone conversations to military threat detection, and from temperature sensing to cable television, all are made possible by optical fibers.

Write an essay on the material aspects of optical fibers.

The two main classes of applications of optical fibers are point-to-point communications links and sensor devices.

A fiber-optic system of the first type consists of three main components: one or more light sources (transmitting devices), a transmission medium (cables containing optical fibers), and one or more light detection devices (Fig. Q57.1). There is little doubt that these systems are attracting an increasing amount of attention,

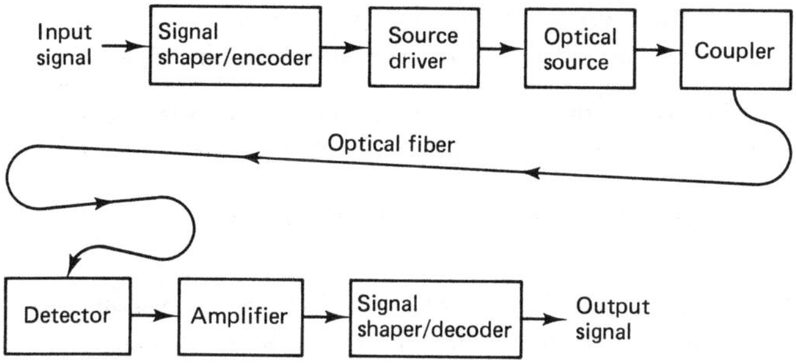

Figure Q57.1 Elements of a fiber-optic communications system.

leading to an expansion of their range and variety of applications. These applications may be grouped into a number of categories (Table Q57.1).

The second class of applications is in the field of sensing devices, with the number and variety of examples (Table Q57.1 and Fig. Q57.2) also increasing rapidly since the early 1980s.

This expanded use of fiber-optic systems is demonstrated by the growth of the market. For example, the fiber and cable sector of this market is projected to grow in sales from $749 million in 1987 to $1.2 billion in 1997.* The total length of optical fibers produced in the United States will increase from 8×10^7 km in 1984 to 3×10^{10} km in 1997.†

Fiber-optic systems have a number of distinct advantages over systems that use metallic media, such as copper cables. These advantages are as follows:

- Low loss. Optical fibers minimize the need for reamplification of signals in long-distance telecommunications applications; signals sent through copper

TABLE Q57.1 SOME APPLICATIONS OF OPTICAL FIBERS

Point-to-Point Communications

- Telephone services[a] (long-distance, intraoffice, interoffice, feeder, local distribution, private networks, and undersea cable systems, such as TAT-8, which provides service between Western Europe and the United States)
- Teleports
- Data transmission systems

Sensors

- Thermometric sensors, used in industry (for measurement of temperature in geothermic walls, in high-voltage electrical equipment, in radio frequency or microwave heating of food or other materials) and medicine (temperature monitoring during removal of tumors)
- Gyroscopes, magnetometers, pressure sensors, hydrophones, liquid level sensors, seismometers
- Remote sensing of a source, as in a military threat warning system in a helicopter
- Location of illegal drugs in concealed compartments and spaces in bulkheads of boats at sea.[b]

Other Applications

- Delivery of high-peak-power laser light to a remote location, as in the U.S. Department of Energy's Advanced Gas Centrifuge Program in Oak Ridge, Tenn.
- Measurement of blood flow and cardiac output by the thermodilution method

[a] 88% of the 1.4×10^6 km of optical files used in the United States in 1985 were for these services, provided by AT&T, the Bell Operating Companies, and the so-called other common carriers (notably MCI and U.S. Sprint).

[b] The device, called Fiberscope, was initially tested off the coast of Florida in 1988.

*Projection based on 1987 and 1988 sales figures provided by Information Gatekeeper, Boston.

†Projection based on 1984 and 1988 production figures provided by Yankee Group, Boston.

wire and cables, for example, have to be "repeated" at about every 2 km, while, with fiber-optic systems currently available, "repeaters" are located about every 30 km or so.

- Higher capacity. Two pairs of copper wire will carry about 24 two-way voice transmissions, whereas one pair of optical fibers can carry up to 1344 of these transmissions.
- Better use of space. The core diameter of a copper wire is about 1.5 mm, whereas that of an optical fiber is about 0.025 mm.

Figure Q57.2 Fiber-optic sensors: (a) temperature sensor; (b) rotation sensor; (c) hydrophone; (d) vibration monitor; (e) seismometer.

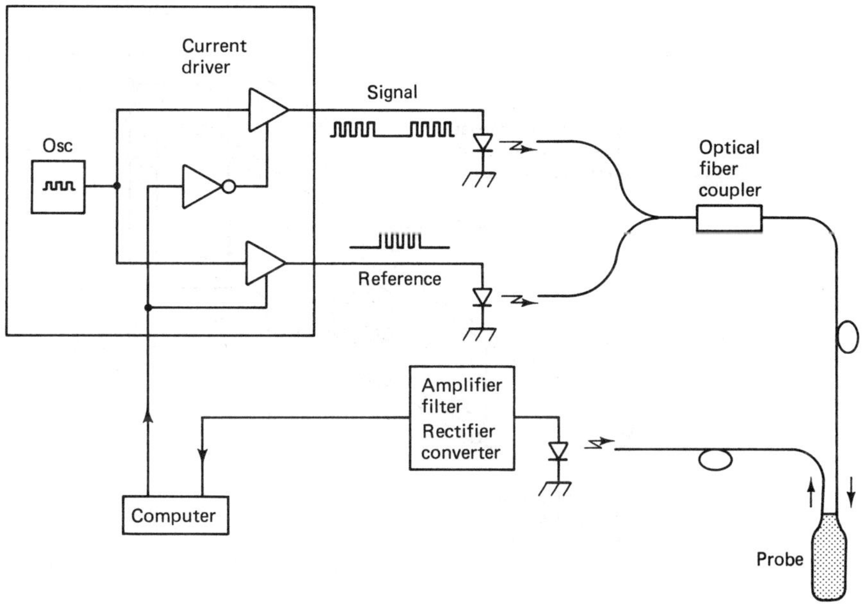

(a)

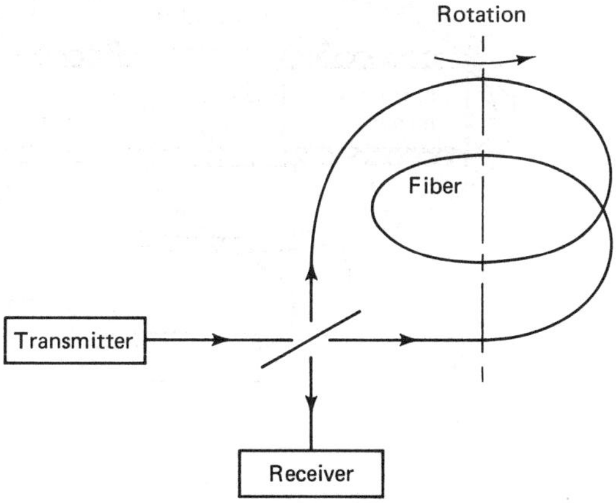

(b)

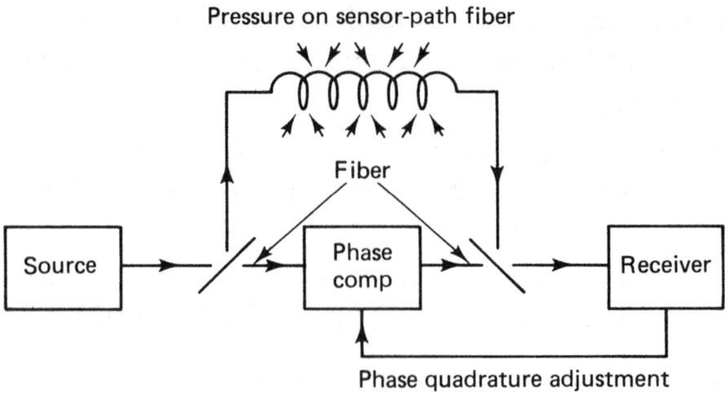

(c)

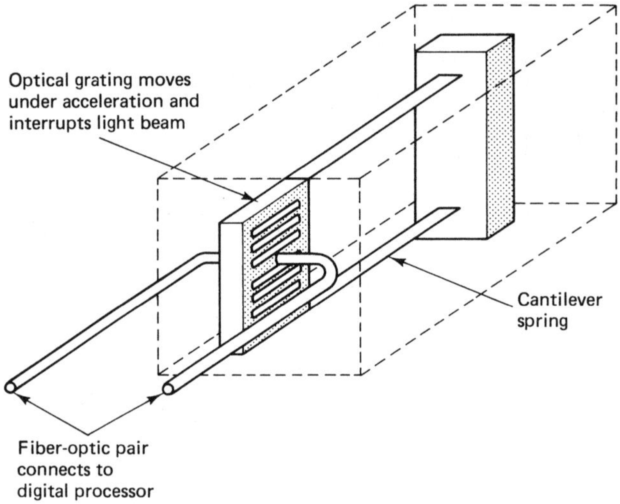

(d)

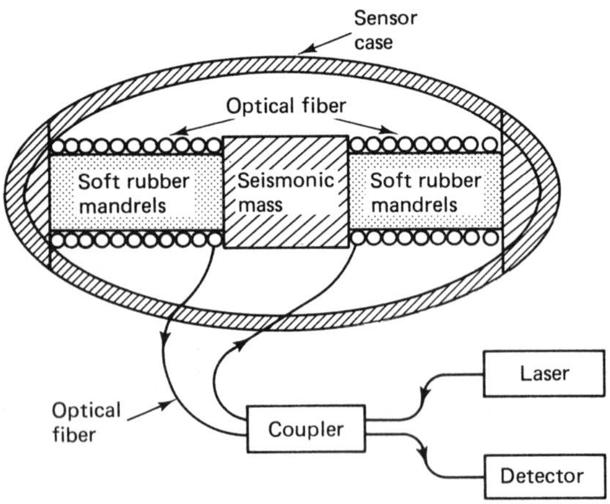

(e)

Figure Q57.2 (Continued)

- Decrease in weight. A single conductor fiber-optic cable weighs about 13 g m^{-1}, whereas a comparable coaxial copper wire cable weighs about 120 g m^{-1}.
- Fiber-optic cables are immune to interference from electromagnetic and radio frequency waves; this translates to security of transmission.
- Optical fibers are noncorrosive, whereas copper is prone to attack by a number of electrolytes.
- Most environmental electrical energy conditions (such as lightning) have little effect on the performance of a fiber-optic system.

Two main disadvantages are associated with fiber-optic systems:

- Delicate physical handling is necessary.
- The systems involve high-quality complex electronic and photonic interface equipment.

The most important aspect of an optical fiber material is its attenuation or loss, which is reflected in the value of the attenuation coefficient $\alpha(\lambda)$, defined as the attenuation per unit length at a given frequency λ, and is expressed in dB km^{-1}. The ratio of the output signal P_o to the input signal P_i is then given by

$$\frac{P_o}{P_i} = 10^{-\alpha L/10}$$

where L is the length. Thus for $\alpha = 1000$ dB km^{-1}, a repeater would be needed every 50 m or so.

The objective, therefore, is to select a material with as low a value of α as possible. Broadly speaking, three categories of materials can be used as optical fibers: monocrystalline fiber materials, polycrystalline fibers, and glass fiber materials. Glasses have attracted the most attention, so they will be considered in some detail in this presentation.

Mention should, however, be made of the other two types of materials. Examples of monocrystalline fibers are those made of CsI, CsBr, KRS-5, and AgBr. To date, the best is CsBr, which combines low attenuation with good mechanical strength. The fibers possess macroscopically smooth surfaces and can be fabricated in approximately circular cross sections from the melt. Examples of polycrystalline fiber materials include KRS-6, KCl, CsI, TeBr, and AgCl. These fibers are prepared by thermal extrusion, and the lower value of α currently available is 120 dB km^{-1}. These fibers have additional scattering loss due to grain boundaries, high toxicity, and insufficient strength, and some of them are hygroscopic.

Let us now examine the main features of the optical glass fiber materials (halide, chalcogenide, and oxide glasses).

Halide Glasses

Halide glasses fall into three subcategories. Single-component halide glasses (SCHGs) have low attenuation due to monovalent ion behavior; for example, BeF_2- and $ZnCl_2$-based halide glasses have α of 10^{-2} to 10^{-5} dB km^{-1}. These glasses have some drawbacks; for example, beryllium in BeF_2 is highly toxic, while $ZnCl_2$ is highly deliquescent.

Single-anion multicomponent or heavy metal fluoride glasses (HMFGs)* have a wide transparency range (from 0.3 μm to about 9 μm, that is, mid-infrared to near ultraviolet). Attenuation losses of these glasses are on the order of 5 to 8×10^{-3} dB km^{-1} at $\lambda = 2.4$ μm. This high loss is mainly because of impurities—such as three-dimensional transition metal additions (mainly iron and nickel), which absorb in the 2–3 μm range, and rare earths—and because of a fundamental stretching of O-H absorption at about 3 μm. These glasses have a tendency to recrystallize from the surface as a result of attack by moisture and oxygen, so special jackets or hermetic coatings are required for fibers made of these glasses. HMFGs have relatively narrow glass-forming regions because viscosities are low (for example, 0.4 Pa at 490°C). HMFGs also have thermal instability tendencies, leading to devitrification; they also have a high susceptibility to attack in aqueous environments (see Fig. C9.2, Chapter C9 in Part III).

Heavier halide glasses are based on chloride, bromide, and iodide mixtures, although the latter two types give rise to glasses of low stability and hence are unsuitable for fiber-optic applications. The chloride-based glasses have major problems with impurity ions and are water-soluble.

Chalcogenide Glasses

Chalcogenide glasses contain the elements Se, S, or Te, together with one or more of Ge, Si, As, Sb, and some others. Attenuation loss is on the order of 10^{-2} dB km^{-1}; this is primarily due to impurity absorption (from H_2S, H_2Se oxides of the original cationic elements, water molecules, etc.). These glasses have large glass-forming regions, high stability against attack by moisture, and reasonably long wavelength cutoff of 0.5 to about 15 μm.

Fused Silica

The losses in this glass amount to about 0.2 dB km^{-1}, making it less attractive than either of the other two types of glass already discussed.

Some characteristics of three examples of these glasses are presented in Table Q57.2 and Figs. Q57.3 and Q57.4.

Sources of Attenuation in Optical Glasses

The optical attenuation in glasses arises from three main sources. Scattering and absorption losses can result from *extrinsic* factors such as mechanical defects (splicing, microdeformations), material inhomogeneities (such as voids and dissolved submicron inclusions), and impurities (such as oxides, hydrides, rare earths, and transition metal ions).

Intrinsic absorption may be due to absorption from the electronic (or Urbach) and edge tails, together with elastic (or Rayleigh) scattering and multiphonon

*For details of these glasses, see Chapter C9 in Part III.

TABLE Q57.2 SOME PROPERTIES OF THREE CANDIDATE OPTICAL FIBER GLASSES

Glass Type, Composition	Density ($kg\ m^{-3}$)	Refractive Index	Glass Transition Temperature (°C)	Thermal Expansion Coefficient ($10^{-6}\ °C^{-1}$)	Vickers Hardness (MPa)	Poisson's Ratio	Young's Modulus (GPa)	Fracture Toughness (MPa $\sqrt{m}$)
Heavy metal fluoride, $57ZrF_4$-$36BaF_2$-$3LaF_2$-$4AlF_3$ (mole %)	4610	1.52	310	18.70	2619	0.30	56	0.31
Chalcogenide (TI 1173 glass), 28Ge-12Sb-60Se (at. %)	4400	2.50–2.65	300	14.10	1570	0.24	21	0.20
Fused silica, SiO_2	2200	1.45–1.47	1100	0.54	6229	0.16	73	0.74

Source: P. Klocek, M. Roth, and R. D. Rock, "Chalcogenide glass optical fibers and image bundles: Properties and applications," *Optical Engineering* *26* (2), pp. 88–95, 1987 (all data except fracture toughness).

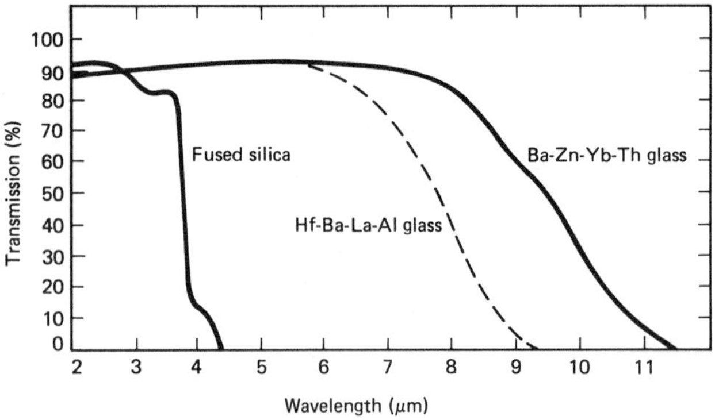

Figure Q57.3 Transmission versus wavelength for various optical glasses.

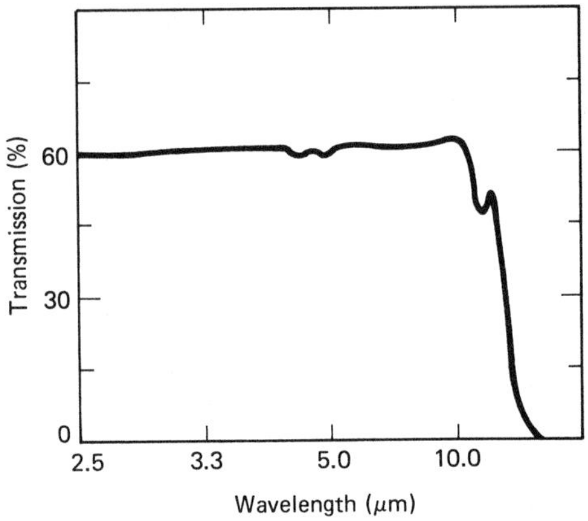

Figure Q57.4 Infrared spectra of $Ge_{28}Sb_{12}Se_{60}$ (TI 1173 glass), 9.08mm thick (courtesy of Texas Instruments, Dallas, Texas).

absorption. All these losses are dependent on wavelength λ. Such dependence for Urbach, multiphonon, and weak absorption tails for some chalcogenide glasses are given in Table Q57.3.

The Rayleigh scattering loss is also dependent on wavelength and is given by B/λ^4, where B is a material constant. The total minimum intrinsic loss in a given glass is obtained by the interaction of the scattering loss and the vibrational edge absorption curves (Fig. Q57.5). This yields the so-called theoretical V curve for the material (Fig. Q57.6).

Attenuation may also stem from a number of material-related *nonlinear effects,* such as Raman and Brillouin scattering, which occur above a critical power level [2]. In the case of Raman scattering, the associated loss is reflected in the value of F_R, which is the ratio of the spontaneous Raman scattering loss to the density fluctuation loss. The values of F_R and the Rayleigh material constant B for three named glasses are given in Table Q57.4.

The trend in reduction in the total optical losses in optical fibers over the period 1972–86 is shown in Table Q57.5. Much of this improvement is directly related to advances in fiber fabrication processing. For mono- or multimode long-distance telecommunications applications, optical fibers are best made by vapor techniques employing energy supplied in a flame, in a heated tube, or in a plasma. Fibers made using these processes have low values of α. Direct melt or liquid phase techniques produce fibers with higher values of α (on the order of 3×10^3

TABLE Q57.3 SOME LOSS CHARACTERISTICS OF THREE CHALCOGENIDE GLASSES (in dB km^{-1})

Glass	Source Of Optical Loss: Urbach Tail	Multiphonon Absorption Tail	Weak Absorption Tail
40 As-60S	$1.6 \times 10^{-11} \exp(23/\lambda)$	$1.8 \times 10^{9} \exp(-95/\lambda)$	$5.9 \exp(4.4/\lambda)$
38 As-5Ge-57Se	$9.3 \times 10^{-8} \exp(25/\lambda)$	$4.7 \times 10^{8} \exp(-124/\lambda)$	$5.7 \exp(6.4/\lambda)$
20Ge-80S	$1.0 \times 10^{-8} \exp(17/\lambda)$	$1.1 \times 10^{8} \exp(-59/\lambda)$	I $66 \exp(3.2/\lambda)$ II $16 \exp(3.2/\lambda)$

λ is in μm; glass composition is in wt %.
Source: T. Kanamori, et al., "Chalcogenide fibers for mid infrared transmission," *J. Lightwave Technology 2* (5), pp. 607–613, 1984.

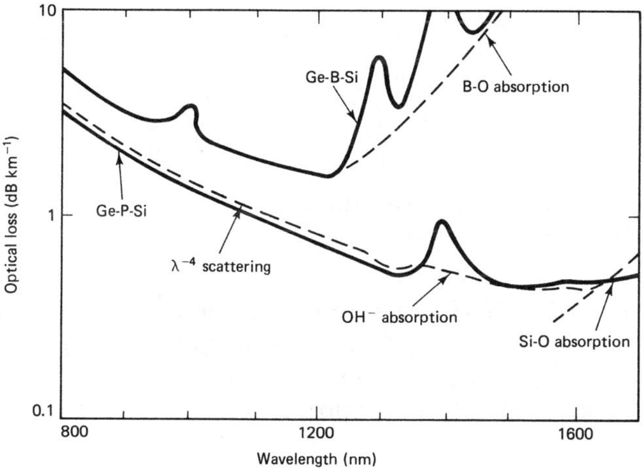

Figure Q57.5 Scattering and absorption optical losses in an optical fiber (graded-index silica glass).

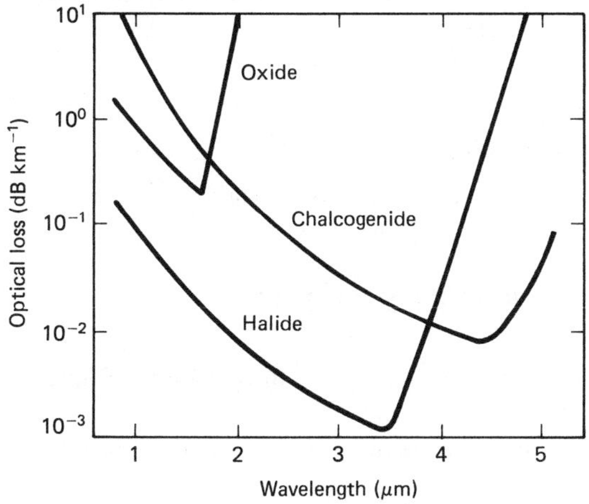

Figure Q57.6 Projected minimum optical losses in glasses [1].

TABLE Q57.4 VALUES OF F_R AND B AT VARIOUS WAVELENGTHS FOR THREE TYPES OF GLASS

Glass Type and Example	λ (μm)	F_R	B (μm^4 dB km^{-1})
Silica (SiO_2)	1.27	0.008	0.470
Chalcogenide (As_2Se_3)	7.60	0.052	0.080
HMFG (Zr_2BaF_{10})	1.63	0.030	0.003

Source: M. E. Lines, "Scattering losses in optic fiber materials: II. Numerical results," *J. Appl. Phys. 55* (11), pp. 4058–4063, 1984.

TABLE Q57.5 REDUCTION IN OPTICAL LOSSES IN OPTICAL MATERIALS, 1972–86

	Optical Loss (dB km^{-1})	
Year	Silica	Fluoride Glasses
1972	10	—
1976	0.5	—
1980	0.2	100.0
1982	0.2	3.0
1984	0.2	1.5
1986	0.2	0.5

Source: N. Poulain, "New glasses for optical fibers," *Endeavour 11* (1), p. 27, 1987.

to 2×10^4 dB km^{-1}). Such fibers are best suited for use in optics systems for short-distance multimode applications. A third fabrication technique, sol-gel, is currently receiving serious research and development attention.

In addition to low attenuation, other desirable properties for an optical fiber material are as follows:

- Low pulse dispersion (or broadening); this determines the information-carrying capacity of the fibers
- High strength
- High chemical durability
- Ease of fabrication in long length
- Absence of toxic and/or exotic starting materials

REFERENCES

1. S. Shibata, et al., "Prediction of loss minima in infrared optical fibers," *Electronics Letters 17* (21), p. 776, 1981.
2. R. H. Stolen, "Nonlinearity in fiber transmissions," *Proc. IEEE 68* (10), pp. 1232–1236, 1980.

QUESTION 58
Cutting Tools

The ideal cutting tool material should have the hardness of diamond, the toughness of high-speed steel, and the chemical inertness of alumina.

In the absence of such an ideal material, comment on the appropriateness of ceramics as cutting tool materials.

There is a clear consensus that ceramics have enormous potential as cutting tool materials. They are hard, retain their hardness at elevated temperatures (Table Q58.1), and have a relatively low reactivity with steel. Thus they can be used at high cutting speeds without deformation or dissolution. Their main drawbacks are low fracture toughness and susceptibility to shock loading.

Six categories of ceramic tool materials are in common use today.

1. Pure oxide ceramics, notably Al_2O_3 containing a small amount of ZrO_2, which improves the toughness of the material, have low thermal conductivities and hence low thermal shock resistance. Thermal effects are particularly evident at high machining speeds or with short machining cycles or variable depths of cut (reduced feeds can be used at the beginning and end of the cut, as is done in computerized numerical control of machining). Figure Q58.1 diagrams a simple cutting operation.

TABLE Q58.1 EFFECT OF TEMPERATURE ON THE HARDNESS OF VARIOUS CERAMICS (INDENTER LOAD: 1 kg; INDENTATION TIME: 10s)

Temperature (°C)	Vickers hardness (kg mm^{-2})		
	Sialon	Black Al_2O_3-TiC	White Al_2O_3
20	1850	2210	2000
200	1700	1850	1650
600	1500	1250	1150
1000	1250	900	780

Source: B. North, "Substitution of ceramics for conventional cutting tools," *Materials and Society,* p. 62, Spring 1984.

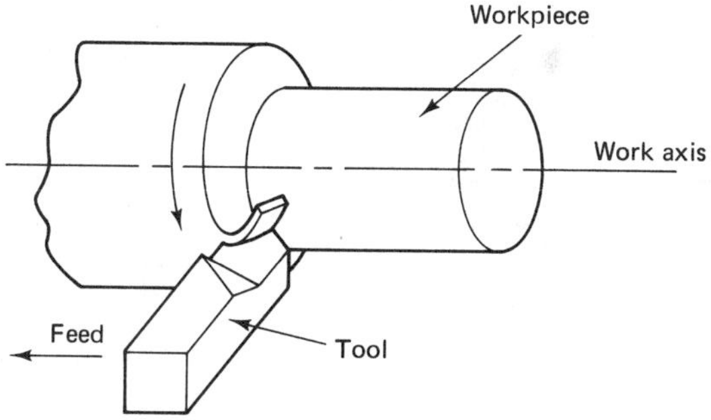

Figure Q58.1 Schematic representation of a simple metal cutting (or turning) operation.

2. Mixed oxides are much better than pure oxides and are obtained by addition of either TiC (because it is usually necessary to hot-press the powder, the shapes of tools made of this oxide are limited) and/or TiN. (This oxide is preferred since this composition can be cold-pressed and sintered). A typical mixed oxide ceramic is thus Al_2O_3-ZrO_2-TiC-TiN.

A recent example of a mixed oxide is Al_2O_3 strengthened with SiC whiskers. The whiskers improve the fracture toughness of the oxide vis-à-vis the unreinforced alumina. This new ceramic is considered the state-of-the-art material for the cutting tool to be used in rough machining of superalloys, especially in aerospace applications. The reactivity of SiC with ferrous materials has, however, limited the use of this cutting tool material to the machining of steel at very low cutting speeds.

3. A sialon is a solid solution of Si_3N_4 and metal oxides such as Al_2O_3, Y_2O_3, MgO, and BeO. The silicon nitride has a very low coefficient of linear expansion, high thermal conductivity, and hence excellent thermal shock resistance. It is not easily sinterable to full density, although this requirement may be achieved by substituting some of the silicon and nitrogen in silicon nitride by aluminum and oxygen. A metal oxide, such as Y_2O_3, is added to improve sintering further.

During sintering, silica on the surface of the Si_3N_4 particles reacts with alumina and yttria to form a low-melting-point liquid. The silicon nitride reacts with this liquid to form sialon plus, on cooling, a glass (Fig. Q58.2). Depending on the relative amounts of the reactants, the sialons so formed can have the atomic arrangement of the β-Si_3N_4 or the α-Si_3N_4. The β sialon has the composition, $Si_{6-z}Al_zO_zN_{8-z}$, where z is the extent of the substitution of Si and N by Al and O_2. The α sialon has the composition $M_x(Si, Al)_{12}(O, N)_{16}$, where M is a metal atom such as yttrium. An outline of the Jänecke triangular prism for the Si-Al-O-N-Y system is given in Fig. Q58.3. In general, the hardness of sialon increases with the amount of the α phase, and this increase is retained at elevated temperatures (Table Q58.2).

The major limitation of sialons is in their use to machine steel, except very hard steels. This limitation is related to their poor resistance to dissolution in the

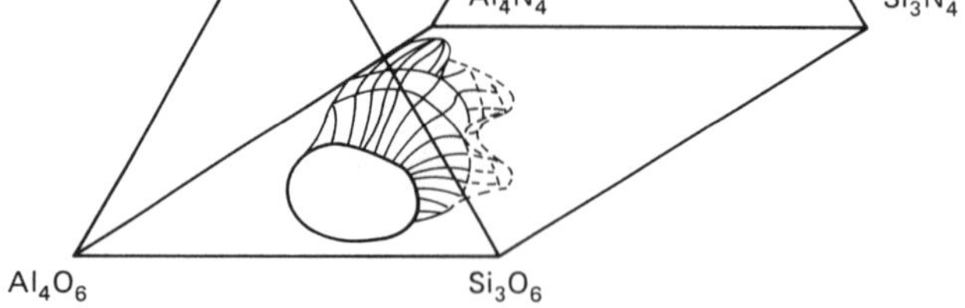

Figure Q58.2 Glass-forming region of the Y-Si-Al-O-N system on furnace cooling from 1700°C [2] (reprinted with permission from the American Society for Metals).

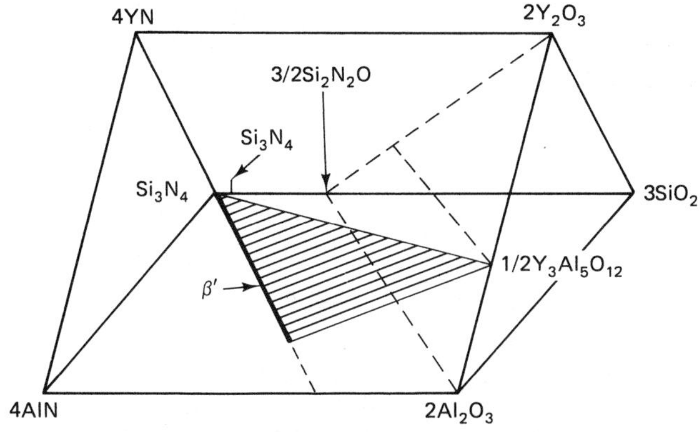

Figure Q58.3 The Jänecke triangular prism for the yttrium-sialon system [2] (reprinted with permission from the American Society for Metals).

TABLE Q58.2 THE VARIATION OF HARDNESS WITH TEMPERATURE FOR A 50α-50β SIALON COMPOSITE COMPARED WITH PURE β SIALON

Temperature (°C)	Vickers hardness (kg mm^{-2})	
	Pure Sialon	50-50α/β Sialon
20	1550	1950
600	950	1450
1000	450	1050

Source: K. H. Jack, "Nitrogen ceramics," *Endeavour 11* (2), p. 85, 1987.

TABLE Q58.3 PREDICTED SOLUTION WEAR RATES OF CERAMIC TOOL MATERIALS IN STEEL AT 1027°C

Material	Log (Predicted Relative Solution Wear Rate)
Zirconium dioxide	0.00014
Aluminum oxide	0.010
Titanium oxide	0.015
Titanium nitride	1.5
Titanium carbide	5.0
Silicon nitride	200.0

Source: D. H. Jack, "Ceramic tool materials," paper presented at the Hi-Tech 6 International Conference, March 1985.

workpiece (Table Q58.3). This tendency is linked to thermodynamic considerations; that is, the more negative the free enthalpy of formation, the greater the chemical stability of the material and the less the degree of reaction that might be expected between the material and the workpiece (Fig. Q58.4). On its own, Si_3N_4 is not very chemically inert, although alloying with aluminum and oxygen to form a sialon does involve stability solubility.

Sialon tools have been partially successful in the machining of alloys that could not be machined with sintered alumina inserts. For example, Brooks [1] reported that with sialon tools, Inconel alloy 718 was worked at 15 m s^{-1} with

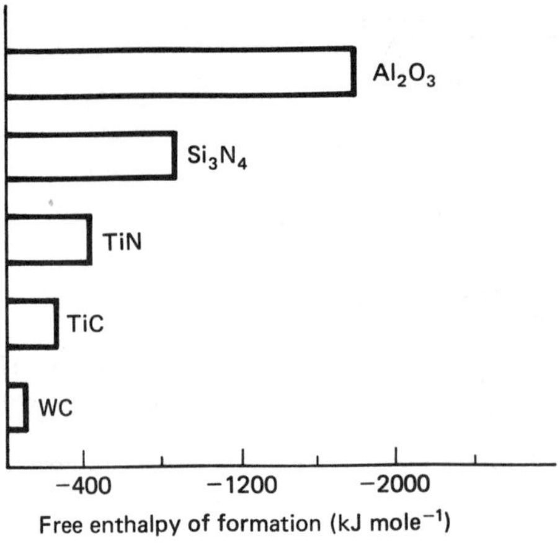

Figure Q58.4 Free enthalpy of formation of various ceramic materials at 260°C. (Data courtesy of Kyocera Feldmuehle, Inc.)

feed rates in excess of 0.305 mm r^{-1} and depths of cut between 3.81 and 5.08 mm. In the same operation using carbide inserts, speeds were limited to 2.3 m s^{-1}. According to Jack, "The lead time for machining some of the RB 211 aircraft engine turbine discs is said by Rolls-Royce to have reduced from 22 weeks to 5 weeks by using the conventional sialon" [2]. Researchers at Ford Motor Company's Manufacturing Process Laboratory in Redford, Mich., have reportedly developed a silicon nitride base material (known as S8) that is claimed to outperform most cutting tool materials currently available. For example, under identical test conditions, Ford engineers claim to have machined a cast-iron direct clutch cylinder in 58s with a conventional tool, and in 18s with an S8 tool [3].

The properties of pure oxide, mixed oxide, and sialon ceramics are compared in Table Q58.4.

TABLE Q58.4 PROPERTIES OF PURE OXIDE (Al_2O_3), MIXED OXIDE (Al_2O_3 + TiC), AND SILICON NITRIDE–BASED (SIALON) CERAMICS

Property	Al_2O_3[a]	Al_2O_3 + TiC	Sialon[b]
Density (kg m^{-3})	3990	4290	3260
Vickers hardness (GPa)			
At room temperature	17.7	21.8	18.34
At 1000°C	7.8	8.8	12.1
Fracture toughness (MPa $\sqrt{m}$)	4.3	4.5	6.5
Modulus of elasticity (GPa)	390	416	304
Thermal conductivity			
At room temperature	29.5	21.6	53.5
At 1000°C	7.6	9.8	8.5
Coefficient of linear expansion[c] (10^{-6} K^{-1})	8.2	8.6	3.1
Bend strength (MPa)	700	910	750
Abrasive wear resistance index[d]	1.00	1.27	1.88

[a] +1 wt ZrO_2.

[b] Values vary with specific grade. Two common grades are Kyon 2000 (Kennametal, Inc., Latrobe, Pa. and CC680 (Sandvik Hard Materials, Coventry, England).

[c] Mean of values at room temperature and 1000°C.

[d] $K_{IC}^{0.5}E^{0.8}H_e^{1.43}$, where K_{IC} = fracture toughness, E = modulus of elasticity, and H_e = hardness at 1000°C; value expressed relative to Al_2O_3.

4. Two types of zirconia-toughened ceramics are common today: phase-transformation-toughened alumina (PTTA) and partially stabilized zirconia (PSZ). The increased toughness of PTTA, compared to conventional sintered alumina inserts, is attributable to a stress-induced phase transformation in the zirconia particles. Metastable grains of tetragonal ZrO_2 are incorporated in the Al_2O_3 mixtures during fabrication. In use, when a crack forms, the metastable tetragonal ZrO_2 transforms to the stable monoclinic structure, thus increasing the energy required for the crack to propagate, and hence its growth is retarded [4]. PSZ is a ceramic material in which a complex microstructure is developed by heat treatment of a zirconia-rich solid solution.

5. Boron carbide has been attracting a lot of attention recently. It is not only harder than alumina but is exceeded in hardness by only two other materials—cubic boron nitride and diamond. Hot-pressed boron carbide has been shown to have excellent wear resistance in machining very abrasive materials such as Al-Si alloys as well as good resistance to cutting tool workpiece welding [5]. One disadvantage of this material, its brittle nature, is now being studied with a view to improving its fracture toughness.

6. Cubic boron nitride (CBN), with a hardness second only to that of diamond, is today a well-established and commercially successful cutting tool material (sold under the trade names Borazon and Amborite, for example). The hardness of CBN is not seriously impaired until subjected to temperatures in excess of about 1000°C. The material also possesses good thermal stability, oxidation resistance (up to ~1000°C), as well as extremely high thermal conductivity. The combination of high hardness and reduced chemical reactivity of CBN makes it attractive as the material for a cutting tool to be used at very high cutting speeds, leading to substantial improvements in productivity, as in the machining of jet engine components [6]. CBN-based tools have also been used successfully to machine cobalt-based superalloys, nickel-based superalloys, very abrasive hypereutectic Al-Si alloys, and ultra-high-strength steels, such as AISI 4130. However, it is not suitable for machining soft, low-alloy steels, possibly because of undesirable chemical reactions. Another disadvantage is its high cost.

REFERENCES

1. R. Brooks, *American Metal Market/Metalworking News,* April 28, 1982.
2. K. H., Jack, "Sialon tool materials," *Metals Technology,* pp. 297–301, July 1982.
3. *Manufacturing Engineering,* p. 21, February 1983.
4. F. F. Lange, U.S. Patent 4,316,964, February 23, 1982.
5. M. Furukawa and T. Kitahuia, *Nippon Tungsten Review 12,* p. 55, 1979.
6. L. E. Hibbs and R. H. Wentorf, *High Temperature–High Pressure 6* (4), p. 409, 1974.

Part III

Latest Developments in the Use of Materials

MATERIALS FOR HYPERVELOCITY VEHICLES

CHAPTER C1

C1.1 THE NATIONAL AEROSPACE PLAN OF THE UNITED STATES

President Reagan, in his State of the Union address in February 1986, announced a nationwide research effort to develop a craft, officially called the National Aerospace Plane (NASP), the Transatmospheric Vehicle (TAV), the X-30 program, or the single-stage-to-orbit (SSTO) vehicle. The Defense Department and NASA define the official NASP goal as "technology development and demonstration to provide the basis for hypersonic flight vehicles leading to space transportation systems, superior U.S. military aircraft, and civil transports that will have technical, cost and operational advantage over existing systems." The near-term NASP objective is "to develop, and demonstrate in an experimental flight vehicle (the X-30), the requisite technologies to permit the nation to develop both military and civil vehicles capable of operating at sustained hypersonic speeds within the atmosphere and/or operating as space launch vehicles for delivering payloads into orbit."

In simple terms, it is envisaged that the NASP will fly from Washington, D.C., to Tokyo in less than 2 h (hence its unofficial nickname, the *Orient Express*), zoom into orbit to battle enemy missiles, and roar off on bombing missions at 25 times the speed of sound (about 27,000 km h^{-1}) at altitudes up to 106 km.

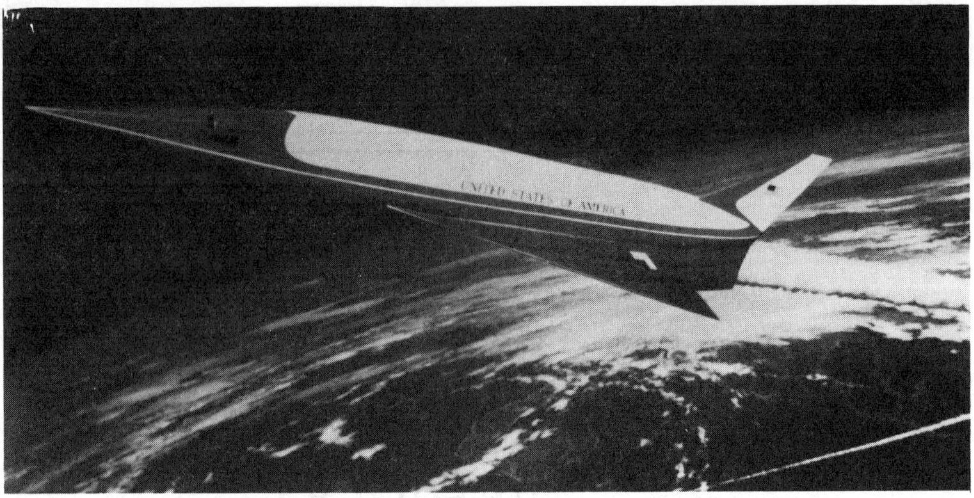

Figure C1.1 Artist's impression of the National Aerospace Plane of the United States. (Courtesy of *Aerospace America*)

An artist's impression of the NASP is shown in Fig. C1.1. The plane will take off from a runway, accelerate to orbital velocities, and descend and land in a conventional manner. It will also probably cruise in the 24–40-km zone, an area that is inaccessible to jet aircraft, which lack the requisite propulsion system. The NASP seeks to be a heretofore elusive entity: a fully reusable space vehicle. Its payload fraction should be about 10 times greater than that of rocket systems. An NASP, it is hoped, would accomplish a multitude of goals, among them these:

- Eliminate conventional rocket launch staging
- Afford greater flexibility by using conventional airfields, in some cases
- Improve system flexibility by eliminating U.S. reliance on a few launch complexes, such as at the Kennedy Space Flight Center in Florida and Edwards Air Force Base in California
- Provide improved mission safety through a combination of new features such as multiengine redundancy, aircraft-type control, and full abort capability
- Greatly reduce the number of ground support staff and eliminate vertical assembly buildings, launch pads, solid boosters, external fuel tanks, and a host of logistics problems
- Operate (with six vehicles at a projected rate of 150 launches per year) at overall costs that are $\frac{1}{10}$ to $\frac{1}{100}$ those of current rocket launch systems.

Two engines would be needed: one to boost the plane to speeds of about Mach 2.5; then a revolutionary new engine, called the supersonic combustion ramjet or scramjet, would take over, powering the plane to hypersonic speeds (Fig. C1.2). In theory, air, heated to extremely high temperatures by the supersonic speeds, would rush into the scramjet engines and be mixed with liquid hydrogen.

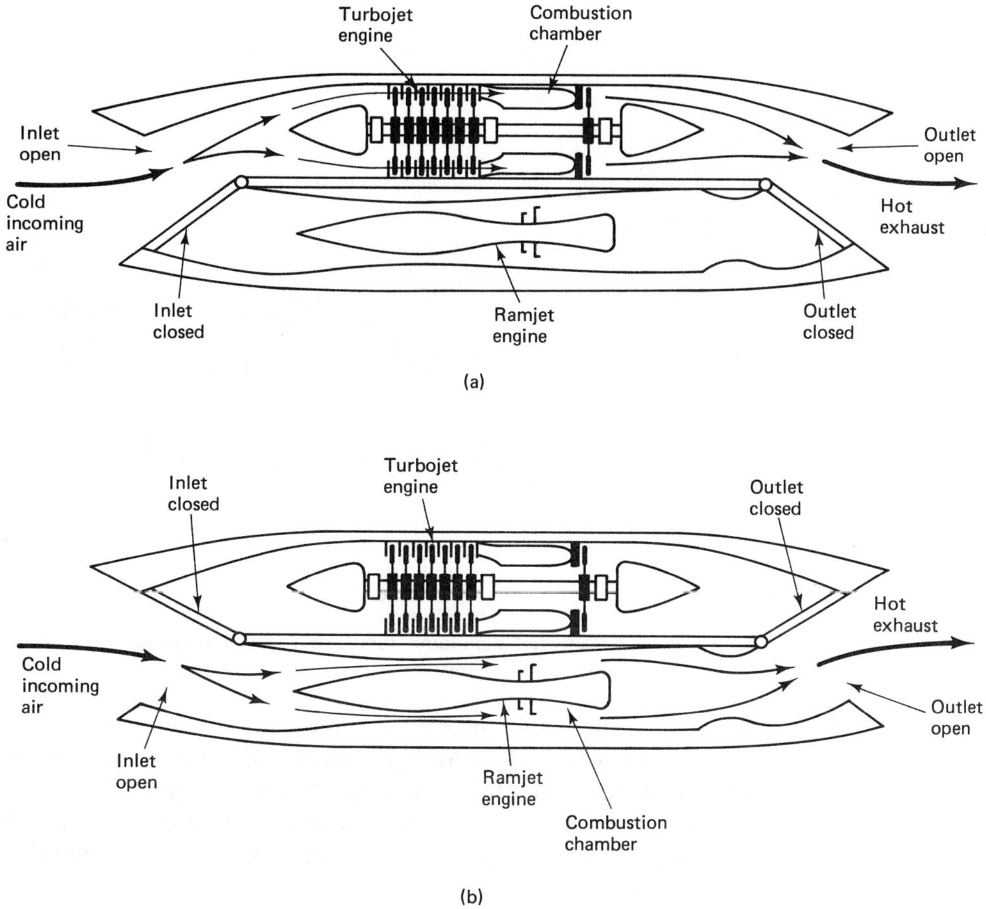

Figure C1.2 The two-in-one engine concept for the National Aerospace Plane: (a) low-speed operation to Mach 2.5; (b) high-speed operation beyond Mach 2.5.

The hot air would ignite the hydrogen, and the explosion would propel the plane forward. Unlike normal jet engines, the scramjet would not have a compressor to pack the air together to make it easier to burn. Instead, the plane's shape would compact it.

However, as air rushes past an airplane at high speeds, it is compressed by shock waves that form along the fuselage. At Mach 25, those shock waves are powerful enough to damage a plane with a circular body or one that is not optimally shaped. Blasting along at Mach 25 could heat the skin of the plane up to 6100°C. Even at lower speeds, to create the massive airflows the plane would need, the entire fuselage would serve as an extension of the air intake. Figure C1.3 shows the intense heat that the airframe of the plane would have to endure. Thus the successful operation of the NASP is contingent on proper methods of containing and/or dissipating this heat. This would be achieved in three ways. One is by shaping—using concepts such as the "spinning leading edge" incorporated in the plane's wings and nose. The nose, for example, would be a ball that would rotate

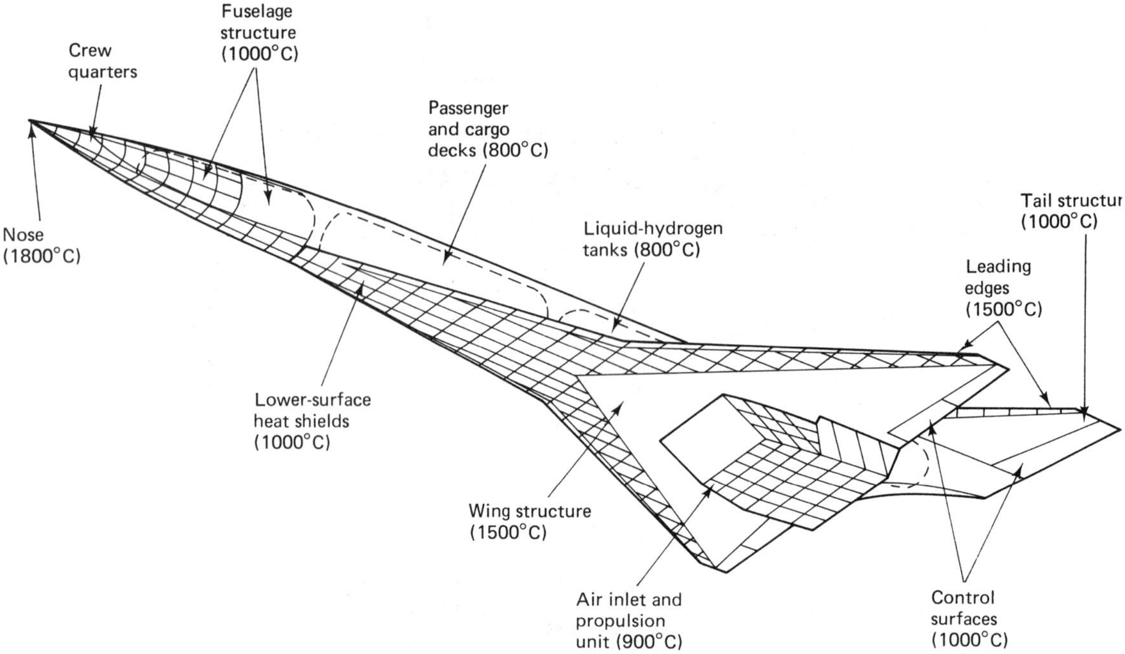

Figure C1.3 Anticipated maximum temperature at various sections on the fuselage of the U.S. National Aerospace Plane.

inside the fuselage to help keep heat from building up at any point, and the wings must use spinning cylinders. A second means is circulating hydrogen through the engine frame and nose, areas that experience the highest heat flux. Third is using high-temperature-resistant, high-strength, lightweight, and fully reusable materials. Preliminary indications are that a variety of advanced materials will be needed, for example:

- Metal-matrix composites (boron fiber reinforced titanium, tungsten fiber reinforced Fe-Cr-Al-Y alloys, silicon carbide reinforced metals, Inconel 718 alloys reinforced with fibers of Al_2O_3 or TiC, for example), to be used in engine cowl lips
- Carbon-carbon composites for use in the nose caps and wing leading edges
- High-temperature polymers reinforced with ceramic or superalloy fibers for the airframe, control surfaces, and engine structure
- Rapid-solidification-rate titanium alloys to be used in engine parts
- Intermetallic compounds, such as titanium aluminides, to be used in the engine's combustion chamber
- Monolithic ceramics, to be used in nose cap and leading edge inserts

A number of contracts for the concept development and airframe design phases of the NASP were awarded in 1987 to various aerospace companies.*

*$25 million each to General Dynamics and Rockwell International for preliminary design of the airframe and $81 million each to Pratt & Whitney and Rocketdyne for technical development of the engines.

Figure C1.4 compares one NASP design with that of other aircraft on the basis of speed (Mach number) and range factor (defined as the product of Mach number and the lift-to-drag ratio divided by specific fuel consumption); Fig. C1.5 compares the specific impulse. The flight envelope of the X-30 encompasses the commercially interesting hypersonic transport regime, which extends well beyond that of even the most advanced jet aircraft (Fig. C1.6).

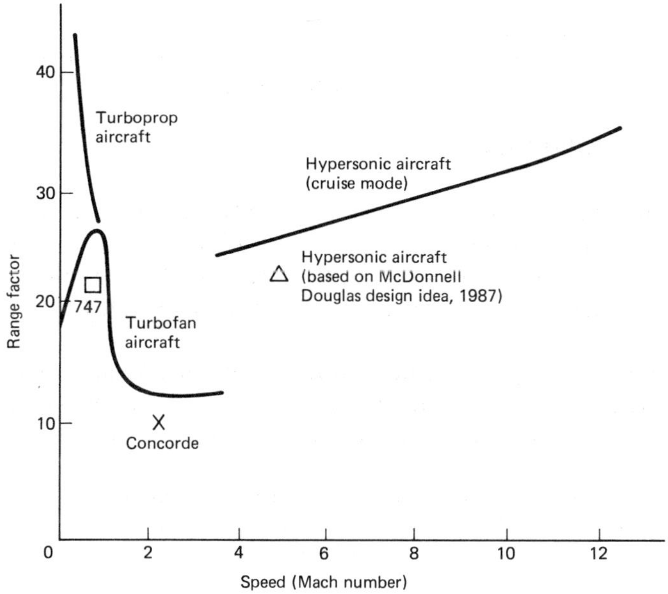

Figure C1.4 Range factor–Mach number envelope of various types of aircraft.

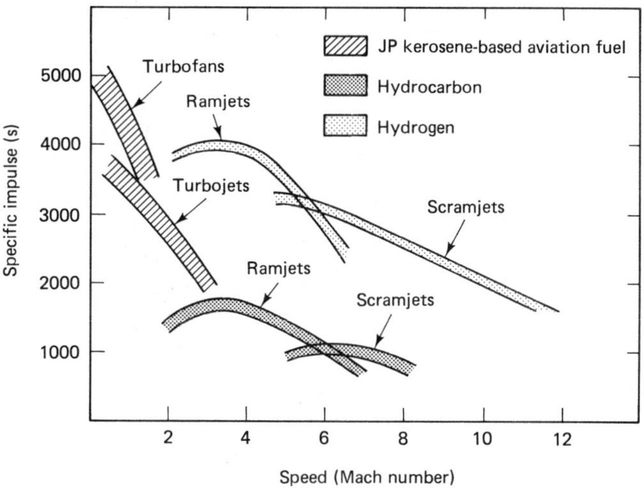

Figure C1.5 Specific impulse of different propulsion systems at various Mach numbers at optimum altitude.

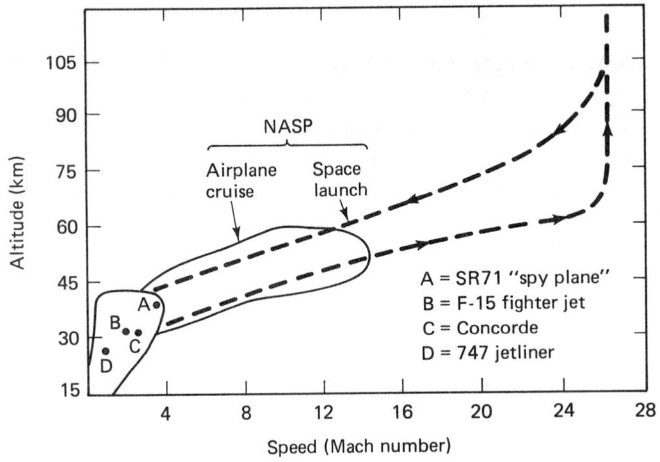

Figure C1.6 Flight envelopes of various advanced aircraft.

C1.2 THE SANGER STS OF WEST GERMANY

The West German government proposed an alternative to the NASP of the United States. The vehicle, known as the Sanger Space Transportation System (STS), would use a large air-breathing first-stage hypersonic aircraft about the size of the Boeing 747 aircraft and be capable of at least Mach 6 and possibly Mach 10 (Fig. C1.7).

The first stage would carry the plane to an altitude of about 30 km, where the orbital vehicle would detach and a conventional cryogenic rocket engine would carry the plane to an orbital altitude of about 40 km. It is envisioned that a

Figure C1.7 A diagram of the artist's impression of the Sanger space transportation vehicle. (Courtesy of *Aviation Week and Space Technology;* details correct as of November 1988)

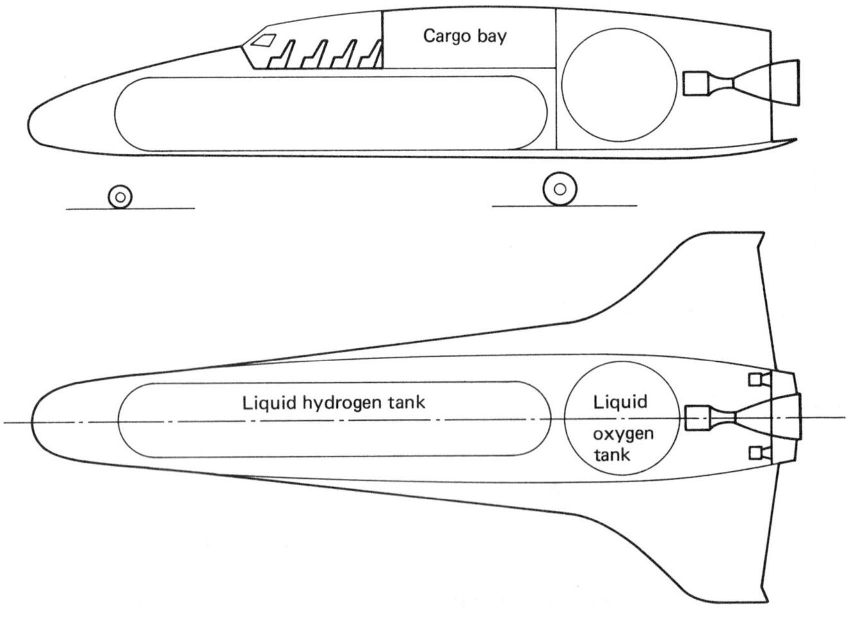

derivative of the first stage could serve as a 200-passenger transatmospheric vehicle, with a range of about 13,000 km.

Further studies are continuing at the aerospace company Messerschmitt Boelkow-Blohm (MBB), with some funding from the West German Aerospace Research Establishment, to refine the plans for this plane.

The material requirements for this plane, particularly for its airframe, would be identical to those of the NASP.

C1.3 THE SPACE TRANSPORTATION VEHICLE

The Space Transportation Vehicle (STV) was proposed by a U.S. engineering firm, TAAS Co., of Atlanta, Ga., as an alternative space transportation system. Its relatively simple design involves replacing the wings of a conventional plane with ones that are more nearly normal to the fuselage (Fig. C1.8), with an aspect ratio near 10. The combination of maximum lift-to-drag ratio and minimum aerodynamic pressure has three advantages: (a) It allows the plane to use a metallic thermal protection system, yielding considerable weight savings and durability (by contrast, the ceramic skin of the U.S. space shuttle weighs about 8.2 t, or 14% of its empty weight, a penalty that is reflected in decreased maneuverability). (b) It permits the STV to take off like conventional aircraft. (c) It results in efficient aerodynamic orbital maneuvering at about 61 km. The vehicle is designed as a low-cost system with great flexibility and a low-earth-orbit payload capacity of 1.82 t. Propulsion will be supplied by rockets using a mixture of liquid oxygen and liquid hydrogen propellants, but more efficient scramjet engines could be used once these became available. According to the firm, the fuselage of the STV would suffer a maximum temperature of 760°C while maneuvering in near-space. (By contrast, parts of the fuselage of the U.S. space shuttle experience temperatures up to 1320°C.)

In addition to high-temperature-resistant superalloys, other suitable materials for the airframe of the STV are the same as those listed for the NASP.

Figure C1.8 Artist's impression of the Space Transportation Vehicle. (Courtesy of TAAS Co.)

C1.4 SUMMARY

The selection of materials for use on external surfaces of hypervelocity vehicles is governed primarily by the need to protect these sections from the very high temperatures created by aerodynamic heating during flight. To keep the weight of these vehicles to the minimum possible, high-temperature-resistant but lightweight materials, rather than thermal protection shields, are preferred. Examples of the former were given in section C1.1, and various aspects of them are presented in more detail in Chapters C5 through C8 and C15 in Part III.

ALUMINUM-LITHIUM ALLOYS

CHAPTER C2

C2.1 BINARY ALLOYS

The ability to alloy large amounts of lithium with aluminum is a fairly recent development, since the highly reactive lithium can explode during alloying. Aluminum-lithium (Al-Li) alloys are made in a controlled atmosphere, then formed like a typical aluminum ingot. It does not take much lithium to achieve the desired result.

Until about 1977, not much was known about the Al-Li alloy system. Phase diagrams were unavailable; mechanisms of precipitation, precipitate morphology, and oxidation were little understood. In addition, data on mechanical properties, such as fracture toughness and endurance limit, was not available. Today, however, these alloys are better understood. For example, there are indications [1] from electron microscopy and x-ray work that precipitation in binary Al-Li alloys proceeds thus:

$$\text{Supersaturated solid solution} \longrightarrow \delta'(\text{Al}_3\text{Li}) \longrightarrow \text{Al-Li}$$

The structure of these alloys has been studied by means of electron energy loss spectroscopy [2], a technique that is particularly suited to examination of these alloys because the low-weight atoms in them absorb little energy and emit low-energy x rays that are usually absorbed before they leave the specimen (thereby

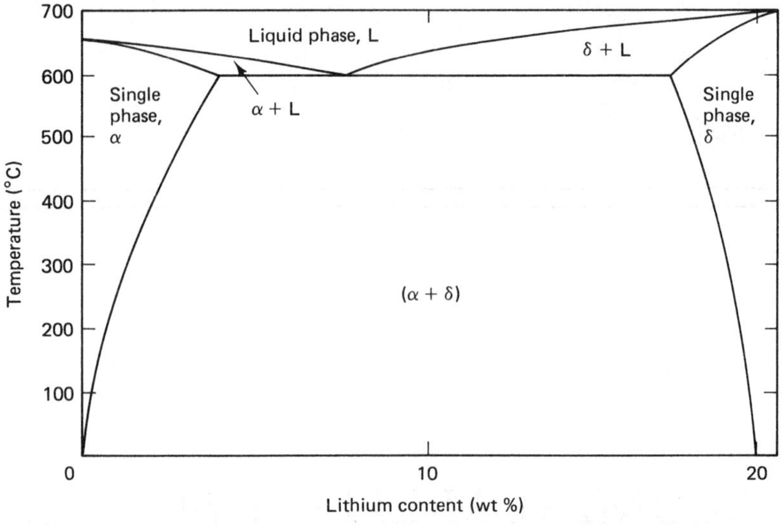

Figure C2.1 Diagram of the aluminum-lithium equilibrium phase.

making specimen structure determination using conventional x-ray spectrometry rather difficult).

The phase diagram for these alloys has been obtained (Fig. C2.1). In addition, the corrosion resistance of several Al-Li alloys has been studied by Sheppard and Parson [3]. Intergranular corrosion was investigated using the standard test method described in ASTM G67-80, and stress corrosion was studied using the C-ring test.

Some examples of aluminum-lithium alloys in commercial production are these:

- Weldalite (Reynolds Metals Co., Richmond, Va.; International Light Metals Corp., Torrance, Calif.)
- Alithalite 2090 (Alcoa, Alcoa Center, Pa.)
- Kalite (Kaiser Aluminum & Chemical Corp., Oakland, Calif.)
- Alloy 8090 (British Alcan)
- Lital A (British Alcan and Royal Aircraft Establishment, England)

All are available in sheet, plate, and extrusion forms and in sizes similar to those of conventional high-strength aluminum alloys.

Al-Li alloys possess the following properties:

- Low density. Although lithium accounts typically for only 2.6% of the weight of these alloys, its volume fraction is much higher (16%).
- They can be shaped by superplastic forming, a process that uses heat and pressure to deform metal plates. Superplastic forming makes it possible to consolidate several small pieces into one large unit, thereby simplifying construction and reducing labor costs. The process also reduces the weight of the final product by eliminating overlapping flanges used to fasten sections together.

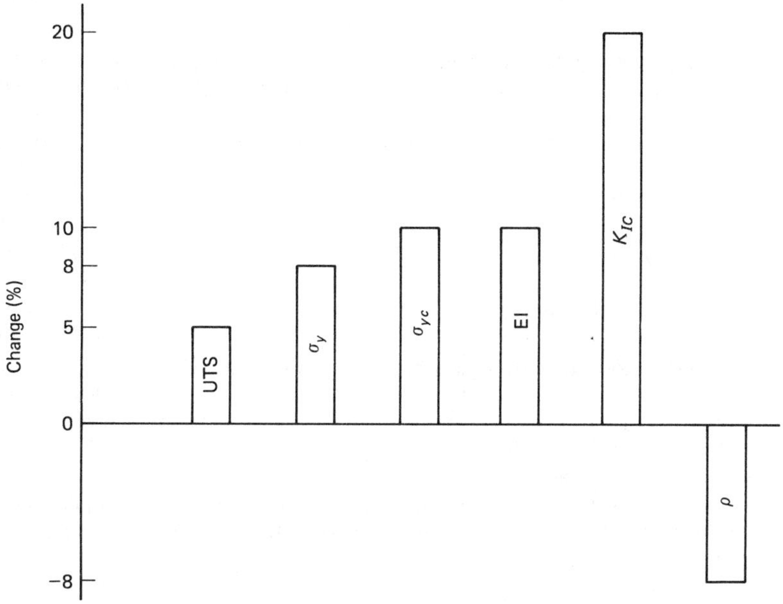

Figure C2.2 Comparison of some mechanical properties of an Al-Li alloy with 7075-T6 Al alloy (σ_{yc} = compressive yield strength; EL = tensile elongation).

- High ductility; typically 6% to 10% for sheet and plate products.
- Good damage tolerance.
- Good corrosion resistance.
- Excellent mechanical properties; typically, modulus of elasticity $E = 79$ GPa, ultimate tensile strength (UTS) = 556 MPa, tensile yield strength (σ_y) = 510 MPa, fracture toughness (K_{IC}) = 33 MPa$\sqrt{\text{m}}$, and density (ρ) = 2685 kg m^{-3}. Some mechanical properties of an Al-Li alloy are compared with those of a competitive conventional aluminum alloy (7075-T6) in Fig. C2.2.
- Ease of fabrication and processing using conventional equipment and methods. However, it must be noted that operations that generate molten metal, such as flame cutting and welding, present serious explosion hazards when moisture is present. These alloys may be solution-heat-treated and aged in most present-day furnace equipment, but salt bath solution heat treatment is not recommended because of possible severe exothermic reactions.

Furthermore, Papazian et al. [4] have shown that the strength of these alloys decreases approximately linearly with the square root of solution treatment time (Table C2.1). They suggest that this is due to lithium depletion in the near-surface region, which produces various other adverse effects, such as local recrystallization and a shift in galvanic potential. Although this loss in strength is only marginal over long solution treatment times, these workers propose that design stress values for thin sheet may have to be lower than those for thicker sections.

Another disadvantage of Al-Li alloys is their easy tendency to oxidize at elevated temperatures [5], leading to decreases in microhardness [6], electrical resistivity [7], and precipitate density [6] in the near-surface zone.

TABLE C2.1 LOSS IN STRENGTH (IN MPa) OF 1.6 mm THICK Al-Li ALLOY SHEETS AS A FUNCTION OF SOLUTION TREATMENT TIME, t (IN MIN.)

Alloy and heat treatment environment	Linear regression fits
8090-T6; air furnace	UTS[a] = 453-2.5$t^{0.5}$ PS[b] = 354-2.5$t^{0.5}$
8090-T6; sodium nitrate salt bath maintained at 530°C	UTS = 456-2.5$t^{0.5}$ PS = 355-1.7$t^{0.5}$
8091-T6; air furnace	UTS = 467-2.2$t^{0.5}$ PS = 383-1.9$t^{0.5}$
8091-T6; sodium nitrate salt bath maintained at 530°C	UTS = 459-1.1$t^{0.5}$ PS = 383-1.2$t^{0.5}$

[a]Ultimate tensile strength

[b]0.2% proof strength

Source: J. M. Papazian, G. G. Both, and P. Shaw, "Effects of lithium loss on strength and formability of Al-Li alloys 8090 and 8091," *Materials Science and Engineering 94*, pp. 219–224, 1987.

On the whole, Al-Li alloys possess enough attractive properties to make them the material of choice in applications such as these:

- Wing sections of the U.S. Air Force F-15 Eagle fighter jets (Fig. C2.3), such as the upper inboard aft wing skin
- Forward area bulkheads and internal fuel tanks of the Advanced Tactical Fighter of the U.S. Air Force
- Forgings for the wheels and brakes of aircraft (Fig. C2.4)
- Precision forging for the wing body attach fitting of a Boeing 757 aircraft

McDonnell Douglas's large commercial order for an Al-Li alloy (Alcoa's Alithalite 2090) in July 1987 signaled the start of the acceptance of this material

Figure C2.3 Aluminum-lithium wing skins, shown in grey, on a U.S. Air Force F-15 fighter jet. (Courtesy of McDonnell Douglas)

Figure C2.4 Inner wheel half of a commercial jet aircraft, the Boeing 757. (Courtesy of BF Goodrich).

in the U.S. aerospace industry. This alloy, used in the floor beams of the MD-11 transport planes, reportedly reduced the weight of the aircraft by 145 kg; 82 kg of this reduction is achieved because the alloy is 7% lighter than Al alloy 7075-T6 (the material that could have been used), and 63 kg of the reduction is gained through redesign that takes advantage of the high strength of the Alithalite alloy.

In Europe, Al-Li alloys have been used in various aerospace applications, including the flaperon skins and carriage floor in the demonstrator for the Eurofighter aircraft and the fuselage skins, wing, and fin attachment areas in the demonstrator for Dassault Breguet's Rafale fighter aircraft.

C2.2 OTHER Al-Li–BASED ALLOYS

Another area of aluminum ingot metallurgy involves ternary, quaternary, and Al-Li-*X*-*Y*-*Z* alloys. They are less dense and much stiffer than conventional aluminum alloys. Two examples of these alloys are Lockalite and Navalite.* The Lockalite alloy composition involves modification in copper and magnesium solute levels to promote δ' phase strengthening. The extruded alloy has an ultimate tensile strength of 552 to 620 MPa, a yield strength of 500 to 572 MPa, and a tensile elongation of 7% to 9%. Navalite has good weldability and corrosion resistance, 3% lower density than most Al-Li alloys, a reasonably high value of the modulus of elasticity of 79 to 81 GPa, a yield strength of 414 to 450 MPa, an elongation at elastic failure of 8% to 10%, and a fracture toughness of 170 to 200 MPa$\sqrt{\text{m}}$.

Potential applications of these alloys include various airframe components of advanced fighter jet aircraft and critical components in next-generation gas turbine engines.

*Composition of Lockalite (International Light Metals, Torrance, Calif.) is Li, 1.6–3.1; Cu, 0.9–3.0; Mg, 0.8–1.8; Zr, 0.14; Al, balance. Composition of Navalite (Naval Surface Weapons Center, White Oak, Md.) is Li, 1.6–2.8; Cu, 0.9–1.4; Mg, 1.7–3.9; Zr, 0.14; Al, balance.

REFERENCES

1. B. Noble and G. E. Thompson, *Metal Sci. J. 5,* p. 114, 1971.
2. "Aluminum-lithium alloys," *Aerospace America 24* (5), p. 22, May 1986.
3. T. Sheppard and N. C. Parson, "Corrrosion resistance of Al-Li alloys," *Mater. Sci. Technology 3,* pp. 345–352, 1987.
4. J. M. Papazian, G. G. Both, and P. Shaw, "Effects of lithium loss on strength and formability of Al-Li alloys 8090 and 8091," *Materials Science and Engineering 94,* pp. 219–224, 1987.
5. D. J. Field et al., in E. A. Starke, Jr. and T. H. Sanders, Jr. (eds.), Aluminum-Lithium Alloys II, Warrendale: Pa., TMS-AIME, pp. 657–667, 1984.
6. S. Fox et al. *Scripta Metallurgica 20,* pp. 71–74, 1986.
7. J. A. Wert and A. B. Ward, Scripta Metallurgica 19, pp. 367–370, 1985.

NEW MAGNETIC MATERIALS

CHAPTER C3

C3.1 NEW PERMANENT MAGNETIC MATERIALS

Although magnetism has been well known for centuries, it was only in 1935 that the first of the modern permanent magnetic materials, isotropic Alnico (Fe-Al-Ni-Co) alloy, was introduced. Since then, development has been gradual, leading to rare-earth-based permanent magnetic materials, neodymium-iron-boron (Nd-Fe-B) alloys. This has led to a widening of the areas of application of permanent magnetic materials, among which applications are these:

- Automotive dc motor drives, starter motors, window winders, wipers, washers, fans, seat activators, speedometers, alternators
- Analog watches, model aircraft, headsets, deaf aids, electric clocks
- Electronic sensors, contactless switches, spherical motors, computer print heads, energy meter disc repulsion bearings, electromechanical transducers
- Aerospace couplings, traveling wave tubes, magnetrons, klystrons, autocompasses
- Dentures, orthopedic components, wound closures, stoma seals, ferromagnetic probes, cancer cell separators
- Magnets in nuclear magnetic resonance (NMR) imagers

TABLE C3.1 SOME PROPERTIES OF PERMANENT MAGNETIC MATERIALS

Material	Year of Introduction	Energy Product ($kJ\ m^{-3}$)	Cost[a] $\$\ kg^{-1}$	Cost[a] $\$\ [kJ\ m^{-3}]^{-1}$	Curie Temperature (°C)
Alnico 8	1935	43.8	0.15	0.48	850
$SmCo_5$	1970	143.3	1.00	1.00	724
Sm_2Co_{17}	1980	222.9	0.85	0.55	673
$Nd_2Fe_{14}B$	1983	254.7	0.81	0.46	320
		294.5	0.81	0.09	—
		358.2	—	—	—

[a]Relative to $SmCo_5$, at 1988 prices.

- Other applications that require a homogeneous magnetic field, such as beam focusing by quadrupole magnets, parts of the SDI systems, magnetic separators
- Brushless generators, wind-power generators, high-torque miniature motors for robots, drive systems for electrical transportation vehicles

The main competition for the material to use in most of these applications, as well as many others, is between Nd-Fe-B and another rare-earth permanent magnetic material, a samarium-cobalt alloy (Sm_2Co_7), which was introduced in 1970. The values of some relevant properties of Alnico, samarium-cobalt, and Nd-Fe-B alloys are given in Table C3.1, and the demagnetization curves at various temperatures for one of these materials are shown in Fig. C3.1. Industry sources predict the supercession of Sm_2Co_7 by Nd-Fe-B by 1992. There are essentially three reasons for this optimism. First and most important, neodymium is more readily available than samarium, so Nd-Fe-B alloys can be used for many high-volume applications. Second, these alloys contain no cobalt, which is a strate-

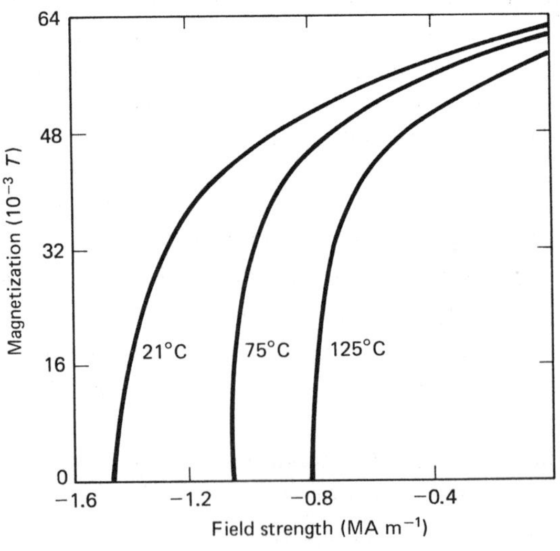

Figure C3.1 Demagnetization curves of hot-pressed $Nd_{0.14}(Fe_{0.91}B_{0.09})_{0.86}$ at various temperatures [1].

gically important material as far as the United States is concerned. Third, these alloys have a high energy product, hence the cost per energy product is low.

The hard magnetic properties of Nd-Fe-B have been attributed to the presence of a highly anisotropic primitive tetragonal $Nd_2Fe_{14}B$ ternary phase (Fig. C3.2). The Curie temperature of this phase is about 300°C, and this leads to a large temperature coefficient of coercivity so that at about 150°C the coercivity becomes very small, thereby limiting the high-temperature range of the Nd-Fe-B alloys.

Two methods have been studied in detail to increase the temperature coefficient of induction of these alloys (and thus improve their elevated-temperature performance). One is raising the Curie temperature and reducing the thermal demagnetization of the magnetic moments. Substitution of cobalt for iron increases the Curie temperature T_c, but the intrinsic coercivity force H_{ci} decreases [3]. Aluminum is found to be effective in increasing H_{ci} of Nd-Fe-Co-B alloys, but it decreases T_c (Table C3.2). Therefore, the amounts of Al, Co, and B in the Nd-

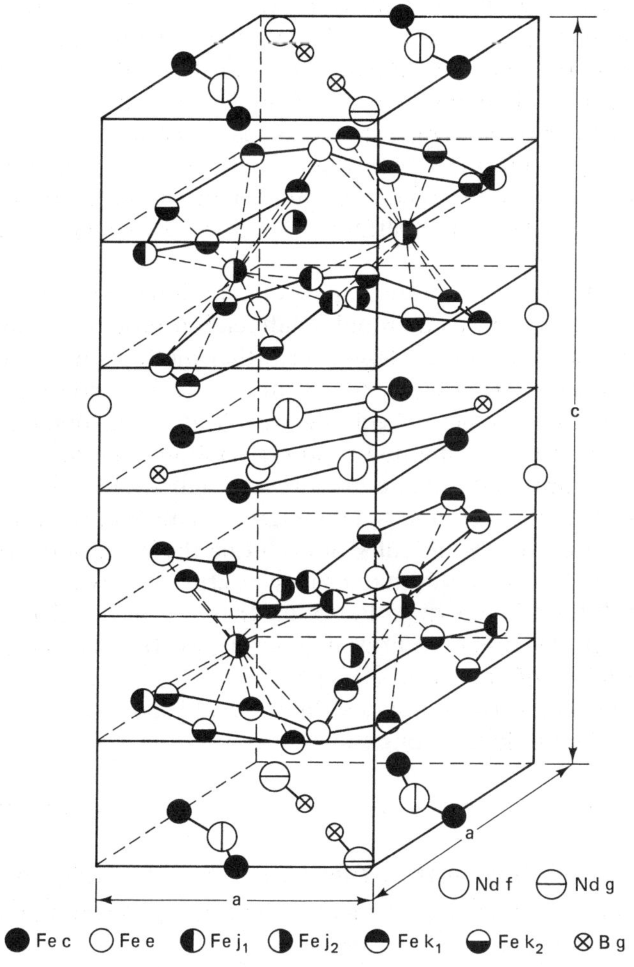

Figure C3.2 Unit cell of $Nd_2Fe_{14}B$ alloy [2]. (reprinted with permission from The American Physical Society)

TABLE C3.2 THE CURIE TEMPERATURE AND INTRINSIC COERCIVITY OF $Nd_{15}Fe_{57-x}Co_{22}Al_xB_6$ ALLOYS

Amount of Aluminum (at. %)	Curie Temperature (°C)	Intrinsic Coercivity ($MA\ m^{-1}$)
0	520	0.56
5	430	0.90
10	375	1.30
12	—	1.55
14	275	1.80
18	250	—

Source: B. M. Ma, and K. S. V. L. Naraisimham, "NdFeB magnets with higher Curie temperature," *IEEE Trans. on Magnetics MAG 22* (5), p. 916, 1986.

Fe-B alloy system have to be carefully controlled in order to optimize performance; that is, the best combination of good intrinsic coercivity, high T_c, and low temperature coefficient must be achieved. Results indicate that 20% Co is the optimum cobalt content [4]. Use of such a high quantity of cobalt in these alloys can, however, compromise their price competitiveness vis-à-vis Sm-Co alloys.

Another method for increasing the temperature coefficient of induction is alloying with heavy rare-earth elements to compensate for the rapid decrease in the neodymium moment with increasing temperature. An example of such an alloy is $(Nd_{0.33}Ho_{0.64}Dy_{0.13})_{15}Fe_{79}B_6$.

These new magnetic alloys can be produced by conventional powder metallurgy techniques, by sintering, or by using rapid solidification processing steps to bulk form by liquid dynamic compaction. However, other methods have been developed. The General Motors (GM) process, for example, involves pouring the molten Nd-Fe-B onto a rotation copper wheel, the resulting rapid quenching producing a fine-grained structure. The method produces ribbons that are either blended with a plastic and compression-molded or pressed and vacuum-impregnated with a plastic. The GM method produces a material with an energy product of only about 50 kJ m^{-3} because of the difficulty of orienting the fine grain structure of the melt-spun ribbon. One way that has proved useful in alleviating this problem is to crush and mill the ribbons or to hot-press them at, say, 725°C and 100 kPa, then to rotate the grains along the pressing directions; this has led to a material with an energy product of 380 kJ m^{-3}.

A number of companies in the United States, Japan, and Europe have commercial Nd-Fe-B alloys on the market, among them:

- Neomax 35 and Neo 30H (Sumimoto Special Metals Co., Egawa, Osaka, Japan)
- Crumax 40, 32, 30A, and 30B (Colt Industries Crucible, Pittsburgh, Pa.)
- Magnequench (Magnetic Products Department, Delco Remy Division, General Motors, Anderson, Ind.)

For example, Magnequench is being used in applications such as starter motors in automobiles, improved power zoom lens motors in cameras, downsized or more powerful hand-held tools, and antilock braking systems in automobiles. The alloy was used in the motor (3.67 kg, 1.5 kW, brushless dc type) in the Sunraycer, the solar-powered car designed and built by a General Motors/AeroVironment team that won the World Solar Challenge, a race for solar-powered cars held in November 1987 over a 3200-km course from Darwin to Adelaide, Australia.

C3.2 NEW SOFT MAGNETIC MATERIALS

Since about 1977, a new soft magnetic material, a rapidly quenched amorphous magnetic alloy (AMA), has been in the vanguard of new developments in the field, challenging the huge market share enjoyed by the traditional soft magnetic materials (grain-oriented 3% Si-Fe electrical steels for use in power transformers and ferrites for high-frequency applications). AMA consists of 75% to 80% transition metals (such as iron, cobalt, and nickel) together with 20% to 25% of metalloid elements (such as boron, phosphorus, and carbon), made into amorphous ribbons.

The main advantage of AMAs in electrical transformer applications is their low coercivity. Although AMAs cost two to three times as much as electrical steels, their use results in huge reductions in power losses. The main disadvantages of AMAs for use in large power transformers have to do with mechanical handling, brittleness, and possible contamination hazards.

To meet the challenge of AMAs in these applications, a number of developments have been reported in oriented electrical steels:

- Nippon Kokan KK of Japan announced in July 1987 that it had successfully developed the world's first highly functional magnetic flat steel, containing 6.5% silicon. The ductility of this material is enhanced by improved methods of rolling and bending.
- Improved forming techniques—such as laser etching or scribing, which is used to prevent the movement of domain boundaries and hence reduce core losses by as much as 10%, and metal spraying—have been developed by various steel producers in the United States, Japan, and Western Europe.

In the field of high-frequency applications (such as high-frequency transformers, filters, television sets, switch mode power supplies for transformers, radio interference suppression chokes, and magnetic heads), ferrites enjoy a huge market share. Essentially ceramic, these materials have unique insulating characteristics—they do not carry eddy currents in response to magnetic fields—which account for their widespread use. There are two broad categories of ferrites; nickel-zinc is more widely used than manganese-zinc.

At levels below 250 kHz, ferrites are being seriously challenged by AMAs. The advantages of AMAs here are twofold: (a) smaller size for the equivalent amount of power losses in applications where head generation should be kept to the minimum, and (b) zero magnetostriction, thus preventing mechanical response. This property also ensures that unlike ferrites, AMAs are not affected by mechanical strains. The major ferrite manufacturers have responded to the threat posed by AMAs by improving the standard ferrite products through, for example, lowering their losses, increasing their permeability, and keeping the concentration of impurities in them, such as CaO, NaO, BaO, and SiO, to the barest minimum. The dominance of ferrites in the high-frequency applications market is likely to continue until at least 1992. The main reason for this is the price differential between ferrites (typical 1988 price: \$2.50 kg^{-1}) and AMAs (typical 1988 price: \$10.00 kg^{-1}).

REFERENCES

1. R. W. Lee, E. G. Brewer, and N. A. Schaffel, "Processing of neodymium-iron-boron melt-spun ribbons to full dense magnets," *IEEE Trans. on Magnetics MAG-21* (5), pp. 1958–1963, 1985.
2. J. F. Herbst, et al., "Relationship between crystal structure and magnetic properties in $Nd_2Fe_{14}B$," *Phys. Rev. B29* (7), pp. 4176–4178, 1984.
3. B. M. Ma and K. S. V. L. Naraisimham, *J. Magnetism and Magnetic Mater. 54–57,* p. 559, 1986.
4. M. Sagawa, et al., *IEEE J. Magnetics in Japan TJMJ-1* (1), p. 48, 1985.

OTHER NEW METALLIC MATERIALS

CHAPTER C4

In spite of the fact that the most interesting developments are taking place in nonmetals, a number of new metal alloys were introduced in the period 1986–88 with the aim of stemming the tide of advanced nonmetallic materials. This is a short list of some of these alloys.

1. A modified 316 stainless steel for the critical labyrinths of tubing for the next generation of fossil-fueled power plants, where steam conditions of 540°C and 24 MPa are likely to be encountered. This alloy, which is based on the 800H steel and the nickel aluminides, contains a fine dispersoid of MC carbides that pins down the dislocations, considerably reducing creep.

2. Another stainless steel, 201L, has corrosion resistance equal to that of 304 stainless steel combined with the high strength of 9% nickel steel. 201L may be used in cryogenic temperature applications, such as in vessels for the liquefaction and transportation of nitrogen, helium, and carbon dioxide. Typical mechanical properties of this steel at −195°C are yield strength of 585 MPa, tensile strength of 1450 MPa, and elongation of 25%.

3. A 60% ferrite, 40% austenite duplex stainless steel called 7-Mo Plus and a nitrogen-strengthened austenitic stainless steel called SCF-19 are recommended for service in aggressive media. SCF-19 alloy has application in drill collars, housings for measurement-while-drilling devices used in directional drilling for oil and gas, and digester and liquor heater tubes for pulp and paper processing mills.

4. A new type of 303 stainless steel, with modified chemistry, has been added to the family of Plus-X free machining steels.

5. Two high-strength stainless steels, RS 329 and RS 323 (Daido Steel Co. Ltd., Japan). RS 329 is a semiaustenitic, precipitation-hardened steel with a tensile strength of 2355 MPa that is recommended for spring applications. RS323 is a nonmagnetic stainless steel that can develop a Vickers hardness of 500kg mm^{-2} after 80% cold work. Magnetic relative permeability, after drawing to this hardness, is reportedly less than 1.01. Uses include microshafts and nonmagnetic springs for electronic components.

6. A second-generation duplex stainless steel, Alloy 2205 (UNS S31803) (22Cr-5.5Ni-3Mo). This steel may be used as a substitute for 316L steel for applications where corrosive environments are very aggressive and, in particular, where chloride stress corrosion cracking is likely to occur.

7. Another second-generation nitrogen-enhanced duplex stainless steel is Ferralium 255. This steel may be used in wellhead components in relatively sweet enhanced-recovery oil wells where fluids are high in carbon dioxide.

8. A low-expansion superalloy, Incoloy 909, with very high tensile and notch rupture strengths. This alloy does not require thermomechanical processing or long heat treatment. This is because 0.4% Si is added to the nickel-iron-cobalt matrix.

9. Mechanically alloyed iron and new nickel-based materials have been introduced for aircraft engine uses. Examples include the 78Ni-20Cr MA 754 alloy as used in General Electric's F404, F110, and F101 aircraft engines and Inconel MA 6000, a 69Ni-15Cr-4W-2Mo-4.5Al-2.5Ti alloy developed as a blade material for advanced gas turbine engines.

10. Research programs are being conducted with the aim of improving the performance of 8% to 14% chromium ferritic steels for high-temperature applications up to 650°C. These alloys should satisfy the requirement for new or improved steels for demanding applications such as those found in advanced power plants. These alloys are the key to the successful operation of future coal-fired power plants and find use in applications such as fuel cladding, wrapper and steam generator tubing for fast breeder reactors, and first-wall materials for a prototypical fusion reactor.

In a program started in 1982, joint research by Kobe Steel and Mitsubishi Heavy Industries of Japan produced modified versions of ferritic TAF steel for operation at 620°C. These new rotor steels, TR 1100, TR 1150, and TR 1200, have higher creep rupture strength compared to the steel that has had wide application in supercritical steam power plant components, General Electric's GE Steel (Table C4.1). TR 1200, with a tungsten content within a molybdenum equivalence (Mo% + 0.5W%) range of 1.2% to 1.5% Mo, was developed for applications at 650°C.

TABLE C4.1 THE CREEP RUPTURE STRENGTH OF ADVANCED 10%–12% Cr STEELS FOR STEAM TURBINE ROTORS

Steel Type	1000,000 h Creep Rupture Strength (MPa)
GE	68
TR 1100	118
TR 1150	157
TR 1200	196

Sources: Kobe Steel Ltd., Tokyo, Japan; Mitsubishi Heavy Industries Ltd., Tokyo, Japan; General Electric Co., Salem, Va.

11. Based on the excellent high-temperature strength of the TR ferritic steel grades, modified grades have been developed for power plant boiler tube applications. Called T1 and T2, these steels contain 0.05% and 0.10% carbon, respectively, 10% chromium, and alloying elements that produce a 20% to 30% delta-ferrite content, resulting in a combination of high-temperature strength and excellent formability and weldability, properties that are desirable in boiler tubing.

12. Type 316 austenitic stainless steel is being used for fuel and wrapper tubing of fast breeder reactors, but type 316 stainless steel with titanium and niobium additions, the so-called D-9 steel (0.05C-14.5Cr-14.5Ni-2Mo-0.25Ti-0.05Zr), and 9% to 12% chromium ferritic steels are preferred for this application. The Power Reactor and Nuclear Fuel Development Corp. in Japan developed one such ferritic steel, TB12, which, reportedly, has a 100,000-h creep rupture strength of 100 MPa at 650°C, when tempered at 760°C. The creep rupture strength is as much as five times that of some conventional ferritic steels (Fig. C4.1).

13. One copper alloy, Incramet 800, a nickel-aluminum bronze, can be used for tooling to be used at temperatures of 315°C to 650°C for thermal processing of resin matrix composite material parts. (Cast aluminum and graphite are among materials normally used for composite tooling.) Two Cu-Fe-Mg-P alloys, C 19700 and C19750, are designed to satisfy the needs of the semiconductor industry for materials that provide better heat dissipation in surface mount devices. These alloys reportedly possess a combination of high electrical conductivity (60% to 75% IACS),* high yield strength (480 to 550 MPa), and excellent bend formability.

14. Many research workers see aluminides (Fig. C4.2) as the main threat to stainless steel and superalloys in the near term. A lot of the initial basic research work focused on nickel and iron aluminides such as these:

- IC-50 (11.3Al-0.02B-0.6Zr), a nickel aluminide
- IC-218 (8.5Al-7.8Cr-0.02B-0.8Zr), a nickel aluminide

*IACS, International Annealed Copper Standard.

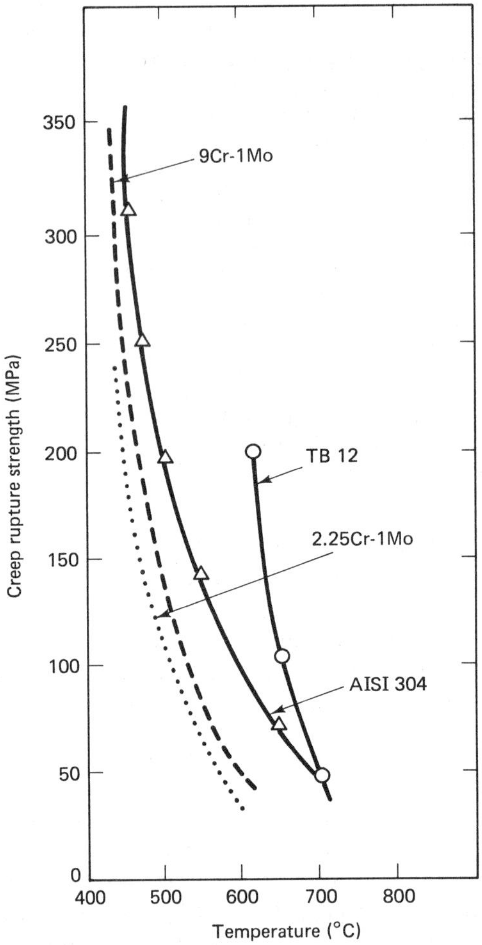

Figure C4.1 Creep rupture strength of four steels for use in boiler tubes.

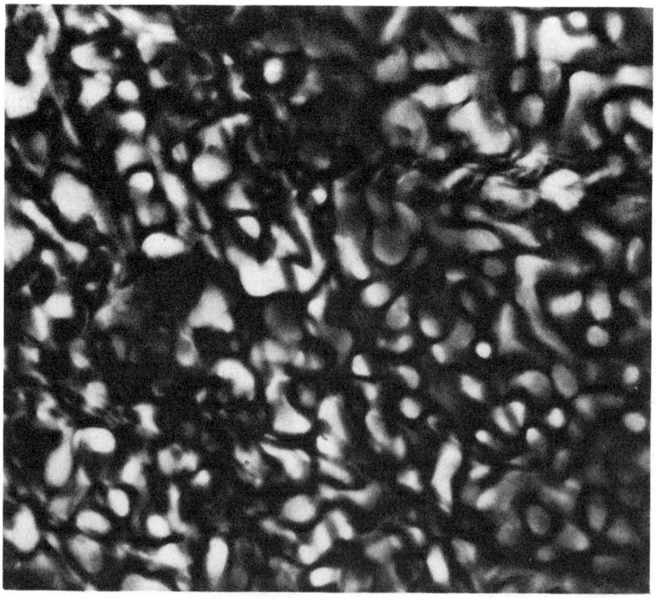

Figure C4.2 Dark-field transmission electron micrograph of an iron aluminide (Fe–24% Al). (Courtesy of Oak Ridge National Laboratory, Oak Ridge, Tenn.)

- IC-221 (8.5Al-7.8Cr-0.02B-1.7Zr), a nickel aluminide
- Fe–40% Al alloyed with minor additions of zirconium, molybdenum, and boron

Considerable research efforts are directed toward improving ductility via microalloying with boron, increasing high-temperature strength by solid solution strengthening with small amounts of hafnium or zirconium, eliminating "dynamic embrittlement" in oxidizing environments at intermediate temperatures (600°C to 800°C) through alloying with chromium, and improving hot fabricability by, for example, isothermal forging. It is expected that by 1994, markets for aluminides will have matured, and they will find widespread use in, for example, gas turbines, aircraft fasteners, pump darts, dental drill bits, dies, Stirling automotive engines, and bellows for expansion components in corrosive media.

15. Alcoa's alloy 2519 has a small, closely controlled amount of magnesium and a larger than usual amount of cold work prior to the precipitation heat treatment. This leads to a refined precipitate and a strength that is about 20% higher than that of alloy 2219-T87.

16. Based on the unfavorable cost penalty of rapid solidification processing compared to wrought-ingot metallurgy products, researchers at BBC Brown Boveri Research Center, in Baden-Dattwil, Switzerland, developed a wrought aluminum alloy (code named T6) with a combination of high-temperature strength and improved creep resistance at temperatures between 150°C and 200°C. In this alloy, small amounts of silver and magnesium are added to promote the formation of the relatively stable Ω phase, which can coexist with the partly coherent Θ' phase. Additions of manganese (0.4% to 0.8%) and zirconium (0.10% to 0.25%) promote dispersion strengthening by forming fine, stable intermetallic compounds. Silicon content is limited to 0.10% to prevent its preferential combination with magnesium to form Mg_2Si (thereby reducing the required interaction of magnesium with silver) and to prevent the formation of low-melting eutectics of Si*X* alloys. With the additon of a small amount of titanium to promote grain refinement, the final nominal composition of the alloy is Al–6.3% Cu–0.5% Mg–0.5% Ag–0.5% Mn–0.2% Zr–0.05% Ti, with maximum iron and silicon contents of 0.15% and 0.10%, respectively. Tests on extruded cylinders solution-heat-treated, quenched, and aged show that the mechanical properties of alloy T6 are superior to those of conventional wrought aluminum alloys at both room temperature and 200°C (Table C4.2).

17. Dow Chemical Co. announced the development of two high-purity versions of the standard AZ91 A, B, and C magnesium alloys. These alloys reportedly are as much as 100 times more corrosion-resistant than the comparable standard alloys and considerably more corrosion-resistant than carbon steel and aluminum alloy 380 (UNS A03800). The improvement is explained by control of both the iron-to-manganese ratio and melt temperatures, yielding minimum solubility of these heavy metal elements. The die-cast alloy AZ91D is used in a number of automotive components, such as the rocker arm cover air cleaner housing, engine grille, headlamp assembly, brake pedal support brackets, and clutch housings.

TABLE C4.2 PROPERTIES OF COMMERCIAL AND EXPERIMENTAL ALUMINUM ALLOYS

	Commercial Alloys				Experimental Alloy
Mechanical Property	7075-T6	2618-T6	2219-T6	2021-T8	T6
0.2% proof strength (MPa)					
At 20°C	495	385	295	435	517
After 30 min at 200°C	235	310	235	275	393
After 100 h at 200°C	125	270	225	240	305
Ultimate tensile strength (MPa)					
At 20°C	565	445	420	500	581
After 30 min at 200°C	285	350	295	320	409
After 100 h at 200°C	140	290	275	305	325
Elongation at fracture (%)					
At 20°C	11.0	7.5	10.0	10.0	7.5
After 30 min at 200°C	12.0	9.0	16.0	16.0	11.0
After 100 h at 200°C	16.0	12.0	20.0	20.0	18.0

Source for T6 data: BBC Brown Boveri, Baden-Dattwil, Switzerland.

The sand-cast alloy AZ91C/HP is used in various military helicopter components. Fansteel Inc., of North Chicago, Ill., reported the development of another high-purity AZ91 alloy with greatly improved salt-fog corrosion performance. Success of the sand-casting alloy AZ91E is linked to a minimized iron-to-manganese ratio, use of pickling after grit blasting to remove surface contaminants, and selection of the proper temper.

18. A tantalum-columbium refractory alloy (named KBI alloy 41) was introduced by the KBI Division of Cabot Corporation of Reading, Pa. This alloy has high yield strength at room temperature and an excellent retention of that strength at elevated temperatures, compared to tantalum (Table C4.3). Complementing its high strength, Alloy 41 shows excellent corrosion resistance and resists hydrogen embrittlement. Thus this alloy finds use, for example, in welded tube bundles for heat exchangers and for patching glass-lined equipment.

19. A titanium-based alloy, TICOREX (Ti–0.05 wt % Ru–0.5 wt % Ni) was introduced by Nippon Mining Ltd. This alloy has excellent resistance to both general corrosion as well as crevice attack (Table C4.4). The corrosion resistance

TABLE C4.3 TYPICAL YIELD STRENGTH OF TANTALUM AND A TANTALUM-BASED ALLOY AS A FUNCTION OF TEMPERATURE

	Yield Strength (MPa)		
	Room Temperature	200°C	500°C
Tantalum	166	83	62
KBI Alloy 41	434	352	310

Source: Cabot Corporation, KBI Division, Reading, Pa.

TABLE C4.4 GENERAL AND CREVICE CORROSION PERFORMANCE OF TITANIUM AND TWO OF ITS ALLOYS

Corrosion Conditions/Indicators	Commercially Pure Titanium	Ti-0.15Pd	TICOREX
Temperature and HCl concentration to produce a general corrosion rate of 0.1 my^{-1}	100°C and 1 wt % or 80°C and 3 wt % or 50°C and 4 wt %	100°C and 2 wt % or 80 °C and 7.5 wt % or 50°C and 14 wt %	100°C and 2.5 wt % or 80°C and 8 wt % or 50°C and 11 wt %
Minimum time for crevice corrosion to occur in boiling 15% NaCl solution[a]	6 h	24 h	24 h

[a] An accelerated test [1] was used.

Source: K. Taki, "Ti-Ru-Ni for corrosive applications," *J. Metals 39* (12), p. 47, 1987.

of this alloy has been explained [2] by the presence of Ti_2Ni, which has a noble potential even in very aggressive media and increases the passivation potential of the alloy.

20. A new tin-containing titanium alloy designated Ti-15-3 (composed of 15% vanadium, 3% tin, 3% aluminum, 3% chromium, and 76% titanium) was developed by the United States Air Force. The alloy has a number of advantages, such as easy melting and excellent weldability, ease of production as a strip (leading to a reduction in labor costs, provision of better yields, and finer gage control), cold formability (thus reducing tooling costs and preventing surface scale formation), and age hardenability (thus allowing fabrication in the malleable solution-treated form before aging to the required best mix of mechanical properties). This alloy is being used by airframe manufacturers, such as North American Aircraft Operations (a subsidiary of Rockwell International), for fabrication of certain components in the Air Force B-1B strategic bomber.

REFERENCES

1. K. Taki and H. Sakuyama, paper presented at the 1986 International Conference on Titanium Products and Applications, San Francisco, Calif., 1986.
2. A. J. Sedriks, J. A. S. Green, and D. L. Novak, *Corrosion 28* (4), pp. 137–142, 1972.

FIBER REINFORCED SUPERALLOY COMPOSITE MATERIALS

CHAPTER C5

It is perhaps uncontestable to state that aerospace materials research has continuously pushed practicable operating temperatures of materials to higher levels. It seems, however, that superalloys, which have traditionally been the material of choice for the most demanding elevated-temperature applications, are fast approaching their maximum temperature limits, about 1000°C for unreinforced superalloys. For further advances in high-temperature performance, the research focus is on materials with unidirectional microstructures, such as fiber reinforced superalloy composites, directionally solidified single crystals and eutectics, oxide-dispersion-strengthened superalloys, iron, nickel or titanium intermetallics or aluminides, and specialty ceramics.

Some of the early advances were obtained with fiber reinforced superalloy (FRS) composite materials. The fibers studied include submicron whiskers, refractory metal wires, and continuous ceramic boron and carbon filaments. These last are easily dissolved by the superalloy matrices at high temperatures, and efforts are being made to develop suitable diffusion barriers to reduce fiber-matrix reactions (such as the formation of solid solution zones and intermetallic compounds, and diffusion-initiated recrystallization of heavily worked fibers). Excellent results have been obtained with refractory metal wires because of their generally greater tolerance of these interfacial reactions. Continuous tungsten alloy wires are widely used because they have high values of ultimate tensile

TABLE C5.1 THE CREEP RUPTURE STRENGTH OF SOME SUPERALLOYS

Material	100-h Creep Rupture Strength (MPa)
Unreinforced superalloys	45–90
Tungsten metal wire-reinforced superalloys	340–1400
Tantalum metal wire-reinforced superalloys	500–550
Molybdenum metal wire-reinforced superalloys	280–300
Niobium metal wire-reinforced superalloys	200–300

strength at elevated temperatures (up to 2 GPa at 1100°C), creep rupture strength (Table C5.1), and specific creep rupture strength. As far as the matrix is concerned, it is selected for oxidation resistance and fiber compatibility. Iron-based superalloys are more compatible with tungsten alloy fibers than are nickel- or cobalt-based superalloys, but the most acceptable matrices for tungsten alloy fibers are the Fe-Cr-Al-Y alloys.

Tungsten alloy fiber reinforced superalloy (TAFRS) composites have an average 100-h rupture strength of about 250 MPa at 1100°C, which is equal to the average strength of unreinforced superalloys at about 950°C, suggesting an improvement in service temperature of about 150°C (Fig. C5.1). At temperatures between 700°C and 1000°C, the high-cycle fatigue strength ratios of TAFRS composites are significantly higher than those of unreinforced superalloys (Fig. C5.2). Similar improvements in low-cycle fatigue properties have been reported. In creep tests, a TAFRS composite containing 20 vol % fiber in a type 713C nickel-based superalloy matrix remains in the secondary creep stage long after the unreinforced 713 superalloy has failed in the tertiary creep stage. The secondary stage creep strain rate, at a given stress, of the TAFRS composite is only 20% that of the unreinforced superalloy (Fig. C5.3). A ductile superalloy matrix alleviates thermal strains (arising from temperature gradients from, for example, rapid heating and cooling cycles and thermal-expansion mismatch between the fiber and the matrix) through plastic deformation. TAFRS composites have reportedly successfully undergone 1000 thermal cycles beween 30°C and 1200°C without any substantial

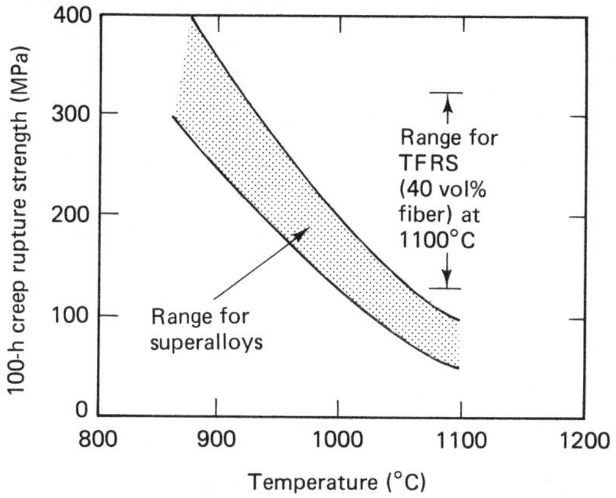

Figure C5.1 The 100-h creep rupture strength of TFRS composites and unreinforced superalloys.

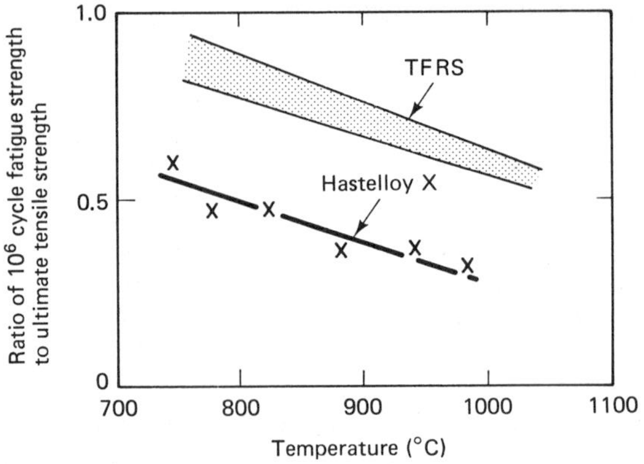

Figure C5.2 High-cycle fatigue strength to ultimate tensile strength ratios of TFRS composites and an unreinforced superalloy.

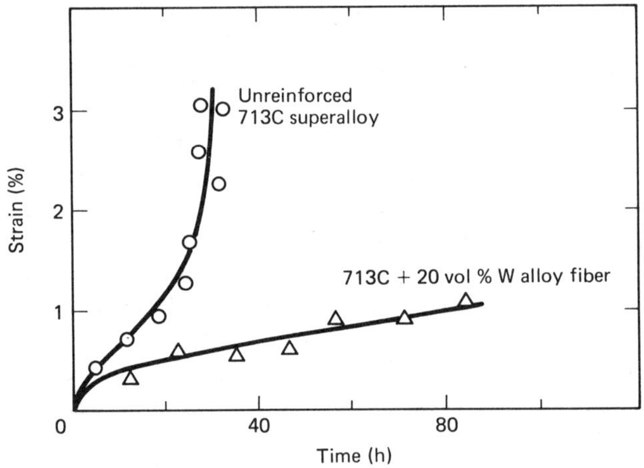

Figure C5.3 Creep behavior of type 713C nickel-base superalloy with and without reinforcement with tungsten alloy fibers, tested at 1200°C and 120 MPa.

structural damage or property impairment. The thermal conductivity of a TAFRS composite is highest in the tungsten fiber reinforcement direction and is much lower in a direction orthogonal to the reinforcement, but even this low value is about twice that of unreinforced superalloys (Fig. C5.4). This advantage becomes very important in the design of turbine engines. With turbine blades and vanes made of TAFRS composites, the turbine inlet temperature can be increased by about 60°C (compared with unreinforced superalloys) while maintaining the same rate of flow of coolant air inside the airfoil and through its walls.

TAFRS composites have a much higher specific tensile strength than unreinforced superalloys and retain this property for a wider high-temperature range (Fig. C5.5).

All these advantages of TAFRS composites have led to their consideration for use in or actual use in applications such as these:

- Turbopump blades in advanced space shuttle engines
- Various automotive parts, such as engine seals and body rails
- Rotating turbine blades in air-breathing and rocket engines
- Various electrical parts, such as bus bars and generator rotors

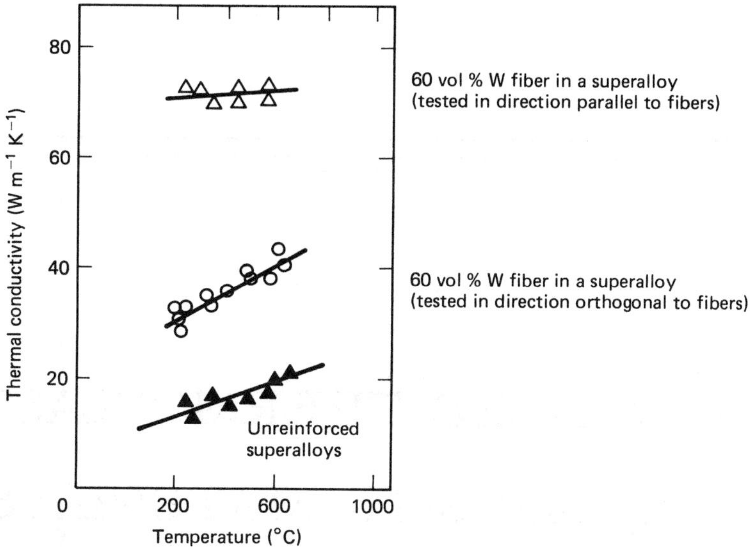

Figure C5.4 Thermal conductivities of TFRS composite materials and unreinforced superalloys.

Figure C5.5 Specific strength versus temperature variation of unreinforced superalloys and a tungsten alloy fiber reinforced superalloy composite material.

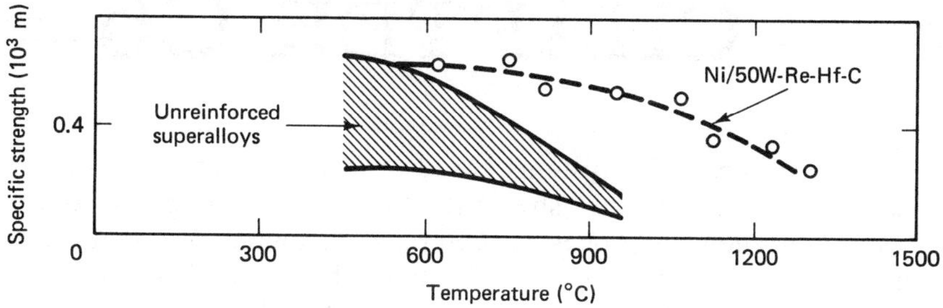

ADVANCED THERMOPLASTIC COMPOSITE MATERIALS

CHAPTER C6

Advanced composite systems based on thermoplastic matrices are prime candidates as replacements for sheet metal and thermoset composites in military, aerospace, and chemical processing components, among their many uses. An advanced thermoplastic composite (ATPC) material may be described as a material containing a high fiber content of a highly heat- and chemical-resistant thermoplastic matrix with continuous, oriented reinforcements. These are usually carbon fibers, although other fibers, as well as random, discontinuous, or chopped reinforcements, have been used.

Polyetheretherketone (PEEK) and polyphenylene sulfide (PPS) are the matrix resins commonly used for thermoplastic composites, although variants on them, as well as other matrices, are being developed. Among the variants are APC-2, APC(HTA), APC(HTX), and APC(HTC), which are developmental grades of PEEK, and polyarylene sulfide, which is a high-temperature form of PPS.

Among the other matrices are polyamide-imide, a very high molecular weight polyamide, a siloxane-modified polymide sulfone, polyether sulfone (PES), polyarylate, improved versions of polyimide, and polyketone.

The structures of PEEK, PES and PPS, are shown in Fig. C6.1. Some properties of four thermoplastic materials are presented in Table C6.1.

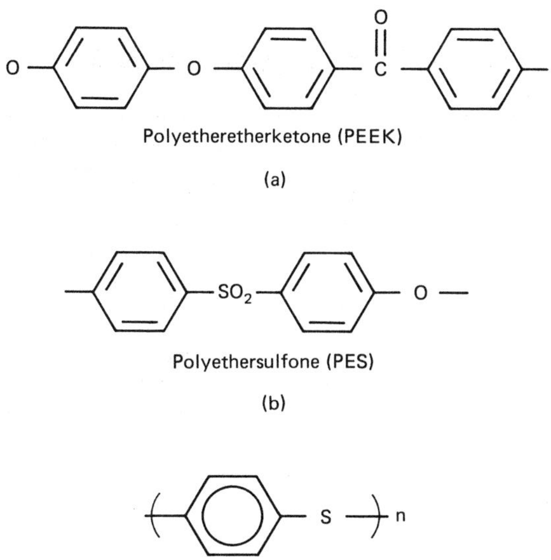

Polyetheretherketone (PEEK)

(a)

Polyethersulfone (PES)

(b)

Polyphenylene sulfide (PPS)

(c)

Figure C6.1 Structure of three widely used thermoplastics.

TABLE C6.1 PROPERTIES OF FOUR HIGH-PERFORMANCE ENGINEERING THERMOPLASTIC MATERIALS[a]

Material Property	Victrex PEEK	Ryton PPS	Torlon PAI	Ardel PAR
Density (kg m^{-3})	1380	1750	1535	1210
Tensile strength (MPa)	152	96	172	70
Tensile elongation at yield (%)	4.9	0.7	14	8.4
Flexural strength (MPa)	245	141	276	76
Flexural modulus (GPa)	8.34	14.13	12.45	2.14
Notched Izod (N·m m^{-1})	90.80	77.40	76.10	213.60
Heat deflection temperature (°C)	232	260	278	174
Dielectric strength (MV m^{-1})	19.02	15.16	26.18	15.75
Water absorption (% in 24 h)	0.55	0.05	0.25	0.27
Price ($ kg^{-1}, June 1988)	59.47	5.18	49.56	5.76

[a]Mean values; PEEK, polyetheretherketone; PPS, polyphenylene sulfide; PAI, polyamide-imide; PAR, polyarylate resin.

Sources: ICI Americas, Wilmington, Del.; Phillips 66, Bartlesville, Ok.; Amoco Performance Products, Chicago, Ill.

ATPC is produced in a number of forms, including these:

- Braided tapes of nylon-glass and carbon fiber with PEEK or PPS
- Tows that consist of three rovings of glass, graphite, aramid, or a combination of these materials impregnated with polysulfone or polypropylene
- Woven fabrics of PEEK slit film and graphite tow
- Prepreg tapes based on PPS, PEEK, polyimide, and polyamide-imide

TABLE C6.2 MECHANICAL AND PHYSICAL PROPERTIES OF THREE ADVANCED THERMOPLASTIC COMPOSITE MATERIALS

Property	PEEK-Glass[a]	PPS-Glass[b]	PEEK-Carbon[c]
Tensile strength (MPa)	200	138	1793
Tensile modulus (GPa)	18	14	—
Flexural strength (MPa)	290	207	1496
Flexural modulus (GPa)	11	11	130
Compressive strength (MPa)	—	172	0.96
Notched Izod impact (J m^{-1})	112	75	—
Heat deflection, at 1.38 MPa (°C)	288	260	—
Coefficient of linear expansion (10^{-6} °C^{-1})	2.2	2.2	—

[a]Volume fraction of glass fibers = 0.30.

[b]Volume fraction of glass fibers = 0.40.

[c]Volume fraction of carbon fibers = 0.56.

TABLE C6.3 ELECTRICAL PROPERTIES OF ADVANCED THERMOPLASTIC COMPOSITE MATERIALS

Property	PC-Glass	PBT-Glass	PPO-Glass
Volume resistivity (Ω·m)	10^{15}	10^{14}	10^{15}
Dielectric strength at 50 Hz (MV m^{-1})	50	50	50
Dielectric constant at 50 Hz	3.3	3.8	3.0
Dissipation factor, tan δ	9×10^{-4}	9×10^{-3}	5×10^{-4}

PC, polycarbonate; PBT, polybutylene terephthalate; PPO, polyphenylene oxide; volume fraction of glass fibers = 0.30.

Some properties of six unidirectional ATPCs are shown in Tables C6.2 and C6.3.

ATCPs find widespread use in applications such as the leading edge of the vertical stabilizer on a hovercraft, flexographic doctor blades used in printing presses, parts of the oil and water pumps and turbo inlet and impeller in a racing car, oil well sucker rod guides, chemical pump impellers, missile housings, sheathing for optical fibers, access covers for armored vehicles, aircraft circuit board covers, and chemical butterfly valves. The aerospace industry has aggressively embraced these materials. For example, Lockheed-Georgia (Marietta, Ga.) used a graphite fiber reinforced PEEK composite material to form the belly skin panel of the U.S. Air Force C-130 transport plane (Fig. C6.2), Westland Aircraft Company (Somerset, England) used carbon fiber reinforced PEEK materials for the tailplane and the roof of the Westland 30 helicopter (Fig. C6.3), and a graphite reinforced PMR14* polyimide air-bypass duct replaced the titanium duct on the GE F-404 engine for the U.S. Navy's F-18 fighter jet aircraft.

*PMR, polymerization of monomer reactants.

Figure C6.2 A graphite-polyetheretherketone composite tape to form the belly skin of the U.S. Air Force C-130 aircraft. (Courtesy of Lockheed-Georgia)

Figure C6.3 Distribution of thermoplastic composites in some components of a Westland 30 helicopter. (Courtesy of Westland Helicopters Ltd.)

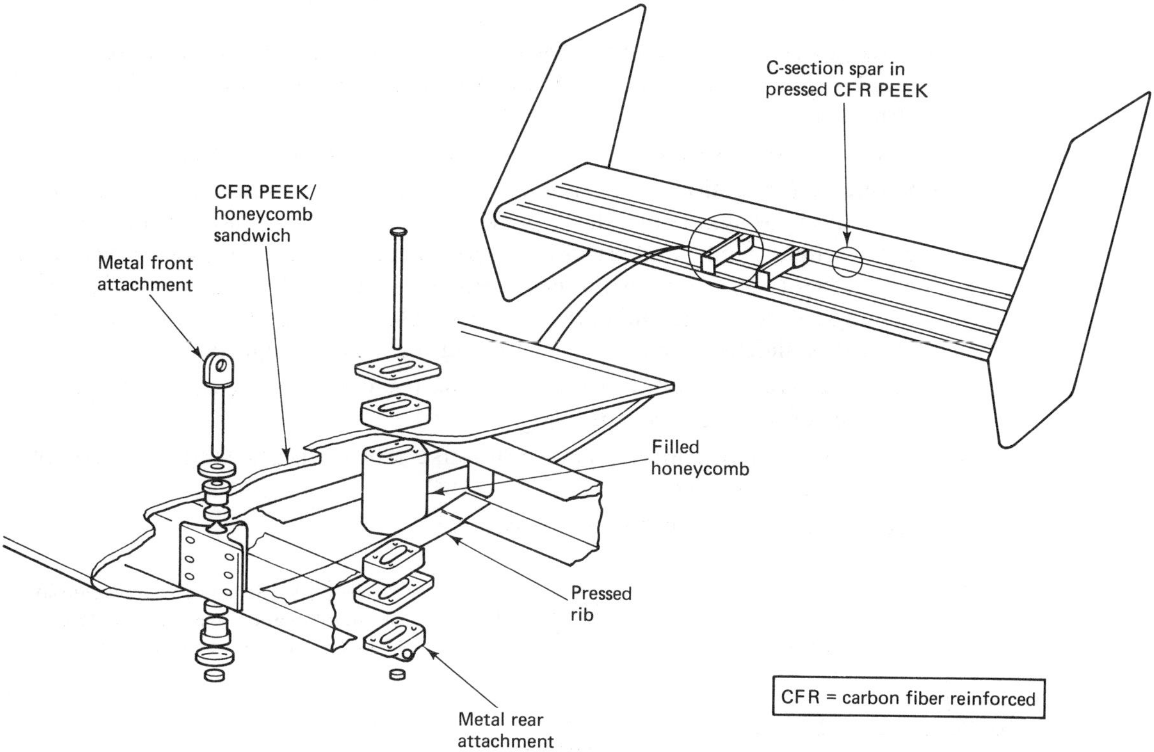

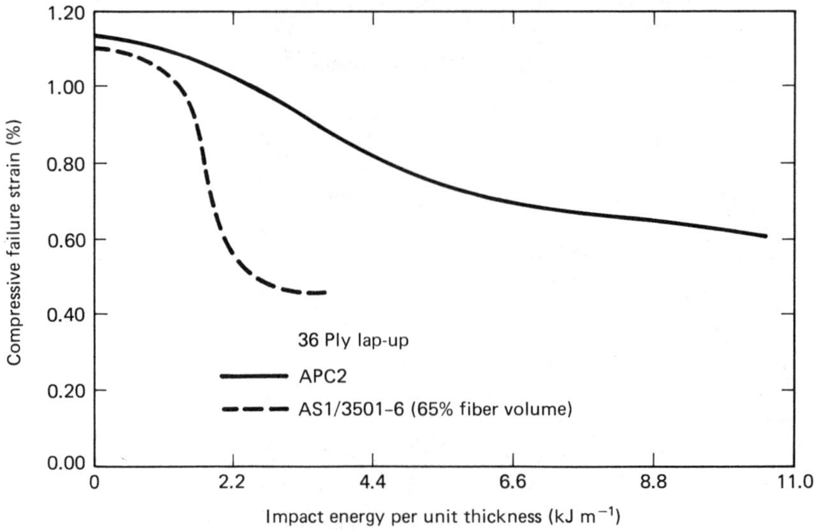

Figure C6.4 Variation of compressive strain to failure with impact energy per unit thickness of APC2 (carbon-PEEK) and AS1/3501-6 Hercules (carbon-epoxy). (Data courtesy of ICI Americas Inc.)

These applications, and many others, are related to a number of advantageous properties of ATPCs. These are some of them:

- A toughness that is several times that of epoxy resin based composite materials (Fig. C6.4); thus an ATPC has good damage tolerance in molded structures.
- Excellent resistance to common solvents, aviation fuels, cleaners, etc.
- Reduced fabrication costs and greater speed of processing; for example, fabrication of stampable sheet materials involves little more than warming precut sheet blanks and stamping them on temperature-controlled matched-metal dies (Fig. C6.5). Although thermosetting ATPCs requires higher processing temperatures and pressures than resin based composites, cycle times are much shorter since no cross-linking reaction is required.
- Lower costs of material storage because the prepregs are cured and do not require refrigerated storage.
- Reduced quality control costs that arise from using a single-component, nonreactive resin system.
- Long- and short-term thermal resistance.

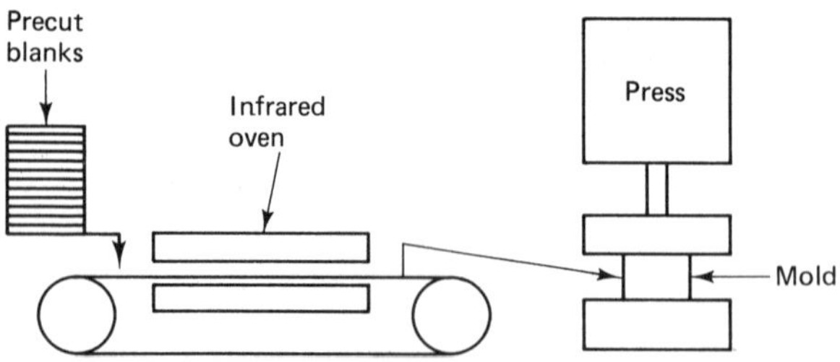

Figure C6.5 Schematic representation of the fabrication process of a stampable thermoplastic composite sheet material.

- Inherent flame resistance.
- Low moisture sensitivity, thereby reducing the likelihood of either failure of the fiber-matrix interface or plasticization.
- Well-consolidated structural changes, profiles, and flat laminates can be produced from ATPC prepregs using conventional pultrusion, stamping, filament winding (Fig. C6.6), diaphragm forming (Fig. C6.7), or automated tape technology (Fig. C6.8).

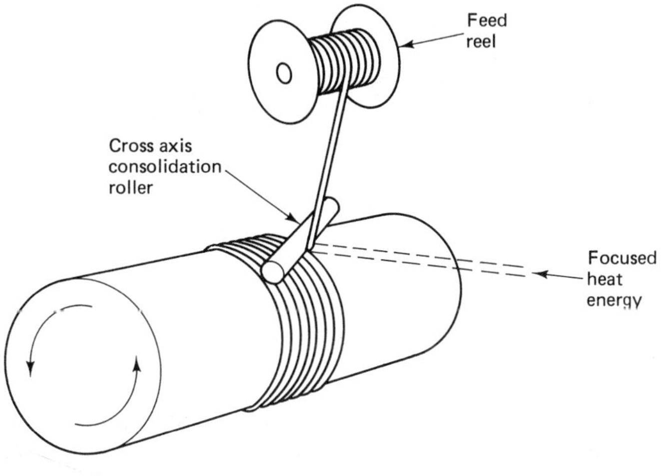

Figure C6.6 Filament winding with continuous consolidation of a thermoplastic composite material. (Courtesy of ICI Americas Inc.)

Figure C6.7 Diaphragm forming of a thermoplastic composite material. (Courtesy of ICI Americas, Inc.)

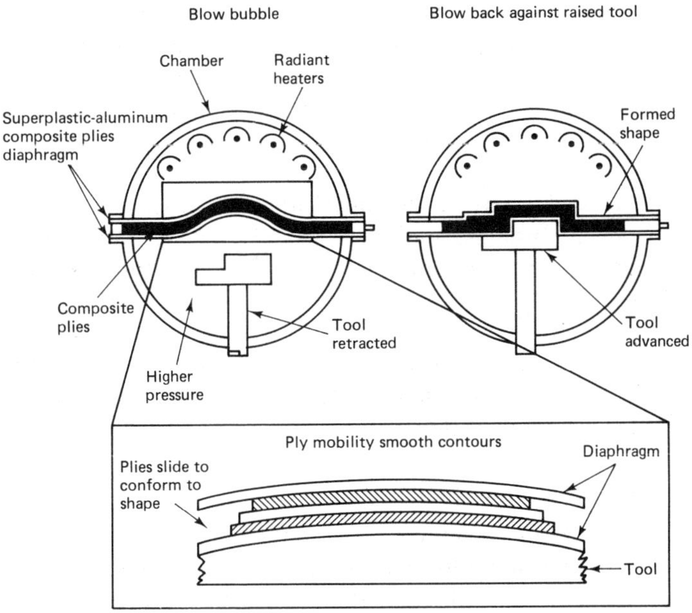

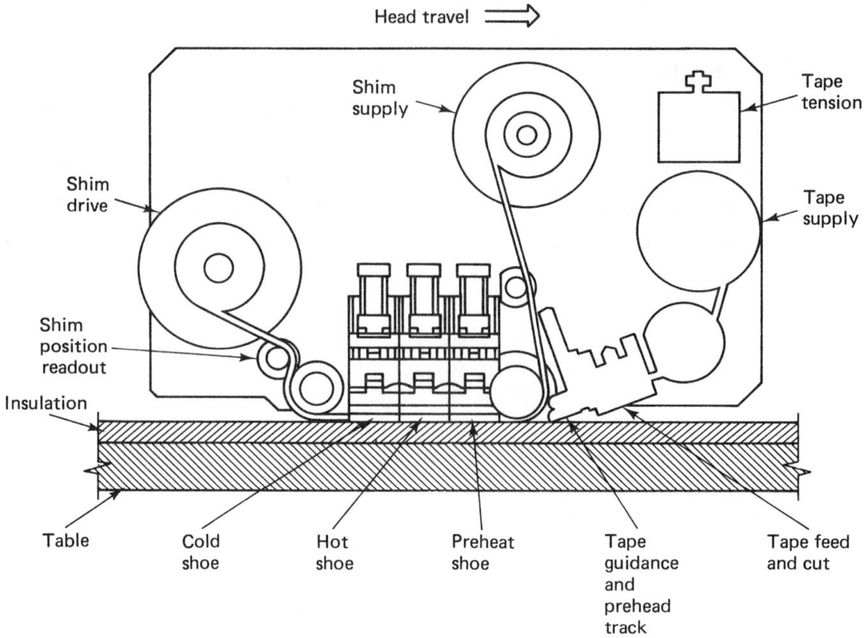

Figure C6.8 The world's first thermoplastic prepreg tape layer. (Courtesy of Cincinnati Milacron)

- A wide range of bonding methods (adhesive bonding, ultrasonic welding, induction heating, fusion bonding, etc.) can be used, making component design very flexible.
- High retention of strength and stiffness at elevated temperatures (Fig. C6.9).
- Excellent damage tolerance characteristics, including high retention of compression strength after impact and resistance to delamination. In general, ATPCs exhibit mode I delamination strain energy release rate (G_{IC}) values that are about an order of magnitude greater than those of conventional thermoset composites. Thus G_{IC} for carbon fiber–PPS, carbon fiber–PEEK, and carbon–epoxy resin composite materials are 0.88–1.41, 1.41–2.00, and 0.14 kJ m^{-2}, respectively [1].

Notwithstanding this long catalog of attractive features, ATPCs have a number of disadvantages, and the search to correct them forms the basis of directions in research and development. Among the most important of these aspects are the following:

- ATPCs have a short experience base, which must be expanded if they are to present a serious challenge to thermoset composites, especially in aerospace applications.
- ATPCs that are currently commercially available have temperature use limitations, to about 120°C.
- The fatigue behavior of these materials has not been thoroughly investigated, although available results point to acceptable performance. But important phenomena, such as microcrack saturation (which occurs in thermoset com-

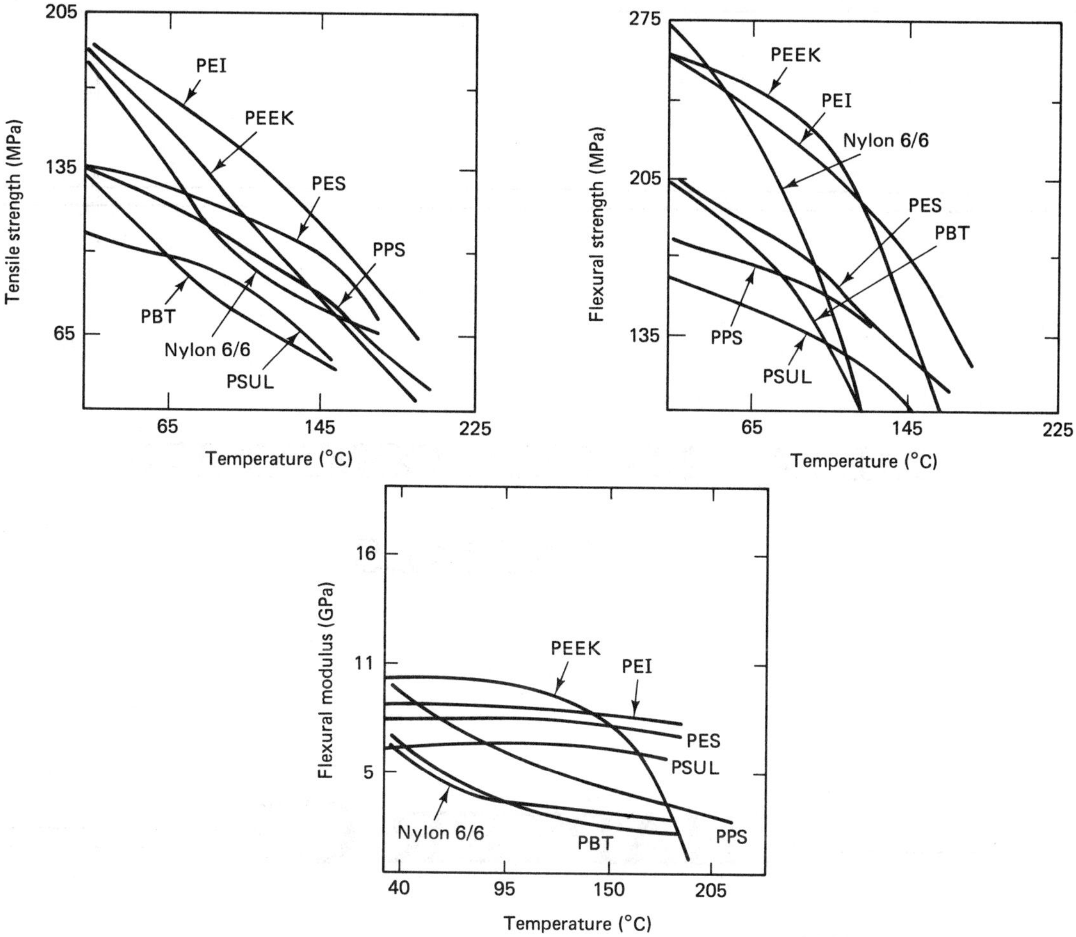

Figure C6.9 Effect of temperature on some mechanical properties of 30% glass fiber reinforced thermoplastic material (PEEK, polyetheretherketone; PEI, polyetherimide; PSUL, polysulfone; PPS, polyphenylene sulfide; PBT, polybutylene terephthalate; PES, polyether sulfone). (Data courtesy of RTP Co.)

posites in the 90° plies in a crossply laminate), which can have a positive influence on fatigue life, are, by and large, unexplored.

- More studies are needed on delamination growth under fatigue loading conditions. Preliminary results showed that the exponent in the Paris-Erdogan law has a value of between 4 and 10 for ATPCs, whereas it is between 2 and 4 for most metals. This indicates unacceptably high fatigue crack growth rates in ATPCs even though, by comparison, they are better in this respect than thermoset composites.
- Characterization of the viscoelastic and creep response of ATPCs is urgently required in order to specify limitations for service temperature.

REFERENCE

1. G. M. Newaz, "Advanced thermoplastic composites," *ASTM Standardization News* *15* (10), pp. 32–35 (1987).

METAL MATRIX COMPOSITE MATERIALS

CHAPTER C7

C7.1 INTRODUCTION

The term "composite material" normally conjures up the image of plastics reinforced with glass or carbon fibers. In such a material, the fibers give strength to an otherwise mechanically weak material. The fibers take the load, and the polymer matrix serves to distribute the load equally between the fibers and the matrix. The same principle is used in another family of composite materials: metal matrix composites (MMCs), in which a metallic phase (the matrix) is reinforced by strong ceramic or metal fibers, whiskers, or particles.

The three basic requirements for a matrix to be used in a metal matrix composite are as follows:

- A high ductility, in order to provide strain accumulation around the brittle fibers
- A low melting point, in order to permit liquid-phase fabrication processes without compromising the properties of the fibers
- A low density, in order to obtain high specific strength and stiffness

Since 1977, a wide selection of high-modulus, high-strength, low-density, and often refractory filamentary materials have become available as candidate reinforcements for metallic matrices. The filaments of major interest are boron, silicon carbide, alumina, graphite, and a silicon-coated boron with slightly lower tensile strength than boron (called Borsic).

C7.2 FABRICATION METHODS

A wide variety of metallurgical processes have been employed for the fabrication of MMCs, including the following:

1. Diffusion bonding, in which heat and pressure are imposed simultaneously (Fig. C7.1). This process has been used, in some form, to fabricate most of the commercially available MMCs, such as B-Al, Borsic-Al, Ti-SiC, and Ti-Borsic.

2. Plasma bonding, which is generally used to prepare monolayer tapes, which are then consolidated by diffusion bonding into thicker sections. The advantages include these:

- ''Locking in'' of good fiber spacing if the initial monofilament winding technique is correct
- Excellent flexibility of monolayer tapes in hardware fabrication
- Production of complex parts, angle-ply configurations, and large panels without prohibitive monofilament handling problems

Disadvantages include these:

- Cost of intermediate processing to a monolayer tape
- Limitations of the method to relatively low-melting-point materials
- Oxygen contamination problems, which eliminate practical consideration of metals like titanium.

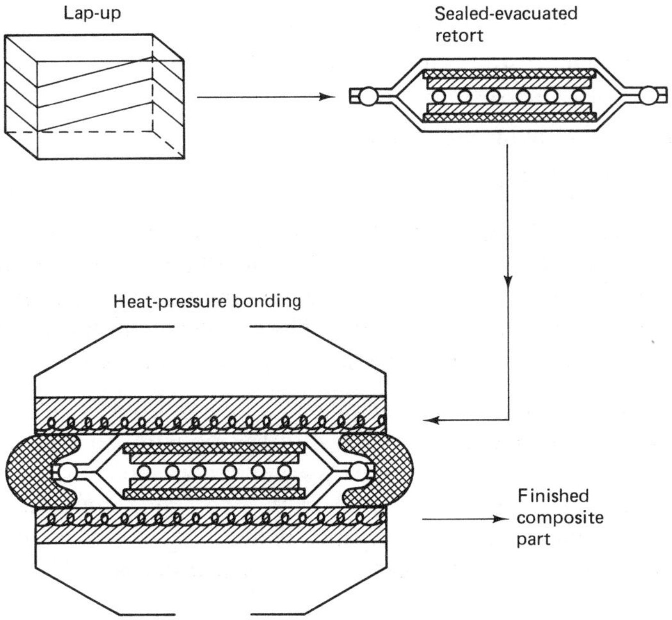

Figure C7.1 Diffusion bonding process of a metal matrix composite part.

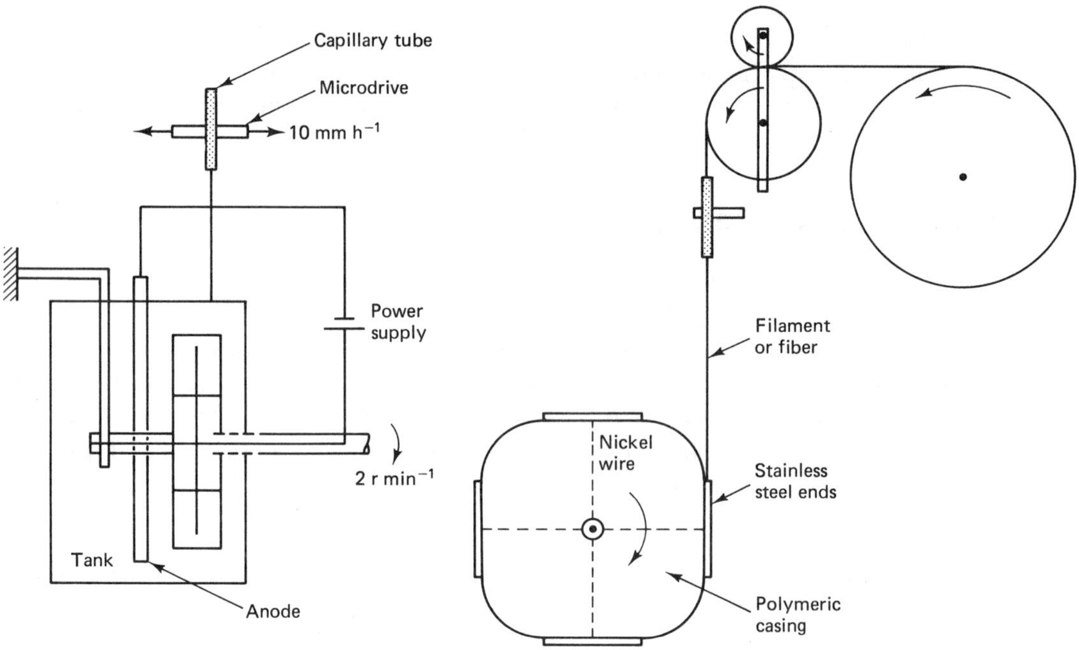

Figure C7.2 Schematic diagram of the electroforming process.

3. Electroforming, which was one of the first methods of fabrication of MMCs studied because the composite could be formed at low temperatures, thereby minimizing the degradation of reactive filaments in metal matrices. A schematic of the electroforming process is shown in Fig. C7.2. The advantages of the process are these:

- Room-temperature fabrication
- A high-density monofilament matrix can be obtained
- Excellent filament-matrix interface contact is possible
- Filament spacing is accurately controlled
- Volume loading of reinforcement is flexible
- A variety of mandrel shapes is possible

Two practical limitations inherent in the electroforming process are the presence of impurities, which are incorporated into the composite from the bath deposition, and the inability to deposit alloys from solution. In all, the method has marginal utility and poses serious problems for most MMCs. The method has, however, been used successfully to prepare carbon fiber reinforced nickel composites.

4. Liquid metal infiltration using a tubular arrangement similar to that shown in Fig. C7.3; the filaments are arranged in parallel with the powder melt in the crucible and are continuously drawn through an exit orifice at the bottom of the crucible and then through cooling coils and a long freeze tube. The technique is limited and has not received much attention because of the few reinforcements that are stable in molten metals. It has, however, been used to prepare Al-SiO_2 composites.

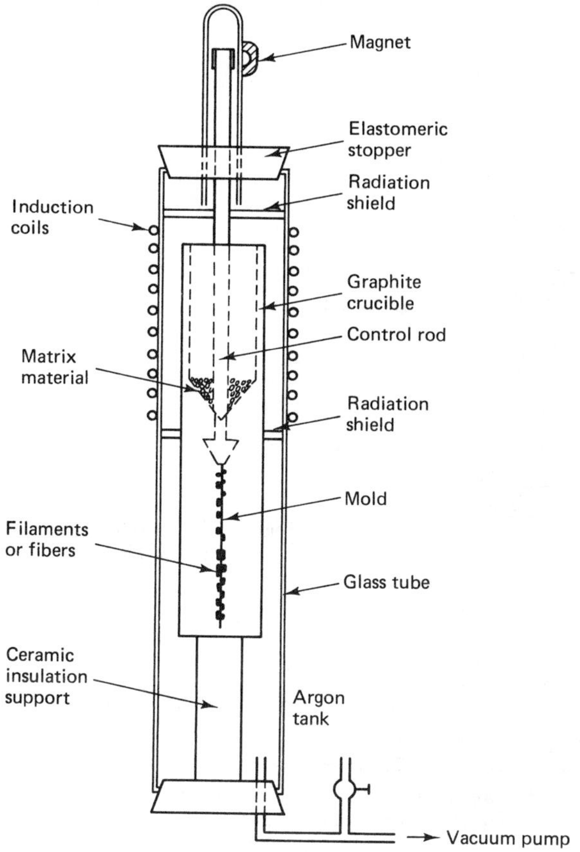

Figure C7.3 Schematic diagram of a vacuum casting apparatus.

5. High-energy-rate forming. One device for this type of fabrication, which has been employed to consolidate whisker reinforced MMCs, such as SiC/Ti-6Al-4V, is shown in Fig. C7.4. The main advantage of the method is the instantaneous application of high pressures (typically about 2.5 GPa), for very short time periods (typically about 10^{-4} s), which prevents interaction of matrix and filament or whisker while providing densification of reasonably complex shapes. The apparatus may use explosive, electromechanical, or pneumatic-mechanical means to achieve the high pressure pulse.

6. Hot-roll bonding. The diffusion bonding process has been modified to provide for a short reaction time during fabrication in a process known as hot-roll bonding (Fig. C7.5). Two problems associated with this method are difficulty in obtaining high volume loading of the filaments and limited possibilities for consolidating thin tapes that have cross- and angle-ply configurations. The method has been used to counter reactivity tendencies in systems such as B-Ti.

7. Mechanical alloying, in which particulates of the reinforcing fiber are enclosed in a powder of the matrix material prior to consolidation. The material is then vacuum-pressed at temperatures below the melting point. The resulting product has a uniform microstructure. Missing from the product are copper aluminides, materials known to be fracture initiators. This method has been used to produce SiC-reinforcing billets out of two proprietary alloys (IN 9021, a 4% Cu–1.5% Mg material, and IN 9052, a 4% Mg material).

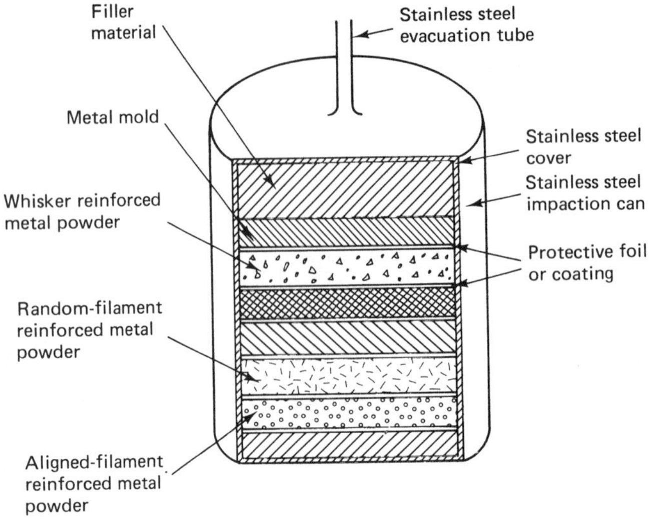

Figure C7.4 Composite billet for high-energy-rate forming.

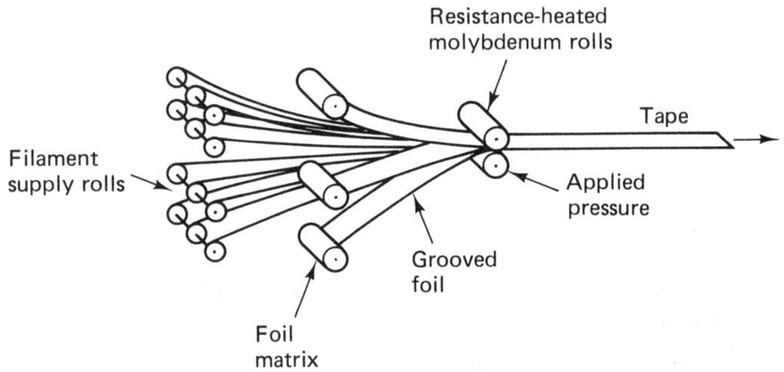

Figure C7.5 Hot-roll bonding of a metal matrix composite material.

8. Squeeze-casting. This is a very attractive fabrication method, especially for fabricating MMCs for automotive parts. Its pressure solidification process eliminates internal porosity, even in thick sections, and provides an improved metallurgical structure. The high pressure used in this process (210 MPa) promotes the desirable wetting of most ceramic fibers. In particular, the process permits selective placement of the reinforcing fibers in the part. Squeeze-cast aluminum alloy reinforced with alumina has been used for the combustion bowls of the pistons of diesel engines.

9. Other methods, such as chemical vapor deposition, cold pressing and sintering, powder slip casting, and hot extrusion. For example, chemical vapor deposition has been employed to form Be-Al composites by infiltrating beryllium filament wound on a heated mandrel with aluminum deposited by thermal decomposition from aluminum alkyls. Hot extrusions of B-Al composite have been obtained in which the filaments in the composite bar to be extruded are aligned manually in the Al powder layer and then cold-pressed into a preform, which is placed into an aluminum alloy can which, in turn, is filled with other preforms and tightly packed with aluminum alloy shins. This can is then heated to about 500°C, and the hot billet extruded with a suitable reduction and at a predetermined rate.

10. Direct-chill casting (DCC). This is a new method, introduced by Dural Aluminum Composites Corporation of San Diego, Calif., in March 1988. The method is used to produce ceramic fiber reinforced aluminum alloy composites in extrusion billet, rolling slab, and foundry pig form. In essence, DCC employs a short, water-cooled mold. Molten metal is poured into the top of mold, and as it is cast, the ingot is lowered by a hydraulically powered platform. The fast casting speed in this method results in both a fine grain size and an improved distribution of the ceramic fibers in the matrix material. The MMC produced using this method is reportedly 60% stiffer and 15% to 40% stronger than the unreinforced aluminum alloy, is very wear-resistant, and has a controllable thermal expansion coefficient.

C7.3 COMPATIBILITY CONSIDERATIONS

Limited compatibility can be a serious problem in MMCs. In a broad sense, compatibility includes not only the familiar chemical interactions between the matrix and the reinforcement during consolidation, during secondary processing, and in service but also mechanical interplay (such as thermal expansion mismatch effects). For the control of chemical interactions that can degrade either the filament or the matrix, several approaches have been taken involving the development of new reinforcements that are thermodynamically stable with respect to the matrix, diffusion-barrier coatings that reduce filament-matrix interactions, and alloy additions that reduce the activity of the diffusing species and enhance mechanical compatibility.

C7.4 PROPERTIES

The properties of metal matrix composites, like all composite materials, are anisotropic (Table C7.1). The ratio of longitudinal to normal strength, for example, varies from 2 to 15 depending on the composite material, but this ratio also reflects the differences in fiber and matrix strengths. In addition, the control of the interfacial reaction between the fiber and the matrix is known to influence the normal properties greatly, and this reaction will also depend on matrix composition. The optimum balance between longitudinal and transverse direction properties can be obtained either by combining fibers or through hybrid selection and stacking sequences, matrix alloy selection, and component design.

Some other mechanical properties of continuously reinforced MMCs are shown in Tables C7.2 and C7.3 and Figs. C7.6 through C7.10. Some mechanical properties of discontinuously reinforced MMCs are shown in Tables C7.4 and 7.5.

TABLE C7.1 PROPERTIES OF SOME METAL MATRIX COMPOSITE MATERIALS

Matrix	Al	Al	Ti	Ti
Fiber Reinforcement/ Volume Fraction (%)	SiC/50	B/50	SiC/50	B/50
Tensile strength (MPa)				
Longitudinal	1590	1310	1724	1069
Transverse	104	103	414	207
Tensile modulus (GPa)				
Longitudinal	207	200	214	262
Transverse	160	124	165	193
Density (kg m^{-3})	2934	—	3931	—

TABLE C7.2 FATIGUE DATA OF THREE VARIETIES OF Lo-Ex[a] ALLOY AFTER 100 h AT 350°C

Lo-Ex Alloy	Cyclical Stress (MPa)	Number of Cycles to Failure at this Stress
Alumina fiber-reinforced (volume fraction of fibers = 0.27)	60	1×10^6
	50	6×10^6
	40	2×10^7
Unreinforced squeeze-cast	38	6×10^5
	30	2×10^6
	20	7×10^7
Unreinforced gravity-die-cast	30–37	6×10^5
	25–36	2×10^6
	15–22	7×10^7

[a]Lo-Ex is a eutectic Al-Si alloy used for pistons in internal-combustion engines.

Source: Anon, "Metal-matrix composites," Technical File No. 144, p. 11, *Engineering 226,* (4), 1986.

TABLE C7.3 WEAR OF SAE 334 ALUMINUM AND ITS METAL MATRIX COMPOSITES (CONTAINING 5 vol% PROPRIETARY CERAMIC FIBERS) AGAINST CAST IRON

Material	Wear Depth (μm)
Base alloy	350
Fiberfrax-reinforced	48
Fibermax-reinforced	40

Source: B. Peters, "Metal matrix composites for automobiles?" *Automotive Engineering 94* (12), p. 64, 1986.

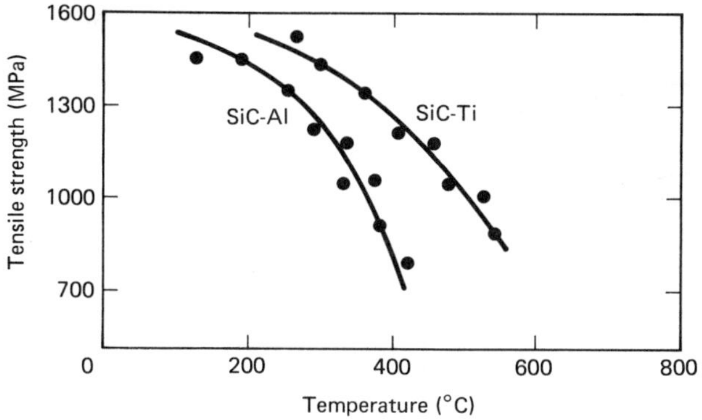

Figure C7.6 Temperature effects on SiC-reinforced metals.

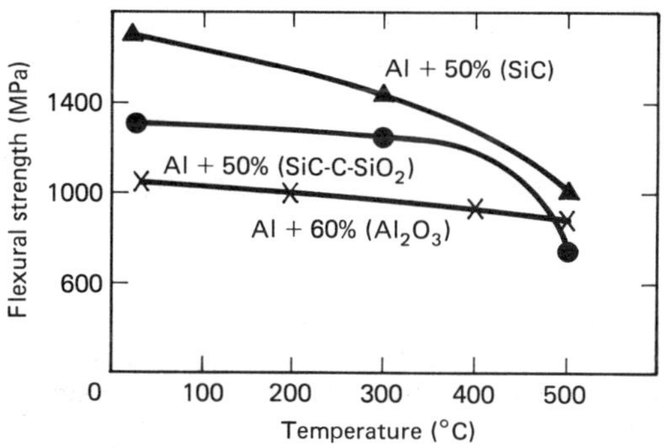

Figure C7.7 Flexural strength of continuously reinforced aluminum composites at various temperatures [1].

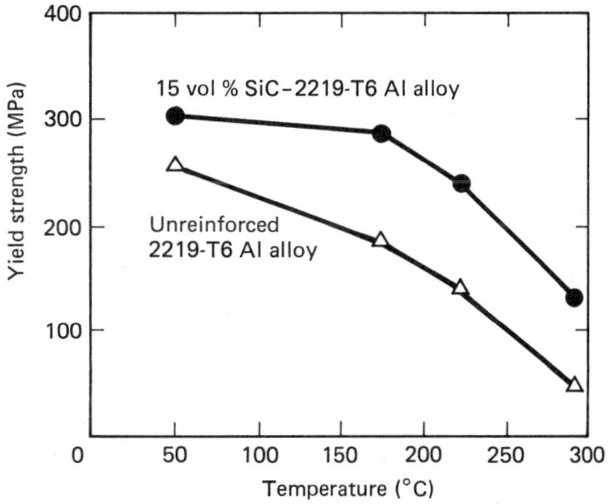

Figure C7.8 Variation of yield strength with temperature of unreinforced and SiC-reinforced 2219-T6 aluminum alloy materials.

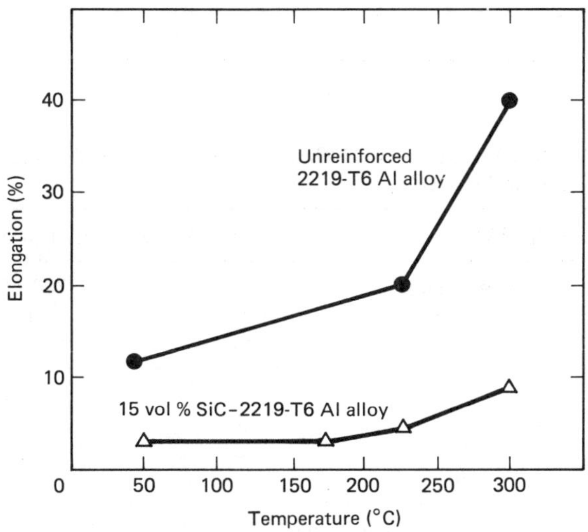

Figure C7.9 Variation of elongation with temperature of unreinforced and reinforced 2219-T6 aluminum materials.

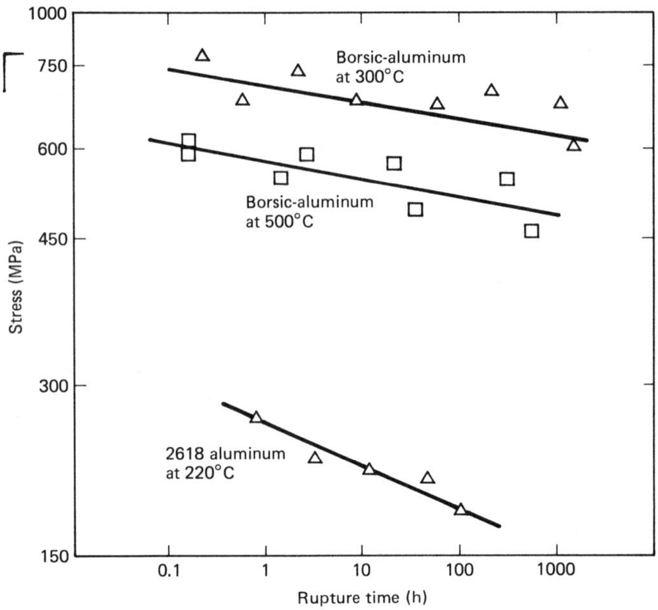

Figure C7.10 Stress rupture properties of Borsic-aluminum metal matrix materials [2] compared with those of an unreinforced aluminum alloy (Borsic content = 50% by volume).

TABLE C7.4 SOME MECHANICAL PROPERTIES OF DISCONTINUOUSLY REINFORCED Al–11% Si–1% Cu–1% Ni–1% Mg ALLOYS

Property	Temperature (°C)	Alloy: Unreinforced	Alloy: +20% Al_2O_3-SiO_2 Short Fibers	Alloy: +20% SiC Whiskers
Modulus of elasticity, E (GPa)	20	72	95	111
Endurance strength, FL, at 10^7 cycles (GPa)	20	80	109	131
Ultimate tensile strength, UTS (MPa)	20	297	312	384
	250	115	198	284
	350	55	112	181

TABLE C7.5 SPECIFIC STRENGTH AND MODULUS OF ALUMINUM AND ALUMINUM ALLOYS REINFORCED WITH SILICON CARBIDE WHISKERS, SiC_w

Temperature (°C)	Specific Strength (10^3 m): 2000 Series Al Matrix + 25% SiC_w	Specific Strength (10^3 m): 7000 Series Al Matrix + 25% SiC_w	Specific Modulus (10^6m): Al Matrix + 15% SiC_w	Specific Modulus (10^6m): 7000 Series Al Alloy + 25% SiC_w
50	225	300	37	42
150	215	230	32	40
250	200	60	27	32
350	—	30	—	—

Source: S. Dermarker, "Metal matrix composites," *Metals and Materials 4* (3), p. 145, 1986.

C7.5 ADVANTAGES AND DISADVANTAGES OF MMCs

ADVANTAGES

- The ratios of the strength (in tension, compression, and shear) to the density and of the modulus (elastic and shear) to the density of these materials are much higher than for the unreinforced metal counterpart. For example, the specific tensile strength and specific tensile modulus of an aluminum-silicon carbide MMC (in the direction of the fiber or reinforcement) are, respectively, about 15 and 2.5 times those of unreinforced aluminum.
- The fatigue strength is improved compared to the unreinforced metal.
- The fibers can be oriented to the principal axis of stress, thus designing the material for the application.
- By changing fiber orientation, harmful natural vibrations can be smoothed out.
- They offer high-temperature capability far above that of polymer matrix composite materials.
- Joining by conventional means is easy.

DISADVANTAGES AND PROBLEM AREAS

- High cost, due to small volume, although both the raw material price and the cost of fabricated products are dropping.
- Insufficient characterization data.
- Lack of standardization of materials and processes.
- Lack of standardized, fully defined test methods and, particularly, nondestructive testing and accelerated testing procedures. In this respect, it is gratifying to note that ASTM Committee D30 on High Modulus Fibers formed four task groups in 1982 charged with directing and guiding efforts to develop test methods and improve basic understanding in the areas of shear, compression, and fatigue fracture of MMCs.

C7.6 APPLICATIONS

AEROSPACE AND AERONAUTICS

- A missile payload adapter, fabricated from B-Al
- Use of continuous SiC-fiber Al and single-crystal SiC-whisker Al for the vertical fins on fighter aircraft
- Spars, skins, undercarriages, propellers, turbine blades, and compressor blades of aircraft
- Gearbox housings, rotor hubs, and skids of helicopters
- Use of SiC-Al for the cover of an inertial measurement unit of the gyroscope guidance system of a ballistic missile
- Torque box for the horizontal tail of the F-15 fighter jet made of an MMC, giving a structure that is 17% lighter than a boron-epoxy counterpart

- A rocket motor in which the Al motor shell is wound with layers of SiC fibers held in place by plasma
- Use of a SiC-whisker Al material in the rotor of a stratified-charge rotary engine for small aircraft
- Airframe of a missile; the weight savings associated with the use of MMC here can be translated into increased velocity, increased range, or increased warhead payload. This in turn means an increase in either lethality or the number of on-board electronic devices that aid guidance, control, and survivability and hence make for the general efficacy of the weapons system
- Radiators in space nuclear power systems
- Blades of advanced turbopumps for the main engines of the next generation of the U.S. space shuttle, made of an MMC comprising of tungsten fibers embedded in an iron-based superalloy matrix.

SATELLITES AND SPACE STRUCTURES

- Deployable antennas
- Solar panels and arrays
- Laser mirrors and x-ray telescopes

AUTOMOTIVE PRODUCTS

- Various areas of the pistons of automotive engines, such as the crown/combustion bowl, the ring grooves, and the gudgeon pin bosses
- Connecting rods
- Leaf springs
- Driveshafts

ELECTRONIC HARDWARE

- Microwave guides
- Radar antennas

OTHER APPLICATIONS

- Tennis racquets, skis, and golf clubs
- Bicycle frames and yacht masts
- Hydrofoils, geothermal well components, pressure vessels, and battery plates and separators

C7.7 CONCLUSION

There is a clear imperative for a number of developments in the fields of matrices,* innovative fiber research, and reductions in undesirable fiber-matrix interfacial reactions.

*As of 1988, fully 90% of MMCs were based on aluminum and its alloys (notably 6061 Al), and a few were based on alloys of magnesium (such as AZ61 and AZ91), tungsten, and titanium.

There have been some spectacular improvements and innovative changes in fibers since the mid-1980s. For example, a new approach for manufacturing silicon carbide was developed by Nippon Carbon Co. of Japan. In involves melting the organic-silicon polymer (polycarbosilane), spinning it into fiber, and baking it. During the process, methane and hydrogen are lost by thermal decomposition to leave silicon carbide fiber. This is then coated with a resin. The resultant silicon carbide fiber, called Nicalon, has a mean diameter of 15 μm, a mean fiber length of 850 μm, a density of 2550 kg m^{-3}, a tensile strength of 2.9 GPa, and a tensile modulus of 188 GPa. Unlike carbon fibers, it wets aluminum alloys well and has limited reaction with metals at elevated temperatures. An even more unusual reinforcement material developed in Japan by Ube Industries is Tyranno. It is a long, inorganic fiber material consisting of silicon, titanium, and carbon made by extracting an organic-metallic polymer from a silane of low molecular weight and a titanium compound, making it a polytitano carbosilane polymer. The resultant fiber has an average diameter of 12 μm, is thermally stable up to 1700°C, and has a density of 2200 kg m^{-3}, a tensile strength of 2.9 GPa, a tensile modulus of 195 GPa, good processability, the ability to wet aluminum, and limited reaction with metals at high temperatures. Details of other SiC whiskers are given in Table C7.6. In addition, it is common to coat the fibers with other materials in order to overcome the problem of high-temperature interaction with the metal matrix. For example, 140-μm-diameter boron fibers are coated with boron carbide before incorporation into the metal matrix.

TABLE C7.6 CHARACTERISTICS OF SOME INNOVATIVE SILICON CARBIDE WHISKER REINFORCEMENTS

Type	Composition	Dimensions	Supplier
Ricehull whisker	Whisker-particle mixture from coked ricehull sources	Average 0.6 μm diameter; 50 μm lengths with irregular particles	Arco (USA)
SCW whisker	Average 95% β silicon carbide, 5% α silicon carbide	Average 0.2–0.5 μm diameter; 10–40 μm length	Tacho Chemical Industries (Japan)
Tokamax whisker	β silicon carbide single crystal	0.1–0.5 μm diameter; 50–200 μm length	Tokai Carbon (Japan)
Silar SC-9	Single crystals of α-SiC	0.6 μm diameter; 10–80 μm length	Arco (USA)

REFERENCES

1. S. Dermarker, ''Metal matrix composites,'' *Metals and Materials 4* (3), pp. 144–146, 1986.
2. K. Kreider, et al., ''Plasma-sprayed metal matrix fiber-reinforced composites,'' Report No. TR-68-119. Wright-Patterson AFB, Ohio: U.S. Air Force Materials Laboratory, July 1968.

CERAMIC MATRIX COMPOSITE MATERIALS

CHAPTER C8

C8.1 INTRODUCTION

Two developments made the emergence of ceramic matrix composites (CMCs) possible. First, fiber reinforced polymer composites are now accepted as being very reliable structural materials, thereby providing the requisite theoretical and fabrication experience needed for the adoption of other fibrous composites. Second, new ceramic fibers of superior quality have become and are becoming available, permitting the creation of composites with highly desirable properties.

Fibers, chosen on the basis of their high specific strength and specific stiffness, include (a) large-diameter (> 50 μm) filaments of boron and silicon carbide (individual fiber diameters 7 to 30 μm); (b) filamentary yarns of graphite, alumina, aluminosilicate glasses, boron nitride, and silicon carbide; and (c) whiskers of silicon carbide and nitride (see Tables C8.1 and C8.2). In addition, there is reinforcement by toughened particles, mainly tetragonal zirconia polycrystals and unstabilized monoclinic zirconium oxides. To date, however, the greatest efforts have been devoted to yarn-reinforced ceramics, particularly those reinforced with graphite, alumina, or silicon carbide yarn. Of the ceramic fibers, Nicalon, a polycarbosilane-derived microcrystalline β-SiC fiber containing oxygen and excess carbon, has received the most attention. This is mainly because of the in

TABLE C8.1 PROPERTIES OF SOME COMMERCIALLY AVAILABLE CERAMIC FIBERS

Designation	Composition	Average Diameter (μm)	Average Ultimate Tensile Strength (MPa)
Nicalon	59Si, 31C, 10O	15	2905
SCS-6	SiC on carbon core	143	3920
Nextel 312	$62Al_2O_3$, $14B_2O_3$, $24SiO_2$	11	1750
Tyranno	Si, Ti, C, O	9	> 2970
MPDZ	47Si, 30C, 15N, 8O	13	1925
PRD-166	Al_2O_3, 15–$25ZrO_2$	20	2275

Designation	Average Tensile Modulus (GPa)	Density ($kg\ m^{-3}$)	Supplier
Nicalon	196	2550	Nippon Carbon (Japan)
SCS-6	406	3000	AVCO (USA)
Nextel 312	154	2700	3M (USA)
Tyranno	> 200	2400	Ube (Japan)
MPDZ	193	2300	Dow Corning/Celanese (USA)
PRD-166	385	4200	DuPont (USA)

TABLE C8.2 PROPERTIES OF OTHER COMMONLY USED FIBERS

Fiber Type	Density ($kg\ m^{-3}$)	Ultimate Tensile Strength (MPa)	Tensile Modulus (GPa)
Carbon fibers			
Thornel 300	2000	2930	234
HM	2000	2700	350
GY-70	2000	2070	516
P-100	2000	2070	654
Al_2O_3 fibers			
Saffil	3300	2000	300
FP	3900	1380	379
Boron monofilament	2500	2750	400
Glass fibers			
Soda lime	2480	50[a]	74
Borosilicate	2230	55[a]	65

[a]Modulus of rupture.

situ formation of a carbon-rich interface layer when processed in a number of ceramic matrices, especially oxides.

Although fibers may appear to be the prime determinants of the characteristics of these CMCs, it is the matrices, with their wide range of compositions and physical properties, that afford the greatest possibility for tailoring a successful

CMC. The matrix composition not only contributes to properties such as modulus of elasticity, coefficient of thermal expansion, and creep resistance but, more important, also provides a means of controlling fiber-matrix interaction, which in turn determines the nature of the fracture mode of the composite material. The matrices in use fall into three broad categories:

- Oxides (such as alumina and zirconia)
- Nonoxides (such as silicon carbide, titanium diboride, and silicon nitride)
- Glasses (such as borosilicate, aluminosilicate, and high-silica glasses) and glass ceramics (such as LASI,* ternary mullite, and hexacelsian)

CMCs are a promising class of structural materials for applications where high strength, high stiffness, low thermal expansion, and high-temperature stability are desirable attributes either severally or jointly. Work in recent years has shown that large gains in fracture strength and toughness are achievable. Progress has been made toward identifying the failure mechanisms in CMCs and understanding the intrinsic fiber-matrix interface properties necessary to take advantage of strengthening and toughening mechanisms. Greatly improved reliability at strengths equivalent to the best of monolithic polycrystalline ceramics has been reported. It is also clear that processing techniques that can produce better fiber-matrix distributions, microstructures, and interfacial properties are necessary for further improvements in the properties and wider applications of CMCs.

C8.2 PROCESSING

Most CMCs, especially those based on glass matrices, are easy to fabricate. The flow characteristics of the glass at high temperatures make it possible to adapt techniques commonly used for the fabrication of near-shape polymer composite materials to the fabrication of CMCs. Fabrication procedures now in use or under serious study belong to four categories.

Category A: Slurry Infiltration Techniques

In slurry infiltration, a fiber preform is impregnated by passage through a slurry mixture containing the matrix material and then gathered and dried (Fig. C8.1). This method is well suited to glass or low-melting-point matrices and is limited to simple (uni- or bidirectional) fiber orientations. It is a costly process and thus economically unattractive for the fabrication of geometrically complex components requiring near-final shapes and dimensions.

Hot matrix transfer into woven preforms (Fig. C8.2) is a method used to fabricate components with complex geometries. The matrix must be injected at temperatures much higher than normal hot-pressing temperatures to aid penetration of the interfiber spaces.

*LASI is a mixture of Li_2O, Al_2O_3, MgO, and SiO_2.

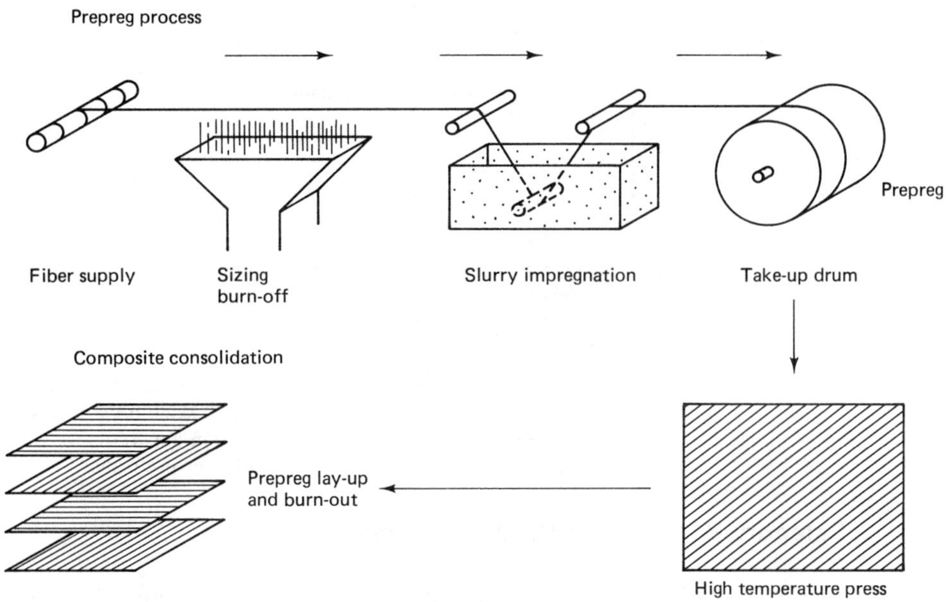

Figure C8.1 Steps in the fabrication of a glass matrix composite material.

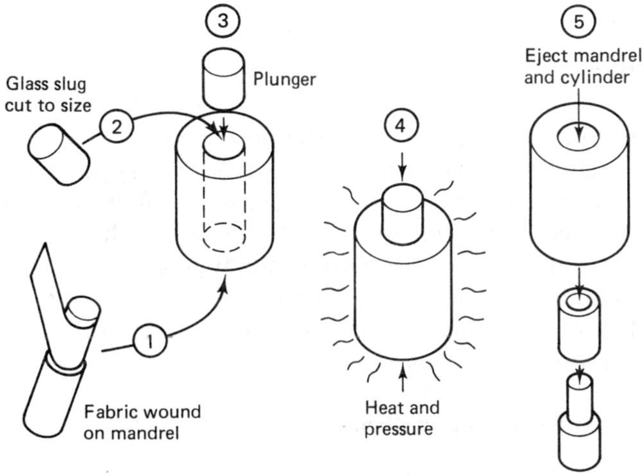

Figure C8.2 Steps in the matrix transfer molding of a glass matrix composite material.

Another technique involves hot injection molding of chopped fiber compounds or preforms. Chopped fiber and matrix grit molding compounds are made by leaving space between the individual yarn strands during the tape-winding operation, as is shown in Fig. C8.1. The strands are chopped into short ($\sim$10 mm) segments, which are then tumbled into a random orientation. The compound (either in the tumbled state or as a preconsolidated slug) is placed in the reservoir section of a double-chambered die and injected at high temperature into a mold cavity by ram action.

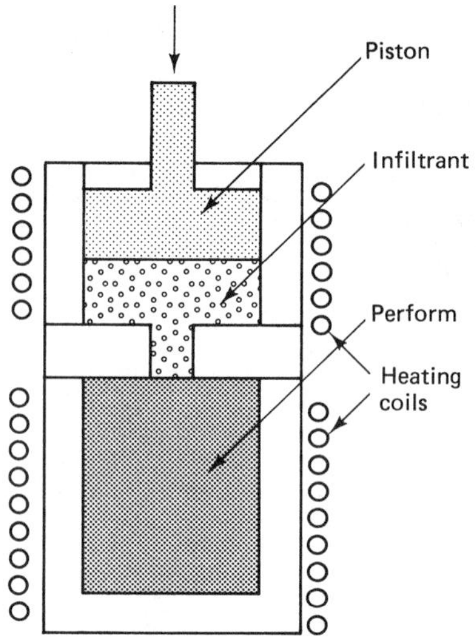

Figure C8.3 Schematic diagram of a small-scale pressure-assisted melt infiltration apparatus for the impregnation of fiber preforms.

Category B: Chemical Synthesis Methods

Examples include sol-gel and polymer precursor processing. They offer a number of advantages, namely good compositional homogeneity in single-phase matrices, excellent potential for forming unique multiphase composites, good ease of infiltration and forming, and low processing temperatures. Among the disadvantages are high shrinkage and low yield.

Category C: Melt Infiltration

This type of procedure is particularly attractive for the high composite densities it affords but is limited to relatively low-melting-point glasses, crystalline ceramics, and glass ceramics. Some advantages are fully dense and flow-free composites formed in a single processing step, minimal dimensional change from preform to final product, and suitability for use with virtually any reinforcement geometry. Disadvantages include fiber-damaging chemical reactions at the high melting temperatures and low rates of infiltration. A schematic representation of a small-scale melt infiltration apparatus is shown in Fig. C8.3.

Category D: In Situ Chemical Reaction Methods

Examples include chemical vapor deposition and chemical vapor infiltration. These methods are especially suited to the processing of nonoxide ceramic matrices.

C8.3 APPLICATIONS

Here is a list of possible applications:

- Radomes and antenna windows of reentry space vehicles
- Aircraft engine components

- Biomedical devices
- Parts of advanced internal combustion engines
- Battery plates and separators, notably in the high-temperature sodium-sulfur traction battery
- Geothermal well components
- Self-lubricated bearings
- Airplane brakes

C8.4 RESEARCH AND DEVELOPMENT DIRECTIONS

Among the key issues to be resolved and the developments to occur as a precondition for the widespread adoption of CMCs are the following:

- Theoretical analyses of the strength of bonding between fibers and matrix that is optimal for toughness and strength of the composites
- Exploration of a variety of coating methods to modify the surface characteristics of reinforcing fibers
- Extension of the slurry infiltration technique to include infiltration of three-dimensional preforms and development of improved infiltration methods for whisker or discrete filament composites
- Investigation of an expanded range of polymers that pyrolyze to ceramics with high yield
- Development of new approaches to the production of in situ composites, with special reference to the production of extremely dense materials without multiple infiltrations and/or slow chemical reactions
- Development of high-performance and cheap ceramic fibers, leading to low-cost CMCs

NEW TYPES OF GLASS

CHAPTER C9

C9.1 METALLIC GLASSES

Metallic glasses, so called because their disordered atomic structure more closely resembles that of glass than the conventional crystalline structure of metals, are a class of materials with exciting potential. They are formed from molten alloys that are quenched very rapidly (at about 10^6 °C s^{-1}), thereby precluding crystallization. This technique is known as rapid solidification technology (RST). Metallic glasses have enormous potential for exceptional strength, corrosion resistance, and magnetization properties.

Although many metal compositions readily form the noncrystalline structure typical of this type of glass, the first commercially available examples were the ferrous metal-metalloid ribbon products called METGLAS from Allied Chemical Corp. of Morristown, N.J.

Researchers at Israel's Ben Gurion University developed a device to produce microcrystalline and amorphous metallic ribbons. Called a centrifuge melt spinning (CMS) machine, the apparatus allows the use of the best combinations of substrate velocities (90 m s^{-1}, for example) and metal ejection pressures (typically 300 kPa). The main disadvantage of CMS is that casting is discontinuous, although the construction of a continuous unit for melting and casting, made under vacuum or controlled atmosphere conditions, is possible.

Rapid solidification is also possible when high pressure is applied instantaneously to a liquid alloy to cause undercooling and hence solidification. This process, known as rapid pressure application (RPA), was used by researchers at the Defense Metallurgical Research Laboratory in India to achieve rapid solidification in several systems of high- and low-melting-point alloys, such as Pb-Sn, Mg-Zn, and Fe-B.

Research at Rensselaer Polytechnic Institute in New York City involved the production of rapidly solidified reactive and refractory alloys in pilot quantities by a unique arc melt spin process—a process that combines an arc furnace (cold copper crucible plus orifice and electrode) with a spinning disc.

Another technique is vapor deposition, developed by the Royal Aircraft Establishment in England. Condensation of an alloy from the vapor produces metastable alloys as well as levels of supersaturation. Scientists at the Université de Neuchâtel in Switzerland used dynamic compaction of powders to consolidate metallic glasses. The potential of this method is good because there is no excessive high-temperature excursion. But compaction leads to many complex shear, pressure, and thermal conditions that produce generalized and sometimes localized microstructural changes.

Several factors are recognized as affecting the ability of an alloy to form a metallic glass; these include atomic size ratio, alloy crystallization temperature and melting point, and heat of formation of compounds. From the collection of experimental data, it seems that atomic size has the greatest influence. For example, work at the Research Institute of Mineral Dressing and Metallurgy in Sentai, Japan, showed that the minimum solute concentration necessary to form an amorphous phase by rapid quenching correlates inversely with atomic size ratio for various binary alloy systems. Notwithstanding the experimental data bank, no satisfactory theoretical predictive relations have yet emerged. In addition, an acceptable correlation between glass-forming factors and metal-metalloid and metal-metal alloy glasses is currently missing.

The most important potential uses of metallic glasses are in applications that take advantage of their unique magnetic characteristics. These glasses are thus used in signal and current transformer cores, switching power-supply units, heads for magnetic tape players, and magnetic amplifiers. In addition, the combination of their physical and mechanical properties suggests many uses beyond power transformers and electrical machinery. Strips of metallic glass can serve as reinforcements for pressure vessels, hoses, tires, belts, pipes, fabrics for shields against electromagnetic interference, pilferproof security tapes for library books, and razor blades. The brazing foils made from these glasses can successfully compete with traditional brass alloys for joining stainless steel and nickel alloys in applications such as aircraft tubes, heat exchangers, food processing equipment, heart pacemakers, and nickel-cadmium batteries.

Some properties of conventional glasses and metallic glasses are compared semi-quantitatively in Table C9.1. The values of a number of properties of three grades of METGLAS are presented in Table C9.2.

TABLE C9.1 SEMI-QUANTITATIVE COMPARISON OF SOME PROPERTIES OF CONVENTIONAL (OXIDE) AND METALLIC GLASSES

Property	Value of Property in Metallic Glass Compared to that in Oxide Glass
Density	2% less
Shear modulus	20%–30% less
Modulus of elasticity	20%–30% less
Fracture toughness	100 times greater

Source: T. R. Anantharaman, "Metallic glasses: an overview," in T. R. Anantharaman (ed.), *Metallic Glasses: Production, Properties and Applications,* Rockport, Mass.: Trans Technical Publications, p. 24, 1984.

TABLE C9.2 SOME PROPERTIES OF THREE GRADES OF METGLAS METALLIC GLASS[a]

Property	2605SC	2605CO	2826MB
Physical Properties			
Density (kg m^{-3})	7300	7560	8020
Crystallization temperature[b] (°C)	480	430	410
Resistivity (10^{-8} Ω·m)	125	130	160
Mechanical Properties (as cast)			
Ultimate tensile strength (GPa)	>0.70	>1.50	1.38
Vickers hardness (GPa)	10.30	10.00	10.50
Magnetic Properties (as cast)			
Saturation magnetization (T)	1.61	1.75	0.88
Coercive force (A m^{-1})	6.37	6.37	7.96
Magnetic Properties (field-annealed)			
Saturation magnetization (T)	1.61	1.75	0.88
Coercive force (A m^{-1})	4.78	3.98	<1.19
Curie temperature (°C)	370	415	353

[a]2065SC is $Fe_{81}B_{13.5}Si_{3.5}C_2$; 2605CO is $Fe_{67}Co_{18}B_{14}Si$; 2826MB is $Fe_{40}Ni_{38}Mo_4B_{18}$.

[b]20°C min^{-1} heating rate.

Source: Allied Chemical, Morristown, New Jersey.

C9.2 HEAVY-METAL FLUORIDE GLASSES

As a general rule, crystalline or polycrystalline materials, such as CaF_2, ZnSe, and various alkali halides, have been used for optical systems and devices that operate in the midinfrared (1–5 μm) position of the spectrum, such as high-energy lasers, windows, optical fibers, and membranes in fluoride-ion-selective electrodes. Although these materials can be fabricated (blown, molded, drawn, stamped, or forged) into practically any shape with high optical homogeneity, vitreous materials have not received much attention because of limitations in their infrared optical behavior. A unique family of nonoxide glasses has emerged that offers the prospect of very low optical absorption in the midinfrared region, together with very high ultraviolet/visible transparency. Known generically as heavy-metal fluo-

TABLE C9.3 COMPOSITIONS AND PROPERTIES OF HEAVY-METAL FLUORIDE GLASSES

	Glass designation						
	ZBLA	HBLA	ZBL	HBL	BZnLuT	BZnLuGT	BZnYbNT
Typical composition (mole %)							
ZrF_4	57	—	62	—	—	—	—
HfF_4	—	57	—	62	—	—	—
BaF_2	36	36	33	33	19	19	14
LaF_3	3	3	5	5	—	—	—
AlF_3	4	4	—	—	—	—	—
ZnF_2	—	—	—	—	27	25	27
LuF_3	—	—	—	—	27	25	—
YbF_3	—	—	—	—	—	—	27
ThF_4	—	—	—	—	27	25	27
GdF_3	—	—	—	—	—	6	—
NaF	—	—	—	—	—	—	5
Density (kg m^{-3})	4610	5880	4790	5780	6200	6200	6200
Glass transition temperature (°C)	310	312	306	312	353	357	343
Crystallization temperature (°C)	390	400	380	395	452	443	446
Refractive index	1.516	1.504	1.523	1.514	1.534	1.534	—
Coefficient of linear expansion (10^{-7} °C^{-1})	187	173	188	177	151	151	—

ride glasses (HMFGs), these glasses contain large amounts of ZrF_4 (in which case they are called fluorozirconate glasses) or HfF_4 (in which case they are called fluorohafnate glasses), plus lesser amounts of alkaline-earth, rare-earth actinide fluorides (Table C9.3). Other HMFGs that contain no ZrF_4 or HfF_4 also exist and exhibit better midinfrared transmission properties than their fluorozirconate or fluorohafnate counterparts. These are called "BaF_2/ThF_4" glasses (Table C9.3).

The following are some of the attractive features of HMFGs:

- Very wide transparency range, extending from 9–11 μm in the infrared to 0.2–0.3 μm in the near ultraviolet (Fig. C9.1)
- Low susceptibility to ionizing radiations in the midinfrared region
- Good mechanical properties (for example, Knoop hardness ~ 2500 MPa)
- Processing without toxic materials
- Easy convertibility to fibers and wide variety of bulk shapes and sizes

Probably their sole major disadvantage is their poor chemical durability in aqueous environments. This aspect has been addressed in detail by Simmons and Simmons [1], whose studies showed that the corrosion of fluorozirconate glasses in water appears to be a complex process involving component solubility, matrix dissolution, ion exchange, hydration, species saturation, and reprecipitation. These glasses leach very rapidly in water and hence show comparatively poor durability in this medium (Fig. C9.2).

Ba-Th-based fluoride glasses have much better chemical resistance than fluorozirconate glasses (mainly because of lower component solubility and much less severe drift in solution pH [2]), but they are still 1000 times less resistant to attack than durable silicates, such as Pyrex glass (Fig. C9.2).

Figure C9.1 Transmission characteristics of a fluorohafnate glass, a BaF_2/ThF_4 glass, and fused silica.

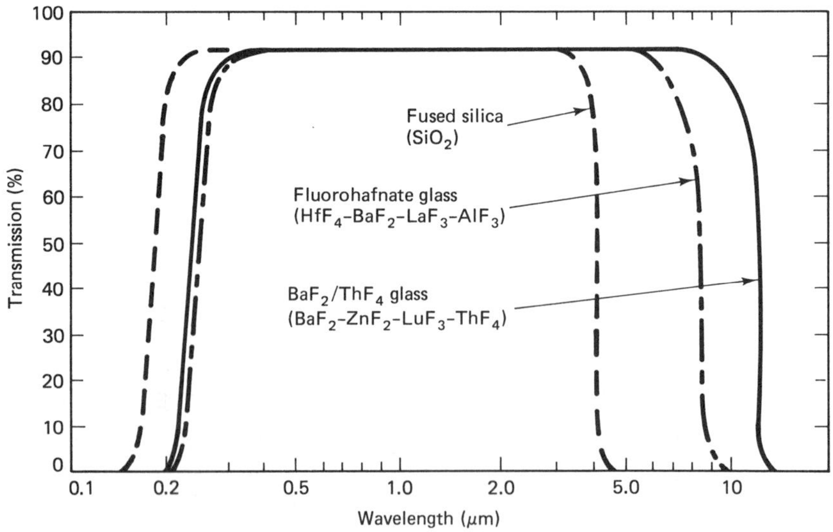

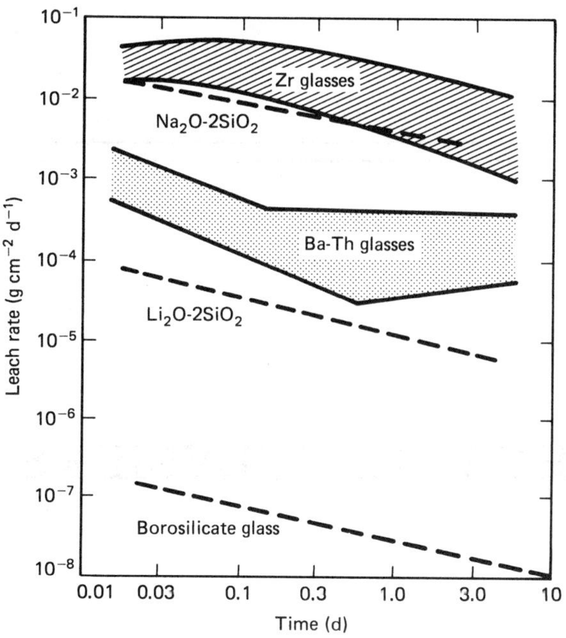

Figure C9.2 Comparison of the corrosion rate of zirconate [1], silicate [2], barium-thorium-based [3], and borosilicate [4] glasses in water at room temperature.

These results emphasize the need for the use of hermetic coatings on these glasses, in order to shield them from severe attack in aqueous media. One promising coating material is diamondlike carbon.

REFERENCES

1. C. J. Simmons and J. H. Simmons, "Chemical durability of fluoride glasses: I. Reaction of fluorozirconate glasses with water," *J. Amer. Ceramic Soc. 69* (9), p. 668, 1986.
2. L. L. Hench and D. E. Clark, "Surface properties and performance prediction of alternate waste forms," Annual Report to the Nuclear Regulatory Commission, Contract No. NRC-04-78-252, 1981.
3. C. J. Simmons, "Chemical durability of fluoride glasses: II. Reaction of barium-thorium-based glasses with water," *J. Amer. Ceramic Soc. 70* (4), p. 299, 1987.
4. A. Barkatt, J. H. Simmons, and P. B. Macedo, "Evaluation of chemical stability of vitrification media for radioactive waste products," *Phys. Chem. Glasses 22* (4), pp. 73–85, 1981.

LIQUID CRYSTAL POLYMERS

CHAPTER C10

C10.1 INTRODUCTION

A class of materials known collectively as thermotropic liquid crystal polymers (TLCPs) appears to be well suited for a growing number of applications in electronics, aerospace, defense, and chemical processing. These polymers are made from paraoxybenzoyl and oxybiphenylene terephthaloyl monomers, and their structure consists of all para-para aromatic groups. They are noted for high-temperature performance, self-reinforcement, an ability to be easily injection-molded, high melting point, and good solvent resistance. These polymers exhibit thermotropism and, in the liquid crystalline phase, may be described as having nematic domains with parallel, one-dimensional order.

The precursors of TLCPs are lyotropic liquid crystal polymers (LLCPs), which are polymers that could be solution-formed only into films or fibers because the stiff amide linkages do not become significantly flexible even under high heat. This makes their melt processing impossible. Examples of LLCPs are Nomex and Kevlar. TLCP is obtained by having more flexible links between the aromatic rings so that melt processing now becomes possible.

The earliest of the TLCPs, Ekkcel, introduced in 1973, incorporates an ester linkage between groups for higher flexibility [Fig. C10.1(a)] and is moldable by

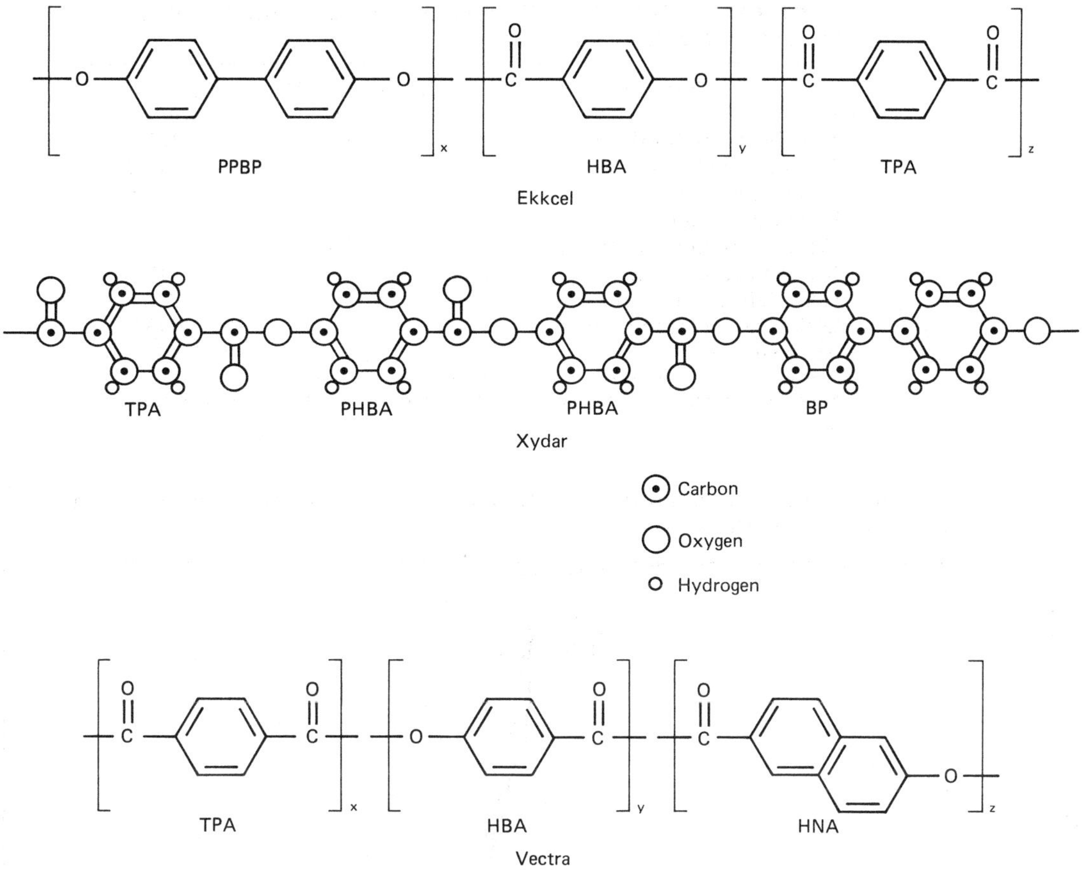

Figure C10.1 Structure of three liquid crystal polymers (PPBP, *p,p′*-biphenol; HBA, hydroxybenzoic acid; TPA, terephthalic acid; HNA, hydroxynaphthoic acid; PHBA, parahydroxybenzoic acid; BP, biphenyl).

injection, compression, or sintering. However, it is difficult (though not impossible) to process because it retains a great deal of orientation in the molded part. Two widely used commercial forms of TLCP are Xydar and Vectra.

C10.2 XYDAR

This TLCP, a self-reinforcing, wholly aromatic polyester [Fig. C10.1(b)], was introduced in late 1984. It is a high-performance material that can be processed rapidly without the need for postcuring and can accept high regrind levels.

Xydar has excellent mechanical and physical properties (Table C10.1). It is inherently flame-retardant (resisting a 1090°C flame for 15 min), producing a swollen-up char that stifles combustion when exposed directly to fire. Under fire conditions, it produces virtually no smoke or noxious fumes and will not melt unless acted on by a kinetic force. The polymer has excellent resistance to thermal

TABLE C10.1 PROPERTIES OF UNFILLED THERMOTROPIC LIQUID CRYSTAL POLYMERS

Property	Xydar SRT 500	Vectra C130
Tensile strength (MPa)	125	165
Tensile modulus (GPa)	8.3	15.9
Elongation (%)	4.8	2.0
Flexural strength (MPa)	136	219
Flexural modulus (GPa)	13.1	14.5
Notched Izod, (J m^{-1})	208	117
Heat deflection at 1.82 MPa (°C)	337	241
Dielectric strength (MV m^{-1})	31	43
Arc resistance (s)	138	137

Sources: Amoco Performance Products, Atlanta, Ga.; Hoechst Celanese, Summit, N.J.

Figure C10.2 Thermogravimetric analysis of Xydar SRT-500. (Data courtesy of Amoco Performance Products Inc.)

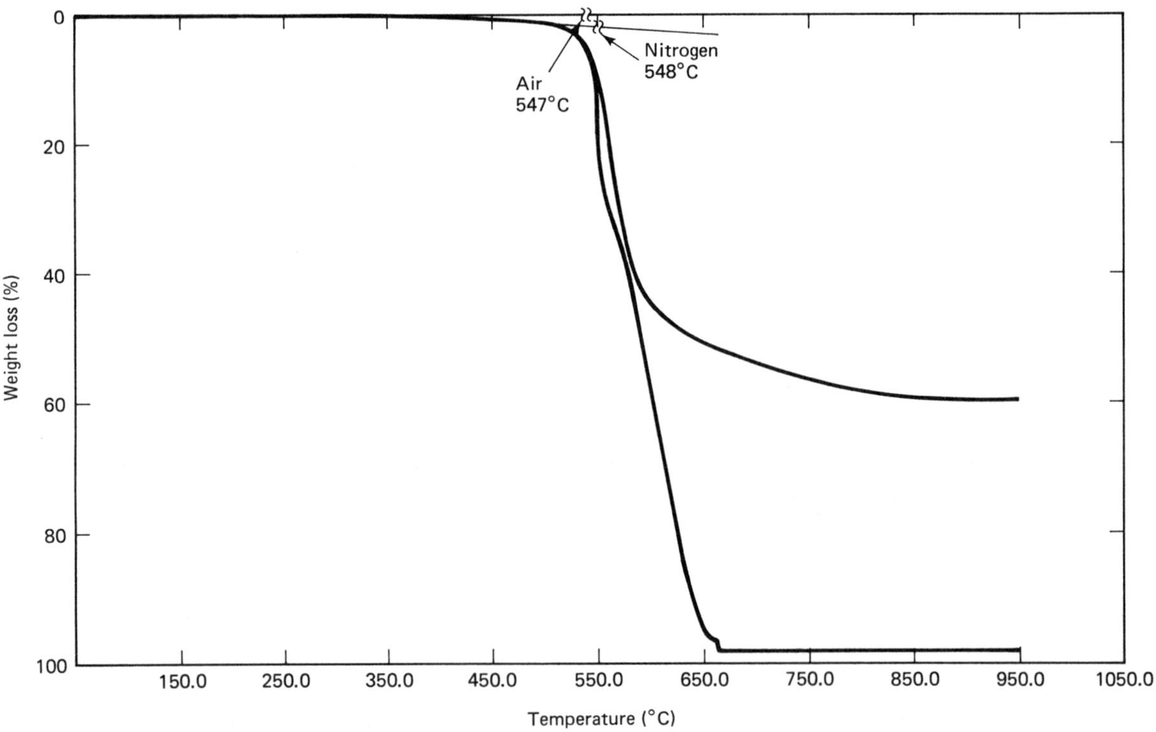

oxidative degradation (Fig. C10.2) and retains a high proportion of its room-temperature properties at elevated temperatures. For example, its flexural modulus is 11.0 GPa at room temperature, 6 GPa at 230°C, and 5.0 GPa at 300°C. Xydar also has excellent creep resistance properties (Fig. C10.3), and its UV and gamma radiation tolerance is exceptionally good. For example, 6700 h of carbon-arc-accelerated UV exposure or 10^9 rad of Co-60 does not produce any significant

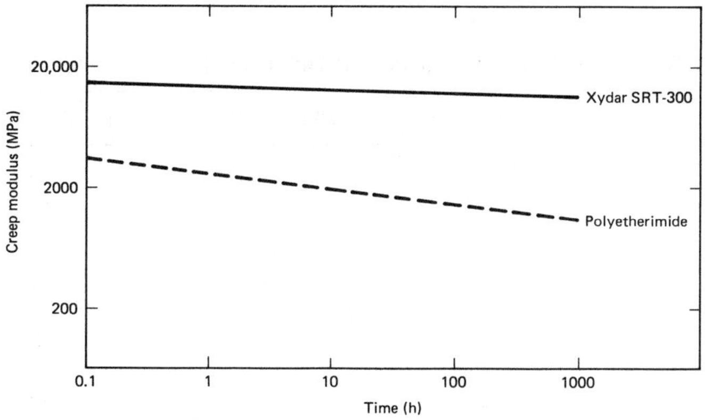

Figure C10.3 Flexural creep of two polymeric materials at 135°C (34 MPa flexural stress). (Data courtesy of Amoco Performance Products Inc.)

deterioration in properties. Because of the high degree of crystallinity, Xydar is not attacked by the common corrosive fluids. Its ultimate tensile strength, after 30 d exposure to 20% H_2SO_4 at 20°C, reduces by 0%, and by only 2% after 11 d exposure to 70% HNO_3 at 20°C. The polymer also resists absorption of water, hydrocarbon fuels, polar and nonpolar solvents, caustic soda, and the like. It also has a low outgassing mass loss (0.03% at 125°C at 1.30 mPa for 24 h on the ASTM E595 outgassing test).

As far as processability is concerned, profiles of melt viscosity as a function of shear rate of Xydar are produced at melt temperatures 230°C to 290°C higher than for the common polyolefin resins. This means that the polymer can be easily molded at high production rates, but typically at a very high melting point (420°C). This entails some processing equipment modifications (such as ceramic heater bands for the barrel, cartridge heaters for the mold, and insulation between the mold and the platens), which increase the cost of such equipment.

C10.3 VECTRA

This is a wholly aromatic copolyester based on a large family of naphthalenic-based copolymers. Room-temperature mechanical and physical properties are outstanding (Table C10.1). The polymer is also highly microwave-transparent, very weatherable, resistant to gamma irradiation, inert to a wide array of chemicals and corrosive fluids, and inherently flame-retardant. In addition, Vectra has one advantage over Xydar as far as processability is concerned. Vectra can be molded using standard equipment with no modifications because it has melt temperatures in the 280°C to 390°C range. Viscosity versus shear rate profiles of unfilled Vectra are much like those of unfilled nylon 6,6. This is achieved by the addition of kinks, as provided by HNA in the structure, giving it more degrees of freedom for easier molecular motion [Fig. C10.1(c)].

C10.4 REINFORCED POLYMERS

Many manufacturers are investigating the possibility of improving some properties of TLCPs through the use of reinforcement and fillers that disrupt the anisotropic character of the properties of the molded parts, thus making the properties of

TABLE C10.2 PROPERTIES OF FILLED THERMOTROPIC LIQUID CRYSTAL POLYMERS

Property	FSR-315	PDX-85194	LCP 3440
Tensile strength (MPa)	81	88	110
Tensile modulus (GPa)	12	—	18
Elongation (%)	3	4	2
Flexural strength (MPa)	112	83	172
Flexural modulus (GPa)	11	9	17
Notched Izod impact (J m^{-1})	75	43	53
Heat deflection at 1.82 MPa (°C)	>316	280	316
Dielectric strength (MV m^{-1})	21	—	>24
Arc resistance (s)	241	—	>180

Sources: Amoco Performance Products, Atlanta, Ga.; LNP Co., Malvern, Pa.; RTP Co., Winona, Minn.

filled or reinforced grades less dependent on flow patterns than those of the unfilled polymers. Among the filled varieties are FSR-315, which is 50% talc filler in Xydar; Thermocomp PDX 85194, which is 15% polytetrafluoroethylene-filled Xydar; and LCP 3440, which is 30% glass fiber reinforced Xydar that incorporates a proprietary heat-resistant coupling agent. The mechanical and physical properties of these reinforced TLCPs are summarized in Table C10.2.

C10.5 PRODUCTION ASPECTS

The properties of parts made of Xydar have been found [1] to be functions of the part thickness and the various molding parameters (such as injection pressure, injection speed, and mold temperature; see Fig. C10.4). The highest tensile and

Figure C10.4 Tensile strength of 50% mineral-filled Xydar as a function of processing parameters [1].

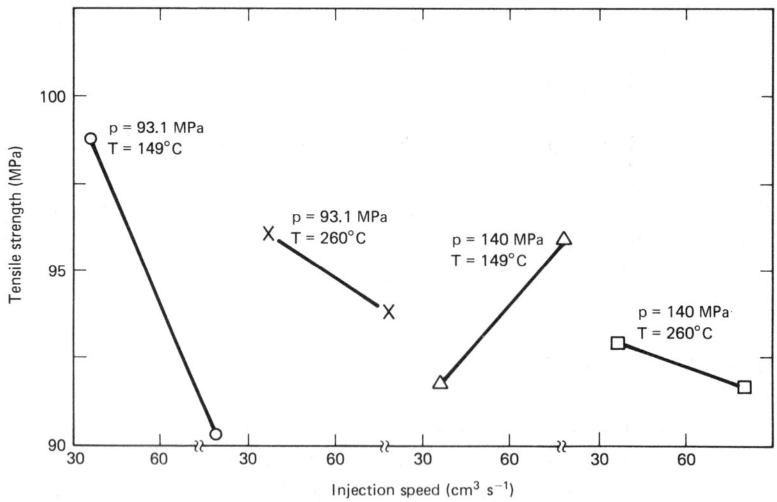

flexural property values are obtained when the parts are thin and the injection pressure, injection speed, and mold temperature are all low [1]. This observation has been explained by the "structural dichotomy of the skin and the core layers of the molded parts" [1]. It seems that the inclusion of fibers reduces the influence of the aforementioned factors and expands the window of processability of thermoformed and/or extruded parts [1]. The thinner the section being extruded, the higher the degree of orientation obtained, and the higher the values of the mechanical properties in the machine direction [1]. There is a proprietary die design that accentuates machine-direction properties in thicker rods and thus enables rods with diameters of about 2 mm to have the same mechanical properties as 0.25 mm-diameter rods. The result of this orientation tendency, though, is that much lower transverse-direction properties result, leading to "splitting" film or to tubing with low burst strength. The use of filled or otherwise modified polymers can alleviate that tendency.

C10.6 RESEARCH RESULTS

A large number of industrial, government, and university research laboratories are working on the development of more commercializable TLCPs and/or improvements to those already on the market. Let us review some of the results from these laboratories.

- The first main-chain thermotropic polyether liquid crystal polymers were produced at Case Western Reserve University, Cleveland, Ohio. These polymers have lower and broader transition temperature ranges and enhanced solubility compared to TLCPs based on polyester. These advantages are reflected in easier processability and greater stability.
- Graft copolymers that exhibit liquid crystal properties, based on unusual macromolecules, were produced at Virginia Polytechnic Institute, Blacksburg, Va., by blending polyethylene terephthalate with a commercially available liquid crystal polymer in order to better higher performance without the constraints of anisotropy.
- High-density polyethylene liquid crystal polymers have been developed with a form of melt drawing that results in an ultimate tensile strength of 145 MPa and a tensile modulus of 7.6 GPa, according to German researchers.
- Aromatic homopolyesters based on *p*-oxybenzoyl have been developed. They display a high degree of crystallinity, excellent wear resistance, inertness to most corrosive fluids, and good thermal stability. One such polymer, Ekonol, may be used either in the unblended form (for plasma spraying with various metal powders for abradable seals for use in jet engines, for example) or blended with, for example, PTFE for use in self-lubricating high-temperature bearings, seals, and rotors.
- Extrudable and thermoformable forms of Vectra (Vectra A900 and Vectra B900) and Xydar have been introduced.

C10.7 APPLICATIONS

Here is a list of some applications that are possible in the near term:

- Industrial uses: mechanical and dimensionally critical components of floppy and hard disc drives; compact disc player parts
- Chemical-resistant uses: pumps, meters, tubing, distillation column packing
- Consumer market uses: dual-use cookware, knobs, handles, microwave oven components
- Aerospace uses: aerodynamic fins, radomes for "smart" bombs and missiles
- Transportation uses: under-the-hood encapsulants, fuel-handling systems
- Electronic and electrical uses: interconnect insulators, high-density connectors, light-emitting-diode cases, fiber-optic connectors, bobbins.

REFERENCE

1. J. J. Duska, "Liquid-crystal polymers: How they process and why," *Plastics Engineering 42* (12), pp. 39–42, 1986.

HIGH-TEMPERATURE SUPERCONDUCTING MATERIALS

CHAPTER C11

C11.1 INTRODUCTION

As most high-purity metals are cooled to temperatures nearing absolute zero, the electrical resistivity decreases gradually, approaching some very small but finite value. There are a few materials, however, for which the electrical resistivity, at a very low temperature, drops precipitously from this small but finite value to practically zero and remains there upon further cooling. Such materials are called superconductors, and the temperature at which they attain superconductivity is called the critical temperature T_c (Fig. C11.1). Other parameters that must attain critical values for superconductivity to occur are the magnetic field H_c and the current density J_c (Fig. C11.2).

The superconductivity phenomenon in metallic alloys has been satisfactorily explained by a well-known theory. To summarize, the superconductive state in metals results from attractive interactions between pairs of conducting electrons; the motions of these paired electrons become coordinated such that scattering by thermal vibrations and impurity atoms is highly inefficient. The electrical resistivity, being directly proportional to the incidence of electron scattering, is therefore very small, almost nonexistent. This is known as the Bardeen-Cooper-Schrieffer (BCS) theory [2].

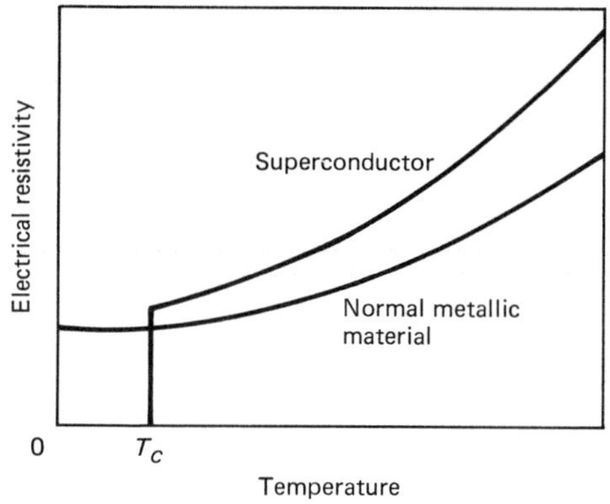

Figure C11.1 Temperature dependence of the electrical resistivity of normally conducting metallic and superconducting materials.

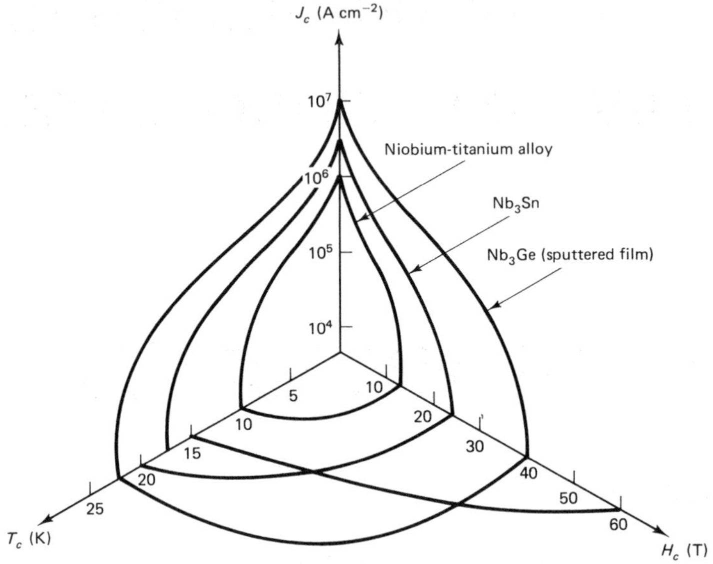

Figure C11.2 Properties of presently available practical Type II superconducting metallic materials as a function of the critical upper field, H_c, the critical temperature T_c, and the critical current density J_c, (on a log scale) [1] (reprinted with permission from John Wiley & Sons, Inc., New York).

TABLE C11.1 CRITICAL TEMPERATURE, T_c, FOR SUPERCONDUCTIVITY IN SELECTED METALLIC SUPERCONDUCTING MATERIALS

Material	T_c (K)
Nb–47% Ti	10.2
V_3Ga	16.5
Nb_3Sn	18.0
Nb_3Ga	20.3
Nb_3Ge	23.2

Practical applications of metallic superconducting materials have been limited to laboratory models, prototypes, and small-scale components. The chief deterrent to the widespread application of these materials has been the difficulty of attaining and maintaining the extremely low values of T_c necessary (Table C11.1).

C11.2 RARE-EARTH-METAL-CONTAINING HIGH-T_c SUPERCONDUCTING MATERIALS (REMCHTCS MATERIALS)

Research and development efforts since about 1965 have been directed toward designing artificially structured high-T_c superconductors. These efforts paid handsome dividends in 1986 and 1987 with the dramatic announcement of the discovery of a ceramic material (La-Ba-Cu-O) with T_c between 30 and 40 K [3].* In early 1987, a research team led by Dr. P. C. W. Chu of the University of Houston, Texas, and another team led by Dr. M. K. Wu of the University of Alabama in Huntsville announced the discovery of another high-T_c ceramic superconductor (Y-Ba-Cu-O) [4]. This report heralded a new dawn in superconductor research. Since then, a host of research teams, from the United States to China, have reported observing superconduction transition at temperatures in excess of 90 K in these systems. Some of these results are summarized in Table C11.2. T_c values for some other ceramic oxide systems are given in Table C11.3. The information in Tables C10.2 and C10.3 is neither comprehensive nor detailed because this exciting field is in a state of flux. It will be a few years yet before an adequate data bank becomes available.

TABLE C11.2 NEW LANTHANUM- AND YTTRIUM-BASED CERAMIC SUPERCONDUCTING MATERIALS

Ceramic Material System	T_c (K)	Research Group
La-Ba-Cu-O	30	IBM, Zurich (Switzerland)
Y-Ba-Cu-O	93	University of Houston (USA); University of Alabama, Huntsville (USA)
Y-Ba-Cu-O	98	Institute of Physics of the Academia Sinica (China)
Y-Ba-Cu-O	100	Science and Technology Agency's National Research Unit for Metals (Japan)

*J. G. Bednorz and K. A. Muller of IBM's Zurich Research Laboratories in Switzerland made the announcement in April 1986. They received the 1987 Nobel Prize in Physics for this discovery.

TABLE C11.3 ONSET CRITICAL TEMPERATURE FOR SUPERCONDUCTIVITY IN SOME RARE EARTH METAL-CONTAINING HIGH-TEMPERATURE SUPERCONDUCTING MATERIALS

Material	Onset T_c* (K)
Yb-Ba-Cu_2-O [5]	91
Tm-Ba-Cu_2-O [5]	91
Sm-Ba-Cu_2-O [5]	85
$Gd_{0.5}Ba_{0.5}CuO_{3-y}$ [6]	95[a]
$Dy_{0.625}Ba_{0.375}CuO_3$ [7]	94
$Ba_2SmCu_3O_7$ [8]	96.5
$Sr_{0.375}Y_{0.625}CuO_{3-y}$ [9]	90.5
$Sm_{1.8}Ba_{1.2}CuO_x$ [10]	82 ± 2
$Eu_{1.5}Ba_{1.5}Cu_2O_x$ [10]	95 ± 3
$Gd_{1.5}Ba_{1.5}Cu_2O_x$ [10]	93 ± 1

[a]Midpoint T_c*

*For a discussion of onset and midpoint T_c, see item 1, section C11.4 and Fig. C11.5.

C11.3 CRYSTAL STRUCTURE AND PROPERTIES OF REMCHTCS MATERIALS

A good deal of information regarding the superconducting properties and other characteristics of these materials is now known, thanks to the feverish pace of research into them during the past few years. Some of the more germane information is summarized in Table C11.4.

Detailed studies of the crystal structure of both the La-Ba-Cu-O and Y-Ba-Cu-O materials have given clear information about this aspect. Thus it is now accepted that Ba-La-Cu-O has a tetragonal K_2NiF_4-type structure at room temperature, consisting of alternate layers along the c-axis of perovskite (CuO_3) and rock salt (La-O) structures [15] (Fig. C11.3). The crystal structure of $YBa_2Cu_3O_7$, an orthorhombically distorted perovskite, also consists of layers of oxygen and copper [16] (Fig. C11.4). Narayan et al. [17] have studied the microstructure and properties of $YBa_2Cu_3O_{9-y}$. They reported resistive transitions at 90 K and near 290 K, by using high-resolution and analytical transmission electron microscopy and Raman scattering techniques. The authors reported the identification of the normal orthorhombic phase associated with T_c = 90 K, together with another phase, which they tentatively assigned to $T_c \sim$ 290 K. The unit cell of this new phase (a) has c = 1.3555 nm (which is about 16% larger than the corresponding value in the unit cell of the orthorhombic phase), (b) grows epitaxially on the {001} planes of the orthorhombic phase, (c) is present in a much smaller (~5%) volume fraction compared to the parent phase, (d) has a higher concentration of

TABLE C11.4 SOME PROPERTIES OF HIGH-TEMPERATURE SUPERCONDUCTING MATERIALS

Material	Property	Value
Single crystal of $YBa_2Cu_3O_x$ [11]	Fracture toughness Hardness	$1.1\ MPa\sqrt{m}$ 8.7 MPa
$La_{1.8}Ba_{0.2}CuO_{4-y}$ [12]	Dominant carrier type (300–28 K)	p
$La_{2-x}(Sr, Ba)_xCuO_4$ ($x = 0.0$–0.3) [13]	Valence of copper	Cu^{II} and Cu^{III}
$La_{1.85}Sr_{0.15}CuO_{4-y}$ [14]	Superconductor gap	12 ± 3 mV

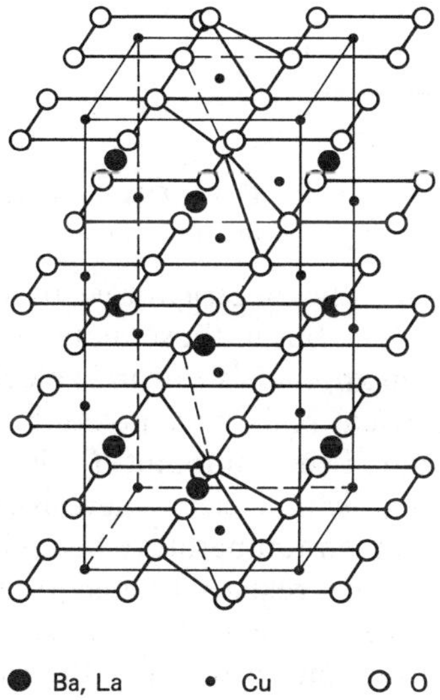

Figure C11.3 Structure of $La_3Ba_3Cu_6O_{14.10}$ [15] (reprinted with permission from American Institute of Physics).

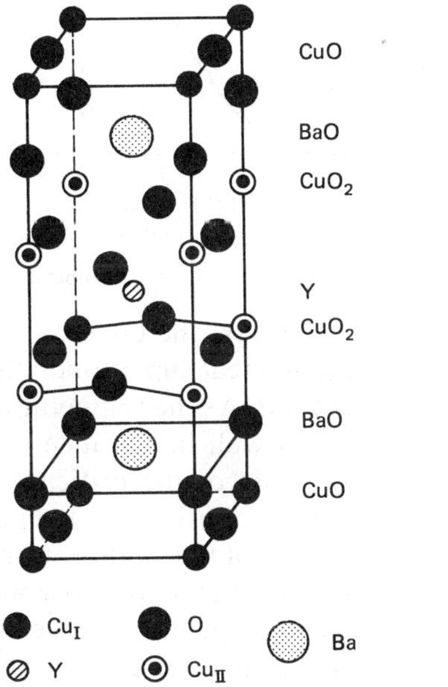

Figure C11.4 Unit cell of $YBa_2Cu_3O_7$ [16] (reprinted with permission from AIME).

copper and oxygen compared to the parent phase, and (e) "is not stable," although stability can be achieved by "inhibiting formation of crystal defects by means well known to materials scientists" [17].

C11.4 RESEARCH CHALLENGES AND DIRECTIONS IN REMCHTCS MATERIALS

Attractive as these high-temperature superconducting materials are, they have a number of unsatisfactory features that need improvement before they can be used in practical applications.

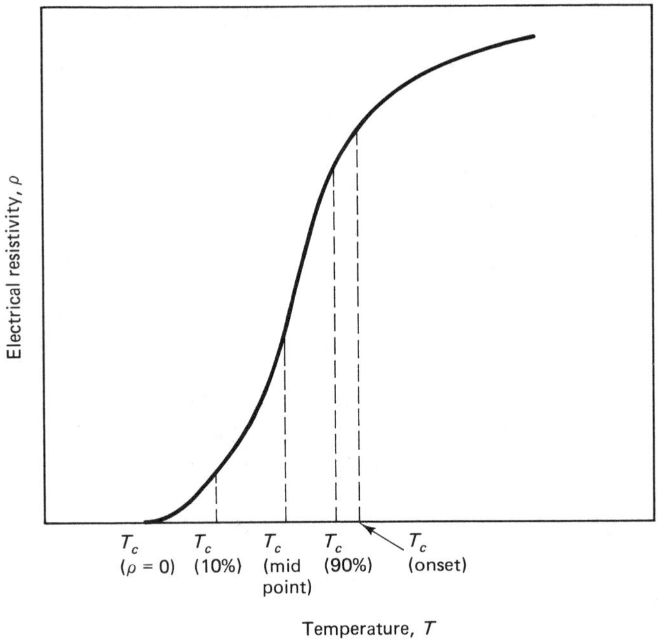

Figure C11.5 A typical superconducting transition.

1. At the onset of superconductivity, the electrical resistivity drops to zero (in an ideal superconductor; see Fig. C11.1), but in these materials, the transition is wide. As the temperature is lowered, an appreciable drop in electrical resistivity starts well in advance of the zero electrical resistivity position. A conservative estimate of the critical temperature, called the midpoint T_c, is the point in the electrical resistivity–temperature curve at which the electrical resistivity has dropped to 50% of the value extrapolated from the high-temperature behavior. Other useful temperatures are those at which the electrical resistivity has dropped by 10% and 90%, respectively, because the difference between them indicates the width of the transition, as well as provides information on whether the transition is symmetrical about the midpoint (note that midpoint $T_c \ll$ onset T_c). [Fig. C11.5] For example, the width of the T_c transition in $Y_{1.4}Ba_{1.6}Cu_2O_x$ is 4 K, with onset T_c being 94 K [18], while for $YBa_2Cu_3O_{7-x}$ onset of superconductivity occurs at 106 K, with the width of the T_c transition being 5 K [19].

2. The critical temperature in the $La_{2-x}Sr_xCuO_{4-y}$ materials is sensitively dependent on the concentration of the strontium [20].

3. Different heat treatments applied to samples with the same chemical composition yield different critical temperatures [20]. It has also been suggested that the oxygen content—the factor y in $(Ba,Sr)_xLa_{2-x}CuO_{4-y}$ and $RBa_2Cu_3O_{9-y}$—has an effect on the superconducting properties of these materials [20].

4. The operational current density associated with the transmission of electricity by these materials is low. For Y-Ba-Cu-O, the critical current density has been given as 168 A cm^{-2} [21], 200 A cm^{-2} [22], and 1100 A cm^{-2} [23]. It is well to note, though, that the critical current density for this material decreases rapidly as both temperature and magnetic field increase (Table C11.5). Some develop-

TABLE C11.5 THE VARIATION OF CRITICAL CURRENT WITH MAGNETIC FIELD OF $YBa_2Cu_3O_7$ SUPERCONDUCTOR

Temperature (K)	Critical Current (mA)	Magnetic Field (T)
65	7.8	3.0
	4.8	4.0
	4.2	5.0
	3.2	6.0
	3.0	7.0
72	11.0	0.8
	4.2	1.0
	2.8	2.0
	2.2	3.0
	2.0	4.0
	1.8	5.0
	1.5	6.0
	1.4	7.0
86	0.5	0
	0.4	1
	0.3	3
	0.2	5
	0.1	7

Source: U. Dai, G. Deutscher, and R. Rosenbaum, "Critical current of $YBa_2Cu_3O_7$ in strong magnetic fields," *Appl. Physics Letters 51* (6), p. 460, 1987.

ments in 1988 give some hope for optimism in raising the critical current density to technologically useful levels. At a temperature of 77 K and fields up to 8 T, a critical current density of 1.8×10^6 A cm^{-2} was reported for $YBa_2Cu_3O_7$ [24]. The material reportedly needs to be crystallographically oriented though, but not necessarily monocrystalline.

5. The performance of these new oxides in high-field situations is uncertain. Although the estimated upper critical magnetic field for the La-Sr-Cu-O materials is reportedly about 85 T [25] and the Y-Ba-Cu-O oxides have an extrapolated upper critical field of 320 T at 0 K [26], the challenge lies in sustaining even one-tenth of this value at near-room temperatures.

6. New ways must be found to deposit thin films of these materials on a variety of substrates. In this respect, the early research results were encouraging. Two research teams reported an excimer ultraviolet laser system that permits the deposition of uniform ceramic superconducting films on aluminum oxide and strontium titanate [27, 28] substrates as well as on silicon dioxide [28]. The latter development is particularly significant as it has wide implications for the use of these high-T_c oxides in integrated circuit chips. Superconducting film (300–400 nm thick) of $La_{2-x}Sr_xCuO_4$ on sapphire have been prepared by a combined evaporation-reaction method [29]. Also, thin films of Y-Ba-Cu-O have been prepared by electron beam evaporation and dc magnetron techniques [30].

7. New fabrication and processing methods must be developed or currently available ones must be adapted to the working of these brittle compounds. A number of developments in 1987 and 1988 gave hope for optimism in meeting this challenge. A 6.35-mm-diameter wire drawn to between 0.25 mm and 1.5 mm in diameter was fabricated from finely pulverized powder of $YBa_2Cu_3O_7$ and reportedly conducted electricity at 80 K [31], and Toshiba scientists reported fabricating a rod (0.6 mm in diameter) and tapes (5 mm wide by 0.1 mm thick) from the same material that conducts at 87 K [32]. Also, workers at the IBM Thomas J. Watson Research Center in Yorktown Heights, N.Y., reported the fabrication of an RF-biased SQUID,* based on bulk Y-Ba-Cu-O, using such techniques as electron-beam vapor deposition, sputtering, ion milling, etching, and lithography [33]. The device reportedly operated with useful signal levels up to 81 K, although "the instability of the break or crack junction may limit the practical application of such a device [33].

Murr and Hare [34] reported some relevant developments in processing—namely, the ability of explosively generated shock waves to consolidate amorphous powders and mixtures of crystalline and amorphous powders while simultaneously bonding them to metal containers in cylindrical shock-loading configurations. This development can form the basis for the processing of these high-temperature superconductors. These workers used this technique to create high-temperature superconducting monoliths and various components of useful shapes and sizes. They also clearly demonstrated the ability to scale up component size rapidly—rings of superconducting $YBa_2Cu_3O_7$ explosively fabricated in a copper matrix (Fig. C11.6) being scaled up by factors of 5 in diameter and 40 in cross section [34].

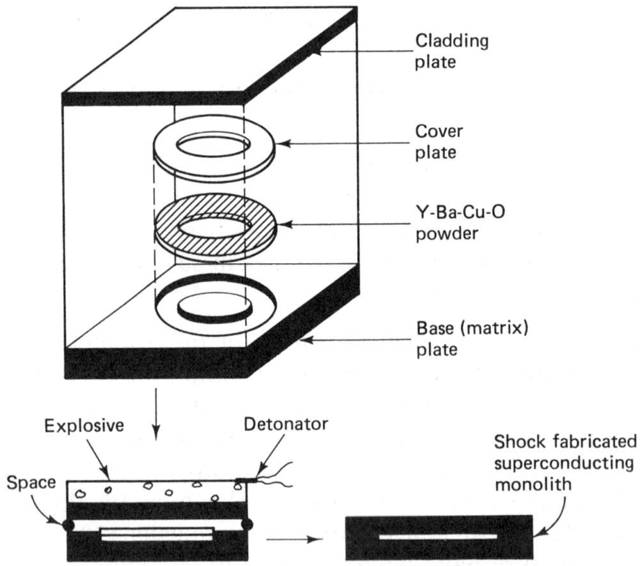

Figure C11.6 Schematic representation of the arrangement for the consolidation and encapsulation of a superconducting powder channel ring into a contiguous, explosively bonded monolith [34]. (reprinted with permission from Plenum Press, Inc.)

*Superconducting quantum interference device; see Question 36 in Part II.

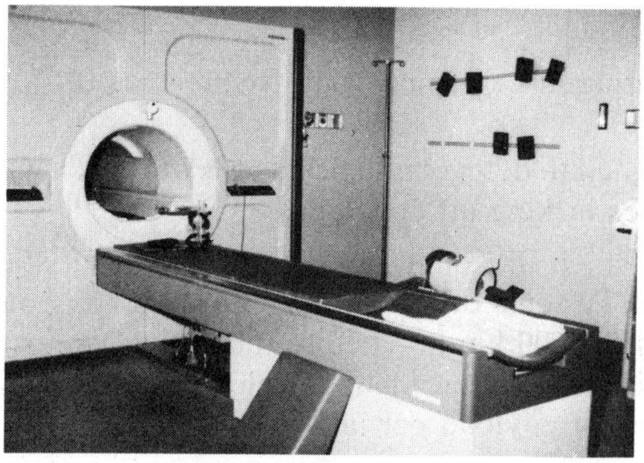

(a)

(b)

(c)

Figure C11.7 Some potential applications of superconducting materials: (a) magnetic resonance imaging instrument (Courtesy of Gil Hayes-University of Tennessee, Memphis); (b) a single gallium arsenide wafer; (Courtesy of National Semiconductor Corp.) (c) magnetically levitated high-speed train (Courtesy of *Design News*).

C11.5 POTENTIAL APPLICATIONS OF REMCHTCS MATERIALS

Once the problems outlined here, and others not mentioned, have been solved, these high-temperature superconductors can be engineered into a number of devices and components and then incorporated into a wide array of systems. Much of the excitement attending research and developments has to do with the wide sweep of potential applications. These are just some of them:

- Advanced nuclear magnetic resonance machines for medical diagnosis [Fig. C11.7(a)].
- Power lines transmitting power over long distances, with low losses, operating at low voltage, thus eliminating the difficulty of providing high-voltage insulation for buried and overhead lines. It has been estimated that only four superconducting transmission lines (diameter 280 mm) would be needed to supply the electrical power demand of New York City.
- New computer circuitry capable of operating at switching speeds of 50 to 100 ps, which is 5 to 10 times faster than present semiconductor devices [Fig. C11.7(b)].
- Magnetohydrodynamic power generation; plasma confinement in fusion reactors.

- Superconducting devices operating at extremely high frequencies for very high resolution radar.
- High-performance sensors designed to detect ballistic missile warheads hurtling through space or submarines lurking off coastlines.
- Electromagnetic accelerators, which are envisaged to replace front-stage rocket boosters for space launch systems of the future. The launcher would accelerate the two upper stages to 3600 km h^{-1} before the rocket ignites to carry the payload into orbit. This would reduce launch vehicle mass by 80%, with attendant reductions in cost of payload per launch.
- Electric motors, which would be one-tenth of their present size; lightweight solenoids, transformers, and power distribution systems.
- The boosting of power of the radio frequency free-electron laser (RFFEL) to levels suitable for Strategic Defense Initiative applications, by way of electron beam accelerator cavities made of superconducting materials. (Experimental superconducting RFFEL facilities operational in 1989, like those at Stanford University in California, used niobium RF cavities at liquid helium temperatures.)
- Magnets for the linear induction motors of magnetically levitated high-speed trains [Fig. C11.7(c)]. One such train is used in short-haul service in England (along a 0.8-km route).
- Magnetic energy storage devices.
- Magnetic separation in hydrometallurgy and levitation melting, which may extend nonequilibrium solidification beyond today's rapid solidification limits.
- The development of SQUIDS. These are the most sensitive magnetic detectors currently available and can be used to study brain waves as well as in paleomagnetic data acquisition and fundamental physics research.
- Improvement of the efficiency of synchronous coaxial accelerators to the point where weapons based on the technology, the so-called coilguns, are practical. One example is an antiarmor projectile accelerator with a muzzle energy of 10 MJ. Coilguns are generally regarded as more efficient than railguns.
- The development of mixers that can operate highly sensitive ultralow noise detectors down to 100-μm wavelengths. Such devices could be used, for example, to monitor the upper atmosphere for trace gases using satellites in the study of ozone layer depletion, as well as to obtain information on the formation of spectral lines, knowledge of which is valuable in astrophysics and astronomy.
- Bearings in ultrafast computer disc drives, high-speed cameras, and gyroscopes made of a superconductor using the Meissner effect. Bearings based on conventional magnets can achieve operating speeds of more than 10^5 r min^{-1} in a vacuum but require control systems in order to maintain stability. Superconducting bearings should eventually operate at 3×10^5 or even 10^6 r min^{-1} and require no elaborate controls since the levitating effect (the Meissner phenomenon) is self-stabilizing.

- Magnets to provide magnetic refrigeration for the hydrogen slush fuel contemplated for the National Aerospace Plane.
- Superconducting magnets in high-energy accelerators that are used in particle physics to produce intermediate vector bosons known as W and Z° particles. These particles play a major role in establishing a "grand verification theory" of forces. Examples of such accelerators are the electron-positron Stanford Linear Collider (Stanford, Calif.), the Tevatron proton-antiproton collider (Fermilab, near Chicago), the electron-positron LEP collider at the European Laboratory for Particle Physics, or CERN (near Geneva, Switzerland), and the HERA proton-electron collider (Hamburg, West Germany). Some 8000 of these magnets (each 17 m long, bending type, cos θ design) will be an integral part of the accelerator of the proposed U.S. Superconducting Super Collider. The 85-km-circumference accelerator will operate at a field of 6.6 T produced by these magnets.

C11.6 A THEORY OF SUPERCONDUCTIVITY IN REMCHTCS MATERIALS

Since there is ample experimental justification for the existence of electron pairing ("Cooper pairs") in REMCHTCS materials, while there is no such evidence for electron-phonon interaction (the mediation agency in low-T_c metallic superconducting materials), considerable research attention by workers in theoretical physics and materials science has focused on specifying the appropriate mediation agency in these materials. The explanations may be grouped into three categories.

The first of these categories are so-called BCS-based hypotheses. There are many variants of this approach. In one, phonon coupling occurs as a result of strong anharmonic effects; that is, lattice ion vibrations that deviate markedly from harmonicity cause electrons to interact. In another variant, pairing between electrons in the metal may be mediated by electron-hole pairs, termed excitons, in the semiconductor. (These materials are essentially metal-semiconductor combinations.) High values of T_c could arise if the conduction-valence band gap in the semiconductor is small. Varma et al. [35] suggested that the high T_c in these materials is due to scattering of electrons from excitonic resonances with large oscillator strengths, rather than from phonons.

Lee and Ihm [36] offered two possible explanations for the high T_c. These are (a) the interband scattering mediated by phonons, and (b) the additional attractive interaction between $d_{x^2-y^2}$ electrons mediated by the d_{z^2} hole-band plasmons and/or $d_{x^2-y^2} \leftrightarrows d_{z^2}$ interband transition. Wang et al. [37] proposed that a BCS-like Hamiltonian with quasi two-dimensional tight-binding electrons on tetragonal lattice describes the high T_c in the La-Ba-Cu-O system. By calculating T_c, assuming, in turn, negligible and nonnegligible weak coupling between neighboring layers, these workers found close agreement with experimental results.

A model of percolation superconduction with exciton exchange was suggested by Tao [38] to explain the high transition temperature in $(La_{1-x}Ba_x)_2Cu_{4-y}$. A bipolaron mechanism proposed by Ting et al. [39] had some difficulties explaining the existence of an energy gap of superconductivity. An electron-phonon theory, put forward by Yu et al. [40], also proved inadequate because it assumed that coupling between the electrons and some phonon modes should be very strong since the density of electron states at the Fermi surface is very low. This would result in the softening of the phonon mode, which is unfavorable for achieving high-T_c superconductivity.

In a modification of the Varma model, Jing and Gosner [41] suggested a model based on electron hopping between octahedral Cu^{2+} and Cu^{3+} ions, the basic evidence for this suggestion being given by the well-known spinel compound, since the perovskite-type structure is similar to the spinel one.

The second category of explanations uses the idea of electric charge fluctuations; that is, charge fluctuations exchanged by electrons lead to a net attractive force. Since electronic interactions can occur at a much higher energy than phonon interactions, the Cooper pairs remain bound at higher temperatures. For example, Li [42] proposed that the electron pairs responsible for high T_c superconductivity in Ba(or Sr)-La-Cu-O compounds fall in the crossover zone from bipolarons to Cooper pairs. Bipolaron condensation could exist in some REMCHTCS materials, while in others BCS-type superconductivity may appear. Chi and Nagi [43] proposed that, in doped compounds like $La_{2-x}Sr_x CuO_4$, the attractive interaction responsible for Cooper pairing of electrons is induced by holes which are created through doping.

The third category of explanations uses the notion of spin-mediated and magnetic interactions; that is, the fluctuations in the spin of electrons can act as quantized entities that exchange energy with electrons, leading to the possibility of electron interaction (and hence attraction) through the fluctuations. Thus Emery [44] suggested that the charge carriers are holes in the O(2p) states and the pairing is mediated by strong coupling to local spin configurations on Cu sites, while Anderson et al. [45] explained the high T_c in these ceramic oxides in terms of a resonating valence bond model, which involves both spin and charge excitations. The main features of the last group of explanations are consistent with the unusual magnetic properties of the high-T_c materials, such as insensitivity to magnetic dopants like gadolinium (which completely destroys superconductivity in metallic low-T_c superconducting materials), antiferromagnetism, and spin glass behavior.

None of these explanations has proved to be satisfactory, so it is necessary for "second-generation" models to be proposed. These models should not be premised on the BCS theory for, as Anderson and Abrahams [46] suggested, "most conventional theories and some unconventional versions of BCS should be eliminated from the running." In the frenzied days of 1987 and 1988, too many explanations were profferred. Indeed, as Sir Nevill Mott remarked in 1987, "there are as many explanations, perhaps, as there are theorists in the field" [47]. This spate of explanations could have been a hurried response to the welter of experimental results published during that period. A period of serious consolidation of ideas is necessary if an acceptable theory of high-T_c superconductivity is to emerge, a theory characterized by predictions that are experimentally verifiable rather than explanations of selected experimental results.

C11.7 RARE-EARTH-METAL-FREE HIGH-T_c SUPERCONDUCTING MATERIALS (REMFHTCS MATERIALS)

A number of high-T_c ceramic materials containing no rare-earth metals were discovered early in 1988. A group of scientists at the Science and Technology Agency's National Research Institute for Metals in Tsukuba, Japan, announced the discovery of one of these materials, a Bi-Sr-Ca-Cu-O compound. It was reported that when annealed for several hours at just under 900°C, this material shows an onset in reduction of electrical resistance at 120 K and zero resistance at 75 K but shows a very sharp drop to almost zero resistance between 105 and 107 K [48]. The research group results suggested that the oxide is composed of two phases, one with T_c of 75 K and another with T_c of about 105 K [48]. Subsequent work [49] on this material revealed the composition of the superconducting phase to be $Bi_2Sr_{3-x}Ca_xCu_2O_{8+y}$, with x ranging from ~0.4 to 0.9, with an A-centered orthorhombic subcell with dimensions of a = 0.5399 nm, b = 0.5414 nm, and c = 0.30904 nm. Using the fact that bond strength calculations point to a mean oxidation state of the copper in this phase to be +2.335, Chippindale et al. [50] have specified the structure of this phase as $Bi_2Sr_{1.67}Ca_{0.99}Cu_2O_x$. One major difference between this bismuth-based material and REMCHTCS ceramics is that the latter have within them chains of copper and oxygen atoms (at one time, researchers thought superconductivity was directly related to the presence of these chains). The bismuth-based materials, by contrast, have double copper and oxygen sheets alternating with double bismuth and oxygen sheets but no chains (Fig. C11.8). It was reported [49] that electrical measurements on a single crystal of

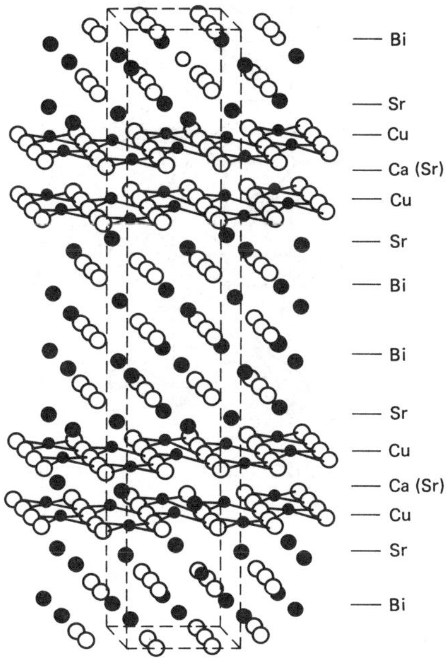

Figure C11.8 The structure of $Bi_2Sr_{3-x}Ca_xCu_2O_{8+y}$ [49] (reprinted with permission from the American Association for the Advancement of Science). Metal atoms are shaded, and only Cu-O bonds are shown.

$Bi_2Sr_{3-x}Ca_xCu_2O_{8+y}$ show onset of superconductivity at 116 K and apparent zero resistance at 91 K.

Two other examples of rare-earth-free high-T_c superconducting materials belong to the Tl-Ba-Cu-O and Tl-Ba-Ca-Cu-O families. They were first reported by a team of physicists from the University of Arkansas. These workers reported onset of superconductivity at 81 K for $Tl_2Ba_2Cu_3O_{8+x}$ [51] and 120 K for $Tl_2CaBaCu_3O_{8+x}$ [52]. Later, scientists at the IBM Research Laboratory in Almaden, Calif., reported a T_c value of 125 K for a proprietary combination of thallium, calcium, barium, copper, and oxygen [53].

In general, the thallium- and bismuth-based ceramics are very similar, with thallium and barium in the former replacing bismuth and strontium in the latter. This observation led to the suggestion that there is a direct correlation between the value of T_c and the number of copper-oxygen layers, that is, the density of states per unit volume n. Thus the 2021, 2122, and 2223 phases* of the thallium-calcium-barium-copper-oxygen materials (Fig. C11.9) become superconducting at about 80 K, 110 K, and 125 K, respectively. The bismuth-based materials show a similar pattern with the values of T_c for the 2021, 2122, and 2223 phases† being about 23 K, 80 K, and 110 K, respectively. These results are consistent with one of the tenets of the BCS theory, which is that T_c increases exponentially with n. This means, though, that T_c reaches a limit of about 200 K at $n = 10$.

Figure C11.9 Structures of three thallium-based superconducting materials [54] (reprinted with permission from *New Scientist*, London, England). A = Cu-O planes; B = Tl-Ba-O planes.

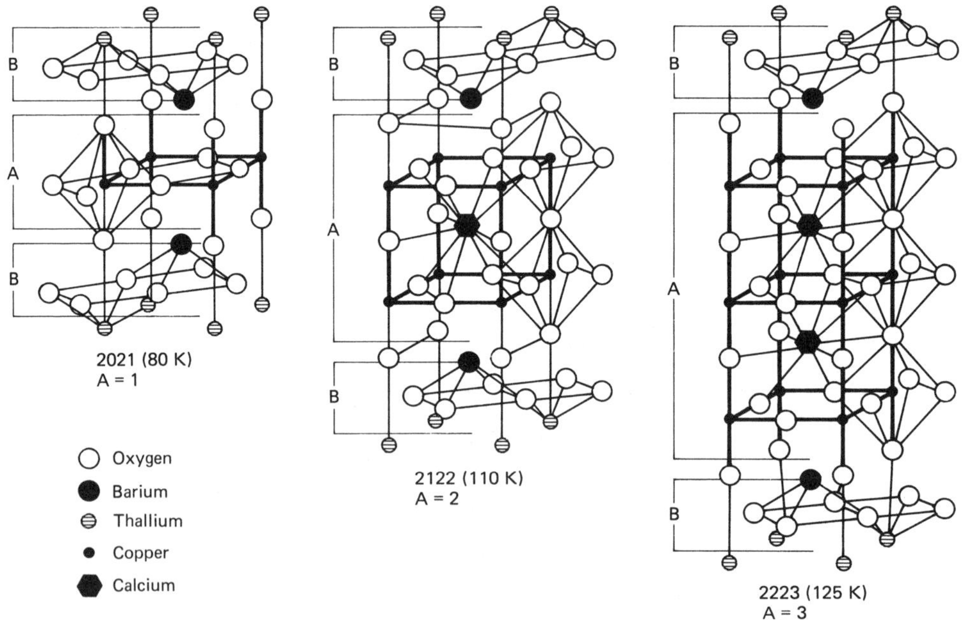

*The digits in each phase designation represent the thallium:calcium:barium:copper ratio.
†The digits denote the bismuth:calcium:strontium:copper ratio.

These rare-earth-metal-free high-T_c superconducting materials have a number of advantages. First, because these materials contain no rare earth, they may be cheaper to manufacture in bulk. Second, they are quite stable; for example, the bismuth-based phases are very resistant to degradation by water. Third, these materials are very tolerant of many variations in composition.

However, these materials have some disadvantages. In particular, because thallium is highly toxic and bismuth is in short supply, superconductors based on materials containing either of these elements may be of scientific rather than technological interest, at least in the near term.

Some of the problems associated with the practical applications of REMCHTCS materials (low critical current density, brittleness, uncertain performance in high-field situations, fabrication and processing limitations, and difficulty of deposition as thin films on a number of substrates), as well as the lack of a satisfactory model to explain their superconductivity, are also applicable to rare-earth-metal-free high-T_c superconducting materials. As far as critical current density is concerned although, for the most part, it is just as low as for REMCHTCS materials, some developments reported in 1988 were particularly important. For example, researchers at one Japanese company reportedly produced a polycrystalline thin film of Bi-Sr-Ca-Cu-O with a critical current density of 1.9×10^6 A cm^{-2} at 77 K [55]. They noted that the crystals need to be aligned so that the copper oxide planes, along which current flows most easily, are parallel to the film.

C11.8 CONCLUSIONS

The initial breakthroughs have been made in the discovery, chemistry, and physics of high-temperature superconducting materials. A period of dedicated research and development lies ahead in various aspects of these materials—the synthesis of more of them, clear understanding of the physics involved, and fabrication for practical, economic applications.

REFERENCES

1. "Superconducting materials," *Kirk-Othmer Encyclopedia of Chemical Technology,* 3d ed., vol. 22. New York: Wiley, 1983, p. 300, fig. 2.
2. J. Bardeen, L. N. Cooper, and J. R. Schrieffer, "Theory of superconductivity," *Physical Review 108* (5), pp. 1175–1204, 1957.
3. J. G. Bednorz and K. A. Muller, *Z. Phys. B 64,* p. 189, 1986.
4. M-K. Wu, et al., "Superconductivity at 93 K in a new mixed-phase Y-Ba-Cu-O compound system at ambient pressure," *Physical Review Letters 58* (9), pp. 908–910, 1987.
5. J. Marcus et al., "Magnetic and transport properties of some Ln-Ba-Cu-O(Ln = Yb,Tm,Sm) superconducting oxides," *Solid State Communications 63* (2), pp. 129–131, 1987.
6. Z. Qi-rui et al., "A new superconducting oxide Gd-Ba-Cu-O with $T_c(\rho = 0)$ above 92 K," *Solid State Communications 63* (5), pp. 415–417, 1987.

7. Z. Qi-rui et al., "Superconducting oxide Dy-Ba-Cu-O with $T_c(\rho = 0)$ beyond 90 K," *Solid State Communications 63* (6), pp. 497–498, 1987.
8. F. Garcia-Alvarado et al., "A new high temperature superconductor: $Ba_2SmCu_3O_{9-x}$," *Solid State Communications 63* (6), pp. 507–510, 1987.
9. Z. Qi-rui et al., "Superconductivity in the Sr-Y-Cu-O system," *Solid State Communications 63* (6), pp. 535–536, 1987.
10. Z. Fisk et al., "Superconductivity of rare earth-barium-copper oxides," *Solid State Communications 62* (11), pp. 743–744, 1987.
11. R. F. Cook, T. R. Dinger, and D. R. Clarke, "Fracture toughness measurement of $YBa_2Cu_3O_x$ single crystals," *Appl. Physics Letters 51* (6), pp. 454–456, 1987.
12. J. T. Chen, et al., "Determination of charge carriers in superconducting La-Ba-Cu-O by thermoelectric measurements," *Physical Review B35,* pp. 7124–7125, 1987.
13. E. E. Alp, et al., "Determination of valence of Cu in superconducting $La_{2-x}(Sr,Ba)_xCuO_4$," *Physical Review B35,* pp. 7199–7201, 1987.
14. S. Pan, et al., "Measurements of the superconductivity gap of La-Sr-Cu-O with a scanning-tunneling microscope," *Physical Review B35,* pp. 7220–7223, 1987.
15. L. Er-Rakho, C. Michel, J. Provost, and B. Raveau, "A series of oxygen-defect perovskites containing Cu^{II} and Cu^{III} oxides $La_{3-x}Ln_x$ $Ba_3Cu^{II}{}_{5-2y}Cu^{III}{}_{1+2y}O_{14+y}$," *J. of Solid State Chem. 37* (2), pp. 151–156, 1981.
16. C. S. Pande, A. K. Singh, and A. DasGupta, "Materials science of high-temperature superconductors," *J. Metals 40* (1), p. 10, 1988.
17. J. Narayan, et al., "Microstructure and properties of $YBa_2Cu_3O_{9-\delta}$ superconductors with transitions at 90 K and near 290 K," *Appl. Physics Letters 51* (12), pp. 940–942, 1987.
18. S. B. Qadri, et al., "X-ray identification of the superconducting high-T_c phase in the Y-Ba-Cu-O system," *Physical Review B35,* pp. 7235–7237, 1987.
19. I. K. Gopalakrishnan, Y. V. Yakhmi, and R. M. Fryer, "Zero electrical resistance at 106 K in $YBa_2Cu_3O_{7-x}$," *Nature 327,* pp. 604–605, 1987.
20. J. M. Tarascon, et al., "Superconductivity at 40 K in the oxygen-defect perovskites $La_{2-x}Sr_xCuO_{4-y}$," *Science 235,* pp. 1373–1376, 1987.
21. D. G. Hinks, et al., "Phase diagram and superconductivity in the Y-Ba-Cu-O system," *Appl. Physics Letters 50* (23), pp. 1688–1690, 1987.
22. I. K. Gopalakrishnan, et al., "Effect of slow cooling rates on the superconducting characteristics of $YBa_2Cu_3O_{7-x}$," *Appl. Physics Letters 51* (17), pp. 1367–1369, 1987.
23. R. J. Cava, et al., "Bulk superconductivity at 91 K in single-phase oxygen-deficient perovskite $Ba_2YCu_3O_{9-\delta}$," *Physical Review Letters 58* (16), pp. 1676–1679, 1987.
24. J. S. Satchell, et al., "High magnetic field critical currents in thin films of $YBa_2Cu_3O_7$," *Nature 334* (6180), pp. 331–333, 1988.
25. S. J. Hwu, et al., "High-T_c superconductivity in regions of possible compound formation," *Physical Review B35* (13), pp. 7119–7121, 1987.
26. T. P. Orlando, et al., "Upper critical fields of high-T_c superconducting $Y_{+2-x}Ba_xCuO_{4-y}$," *Physical Review B35* (13), pp. 7249–7251, 1987.
27. D. Dijkkamp, et al., "Preparation of Y-Ba-Cu-oxide superconductor thin films using pulsed laser evaporation from high T_c bulk material," *Appl. Physics Letters 51* (8), pp. 619–621, 1987.
28. N. Biunno and R. Singh, "NCSU announces new superconductor developments," *J. Metals 39* (9), pp. 5–6, September 1987.

29. O. F. de Lima, et al., "Superconducting thin films based on $La_{2-x}Sr_xCuO_4$," *Appl. Physics Letters 51* (5), pp. 369–370, 1987.

30. R. B. Laibowitz, et al., "Thin superconducting oxide films," *Physical Review B35,* pp. 8821–8823, 1987; B. Y. Tsaur, M. S. Dilorio, and A. J. Straws, "Preparation of superconducting $YBa_2Cu_3O_x$ thin films by oxygen-annealing of multilayer metal films," *Appl. Physics Letters 51* (11), pp. 858–860, 1987; R. E. Somekh, et al., "High superconductivity transition in sputter deposited YBaCuO thin films," *Nature 326,* pp. 857–859, 1987.

31. S. Jin, et al.,"High T_c superconductors: Composite wire fabrication," *Appl. Physics Letters 51* (3), pp. 203–204, 1987.

32. D. Swinbank, "Japanese poised to dominate in superconductors as well?" *Nature 327,* pp. 356–357, 1987.

33. J. E. Zimmerman, et al., "Operation of a Y-Ba-Cu-O rf SQUID at 81 K," *Appl. Physics Letters 51* (8), pp. 617–618, 1987.

34. L. E. Murr, et al., "Interfacial phenomena and microstructural connectivity in explosively fabricated Y-Ba-Cu-O superconductors," *J. Superconductivity 1* (1), p. 5, 1988.

35. C. M. Varma, S. Schmitt-Rink, and E. Abrahams, "Charge transfer resonances and superconducting pairing in the new oxide metals," in *Novel Superconductivity,* S. W. Wolf and V. Kresin (eds.), New York: Plenum Press, pp. 355 365, 1987. C. M. Varma, S. Schmitt-Rink, and E. Abrahams, "Charge transfer excitations and superconductivity in 'ionic' metals," *Solid State Communications 62* (10), pp. 681–685, 1987.

36. D. H. Lee and J. Ihm, "Two-band model for high T_c superconductivity in $La_{2-x}(Ba,Sr)_xCuO_4$," *Solid State Communications 62* (12), pp. 811–815, 1987.

37. Z. Wang et al., "A possible reason for high T_c in oxide superconductors," *Solid State Communications 64* (4), pp. 531–533, 1987.

38. R. Tao, "A possible mechanism of exciton exchange percolated superconductivity for the oxide Balacuo," *Physics Letters A123* (5), pp. 254–256, 1987.

39. C. S. Ting, D. Y. Xing, and W. Y. Lai,

40. J. Yu, A. J. Freeman, and J. H. Xu, "Electronically driven instability and superconductivity in the layered $La_{2-x}Ba_xCuO_4$ perovskites," *Physical Review Letters 58* (10), pp. 1035–1037, 1987.

41. J. Jing and U. Gosner, "A model for high-T_c superconductivity based on electron hopping," *Appl. Physics A44,* pp. 101–104, 1987.

42. Y-M. Li, "A possible picture for high T_c superconductivity in Ba(or Sr)-La-Cu-O compounds," *Solid State Communications 62* (10), pp. 677–680, 1987.

43. H. Chi, and A. D. S. Nagi, "A possible mechanism of high T_c superconductivity," *Solid State Communications 65* (12), pp. 1563–1565, 1988.

44. V. J. Emery, "Theory of high-T_c superconductivity in oxides," *Physical Review Letters 58* (26), pp. 2794–2797, 1987.

45. P. W. Anderson, et al., "Resonating-valence bond theory of phase transitions and superconductivity in La_2CuO_4-based compounds," *Physical Review Letters 58* (26), pp. 2790–2793, 1987.

46. P. W. Anderson and E. Abrahams, "Superconductivity theories narrow down," *Nature 327,* p. 363, 1987.

47. N. Mott, "Superconducting ceramics: Is there an explanation?" *Nature 327,* pp. 185–186, 1987.

48. H. Maeda, et al., *Japan J. Appl. Physics Letters 27,* p. L209, 1988.

49. M. A. Subramanian, et al., "A new high-temperature superconductor: $Bi_2Sr_{3-x}Ca_xCu_2O_{8+y}$," *Science 239,* pp. 1015–1017, 1988.

50. A. M. Chippendale, S. J. Hibble, J. A. Hriljac, L. Cowery, D. M. S. Bagguley, P. Day, and A. K. Cheetham, "Chemical characterization and superconductivity of two phases in the Bi-Sr-Ca-Cu-O system," *Physica C 152* (2), pp. 154–156, 1988.

51. Z. Z. Sheng and A. M. Hermann, "Superconductivity in the rare-earth-free Tl-Ba-Cu-O system above liquid-nitrogen temperature," *Nature 332,* pp. 55–58, 1988.

52. Z. Z. Sheng and A. M. Hermann, "Bulk superconductivity at 120 K in the Tl-Ca/Ba-Cu-O system," *Nature 332,* pp. 138–139, 1988.

53. M. M. Waldrop, "Thallium superconductor reaches 125 K," *Science 239* (4845), p. 1243, 1988.

54. Ghosh, P. "Thallium puts superconductivity on the road to 200 K," *New Scientist 118* (1607), p. 28, 1988.

55. D. Swinbanks, "Thin films carry more current," *Nature 334* (6180), p. 280, July 28, 1988.

MOTION MEMORY ALLOYS

CHAPTER C12

One disadvantage of most shape memory alloys* is that their maximum use temperature is about 180°C. A copper-based alloy has been developed [1] that shows shape change upon heating to a high temperature below some critical value and beyond; the alloy is unique in its ability to remember the motion, rather than the shape, with thermal cycling. This new type of alloy is referred to as a "motion memory" alloy.

The new alloy consists of copper, zinc, and minor additions of aluminum, silicon, or tin. One such alloy is Cu-27.6Zn-3.8Al. This alloy may be processed by melting, annealing, or forging. Processing is followed by heating in the β phase region and cooling at as high a rate as possible to allow for the transformation of the metastable β phase and/or martensitic phase.

Deforming, and preheating up to a critical temperature T_c, allows the alloy to change shape during subsequent thermal cycling, provided that T_c is less than the temperature corresponding to the $(\alpha + \beta)/\beta$ boundary. The extent of the spontaneous shape change increases with T_c.

By way of illustration of this phenomenon, the degree of spontaneous shape change of the copper-based alloy (originally in a U configuration, bent at an angle of 200°, at 20°C) during two series of thermal cycles is shown in Fig. C12.1.

*See Chapter 11, Part I

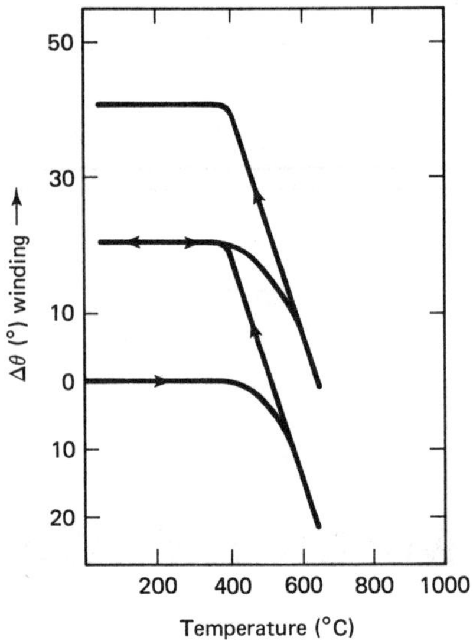

Figure C12.1 Relationship between spontaneous shape change, ΔΘ, and temperature of Cu-27.6Zn-3.8 Al alloy [1] (reprinted with permission from AIME).

REFERENCE

1. Y. Hanatate, et al., "Motion memory alloys," *J. Metals 39* (3), pp. 60–61, 1987.

CONDUCTING POLYMERS AND MOLECULAR METALS*

CHAPTER C13

C13.1 CONDUCTING POLYMERS

The structures of four prototypical conducting polymers are shown in Fig. C13.1. Polyacetylene shows the zigzag arrangement of carbon atoms common to trans structures; polyphenylene has a quinoid structure; polypyrrole and polythiophene show the crenulated arrangement typical of trans-cisoid structures.

These and other conducting polymers have been the subject of research since 1977, and it is expected that large-scale commercialization of these materials will occur by 1992. This optimism is based on developments such as the following:

- The grafting of copolymers based on polysaccharide and derivatives, as well as water-soluble copolymers, cyclic polymers, and star polymers (University of Florida, Gainesville).
- Experimentation with organic conducting spines that contain macrocyclic dyes. The polymers consist of stacked metalloporphyrin or metallophthalocyanine complexes (Stanford University, Stanford, Calif.).

*The proper designation of these materials is integral-oxidation-state complexes with metallic properties.

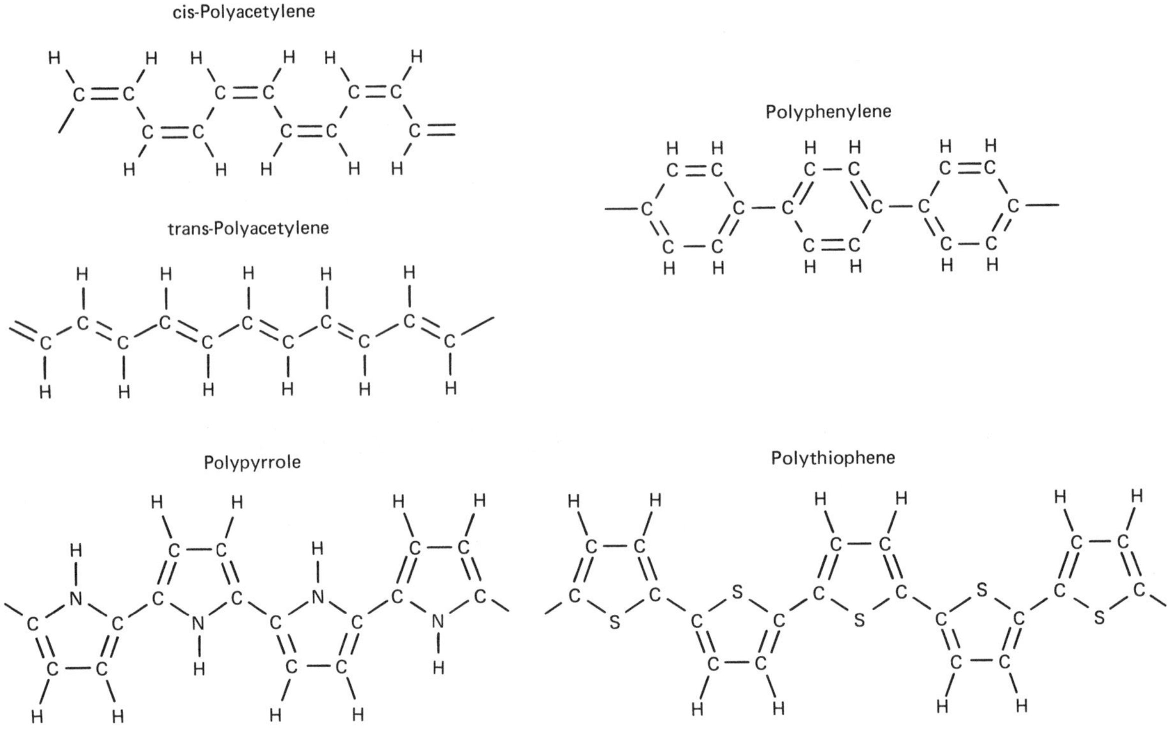

Figure C13.1 Structure of four conducting polymers.

- The synthesizing of chainlike molecular crystals that can conduct electricity at near-metallic levels. Researchers doped electronically active polymers with conjugated backbones, such as polyacetylene. Doping traps electrons or vibrons within the backbone, enabling electrical charges to move from atom to atom (Los Alamos National Laboratory, Los Alamos, N.M.).
- The preparation of very thin polymer films with highly oriented structures. The ordered crystalline polymers can be used in electronic devices, replacing inorganic materials (Case Western Reserve University, Cleveland, Ohio; University of Akron, Ohio).
- The creation of conducting polymers based on linear polyethers, polypyrrole, and polysilane (Brookhaven National Laboratory, New York; Polytechnic Institute of New York).
- The development of organic conductors and superconductors through a better understanding of band structure, doping, and carrier transport mechanisms. Experimenting at temperatures as low as 0.05 K, fields up to 9.5 MA m^{-1}, and hydrostatic pressures up to 1 GPa, researchers have grown high-purity organic conductors (Sandia National Laboratories, Albuquerque, N.M.).
- The building of organometallic polymers based on rigidly held five- and six-member aromatic rings. Five-ring polymers are derivable from cyclopentadienylnaphthalene, and six-ring polymers may be obtained from paracyclophenes. The aromatic ring is held in a framework such that the electron

or hole conducts through the overlap of π orbitals on adjoining aromatic rings (Brandeis University, Waltham, Mass.).

- The development of soluble conducting polymers that employ heterocycles as the repeating units. The program involved a study of monomers such as thiazole, indole, pyrazole, and thienylpyrrole. The method of formation involved an electrochemical polymerization technique, which yields conducting materials by doping during the anodic oxidation (Los Alamos National Laboratory, Los Alamos, N.M.).
- The anodic electropolymerization of long-chain 3-substituted pyrroles and thiophenes, some of which are soluble in common organic solvents in the conducting state (Department of Chemistry, University of Durham, England; Cookson Group PLC, Middlesex, England). The structures of both starting molecules are built on pentagonal rings of atoms. In each pentagon, four of the atoms are carbons. The extra atom in the pentagon is nitrogen in pyrroles, sulfur in thiophenes. The significance of this work is that molecular structures of the monomers have been successfully altered to make them more stable once polymerized. The polymers obtained are flexible, highly conducting (greater than $10^3\ \Omega^{-1}\ m^{-1}$; one polymer, polythiophene, reportedly has a conductivity of $1.05 \times 10^5\ \Omega^{-1}\ m^{-1}$), and typically stable up to 210°C.
- Researchers at Lockheed Aeronautical Systems developed a proprietary air- and water-soluble polymer called APA, reportedly based on polyaniline. This polymer may be melt-processed or solution-cast from common solvents, such as acetone and methylene chloride. Electrical conductivity is in the near-metallic range, about $5 \times 10^3\ \Omega^{-1}\ m^{-1}$, and does not degrade with time. Lockheed also stated that the conductivity can be varied from insulator values to near-metal values in the same structure, something that cannot normally be done with doped inherently conducting polymers.
- Researchers at the Centre d'Études Nucléaires at Grenoble University in France discovered a process of preparation of polyaniline that produces excellent electrical conducting properties, very good charge-recharge cycling, and low self-discharge from an organic medium.
- A team of chemists at BASF, Ludwigshafen, West Germany, and at the University of California in Santa Barbara produced a polyacetylene polymer that demonstrates a higher conductivity than copper metal. The breakthrough involved making polyacetylene as crystalline as possible, which improves its stability in air. The conductivity of the polyacetylene is increased by doping it with iodine, yielding a conductivity of $2 \times 10^6\ \Omega^{-1}\ m^{-1}$ [1]. One of the members of the research team at BASF reported further improvements in the preparation of oriented polyacetylene, leading to a room-temperature conductivity of about $1.5 \times 10^7\ \Omega^{-1}\ m^{-1}$ [2]. The effect of doping, with bromine, iodine, or arsenic fluoride, on the electrical conductivity of *trans*-acetylene is given in Table C13.1. The values of electrical conductivity for some conducting polymers, both undoped and doped (in some cases), are given in Fig. C13.2. The only problem with the iodine doping is that it reduces the conductivity of polyacetylene by 30% over a week. This is because the light activates the iodine, causing it to attack the

TABLE C13.1 ELECTRICAL CONDUCTIVITY OF DOPED *TRANS*-ACETYLENE[a] AS A FUNCTION OF DOPANT CONCENTRATION

	Dopant Concentration (wt %)	Electrical Conductivity (Ω^{-1} m^{-1})
$[CH(AsF_5)_y]_n$	0.00	1.0×10^{-3}
	0.05	1.0×10^{4}
	0.10	1.5×10^{4}
$[CHI_y]_n$	0.00	1.0×10^{-3}
	0.05	1.5×10^{2}
	0.10	2.0×10^{2}
	0.20	3.0×10^{2}
$[CHBr_y]_n$	0.00	1.0×10^{-3}
	0.05	6.0×10^{0}
	0.22	5.8×10^{0}

[a]System represented as $[(CH)D_y]_n$, where D is dopant which may be AsF_5, or I_2, or Br_2.

Source: W. Hayes, "Conducting polymers," *Contemporary Physics 26* (5), p. 430, 1985.

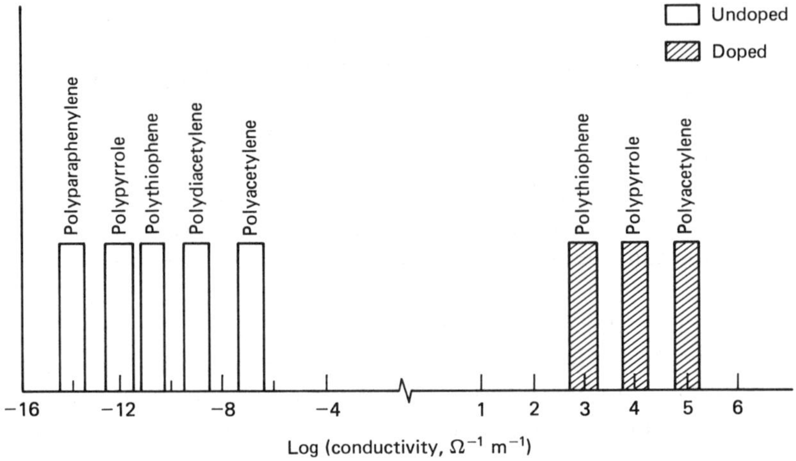

Figure C13.2 Values of the room-temperature electrical conductivity of undoped and doped conducting polymers. (By comparison, electrical conductivity of polytetrafluoroethylene, a polymeric insulator, is 10^{-18} Ω^{-1} m^{-1}, and that of copper, an excellent metallic conductor, is 5.9×10^7 Ω^{-1} m^{-1}.)

double bonds of the polymer. As soon as this happens, the "conduction band" of the polymer is interrupted, and the conductivity falls away.

- Polymer scientists at the University of California at Santa Barbara (UCSB) synthesized two water-soluble electrically conducting polymers, ethane- and butane-tiophene sulfonic acid (P3-ETS and P3-BTS). The materials are self-doping. A small positive voltage causes hydrogen nuclei to escape from the molecules, leaving sulfonate groups behind. The negative charges of the sulfonates balance the positive charges created in the backbone by the initial positive voltage. This property of self-doping has created interest among battery makers. The compounds have another unusual attribute—electrochromicity. They can be electrically switched from an orange color to blue and back again. Tests suggest that this switching can be done very quickly, making these polymers candidates for color components in computer displays and digital watches. UCSB scientists reported systems that could be fabricated into commercially useful devices using readily available solvents. For example, these workers showed that conjugated poly(3-alkylthienylenes) can be solution-processed.

- The development of an acetylene-terminated oligomer by National Starch and Chemical Corporation (Bridgewater, N.J.). This polymer has a high viscosity, so it is better suited to injection-molding applications. It is envisaged that this polymer may replace silicon in circuit boards since they have about the same electrical conductivity.

C13.2 CONDUCTING ORGANIC METALS

One major development in this area was the discovery, by scientists at the University of Cambridge in England and at the University College of North Wales, of a nonmetallic electrical conductor that is electrically conducting in two directions. The material, an organometallic complex of palladium, $Cs[Pd(S_2C_2(CN)_2)_2]\cdot 0.5H_2O$ (Fig. C13.3), has an electrical conductivity of

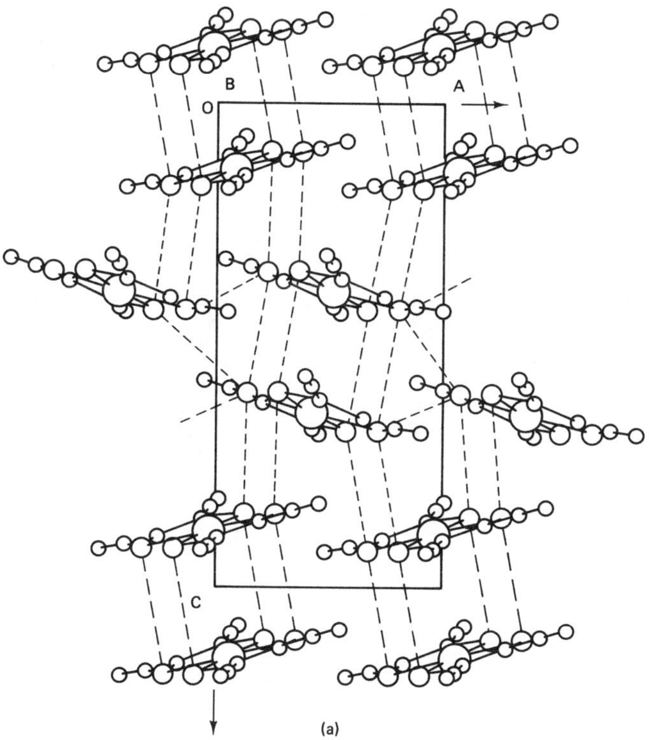

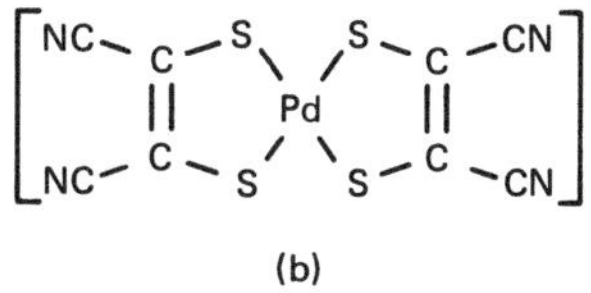

Figure C13.3 A new nonmetallic conductor: (a) *b*-axis projection of the crystal structure of $C_s\ [Pd(S_2C_2(CN)_2)_2]\cdot 0.5H_2O$; the broken lines indicate the S–S contacts; (b) molecular structure of $[Pd(S_2C_2(CN)_2)_2]^-$ [3] (reprinted with permission from Macmillans Magazines Ltd., London, England).

500 Ω^{-1} m^{-1} at ambient pressure and room temperature but is semiconducting at low temperatures under pressure.

If these and other research efforts are successful, conducting polymers and so-called molecular metals could be used in switching and memory devices, various battery systems, lightweight shielding of electronic equipment from electromagnetic impulses, the fabrication of molecular electronic devices, chemically modified electrodes, and sensors.

REFERENCES

1. N. Basescu, et al., ''High electrical conductivity in doped polyacetylene,'' *Nature 327,* pp. 403–405, 1987.
2. H. Naarman, paper presented at the American Chemical Society meeting and symposium on conducting polymers, *Future,* Denver, April 1987.
3. I. D. Parker, et al., ''A new low-dimensional metal, $Cs[Pd(S_2C_2(CN)_2)_2]\cdot 0.5H_2O$,'' *Nature 324,* pp. 547–549, 1986.

ORGANIC MAGNETIC MATERIALS

CHAPTER C14

Dedicated research at various institutions in the United States, notably at Du Pont de Nemours, Ohio State University, and Northwestern University resulted in the world's first organic magnet. It is a ferromagnetic material that is truly organic because it contains more than three-quarters carbon and hydrogen and dissolves in organic solvents. The material becomes ferromagnetic at temperatures below 4.8 K. When fully magnetized, its magnetic strength compares favorably with that of iron.

The compound is an organometallic salt composed of pairs of decamethyl ferrocene cations and tetracyanoethylene anions, $[Fe^{III}(C_5Me_5)_2]^+[TCNE]^-$ (see Fig. C14.1). Each layer is made of discrete pairs of donor and acceptor radical ions, with each layer of donor-acceptor pairs out of sequence with neighboring layers by one-half unit. The salt thus has a valid three-dimensional structure.

This compound is the product of some chemical engineering of the electronic properties of charge transfer or double radical compounds, which are brightly colored salts that have an unpaired electron on both the cation and the anion. These unpaired electrons are responsible for both paramagnetic and ferromagnetic interactions. Charge transfer compounds are usually paramagnetic rather than ferromagnetic. Because the ions are constantly in motion, the spins of the unpaired electrons do not impinge on each other. Each spin behaves like a very small,

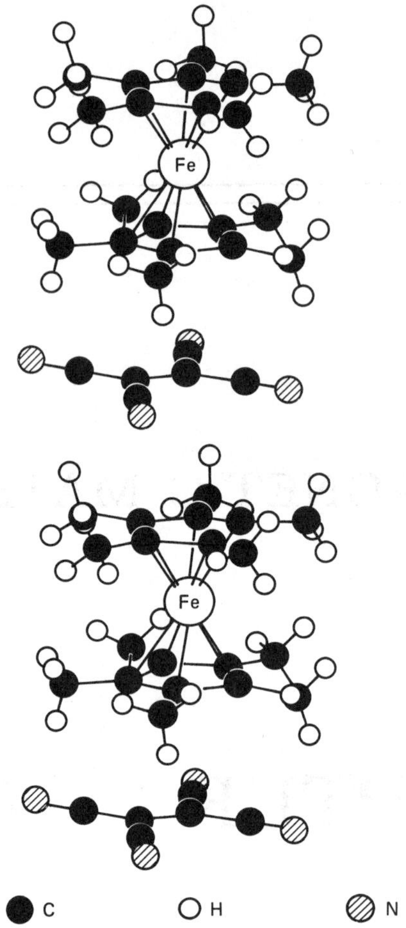

Figure C14.1 One-dimensional structure of the organic magnet $[Fe^{III}(C_5Me_5)_2]^+$ $[TCNE]^-$ [1] (reprinted with permission from *New Scientist,* London, England).

more or less independent magnet. Ferromagnetism occurs only when the magnetic force between the spins is strong enough to overcome the effects of thermal agitation of the atoms or ions. Below the Curie temperature, the electron spins align simultaneously throughout the solid, producing a bulk magnetic effect.

Some magnetic properties of this material are given in Table C14.1, and its magnetization hysteresis behavior is shown in Fig. C14.2.

The next stage in the development of an organic magnetic material was a ferromagnet that is totally organic—a polymer, for example, a material that has no metal in its structure. This step was achieved by Soviet scientists [3]. The compound synthesized by these workers is a member of the polydiacetylene group of materials, which are essentially long chains of carbon atoms with alternate single, double, and triple bonds. These bonds have extra electrons that stretch along the entire chain to produce an extended π bond. This type of bonding bestows interesting electronic properties. Some of these long-chain compounds conduct electricity, and some have unusual nonlinear optical characteristics. Furthermore, the π bonding means that any unpaired electrons introduced along the chain can align their spins and hence produce ferromagnetism.

The Soviet workers synthesized an unusual diacetylene in which the R groups are substituted by nitroxyl groups, each with an odd electron, a biradical (Fig.

TABLE C14.1 SOME MAGNETIC PROPERTIES OF $[Fe^{III}(C_5Me_5)_2]^+$ $[TCNE]^-$ MOLECULAR COMPOUND

Magnetic Property	Value or Comment
Magnetic behavior	Ferromagnetic
Magnetic susceptibility (emu mole^{-1})	
At 30 K	25.0×10^{-3}
At 300 K	1.32×10^{-3}
Effect of field and field gradient product	Attractive
Effect of temperature on magnetic susceptibility	Complex[a]

[a]The high-temperature ($T \geq 50$ K) magnetic susceptibility χ can usually fit the Curie-Weiss expression χ = constant $(T - \Theta)^{-1}$, with Θ (the Curie-Weiss constant) < 0 (see Table C14.2).

Source: J. S. Miller, A. J. Epstein, and W. M. Reiff, "Ferromagnetic molecular charge-transfer complexes," *Chem. Rev. 88* (1), p. 204, 1988.

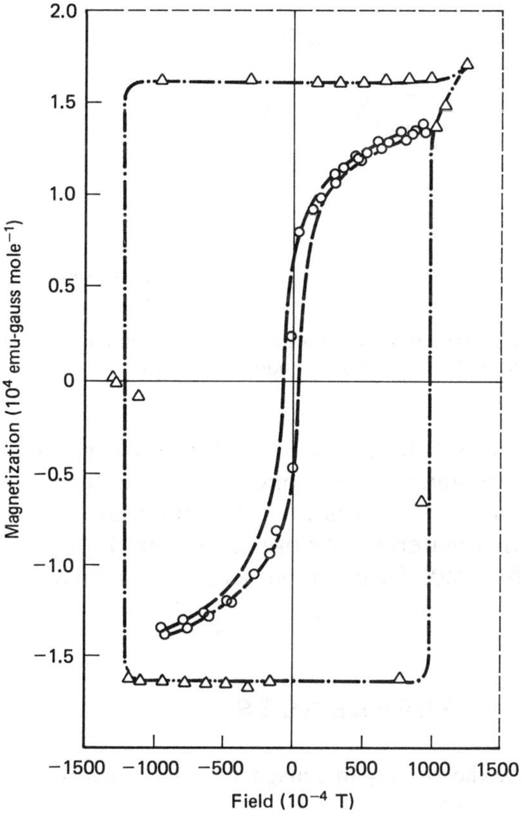

Figure C14.2 Field-dependent magnetization hysteresis behavior of a single crystal of $[Fe^{III}(C_5Me_5)_2]^+$ $[TCNE]^-$ oriented with the stacking axis parallel to the applied field (△: temperature = 2 K, coercive field = 0.1 T; ○: temperature = 4.7 K, coercive field = 0.003 T) [2] (reprinted with permission of the American Chemical Society).

C14.3). The biradical is an antiferromagnetic species because the spins on the two odd electrons pair off. But when the biradicals combine to form chains under the action of light or heat, the spins align to produce the ferromagnetic material. The Soviet workers drew attention to the fact that only a small portion of the polymerized material is magnetic, but it has a very high Curie temperature, remaining ferromagnetic at temperatures as high as 470 K with a magnetization of 10^{-4} T.

TABLE C14.2 RECIPROCAL MAGNETIC SUSCEPTIBILITY AS A FUNCTION OF TEMPERATURE FOR A SINGLE CRYSTAL OF $[Fe^{III}(C_5Me_5)_2]^+[TCNE]^-$ ALIGNED PARALLEL TO THE MAGNETIC FIELD

Temperature (K)	Reciprocal Magnetic Susceptibility, χ^{-1} (mole emu^{-1})
0	0
100	45
200	98
285	150

Source: J. S. Miller, A. J. Epstein, and W. M. Reiff "Ferromagnetic molecular charge-transfer complexes," *Chemical Reviews 88* (1), p. 206, 1988.

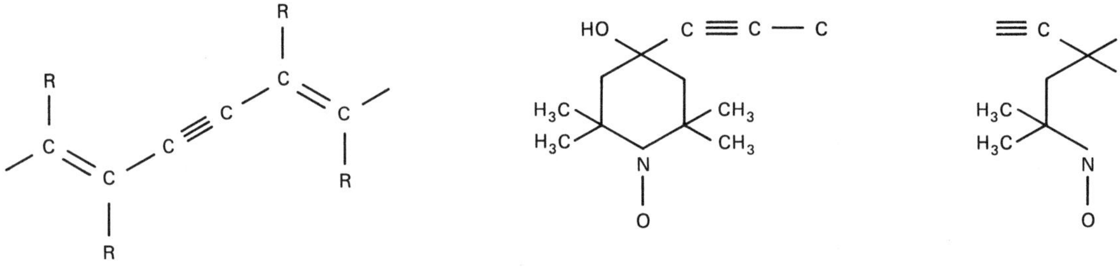

Diacetylene

Biradical with substituted nitroxyl groups

Figure C14.3 One-dimensional structure of the organic magnet group polydiacetylenes [4] (reprinted with permission from *New Scientist,* London, England).

The developments in this field are very exciting for a number of reasons. First, they facilitate an understanding of cooperative phenomena such as ferromagnetism and superconductivity. Second, organic materials are easy to make, especially as coatings on other materials. An organic material that is ferromagnetic at room temperature can be used, for example, as a coat for magnetic recording tape.

REFERENCES

1. N. Hall, "Americans make the first organic magnet," *New Scientist 113* (1546), p. 41, 1987.
2. J. S. Miller, A. J. Epstein, and W. M. Reiff, "Ferromagnetic molecular charge-transfer complexes," *Chemical Reviews 88* (1), pp. 201–204, 1988.
3. Y. V. Korshak, et al., "Organic polymer ferromagnet," *Nature 326,* pp. 370–372, 1987.
4. Anon, "The plastic magnet makes its debut," *New Scientist 114* (1557), p. 27, 1987.

ADVANCED CARBON-CARBON COMPOSITE MATERIALS

CHAPTER C15

Carbon-carbon (C-C) composite materials consist of carbon fibers dispersed in a carbonaceous matrix. However, because carbon takes a number of different forms (such as graphite and turbostratic) and a wide variety of processing methods (weaving patterns and lay-up procedures) may be used in fabricating these materials, their properties exhibit a broad range of values.

A typical manufacturing program for a two-dimensional C-C composite part is shown schematically in Fig. C15.1.

Carbon-carbon composite materials have found their most extensive use in aerospace applications, notably the external surfaces of space vehicles (the nose cone and leading edges of the U.S. space shuttle, for example). There are a number of reasons for this choice:

- For a given weight, C-C can endure higher temperatures for longer periods of time than other ablative materials. For example, the U.S. space shuttle speeds into orbit at velocities exceeding 27,000 km h^{-1}. The heat generated is intense: up to 1370°C at the nose cone and wing leading edges, with reentry temperatures approaching 1700°C. Such temperatures are clearly beyond the capability of metals.

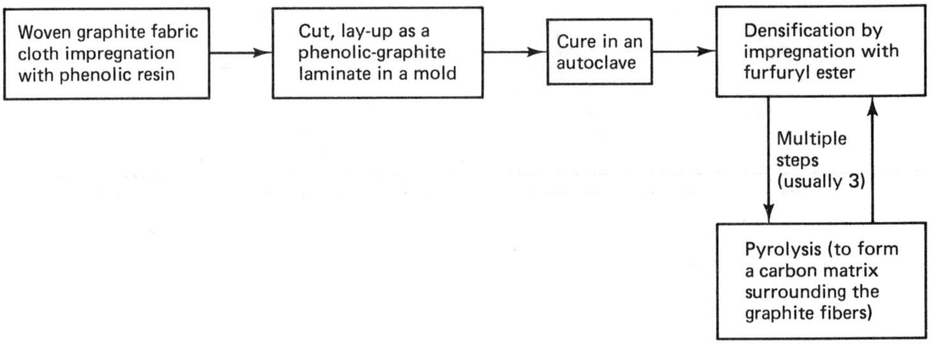

Figure C15.1 Steps in the manufacture of a two-dimensional carbon (graphite)-carbon composite material.

- Not only does C-C withstand such high temperatures, but its mechanical properties, such as specific tensile strength, are not degraded by these temperatures, in contrast to metals and ceramics (Fig. C15.2).
- C-C has a low value of the coefficient of thermal expansion α. At 100°C, α is 10^{-6} °C^{-1} in the r and y directions and 5×10^{-6} °C^{-1} in the z direction; at 800°C, these values are 2×10^{-6}, and 10^{-5}, respectively. These low values of α make C-C parts dimensionally stable over a wide temperature range.
- The material has a high thermal shock resistance, thereby permitting rapid transition from −150°C in outer space to nearly 1700°C during reentry.

Figure C15.2 Variation of specific tensile strength with temperature of high-temperature-resistant materials.

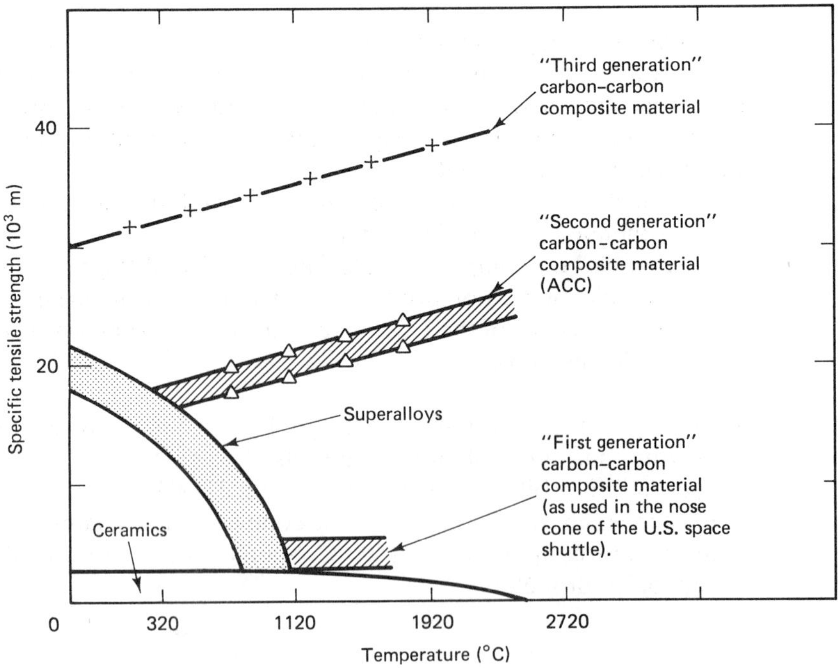

- It has high specific strength and specific stiffness. C-C composites are 20 times stronger than conventional graphite, yet are 30% lighter, with a density of 1380 kg m^{-3}.
- It is very stiff and resists buckling, thus maintaining the aerodynamic shape of a component made of it over a wide temperature range.
- When subjected to thermal cycling, the composite has a long fatigue life. It can withstand the impact of meteors without fracturing, for example.

C-C composite materials find use in a large number of other applications, such as these:

- Brake discs in racing cars, aircraft, trains, motorcycles, and tanks. C-C is used in this application for several reasons. Discs made of this material stop faster and in shorter distances than those made of metals such as beryllium. C-C brakes are less susceptible to warping than metal ones. They are light, thereby permitting a savings in disc weight. C-C has a high heat capacity, making for a good heat sink. C-C has good friction characteristics; C-C brakes have a wear life several times that of metal brakes. C-C can withstand the thermal shock arising from rapid heating and withstand a large number of thermal cycles without thermal fatigue occurring. The McLaren Grand Prix racing car, equipped with brakes made of C-C, has done very well in Formula I races. Figure C15.3 shows the brake of the Gulfstream IV business jet, which has alternating rotating and stationary discs that are squeezed together for braking.
- Clutches in racing cars (Fig. C15.4). Compared to a clutch made of metal, the use of carbon-carbon composite material leads to a reduction in the size of the clutch (which helps maintain a low center of gravity, a necessity in modern racing vehicles), high resistance to attack by contaminants such as engine oil and brake fluid, and better tolerance of intermittent slippage without compromising the integrity of the facing material. One carbon-carbon clutch unit installed in a Lotus Formula I racing car had reportedly accumulated about 3200 km by the end of 1987.

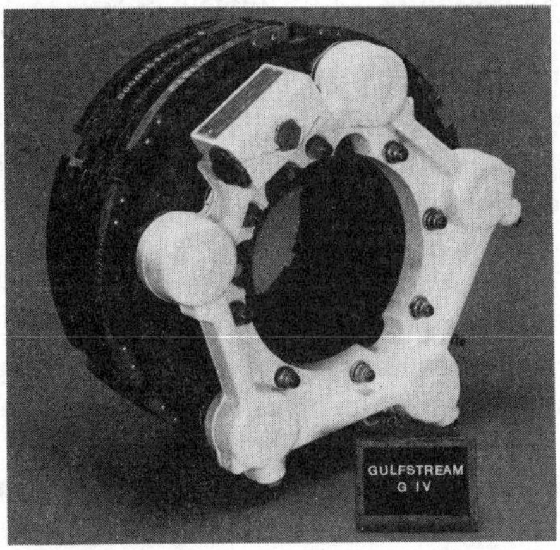

Figure C15.3 Brake of the Gulfstream IV business jet. (Photo courtesy of Loral Aircraft Braking Systems)

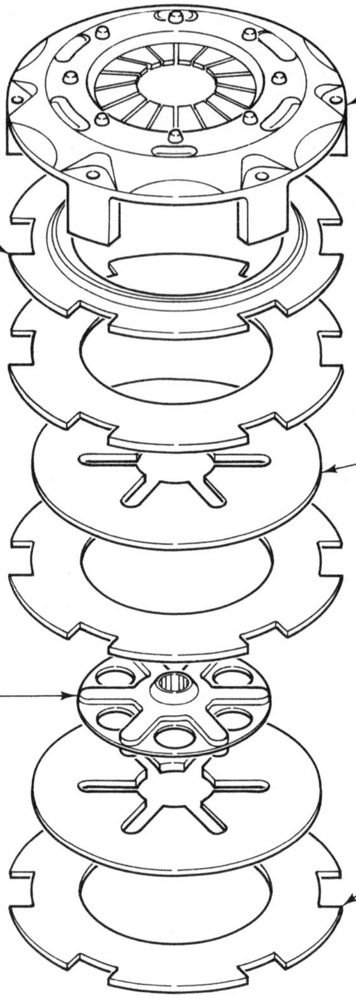

Figure C15.4 Various parts of a carbon-carbon clutch for a racing car. (Courtesy of Tilton Engineering, Inc.)

- Jet engine afterburners and nozzles for missiles. C-C resists stress and erosion from hot gases moving at very high velocity. In such applications, C-C endures flame temperatures up to 3300°C. A high-performance C-C rocket nozzle gives a very long range to the Pershing II missile (Fig. C15.5).
- Various components in a gas turbine engine: rotor, exhaust nozzle flaps and seals, augmenters, combustors, and acoustic panels.

To extend the life of parts made of C-C materials, in the listed applications and others where operational temperatures are very oxidizing, it is usual either to use C-C in a coated and sealed version (Fig. C15.6) or to compound the material with additives that produce oxidation-protective layers in the parts through interaction with oxygen.

The demands of hypersonic flight have led to the development of an improved carbon-carbon composite material, called advanced carbon-carbon (ACC) com-

Figure C15.5 Pershing II missile during flight testing. (Courtesy of Martin Marietta)

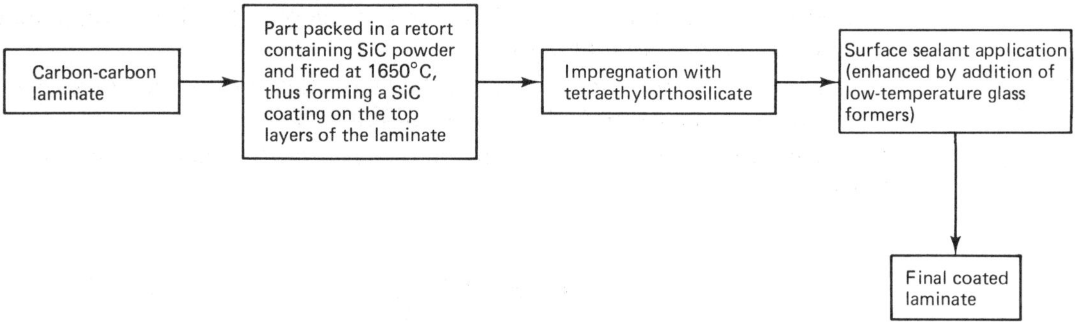

Figure C15.6 Steps in the coating of a carbon-carbon laminate.

posite, for use in various parts of aircraft that will reach at least Mach 11.* ACC is much stronger and stiffer than conventional C-C composites and has a higher specific strength (Fig. C15.2). To increase the strength, continuous polyacrylonitrile (PAN) fibers are used as reinforcement, and the ACC is impregnated with resin and pyrolyzed four times compared to three for conventional C-C composites. ACC has a tensile strength of about 2.5 GPa, two to five times that of most steels, and with an elastic modulus of 340 GPa, it is nearly three times as stiff as titanium.

*Such as the HOTOL (horizontal takeoff and landing space launch vehicle) now in research and development stages (by British Aerospace in close cooperation with Rolls-Royce Ltd.). It is envisaged that by the year 2000, this plane (Fig. C15.7) would be used instead of the Ariane rocket system for launching of satellites for the European Space Agency.

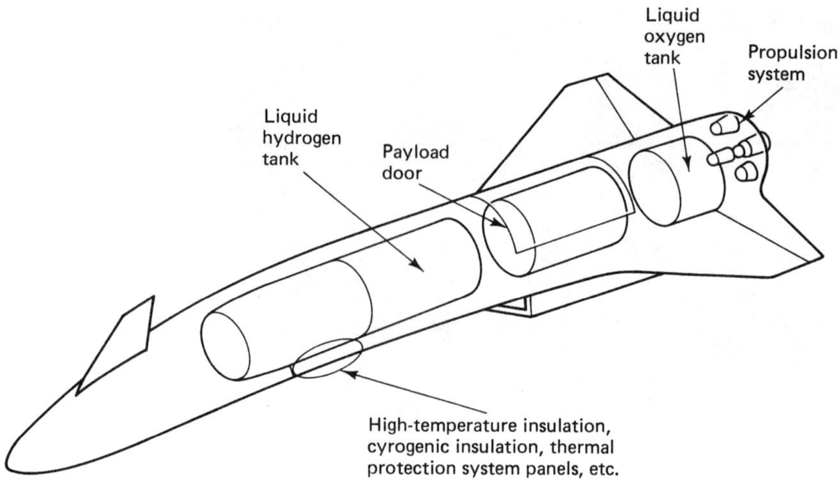

Figure C15.7 Schematic representation of the HOTOL hypersonic plane [1].

Among the potential applications of ACC in hypersonic aircraft are these:

- Engines that can be run at very high temperatures; this results in improvements in the thermodynamic efficiency of the engines.
- Engine rotors that can be made in one piece, thereby leading to reductions in fabrication and assembly costs.
- The airframe; the attractions here are the high values of the specific strength and specific modulus of ACC at elevated temperatures.
- Hot structures (e.g. nose cap, leading ledges) and thermal protection systems.

REFERENCE

1. D. Baker, "US prepares for a hypersonic future," *New Scientist 116* (1589), p. 43, 1987.

OTHER NEW MATERIALS

CHAPTER C16

C16.1 ARALL

This is the acronym for the generic name of a laminate consisting of a number of layers of a high strength aluminum alloy (the core) adhesively bonded to a number of layers of prepregs, which consist of an epoxy resin matrix which has been strengthened by unidirectional high modulus (aramid) fibers, in what has been colloquially dubbed "metallic plywood". The original studies of this material were carried out at Delft University of Technology in Holland. ARALL is an engineered material, and thus can be tailored to meet stated needs; that is, it can be developed in a variety of configurations which can be used as "building blocks" for specific structures. For example, the primary ARALL-1* laminate or lapup consists of two layers of prepregs, each 0.20 mm thick and three sheets of aluminum alloy 7075-T6, each 0.30 mm thick, giving a final product thickness of 1.30 mm (Fig. C16.1).

ARALL has a number of attractive material properties (see Table C16.1), the main ones being

- high specific strength and stiffness (which lead to reduced cross-sectional

*Produced and marketed by Alcoa under license from Enka of Holland.

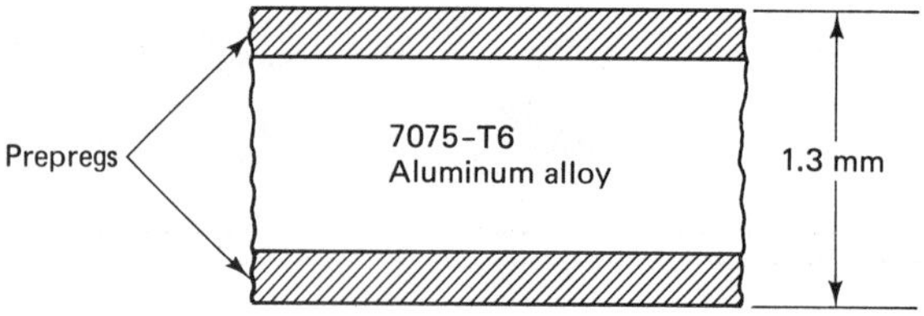

Figure C16.1 The crossection of the primary ARALL-1 lap up.

TABLE C16.1 COMPARISON OF SOME PROPERTIES OF ARALL-1[(a)], 7075-T6 ALUMINUM ALLOY, AND A GRAPHITE FIBER-EPOXY RESIN COMPOSITE

Material property	Test direction	ARALL -1	7075 -T6	Test direction[(b)] (°)	Composite[(c)]
Ultimate tensile strength (MPa)	L[(d)]	765	572	0	655
	LT[(d)]	414	572	90	276
Tensile yield strength (MPa)	L	593	510	0	–
	LT	352	496	90	–
Compressive yield strength (MPa)	L	372	503	0	–
	LT	393	524	90	–
Ultimate compressive strength (MPa)	L	–	–	0	655
	LT	–	–	90	290
Elongation to failure (%)	L	1	11	0	–
	LT	9	11	90	–
Tensile modulus of elasticity (GPa)	L	68	72	0	76
	LT	52	72	90	34
Density (kg m^{-3})		2350	2790		1550

[(a)]Extracted from respondents' data to a round robin evaluation of ARALL-1 properties organized by Alcoa under the ARALL-1 Laminate Test Program, 1985-1988.

[(b)]With respect to the undirectional fibers

[(c)]60% volume fraction of fiber in epoxy resin (42% of the fibers are in the 0° direction, 50% of the fibers are in 45° directions, and 8% of the fibers are in the 90° direction)

[(d)]L: longitudinal; LT: longitudinal-transverse

area of component for a given applied load, or increased load for a given cross-sectional area)

- excellent resistance to fatigue loading; in fact, ARALL is largely insensitive to these loads because fatigue cracks will not propagate in this material (for a comparsion of the fatigue resistance behavior of ARALL and aluminum alloy 7075-T6, see Fig. C16.2).
- ease of fabrication of parts from it; unlike filamentary composite materials, fabrication methods for parts using metallic or ARALL materials are practically identical.
- good ability to withstand Zone II lightning; this is an especial attraction for use of this material in aircraft fuel tank areas.
- good damping properties, which are important for sonic or acoustic fatigue resistance considerations.

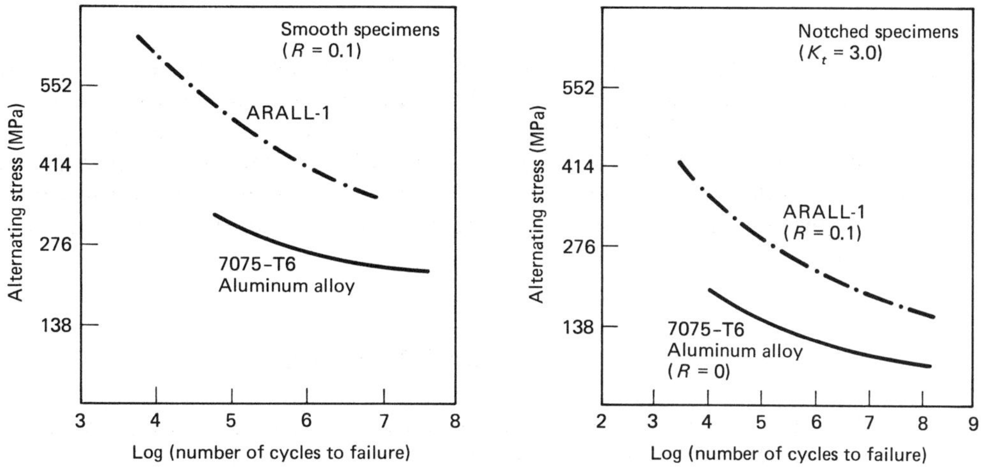

Figure C16.2 The fatigue behavior of ARALL-1 and aluminum alloy 7075-T6 (R: stress ratio = minimum stress/maximum stress in the stress cycle; K_t: stress concentration factor).

ARALL is to be used, at least in the initial stages of development, on aircraft—almost exclusively in tension-loaded, fatigue-prone structures, such as the fuselage pressure cabin, lower wing and tail skins, and wing underside panels. A number of aircraft manufacturers, both in the United States and Western Europe, have vigorous research and development programs involving ARALL. The most advanced of these are based at Fokker Aircraft of Holland. At the end of 1988, this company was flight-testing ARALL in five different designs in the underwing panels of the Fokker 50 aircraft as a prelude to having production parts in the material once airworthiness has been certified.* Preliminary results are very encouraging.**

C16.2 MISCELLANEOUS NEW MATERIALS

Some of these were discussed in two wide-ranging surveys of developments in materials science and engineering (MSE) that appeared in the literature in early 1987 [1,2]. Also, the journal "Advanced Materials & Processes" has regularly published state-of-the-art reviews in its first issue of the year. For example, in 1987, the materials discussed included advanced metals, plastics, and ceramics [3–5], while in 1988 biomedical materials, tough ceramics, automotive materials, aerospace materials, high-temperature superconductors, and other "cutting edge" materials were discussed [6–11].

*By Riiks Luchtvaart Dienst (RLD), the Dutch airworthiness certification authority.

**270,000 simulated flight cycles (including loads simulating the takeoff run, typical enroute turbulence, touchdown and landing runs) at lg stress level (corresponding to mean flight stress) were imposed on a 1.3m × 2m ARALL test panel (which formed an integral part of the under surface of the wing torsion box structure and which was mounted on a dedicated wing panel test rig at Fokker Aircraft's Experimental Department at Schiphol Airport, Amsterdam). The test panel survived. It should be noted that the economic repair life of these panels installed in the Fokker 50 aircraft in current service is 90,000 flight cycles (Source of data in this footnote: Fokker Aircraft BU, Amsterdam, Holland).

Of course, no survey (no matter how comprehensive it is) can realistically hope to capture all the trends in MSE since this field is probably the most exciting and fastest-growing in all of engineering today. New materials and innovative applications of both established and emerging materials are reported on a daily basis, reflecting the vibrancy of research and development efforts in industry, universities, and government agencies in the United States, Japan, Western Europe and, indeed, all over the world.

REFERENCES

1. A. Kelly, "An outline of trends in materials science and processing," Materials Science and Engineering *85*, pp. 1–13, 1987.
2. P. A. Psaras and H. D. Langford, (Editors) *Advancing Materials Research*, National Academy Press, Washington, D.C. 1987.
3. Anon. "A new age of metals," Advanced Materials & Processes *131* (Issue 1), pp. 6–8; pp. 17–26, 1987.
4. Anon. "Plastics: a bird's eye view into the future," ibid, pp. 32–43, 1987.
5. Anon. "The promise of ceramics," ibid, pp. 44–50, 1987.
6. Anon. "Forecast 1988: a staff report," Advanced Materials & Processes *133* (Issue 1), p. 8; pp. 13–38, 1988.
7. J. W. Graham, "Biomedical materials emerge through teamwork," ibid, pp. 41–42, 1988.
8. J. Wachtman, "Moving toward tougher ceramics," ibid, pp. 45–46, 1988.
9. J. G. Simm, "Directions in automotive materials," ibid, pp. 63–65, 1988.
10. R. A. Sprague, "Future aerospace-materials directions," ibid, pp. 67–69, 1988.
11. F. M. Mueller, "Superconductor landscape: view from Los Alamos," ibid, pp. 79–81, 1988.

Part IV

Exercises

1. *Traditionally, friction materials for automotive braking systems have been based on chrysotile asbestos fibers. It is now known, however, that these fibers pose a serious health hazard to workers in the field. The search is on, therefore, for alternative fibers.*

   By specifying the main requirements in a braking material, comment on the merits and drawbacks of steel wool, glass fiber, wool slag fiber, and aramid fiber relative to asbestos for this application.

2. *The sodium-sulfur battery is a high-energy-density, solid-electrolyte battery system that is being developed for use in electric utility load-leveling and electric vehicle applications. It is recognized that the major considerations in material selection in this system are the chemical stability of the chosen materials in molten chemical reactants (such as sodium, sulfur, or Na_2S_x) and the electrochemical stability of the chosen material as an electrode under cycling.*

   Discuss the suitability of named uncoated metals, coated metals, and nonmetals for use as containers for this battery from the viewpoints of chemical and electrochemical stability.

3. *Pipelines are the most important link between offshore oil and gas reserves and onshore facilities.*

   Stating the most important material selection considerations in this case, comment on the suitability of certain steels, such as C-Mn, austenitic stainless, duplex, and high-nickel alloy.

4. *Generally, electrostriction, a second-order electromechanical coupling phenomenon, has been considered a small, negligible effect and has until recently received little attention. Since about 1986, however, following the discovery of a large electrostrictive effect in newer ceramic materials, notably lead-magnesium-niobate-based ceramics, the phenomenon of electrostriction has attracted much research and development attention.*

   Write a short essay on electrostriction, emphasizing the main electrostrictive properties of suitable materials. Suggest some suitable materials for an electrostrictive actuator, such as a deformable mirror to control the phase of the light wave in optical communications.

5. *Excellent biocompatibility and vibratory properties are probably the most important requirements in a material to be used in the artificial tympanic membrane for the attachment of the artificial ossicular chain to the canal wall of a total alloplastic middle ear prosthesis.*

   From the stated requirements, comment on the advantages and disadvantages of each of the following candidate materials for this application: silicone rubbers, polyether and polyester urethanes, a polyether-polyester copolymer, and polypropylene oxide.

6. *A basic understanding of the links between the thermal properties of materials and the conditions that produce the physiological condition of thermal injury is central to the development of standards to protect consumers from heated surfaces.*

List the main thermal properties of materials having hot surfaces that a consumer might contact. Using some derived material property criteria, compare oak wood, a reinforced epoxy resin, and aluminum as candidate materials.

7. Discuss the use of ceramics for the recuperator module on an aluminum remelt furnace (heat exchanger) from the viewpoints of thermal cycling, thermal shock, and corrosion resistance.

8. List the main requirements in a material to be used for the fabrication of the leaf spring of a race car. Compare metals, filamentary composite materials, and metal matrix composite materials as candidates for this application.

9. *American cyclists in the 1988 Summer Olympic Games in Seoul, South Korea, rode "state of the art" bicycles.*

   By deriving an expression for the elastic deflection of the frame of a bicycle, under appropriate loading, obtain the combination of material properties that defines the mass of the bicycle for a given frame stiffness. Comment on the suitability of boron fiber reinforced epoxy resin, 6061-T8 aluminum alloy, and high-purity magnesium as candidate materials for this application.

10. *Blood glucose concentration, arterial blood gas tensions (and pH), and potassium ion concentration are probably the physiological parameters that can be most suitably monitored with the aid of in vivo biosensors.*

    List the main requirements in a material to be used for such a sensor.

11. *The design of a helmet presents a formidable engineering challenge in that it requires knowledge of both human injury tolerance and the mechanical behavior of the helmet itself.*

    State the main desirable properties in a material to be used to manufacture a football helmet. Present a critical comparison of polycarbonate and acrylonitrile-butadiene-styrene as candidate materials.

12. *Ring laser gyroscopes are one example of inertial guidance systems in a missile that require dimensionally stable parts to minimize deviation from established reference coordinates.*

    Discuss the main requirements of a material to be used in these gyros, and comment on the suitability of beryllium, stainless steel, and silicon carbide whisker reinforced aluminum alloy composite material.

13. Comment on the use of ceramics in the following applications: a quench block, a sandblast nozzle, furnace liners, and a gas turbine stator.

14. *The main life- and reliability-limiting components in a gas turbine engine are the discs.*

    State the service conditions under which discs operate in this application, and list the requirements in a suitable disc material.

15. *Springs are an energy storage device for which two of the most important design requirements are the maximization of the values of the density of energy storage per unit volume and per unit weight.*

    Show that each of these energy density terms is a function of material properties. Compare the effectiveness of AISI Q260 steel, titanium, Ti-6Al-4V alloy, and Beta-C (a new β-type alloy, with composition Ti-3Al-8V-6Cr-4Mo-4Zr, made by RMI Co., Niles, Ohio) as the material for the nose and main landing gear down-lock springs and elevator and aileron control springs in the McDonnell Douglas DC-10 aircraft.

16. Discuss the suitability of rock maple wood, pine wood, and a glass fiber mat/polyester resin composite material for the slide approach area, ball impact area, ball roll area, and pin deck of a bowling alley (see Fig. D1).

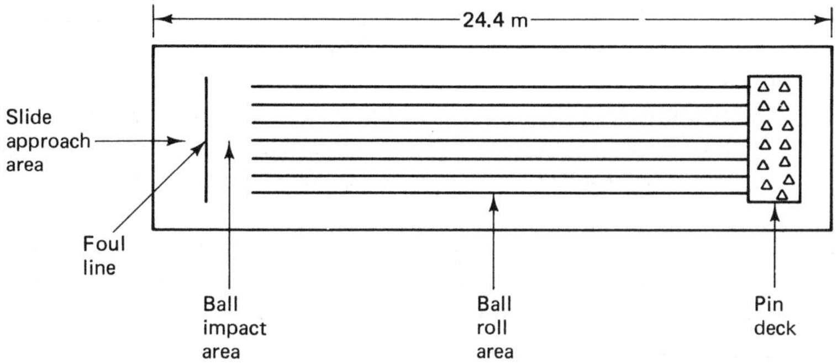

Figure D1 Parts of a bowling alley.

17. *One important objective in radome design is to minimize damage sustained by a vehicle in flying through rain, thereby ensuring adequate function of all essential components.*

Write an essay on rain erosion, with special reference to the use of polymer composite materials for the radome of a fighter jet.

18. *The major life-limiting factors for the hot-gas path components in industrial gas turbines (such as rotor blades and nozzle guide vanes) have been determined to be creep, corrosion, and fatigue (in that order of frequency; see Table D1).*

Discuss the selection of materials for these applications specifically from the viewpoint of these degrading mechanisms.

19. *Skirts are the inflated structures that extend around a hovercraft and permit the vehicle to operate in rough seas and over uneven ground.*

State the main design considerations for a skirt, and comment on the suitability of (a) a nylon woven fabric coated with either neoprene or natural rubber and (b) Kevlar 49 fiber reinforced polyester laminates as candidate materials for this application.

20. *To double the turboprop capabilities of commercial aircraft by the year 2000, a three-pronged approach is being taken. This involves three-dimensional aerodynamic designs, use of new high-temperature and lightweight materials, and structural design innovations.*

Discuss the introduction of carbon-carbon composites, metal matrix composites, and aluminides for use in various hot-section aircraft engine components, such as the combustor-turbine and augmenter-nozzle.

TABLE D1 PRINCIPAL LIFE-LIMITING FACTORS IN AN INDUSTRIAL GAS TURBINE AND THEIR RELATIVE OCCURRENCE

Life-limiting mode	Relative occurrence (%)
Creep	46
Corrosion	35
Fatigue	15

Source: A. S. Radcliff, "Factors influencing gas turbine use and performance," *Materials Science and Technology 3* (7), p. 555, 1987.

21. *Transducers that incorporate a component made of a piezoelectric material are used extensively in applications that range from vending machines to digital weighing machines.*

Write a brief essay on the phenomenon of piezoelectricity, including the properties that define the performance of a piezoelectric material. Compare the performance of a cartridge in a stereo system made of quartz, barium titanate, or lead zirconate titanate.

22. List the main requirements in a material to be used to manufacture a windshield wiper blade. Comment on the suitability of the following candidate materials: polyphenylene-ether, polyester, polycarbonate, and polyacetate.

23. *Stiffness, weight, cost, fire safety, and ballistic protectiveness are probably the most important requirements in materials to be used in naval defense vehicles.*

From these viewpoints, discuss the choice of composite materials for the hatch doors of the U.S. Navy's Aegis class of ships.

24. The following is an extract from an advertisement from a leading manufacturer of beryllium-based products (Brush Wellman, Elmore, Ohio):
Beryllium keeps America's eye on the target. . . . Even the slightest movement in target destination systems means wasted firepower and vulnerability to return fire.

Discuss the choice of beryllium for the gimbaled optical component in the low-altitude infrared targeting system on fighter jets.

25. *On the final day of the 1988 Winter Olympic Games in Calgary, Canada, Ekkehard Fasser of Switzerland won the closest four-man bobsleigh competition in Olympic history. The bobsleigh he drove was claimed to be the first to be manufactured using advanced polymer composite materials.*

Discuss the operational conditions of a bobsleigh, and outline the main requirements in a material to be used in this application. Comment on the advantages of polymer composite materials for use in a bobsleigh.

26. *One frequent complaint leveled against manual lawn mowers is that they are very noisy.*

Discuss the sound-dampening properties of various materials for use in lawn mower motorcases and shrouds.

27. *Recently it was suggested that a viable approach to the solution of the problem of the control of rigid body and elastic deformations of large spare structures is through the use of so-called intelligent structures. These are designed as structures with a dense network of sensors, actuators, and processors through which the software adjustments needed to modify and tune the system's closed loop behavior could be effected.*

Discuss the relative merits and drawbacks of piezoelectric ceramics and polymers as the material of choice for the actuator elements in an "intelligent structure."

28. *Low acoustic velocity, low acoustic attenuation, small acoustic nonlinearity, good optical quality, and a high figure of merit are the desirable properties of a suitable material for a large-time-aperture Bragg cell.*

Compare the following candidate materials for this application on the basis of the stated desirable properties: tellurium dioxide (TeO_2), lead molybdate ($PbMoO_4$), and gallium phosphide (GaP).

29. *Two of the most important properties in a material to be used as a high-density medium in longitudinal video recording are remanence and the ratio of remanence to coercivity.*

On the basis of these criteria, compare the performance of the following candidate materials: γ-Fe_2O_3, Co-γ-Fe_2O_3, CrO_2, iron metal, and cobalt-modified barium ferrite.

30. *The maintenance of resilience of the insole of a running shoe, over a period of prolonged use, is a most desirable requirement.*

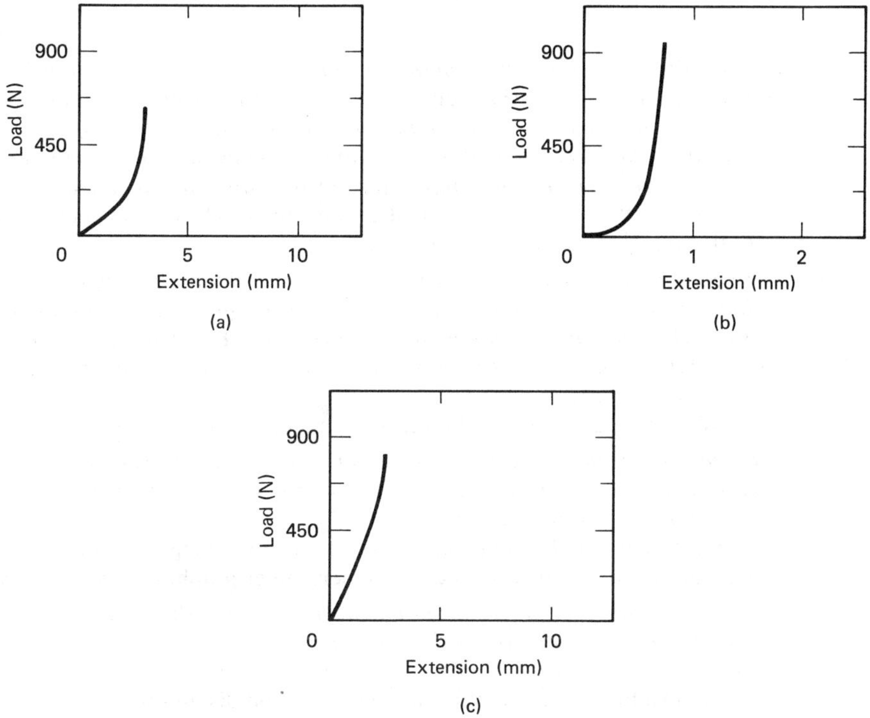

Figure D2 Load extension curves for three types of insole materials: (a) a thermoplast; (b) an elastomer; (c) a urethane.

Present a suitable derivation for resilience in this case, and compare the resilience of the candidate materials whose load extension curves are given in Fig. D2.

31. *The 50,000 or so gallium arsenide chips are the main reason why the next-generation supercomputer Cray 3 will be better than its predecessor the Cray 2, which contains silicon chips.*

Discuss the use of gallium arsenide for this application, specifically from the viewpoints of speed of computation, system power, integration level, density of packing, size, speed of logic, and frequency.

32. *In Nordic or cross-country skiing, the ski pole plays a very important role.*

Comment on the advantages and disadvantages of the following candidate materials for this application: high-modulus graphite fiber reinforced epoxy resin, aluminum alloy, and bamboo.

33. *In much the same way that the wings of an aeroplane are far more crucial to flight than the fuselage, the shaft is the most important part of a golf club because it transfers energy generated by the golfer's body to the club head.*

By specifying the operational conditions of a golf club shaft and hence the main requirements in a material to be used for its construction, comment on the suitability of steel alloy, titanium, and boron fiber reinforced epoxy resin.

34. *Technology has radically changed the face of tennis racquets and their frames by the introduction of new materials, new methods of stringing, and innovative frame shapes.*

List the main requirements in a material for the frame of a tennis racquet, and make some comparisons between filamentary polymer, metal matrix, and natural fibrous (such as wood) composite materials.

35. *The discoidal feed-through capacitor, an innovative form of a ceramic-based multilayer chip, is used as a suppression device and as a filter assembly element in electromagnetic interference applications. The choice of materials for such a capacitor is dependent on a tradeoff between capacitance volume efficiency and the variation in capacitance with the range of operating temperature of the capacitor installed in a circuit.*

Discuss the use of some named ceramic materials for such a capacitor, especially from the viewpoint of the tradeoff stated.

36. *The continual demand for increased performance in modern signal processing applications has led to novel techniques for analog signal identification, control, and processing. One such innovation is magnetostatic wave (MSW) technology. Yttrium iron garnet (YIG) is at present the only epitaxially grown ferrimagnetic material for MSW device applications.*

Write a brief essay on the reasons for this choice.

37. *The most important property of contacts for electric controls is the contact resistance. This resistance is in turn the sum of the constriction resistance of the contact material and the tarnish film resistance.*

Comment on the relative merits of the following contact materials from the viewpoint of these two resistances: copper, silver-graphite, tungsten, and rhodium.

38. *Perhaps the most important requirement in a material to be used for packaging applications, as for motor oil and cleaning products, is its resistance to a degrading mechanism known as environmental stress cracking.*

Write a brief essay on this phenomenon, and discuss the performance of general-purpose polyethylene and high-density polyethylene on the basis of their resistance to this effect.

39. *Under the Apennine Mountains in Italy, a radiation detector is to be employed to study cosmic rays by measuring light flashes in a scintillator liquid. The detector consists of three 12-m-long tanks for the scintillator fluid, strata of stone, and electronic grids. As the cosmic radiation particles pass through the scintillator fluid, flashes of light are produced. The intensity and duration of these flashes are measured and analyzed. The flashes must be reflected by the tank liner in order to ensure accurate detection and measurement.*

Comment on the use of fluorinated ethylene propylene (FEP) for the liner in this application.

40. *In an era of satellites and space shuttles, balloons continue to be a key tool for astronomers and atmospheric scientists because they offer a gentle, low-cost ride to the edge of space. However, these balloons are prone to rupture in the tropopause (at between 18 and 22 km) due to a phenomenon known as cryogenic embrittlement.*

Write a brief essay on this phenomenon, and comment on the relative merits of some named polymers for the balloon to place ozone monitors in the upper atmosphere.

41. *Biphasic culture bottles, myringotomy tubes, and baby bottle nipples are all examples of applications of thermoplastic elastomers in the biomedical field.*

List the major requirements in such elastomers, and give some specific examples.

42. *In the field of telecommunications, a major design requirement is the protection of circuits and components against the effect of nearby electrical and magnetic fields. For example, it is necessary to prevent strong field radiators, like transformers, from affecting their surroundings, and sensitive circuits, such as filter coils, must be protected from stray fields. Therefore, it is common practice to provide magnetic shielding for such critical components.*

Briefly outline the considerations in the selection of a suitable material for such a shield.

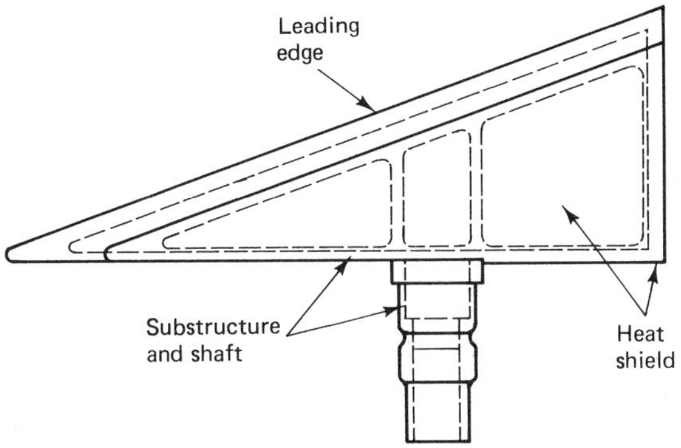

Figure D3 An antiballistic missile air vane control structure.

43. *An air vane control structure (Fig. D3) in an antiballistic missile is fabricated typically of a one-piece high-strength steel casting protected from overheating by a silica-phenolic heat shield and a quartz-phenolic leading edge.*

Discuss, in some detail, the operational conditions of this component, and comment on the choice of material for each of the three sections mentioned.

44. Discuss the choice of beryllium for the following applications:

(a) The mirror in the infrared astronomical satellite, which was used in the identification of a young solar system around Vega in 1983

(b) The cantilever to which the diamond tip in a stylus, in a stereophonic audio system, is attached

45. Justify the use of the named material in the corresponding named application:

(a) High-density polyethylene for the fuel tank in the 1988 Chrysler Premier automobile

(b) Glass fiber reinforced unsaturated polyester for the upper and middle exterior body panels in Greyhound's redesigned 06 Series RTS bus

(c) Polyetheretherketone reinforced with continuous fibers of graphite for the main landing gear strut door on the U.S. Air Force F-51 fighter jet

(d) Carbon fibers reinforced epoxy resin for the main beams of the Heavy Assault Bridge, a prefabricated ground warfare support structure made for the U.S. Army

(e) Toughened molding compound for the external parts (such as the fenders, side panels, and engine access door) of the cab on the Magnum farm tractor

(f) A coat of polytetrafluoroethylene on waxed wood for skis on aircraft to be used in winter resorts in Austria

(g) Polyethylene terephthalate for so-called see-through 0.34-l containers for soft drinks

(h) Silicon carbide for the wall lining of the combustion chamber in a waste-to-energy plant

(i) Nitinol shape memory alloy for the wire supports in a new line of brassieres

(j) PEBAX, a thermoplastic elastomer, for the soles of Air Jordan basketball shoes, Pegasus athletic shoes, and Adidas soccer shoes

(k) AISI 316 steel for the superheater tubing in a fossil-fueled boiler

(l) Silicone rubber reinforced with polyester fiber for the inflatable cockpit seal in the British Defense Forces' Hawk combat aircraft

(m) Aluminum sheet, coated on both sides with solder, for the radiator of a sports motorcycle

(n) A nickel-based alloy for the scrubber of a flue gas desulfurization system

(o) Zytel ST 901, a supertough nylon, for the frame of a backpack used by mountain climbers
(p) A tektite glass for a nuclear waste storage container
(q) Aluminum for the hull of the United States' boat *Stars and Stripes,* which won the America's Cup in 1987
(r) Hot-pressed silicon nitride for the recuperative heat exchanger in an experimental vehicular engine
(s) A high-molybdenum stainless steel for the condenser in a seawater-cooled power station
(t) Magnesia-chrome-titanium for the humidity sensor in an industrial air conditioner
(u) High-density polyethylene for the body of a kayak used in a U.S. white-water expedition over the Braldu River in Pakistan
(v) Aluminum-lithium for the extruded spar cap in an aircraft
(w) Carbon-carbon composite for the liner of the magnetic confinement field of an experimental fusion reactor
(x) A proprietary resin for the ball used in "Cayman Golf," a new short-course form of golf
(y) Zincrometal, an electrogalvanized steel, for the underbody, door panels, deck lid, and hood in the 1987 Buick Somerset automobile
(z) A samarium-cobalt alloy for the support of a swing in a children's playground
(aa) Aluminum for the body of the Hummer (High-Mobility Multipurpose Wheeled Vehicle), a light military truck designed by the AM General Divison of the LTV Aerospace and Defense Co.
(bb) Santoprene, a thermoplastic vulcanizate, for the molded bellows in a sterilizable emergency resuscitator
(cc) Magnesium for the disc drive carriage in a computer printer
(dd) A modified asphalt-concrete, containing latex rubber particulates, for the runways at Boston's Logan International Airport
(ee) Silicide-coated niobium for the afterburner in the engine of a fighter jet
(ff) Polycarbonate for a cardiotomy reservoir, a critical component in heart-lung bypass surgery
(gg) Graphite fiber reinforced aluminum composite material for the hollow radio frequency waveguide on NASA's Hubble Space Telescope
(hh) Aluminum for the facade and internal columns of Infomart (the International Information Processing Market Center, located in Dallas, Texas), which is a re-creation of the Crystal Palace in London
(ii) Stainless steel for the casing of the International Fusion Superconducting Magnet Test Facility at Oak Ridge National Laboratory, Oak Ridge, Tenn.
(jj) Epichlorohydrin rubber for a gasoline pump hose
(kk) A hybrid aramid–graphite–epoxy resin composite material for the wing-to-fuselage fairing, undercarriage doors, engine cowlings, and fixed trailing-edge panels in the Boeing 767 aircraft
(ll) Incoloy MA 956 for parts of the exhaust ducting in an aircraft gas turbine engine
(mm) Fiberil TN, a toughened nylon containing 36% loading of mineral fillers, for automobile mirror housings

APPENDICES

MATERIAL PROPERTY DATABASES

APPENDIX A

A number of databases for the properties of materials are currently available. Among them are the following:

1. CorSur, a corrosion data software package from the National Association of Corrosion Engineers (NACE) of Houston, Texas, covers the effects of 1000 corrodents on more than 25 metallic materials. The data can be retrieved in one of three ways: (a) by tabular listing of materials exhibiting a specified range of corrosion rates in the corrodent conditions of interest (temperature, concentration, etc.), (b) by graphical presentation of corrosion rate ranges for a given material in a series of environmental temperatures and concentrations for a specified medium, and (c) by tabular listing of corrosion rate data for a given material in a given corrodent, as a function of temperature and concentration.

2. The Society of Automotive Engineers (SAE), Inc., of Warrendale, Pa., has a software package on metals and alloys called UNSearch that provides a way of cross-referencing the many different numbering systems used to identify the thousands of metallic materials in commercial use. More than 3500 of them are included in this package. UNSearch has a menu format, thereby allowing the user to organize the data to meet specific design needs. From a series of menus, the selection criteria can be established, and UNSearch then scans the database for the most suitable alloy.

3. A software selection package for powder metals is available from MPR Publishing Services Ltd., Shrewsbury, England. Featured in the company's Ferrous P/M Materials Selector are data on the mechanical properties of 140 grades of as-sintered materials drawn from plain iron as well as 13 different ferrous-based alloy families, that is, various steels (carbon, phosphorus, copper-lead, copper-carbon, nickel, copper-nickel-molybdenum, low alloy, and stainless).

4. MTS Systems Corp., of Minneapolis, Minn., has structural analysis software programs that use data obtained from the company's extensive in-house testing results. One such program is the Life Analysis Program (LAP), a fatigue life package that, among other things, will allow fatigue damage contours to be plotted on finite element models at early stages of design.

5. MetSel/2 is a basic software package from ASM International, Inc., of Metals Park, Ohio. Included is data on various materials such as carbon steel, aluminum, titanium, and iron and steel castings. MetSel/2 can be used to search for data on specific topics such as chemical composition, temper conditions, and UNS (Unified Numbering System) number and processing characteristics.

6. The National Materials Property Data (MPD) Network, Inc., of New York City, has as its objective the establishment of a comprehensive numeric/factual material property data bank. The main feature will be a "gateway" that provides single-point access by common command language to many sources of well-focused data.

7. The Steeltuf database on steel toughness from the Electric Power Research Institute of Palo Alto, Calif., and the Materials Property Council of New York City consists of about 30,000 individual test results on more than 80 steels.

8. PLASPEC, an engineering and marketing data bank from *Plastics Technology* magazine in New York. This package includes a complete plastic materials databank selectable on over 50 specifications, a complete plastics machinery data bank, complete supplier directories, on-line pricing information, and a product industry and pricing news service. The material selection menu, for example, covers 44 processing/physical, mechanical, thermal, electrical, and optical properties on over 100 thermoplastics, thermosets, and elastomers. This databank can be accessed by subscribers using a terminal and modem via a local telephone number.

9. Thermofil, Inc., of Brighton, Mich., offers a computerized databank containing mechanical and electrical properties and pricing information on about 450 thermoplastic resins and compounds the company supplies. The program covers 53 base resins and compounds ranging from polyolefins to polyetheretherketone and grades containing any of 12 types of particulate fillers and/or fiber reinforcement and a wide range of additives.

10. GE Plastics, of Pittsfield, Mass., has developed the Engineering Design Database (EDD), a system that incorporates data acquisition, storage, and manipulation into an interactive computer network. EDD provides tensile stress-strain, short-term creep, rheology, and fatigue and failure property data about engineering thermoplastic materials that the company supplies, such as Noryl (a modified polyphenylene oxide), Lexan (a polycarbonate), and Lomod (an elastomer). EDD also includes data on filled and unfilled structural foams. All data generated for EDD is collected using appropriate ASTM guidelines for plastics testing.

11. Polyfacts, an on-line menu-driven, interactively-formatted database, provided by DuPont, contains seven modules, one of which (the plastics selection and graphics module) has data on mechanical, thermal, and electrical properties, as well as pricing information on about 170 products.

12. Plastivision, an on-line database offered by Borg-Warner, through Compuserve, has data not only about quantitative properties of materials but also about qualitative features, such as transparency and heat resistance.

13. The High Temperature Materials Properties (HTMP) database, operated by the Center for Information and Numerical Data Analysis and Synthesis (CINDAS) of Purdue University, in conjunction with the U.S. Department of Defense, contains data on about 600 varieties of aerospace structural composite materials, as well as metals and infrared detector/sensor materials. CINDAS also operates an Engineering Materials Properties database which consists of data on more than 7000 materials, ranging from metals to biological materials.

14. ICI developed the Engineering Plastics On Screen (EPOS) database, which has been successfully used in many countries in Europe. In the United States, it contains data on mechanical, thermal, electrical, and other properties of injection-moldable materials.

15. CenCad Services developed a database called CenBase, which contains data on 7500 polymeric, 2000 thermoset, 1000 elastomeric, 3000 composite, and 1400 metallic materials.

16. Plascams, a package developed by RAPRA in England, and marketed by ESDU International in the United States, has data on mechanical and electrical properties of about 400 polymers.

17. Various new databases include one from ASM International on phase diagrams for alloys, one from the American Ceramic Society on phase diagrams for ceramics, one from the American Welding Institute on weld properties, and one from Iowa State University on fatigue properties of aluminum alloy weldments. Also, fully configured data discs are available from some suppliers of materials such as Carpenter Technology Corp., Brush Wellman, AMAX Specialty Metals, Haynes International, Bethlehem Steel, INCO Alloys International, and Eastman Chemicals.

18. These are also a number of reliable and comprehensive printed data sheets or bound books containing material property data. Among the best of these are the International Plastics Selector (which contains data on over 10,000 polymers), and Materials Selector, published annually by the magazine, *Materials Engineering* (which contains data on all classes of materials).

MATERIAL PROPERTY WEIGHTING FACTORS AND COST ELEMENTS

APPENDIX B

B.1 MATERIAL PROPERTY WEIGHTING FACTORS

The case studies in this book have illustrated the methodology of selection of materials based only on engineering performance considerations for a given application. This approach has not taken into account at least two other sets of considerations: the relative importance of the required material properties and the final cost of the component. In many cases, the best engineering and the best economic solutions for a given design do not coincide. A compromise is thus a design that yields the optimal mix of these two sets of requirements. Let us briefly examine a methodology that may be used to incorporate the influence of these two factors on the final choice of a material.

In essence, in a systematic and holistic material selection procedure, the steps are (a) identification of the desirable material properties; (b) identification of the important costs of the component, mainly raw material and fabrication costs; and (c) an ordering of the identified material properties in item (a) based on certain proven methods.

This ordering or assignment of the so-called material property weighting factors is far from trivial. For example, digital logic analysis techniques have been used to obtain these factors for a number of applications (Table B.1).

TABLE B.1 MATERIAL PROPERTY WEIGHTING FACTORS FOR FOUR APPLICATIONS

Cutting Tools		**Rolling Element Bearings**	
Hardness at service temperature	0.40	Hardness	0.33
Room temperature hardness	0.25	Flexural strength	0.27
Toughness	0.15	Corrosion resistance	0.20
Raw material cost	0.10	Fracture toughness	0.13
Fracture resistance	0.10	Modulus of elasticity	0.07
Gas Turbine Blades		**Artificial Hip Joints**	
Specific creep rupture strength	0.27	Fatigue strength	0.15
Thermal fatigue resistance	0.23	Corrosion resistance	0.15
Specific tensile strength	0.20	Biocompatibility	0.15
Oxidation resistance	0.17	Hardness	0.11
Raw material cost	0.13	Modulus of elasticity	0.08
		Ultimate tensile strength	0.06
		Density	0.06
		Raw material cost	0.06

Source: A. Jebb and J. M. McCullough, "Database impedes ceramics development," *Chartered Mechanical Engineering 34* (10), pp. 49–50, October 1987.

Once these weighting factors have been assigned, the overall rating *OR* may be obtained, and the candidate materials can then be compared on the basis of their *OR* values. One simple and well-tried expression for *OR* is given by

$$OR_i = \sum_{i=1}^{n} \frac{\text{value of property } j \text{ for material } i}{\text{maximum value of property } j \text{ in candidate materials set}} \times W_{ij} \qquad \text{(B.1)}$$

where W_{ij} is the weighting factor for property j for material i. This approach to obtaining *OR* has been taken by a number of workers, such as Waterman, and Appoo and Alexander [1]. Other expressions for *OR* have been given; for example, the mean weighted property index, which uses the root mean square deviation. This index, due to Hobson and Hanley [2], measures how closely the overall weighted properties meet requirements, and thus the nearer the index is to unity, the better the material. The root mean square deviation indicates how each property deviates from the mean weighted property index, and thus the nearer the index is to zero, the better the material.

The use of the overall rating, as given in equation (B.1), is illustrated for the selection of a material for the wings of a military supersonic aircraft. The desirable properties of a suitable material for this application and the relevant weighting factors are as follows:

- Specific yield strength (0.20)
- Maximum crack tolerance (0.20)
- Structural efficiency (0.25)
- Homologous temperature; that is, the temperature above which creep is a possibility (0.30)
- Specific raw material cost (0.05)

TABLE B.2 OVERALL RATING OF FIVE CANDIDATE MATERIALS FOR THE WING OF A MILITARY SUPERSONIC AIRCRAFT

(1) Material	(2) σ_y (MPa)	(3) ρ (kg m^{-3})	(4) K_{IC} (MPa$\sqrt{m}$)	(5) E (GPa)	(6) T_H (°C)	(7) C (\$ kg^{-1})	(8) ν	(9) σ_y/ρ[a]	(10) $(K_{IC}/\sigma_y)^2$[b]	(11) SE[a]	(12) C[a]	(13) T_H[c]	(14) OR[d]
2014 Al alloy	550	2796	25	73	200	2.0	0.35	0.31	0.17	0.48	0.03	0.33	0.362
Ti-6Al-4V alloy	880	4540	60	116	600	10.0	0.31	0.31	0.37	0.35	0.17	1.00	0.565
301 stainless steel	900	8027	100	193	550	1.5	0.28	0.18	1.00	0.23	0.03	0.92	0.618
Alithalite Al-Li alloy	510	2685	33	79	200	19.8	0.30	0.30	0.34	0.51	0.33	0.33	0.430
60 vol % carbon fiber reinforced epoxy resin	1050	1661	39	138	120[e]	60.0	0.30	1.00	0.11	1.00	1.00	0.20	0.532

σ_y, yield strength; ρ, density; K_{IC}, fracture toughness; E, modulus of elasticity; T_H, homologous temperature; C, raw material cost; ν, Poisson's ratio; SE, structural efficiency $= \dfrac{[E/(1-\nu^2)]^{0.33}}{\rho}$.

[a]Relative to carbon fiber reinforced epoxy resin.

[b]Relative to 301 stainless steel.

[c]Relative to Ti-6Al-4V alloy.

[d]OR = 0.20 (column 9) + 0.20 (column 10) + 0.25 (column 11) + 0.30 (column 13) + 0.05 (1 − column 12).

[e]Maximum use temperature.

The results of the computations for values of *OR* for the candidate materials are given in Table B.2. The best material is 301 stainless steel. But use of steel gives rise to considerable manufacturing problems. For example, for wing skins, steel has to be corrosion-resistant, and the steels with the best corrosion resistance and also the best formability are the austenitic stainless steels, such as the 18Cr-8Ni type. These steels are low-strength materials and are usually strengthened by cold work, which severely reduces their formability. In addition, the high temperature causes considerable distortion, so that wing panels cannot be manufactured with a high degree of flatness. Nowadays, the main materials for use in airframes are the titanium alloys, notably the Ti-6Al-4V alloy fabricated by superplastic forming and diffusion bonding. It is envisaged that filamentary and metal matrix composite materials will begin to find increased use in these applications (see Chapter 2 in Part I).

It should be pointed out that some alternative procedures for the ordering of the identified material properties, other than the use of weighting factors, have recently been introduced. For example, Field and de Neufville [3] have presented a method based on the operational research concept known as utility maximization which represents the relative values of different properties as they vary throughout their range. They showed that the methodology is neither sensitive to arbitrary choices of normalizing constants nor limited to linear representations of preference for a material(s). However, the method is rather involved. In time, though, it should be more easily and widely used, once interactive computer programs that calculate utility become available. Demaid and Zucker [4] have applied 'knowledge engineering' principles to material selection. Arguing that descriptions are at the heart of the material selection problem, they suggest that what is needed in a material selection system is the ability to reason with a wide range of levels and types of descriptions. These workers suggested the use of an inheritance graph as the appropriate tool in material selection. Again, their approach suffers from being conceptually involved and mathematically abtruse.

B.2 COST ELEMENTS

A useful approach to the evaluation of a material for a particular application is the overall savings accruing from its use as a replacement for the material currently being used. Taggart and Schwan [5] recently presented a very lucid expression for this parameter:

$$\text{Overall cost savings penalty } X = (w_o m_o - w_n m_n) + (w_o f_o - w_n f_n) + P(w_o - w_n) \qquad \text{(B.2)}$$

where m_o and m_n are the raw material costs per unit mass of the original and the new materials, respectively; w_o and w_n are the masses of the original and the new materials; respectively; f_o and f_n are the fabrication costs per unit mass of the original and the new materials, respectively; and P is the performance premium; for automotive components, P is given by [6]

$$P = \frac{MDNG}{B(B + DR)}$$

where M is the number of kilometers driven per year, D is the increase in fuel economy (km 1^{-1}) per unit mass of component saved; N is the design life of the automobile, G is the unit price of gasoline at the time the automobile is bought (\$ 1^{-1}), B is the fuel consumption (km 1^{-1}) of the automobile containing one or more components made of the original material, and R is the mass savings arising from the substitution of the new material for the original one in the components. The performance premium is the monetary penalty to be incurred for every unit mass of the component saved by the use of the new material.

It has been suggested [5] that if X is positive, the material substitution is justified. Otherwise, the fuel cost savings over the design life of the vehicle arising from the use of the new material in the component will not pay back the additional raw material and fabrication costs associated with this new material.

Let us apply equation (B.2) to a specific example: an investigation of the advisedness of replacing a metallic (steel) driveshaft in a station wagon with one made of a filamentary composite material (vinyl ester resin reinforced with fibers of carbon and glass).* The values of the relevant parameters are as follows: M = 24,000 km y^{-1}; N = 7 y; B = 10.0 km l^{-1}; G = \$0.26 l^{-1}; D (according to Fig. B.1) = 0.0050 km l^{-1} kg^{-1}; R = 1 kg. With these values, P is calculated to be \$2.18 kg^{-1}. The various cost items have the following values [8]: m_o = \$0.60 kg^{-1}; m_n = \$0.90 kg^{-1}, f_o = \$0.20 kg^{-1}; f_n = \$0.30 kg^{-1}; w_o = 14.5 kg; w_n = 8.4 kg. The value of X in this case is \$14.82, thereby providing economic justification for the material substitution.

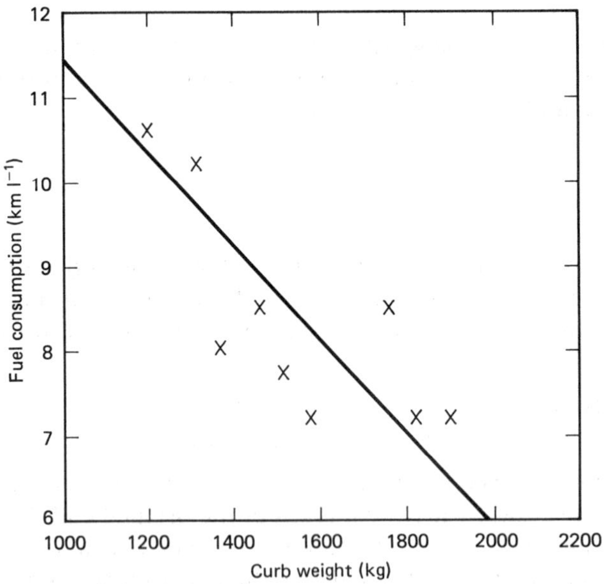

Figure B.1 Dependence of engine fuel consumption on curb weight of station wagons [7].

*Such replacements have been successfully carried out, as in the 1985 Ford Econoline van.

REFERENCES

1. N. A. Waterman, "Materials optimization," *Int. Mech. Conf. 22,* p. 15, 1973; P. M. Appoo and W. O. Alexander, "Computer analysis speeds assessment of engineering materials," *Metals and Materials 00* (0), p. 42, 1976.
2. E. Hobson and D. P. Hanley, "Computerized materials selection," *J. Eng. Mater. Technology,* Paper No. 73-Mat-cc, pp. 147–201, 1973.
3. R. F. R. Field and R. de Neufville, "Materials selection-maximizing overall utility," *Metals and Materials 4* (6), pp. 378–382, 1988.
4. A. Demaid and Zucker, "A conceptual model for materials selection," *Metals and Materials 4* (5), pp. 291–297, 1988.
5. D. G. Taggartt and F. J. Schwan, "Evaluation of woven fabric composites for automotive applications," *J. Mater. Shaping Technology 5* (1), pp. 41–51, 1987.
6. "Woven fabric reinforced composites for automotive applications," Tech. Final Report, NSF Grant No. DMR-8212867, MSC TPR 1605/8102, December 1985.
7. "The right car at the right price," *Changing Times 41* (12), pp. 29–61, 1987.
8. Estimates based on figures provided by John C. Soroka, Ford Motor Co., Transmission and Chassis Division, Van Dyke Plant, Sterling Heights, Mich., January 1988.

SAMPLING OF COURSES IN MATERIAL SELECTION AT VARIOUS UNIVERSITIES IN THE UNITED STATES

APPENDIX C

University of Florida, Gainesville
Course Number and Title: EMA 4714, Material Selection and Failure Analysis
Course Description: Philosophy and practice of engineering selection of materials. Case studies in product liability and failure analysis.

Clemson University, Clemson, S.C.
Course Number and Title: ME 459, ME 659, Material Selection in Design
Course Description: Introduction to the systematic material selection process with emphasis on metallic materials for specific applications. Polymeric, ceramic, and composite materials are also considered. Case studies are the primary mode of instruction, supplemented with lab demonstrations.

Vanderbilt University, Nashville, Tenn.
Course Number and Title: MME 281, Material Selection and Design
Course Description: Applications of the principles of materials science to the selection, design, and analysis of materials for engineering structures. Emphasis on the role of microstructure in applications, design, and failure analysis. A design project and the analysis of engineering failures are included.

University of California, Berkeley
Course Number and Title: ME 130, Materials Engineering
Course Description: Case study course concerned with development and selection of materials and failure analysis. Examples relate to mechanical properties, electrical properties, and corrosion.

Oregon State University, Corvallis
Course Number and Title: ME 481, Material Selection Design Considerations
Course Description: The use of material properties (elastic modulus, strength, etc.) in mechanical design, covering the major families of materials, their properties, and control of those properties. Case studies and projects in material selection are included.

Massachusetts Institute of Technology, Cambridge
Course Number and Title: 3.555, Principles of Materials Engineering
Course Description: Synthesis of material selection, design, and manufacturing from the viewpoint of the materials engineer. Includes fundamentals and applications of manufacturing processes, failure modes, material systems analysis, and cost engineering. Considers a wide variety of materials. Uses lectures and case studies.

University of Alabama, Birmingham
Course Number and Title: MTE 410, Materials Science and Failure Analysis
Course Description: Principles and applications associated with the selection of engineering materials. Case studies involving failure analysis and product liability.

REFERENCES FOR ASPECTS OF MATERIAL BEHAVIOR IN SERVICE

APPENDIX D

Creep

F. GAROFALO, *Fundamentals of Creep and Creep Rupture in Metals*. New York: Macmillan, 1965.

R. W. EVANS and B. WILSHIRE, *Creep of Metals and Alloys*. London: Institute of Metals, 1985.

J. E. DORN (ed.), *Mechanical Behavior of Materials at Elevated Temperatures*. New York: McGraw-Hill, 1961.

Corrosion and Oxidation

U. R. EVANS, *The Corrosion and Oxidation of Metals*. London: Edward Arnold, 1968.

J. M. WEST, *Electrodeposition and Corrosion Processes*. London: Van Nostrand Reinhold, 1971.

N. BIRKS and G. H. MEIER, *Introduction to High-Temperature Oxidation of Metals*. London: Edward Arnold, 1983.

Fatigue

M. KLENSNIL and P. LUKAS, *Fatigue of Metallic Materials*. New York: Elsevier, 1980.

S. T. ROLFE and J. M. BARSOM, *Fracture and Fatigue Control in Structures*. Englewood Cliffs, N. J.: Prentice-Hall, 1977.

B. J. SANDOR, *Fundamentals of Cyclic Stress and Strain*. Madison: University of Wisconsin Press, 1972.

Fracture

J. F. KNOTT, *Fundamentals of Fracture Mechanics*. London: Butterworths, 1973.

R. W. HERTZBERG, *Deformation and Fracture Mechanics of Engineering Materials*. New York: Wiley, 1976.

F. M. BROEK, *Elementary Engineering Fracture Mechanics*, 3d ed. The Hague, Netherlands: Nijhoff, 1982.

Wear and Erosion

L. E. SAMUELS, E. D. DOYLE, and D. M. TURLEY, *Fundamentals of Friction and Wear of Materials*. Metals Park, Ohio: American Society of Metals, 1981.

C. M. PREECE (ed.), *Treatise on Materials Science and Technology: Vol. 16, Erosion*. London: Academic Press, 1979.

G. S. SPRINGER, *Erosion by Liquid Impact*. Washington, D.C.: Scripta Technica, 1976.

TRADEMARK INFORMATION

Adidas (Adidas USA, Inc., Kutztown, Pa.)
Air Jordan (Nike International Inc., Beaverton, Oregon)
Air Sandwich (Philips Research Laboratories, Eindhoven, Netherlands)
AN-74 (Emerson & Cuming, Inc., Canton, Mass.)
Avcothane 51 (Avco Everett Co., Everett, Va.)
Beta-C (RMI Co., Niles, Ohio)
Biomer (Ethicon Corp., Somerville, N.J.)
Bio Metal Fiber (Toki Corp., Japan)
Borsic (Hamilton Standard Div., United Technologies Corp., Windsor Locks, Conn.)
Clearite (Fourco Glass Co., Bridgeport, WV)
CLEARTRAN (CVD Inc., Woburn, Mass.)
Conolene II (Continental Can Co., Stamford, Conn.)
Cray X-MP/2 (Cray Research Inc., Minneapolis, Minn.)
Cryocon (Raychem Corp., Menlo Park, Calif.)
Cryofit (Raychem Corp., Menlo Park, Calif.)
CryoTact (Raychem Corp., Menlo Park, Calif.)
DC 6-1104 (Du Pont de Nemours, Wilmington, Del.)
Delrin 150 (How Medica International Inc., Limerick, Ireland)
Drexon (Drexler Technology Corp., Mountain View, Calif.)
E6 (Ohara Optical Glass, Inc., Watchung, N.J.)
Ebanol (Asarco Inc., Zinc Oxide Dept., Columbus, Ohio)
Ecco-Coat (Emerson & Cumming, Inc., Canton, Mass.)
Eccosorb (Emerson & Cumming, Inc., Canton, Mass.)
Ekkcel (Carborundum, Falconer, N.Y.)
Ekonol (Kennecott Corp., Salt Lake City, Utah)
EMI-X-DA-40 (LNP, Malvern, Pa.)
Enersorb (De Soto, Inc., Fort Worth, Texas)
Epron (Shell Chemical Co., Houston, Texas)
Etrema (Edge Technologies, Inc., Ames, Iowa)
Fiberfrax (Standard Oil Engineered Materials, Niagara Falls, N.Y.)
Fiberil TN (Wilson Fiberfil International, Neshanic Station, N.J.)

Fiberlam (Ciba-Geigy, Basel, Switzerland)
Fibermax (Standard Oil Engineered Materials, Niagara Falls, N.Y.)
Fiberscope (Olympus Corp., Lake Success, N.Y.)
GE RTV 655 (GE Plastics, Pittsfield, Mass.)
Gulflite (Boyle Midway Household Products, Inc., Cranford, N.J.)
Hexsyn (Goodyear Tire & Rubber Co., Akron, Ohio)
Hutton Unit Partners [Conoco (UK) Ltd., Britoil plc, Gulf Oil (UK) Ltd., Amerada Hess Ltd., Amoco (UK) Exploration Co., Enterprise Oil plc, Mobil North Sea Ltd., Texas Eastern North Sea, Inc., and Oil and Pipelines Agency]
IBM System/36 (IBM, New York, N.Y.)
Jarvik 7 (Kolff Medical, Inc., Salt Lake City, Utah)
Kevlar (Du Pont de Nemours, Wilmington, Del.)
Killmaster (Vesoco Chemical Co., Tuscon, Arizona)
LCP 3440 (RTP Co., Winona, Minn.)
Lexan (General Electric Plastics Group, Pittsfield, Mass.)
Lexan 141 (General Electric Plastics Group, Pittsfield, Mass.)
Lomod (General Electric Plastics Group, Pittsfield, Mass.)
Memorite (Beta Phase Inc., Menlo Park, Calif.)
Memrytec (Memory Metals, Inc., Stamford, Conn.)
Merlon CD-2000 (Mobay Chemical Corp., Pittsburgh, Pa.)
METGLAS (Allied Chemical Corp., Morristown, N.J.)
Microvax II (Digital Equipment Corp.)
MM-1000 (Spaulding Fiber Co., Inc., Tonawanda, N.Y.)
MPDZ (Dow Corning/Celanese, Midland, MI)
Nextel (3M Co., Minneapolis, Minn.)
Nextel 312 (3M Co., Minneapolis, Minn.)
Nicalon (Nippon Carbon Co., Tokyo, Japan)
Nomex (Du Pont de Nemours, Wilmington, Del.)
NORYL (General Electric Plastics Group, Pittsfield, Mass.)
PEBAX (Atochem, Inc., Glen Rock, N.J.)
Pegasus (Nike International, Inc., Beaverton, Oregon)
Pellethane (Upjohn Co., Torrance, Calif.)
PineSol (American Cyanamid Co., Household Products Div., Wayne, N.J.)
PRD-166 (Du Pont)
Pyrex (Corning Glass Works, Corning, N.Y.)
Pyroceram 9608 (Corning Glass Works, Corning, N.Y.)
Quartz T08E (Heraeus-Amersil Inc., Sayreville, N.J.)
RAYTRAN (Raytheon Co., Lexington, Mass.)
Ronabond (LeaRonal, Inc., Freeport, N.Y.)
Ryton (Phillips Petroleum, Bartlesville, Okla.)
Santoprene (Monsanto Co., Dayton, Ohio)
Scotchcast 281 (3M, Minneapolis, Minn.)
SCS-6 (AVCO Specialty Materials Div., Textron, Inc., Lowell, Mass.)
7070 (Corning Glass Works, Corning, N.Y.)
7740 (Corning Glass Works, Corning, N.Y.)
Solarex (AFG Industries, Kingsport, Tenn.)
Solar RB (Du Pont de Nemours, Wilmington, Del.)
STP (STP Corp., Boca Raton, Fla.)
Sunadex (AFG Industries, Kingsport, Tenn.)
SunLite (Kalwall Corp., Manchester, New Hampshire)
Tedlar (Du Pont de Nemours, Wilmington, Del.)
Teflon (Du Pont de Nemours, Wilmington, Del.)
Tempax (Schott Glass Technologies, Duryea, Pa.)
Thermocomp PDX 85194 (LNP Co., Malvern, Pa.)
TICOREX (Nippon Mining Ltd., Tokyo, Japan)
TINEL (Raychem Corp., Menlo Park, Calif.)
Tinel-Lock (Raychem Corp., Menlo Park, Calif.)
Tivanium (Zimmer Inc., Warsaw, Indiana)
Torlon (Amoco, Chicago, Ill.)
Tyranno (Ube Industries, Japan)
UF-1006 (LNP, Malvern, Pa.)
ULE (Corning Glass Works, Corning, N.Y.)
Unisys PC/IT (Unisys Corp., Detroit, Michigan)
Urolane LV5753 (Furnane, Los Angeles)
Vease (Raychem Corp., Menlo Park, Calif.)
Vectra (Hoechst Celanese, Summit, N.J.)
Visions (Corning Glass, Corning, N.Y.)
VITUF 5172 (Goodyear, Akron, Ohio)
Vycor (Corning Glass Works, Corning, N.Y.)
Xydar (Amoco Performance Products, Augusta, Ga.)
Zerodur (Schott Glass Technologies, Duryea, Pa.)
Zincrometal (Inland Steel Industries, Inc., Chicago, Ill.)
Zytel ST 901 (Du Pont de Nemours, Wilmington, Del.)

INDEX

B

C

D

E

F

G

H

I

J

N

O

P

Q

R

S

U

V

W

X

Y

Z